Food Safety and Toxicology

Food Safety and Toxicology

Present and Future Perspectives

Edited by
Oluwatosin Ademola Ijabadeniyi
and Omotola Folake Olagunju

DE GRUYTER

Editors
Oluwatosin Ademola Ijabadeniyi
Department of Biotechnology and Food Science
Durban University of Technology
South Africa
tosynolu@yahoo.com

Omotola F. Olagunju
Department of Science
School of Health and Life Sciences
Teesside University
Middlesbrough
UnitedKingdom
tolaolagunju75@gmail.com

ISBN 978-3-11-074833-8
e-ISBN (PDF) 978-3-11-074834-5
e-ISBN (EPUB) 978-3-11-074851-2

Library of Congress Control Number: 2023947353

Bibliographic information published by the Deutsche Nationalbibliothek
The Deutsche Nationalbibliothek lists this publication in the Deutsche Nationalbibliografie;
detailed bibliographic data are available on the Internet at http://dnb.dnb.de.

© 2024 Walter de Gruyter GmbH, Berlin/Boston
Cover image: LightFieldStudios/iStock/Getty Images Plus
Typesetting: Integra Software Services Pvt. Ltd.
Printing and binding: CPI books GmbH, Leck

www.degruyter.com

Foreword

I would like to thank Prof. Oluwatosin Ademola Ijabadeniyi of the Department of Biotechnology and Food Technology, Durban University of Technology, Durban, South Africa, and Dr. Omotola Folake Olagunju of the School of Health and Life Sciences, Teesside University, Middlesbrough, United Kingdom, for the honor given me to write the foreword of this book on food safety.

This book, entitled *Food Safety and Toxicology: Present and Future Perspectives*, comprehensively covers the essential knowledge required by food safety experts to excel in their profession. At a time when food safety takes the center stage in food processing and when consumers are becoming increasingly aware of the challenges associated with the consumption of contaminated and adulterated foods, this book should find its place in all universities and professionals' libraries in the world. The dynamics among food safety professionals are intricate, with laboratory analyses often insufficient in managing health issues caused by chemical and biological contaminants. Experts' assessments, carried out collaboratively by independent scientific experts and national health agencies, are crucial to understanding the effects of these contaminants on human health, especially in cases that could lead to potential hazards.

Toxicology, once phased out of university curricula, has made a comeback due to heightened consumer demand for risk-free foods. Experts now rely on toxicology to shape national policies. The quantification of health hazards is indeed important but not sufficient. The expert must have a perfect understanding of global standards and toxicologically dangerous doses, which are two strictly different parameters. At the industry level, it is necessary to know how to prevent the hazards by a detailed analysis of the risks. The normal quality analyses and the implementation of the HACCP system, compulsory in some countries, have been of great help in controlling hazards. For example, the presence of biofilms on food processing equipment is now being addressed and has led to the development of specific cleaning techniques that previously were not considered. In recent years, analytical methods of monitoring and evaluating risks in food processing has evolved. DNA analysis of pathogens by sequencing is one of them, although requiring improvement in the protocols. Another recent development is in the application of coupled mass spectrometry/chromatography for mycotoxin assay.

This book presents all these themes written by eminent specialists in their respective fields in food safety and toxicology. It evaluates food biological hazards using conventional methods, molecular techniques, and other rapid methods. It proposes also to identify and implement processes to prevent food hazards and toxicants contaminations in the food system, while providing mitigating strategies to prevent public health risks. Reading this book promises to be interesting and informative. It will also provide a network of seasoned experts whose knowledge and experiences may help

https://doi.org/10.1515/9783110748345-202

to navigate through identification and monitoring of health risks in the food industry and in the public space.

Dr. Didier Montet
Montpellier, France
October 2023

Contents

List of contributors

Abidemi Ojo
Centre for Applied Food Sustainability and
Biotechnology
Central University of Technology
Bloemfontein 9301
Free State
South Africa

Abimbola Motunrayo Enitan-Folami
Department of Biotechnology and Food
Technology
Durban University of Technology
Durban 4000
KwaZulu-Natal
South Africa

Abiola Ezekiel Taiwo
Faculty of Engineering
Mangosuthu University of Technology
Durban
South Africa

Abiola Folakemi Olaniran
Department of Food Science and Nutrition
College of Pure and Applied Sciences
Landmark University
Omu-Aran, Kwara State
Nigeria

Abisoye Solomon Fiyinfoluwa
Department of Biological Sciences
Olusegun Agagu University of Science and
Technology
Okitipupa, Ondo State
Nigeria

Aboi Igwaran
Department of Biochemistry and Microbiology
University of Fort Hare
Alice 5700
South Africa

Adebola O. Oladunjoye
Department of Food Technology
University of Ibadan
Ibadan
Nigeria

Adeoluwa I. Adetunji
Research and Development
Labworld
Division of Philafrica Foods
Johannesburg
South Africa

Adeoye John Kayode
Department of Viticulture and Oenology
Stellenbosch University
Durban 4100
KwaZulu-Natal
South Africa

Adeyemi Ayotunde Adeyanju
Department of Food Science and Microbiology
Landmark University
Omu-Aran, Kwara State
Nigeria

Ahmad Cheikhyoussef
Science and Technology Division
Multidisciplinary Research Services
Centre for Research Services (CRS)
University of Namibia
Windhoek
Namibia

Aina A. Daniel
Department of Microbiology
School of Science and Technology Sciences
Babcock University
Illishan-Remo, Ogun State
Nigeria

Ajibola B. Oyedeji
Research and Development (Commercial
Product Development)
Labstat International Inc.
Edmonton
Alberta
Canada

https://doi.org/10.1515/9783110748345-204

Alaofin Sefunmi
Department of Plant Pathology
North Dakota State University
Fargo
USA

Angela Parry-Hanson Kunadu
Department of Animal Science
Texas A&M University
College Station, Texas
USA
And
University of Ghana
Accra
Ghana

Ayodeji Charles Osunla
Department of Microbiology
Adekunle Ajasin University
Akungba Akoko
Ondo State
Nigeria

Bartholomew Saanu Adeleke
Food Security and Safety Focus Area
Faculty of Natural and Agricultural Sciences
North-West University
Mmabatho 2735
South Africa
And
Department of Biological Sciences
Olusegun Agagu University of Sciences and
Technology
Okitipupa
Ondo State
Nigeria

Bhekisisa C. Dlamini
Department of Biotechnology and Food
Technology
University of Johannesburg
Johannesburg
Gauteng
South Africa

Celestina Gosu-Attapkah
Department of Nutrition and Food Science
University of Ghana
Accra
Ghana
And
Ghana Standards Authority
Ghana

Celestina Omohimi
Department of Food Science and Technology
College of Food Sciences and Human Ecology
Federal University of Agriculture
Abeokuta
Nigeria

Christiana Eleojo Aruwa
Department of Biotechnology and Food Science
Faculty of Applied Sciences
Durban University of Technology
South Africa

Clinton Emeka Okonkwo
Department of Food Science
College of Food and Agriculture
United Arab Emirates University
Al Ain
United Arab Emirate

Ebunoluwa J. Atolagbe
Department of Public Health Sciences
Faculty of Basic Medical Sciences
Adeleke University
Ede, Osun State
Nigeria

Elizabeth Sivhute
Centre for Innovation and Development
University of Namibia
Windhoek
Namibia

Elizabeth Toyin Akinleye
Department of Microbiology
Federal University of Technology
Akure, Ondo State
Nigeria

Elizabeth U. Awani-Aguma
Department of Food Technology
University of Ibadan
Ibadan
Nigeria

Emmanuel G. Abere
Department of Economics
Faculty of Social and Management Sciences
Southwestern University
Okun-Owa, Ogun State
Nigeria

Ezekiel Green
Department of Biotechnology and Food
Technology
Faculty of Science
University of Johannesburg
Gauteng
South Africa

Fakoya Soji
Department of Biological Sciences
Olusegun Agagu University of Science and
Technology
Okitipupa, Ondo State
Nigeria

Fatai Oladunni Balogun
Department of Biotechnology and Food Science
Faculty of Applied Sciences
Durban University of Technology
Durban 4001
KwaZulu-Natal
South Africa

Feroz Mahomed Swalaha
Department of Biotechnology and Food
Technology
Durban University of Technology
Durban 4000
KwaZulu-Natal
South Africa

Folasade Banji-Onisile
Department of Microbiology
School of Life Sciences
College of Agriculture
Engineering and Science
University of KwaZulu-Natal
Durban 4000
South Africa

Frank Abimbola Ogundolie
Department of Biotechnology
Faculty of Computing and Applied Sciences
Baze University
Abuja
Nigeria

Funmilola Oluyemi Omoya
Department of Microbiology
Federal University of Technology
Akure, Ondo State
Nigeria

George Ooko Abong
Department of Food Science
Nutrition and Technology
University of Nairobi
Nairobi
Kenya

James A. Elegbeleye
Department of Consumer and Food Sciences
Faculty of Natural and Agricultural Sciences
University of Pretoria
Pretoria
South Africa
And
Department of Microbiology
Faculty of Pure and Applied Sciences
Southwestern University
Okun-Owa, Ogun State
Nigeria

Jamiu Olaseni Aribisala
Department of Biotechnology and Food Science
Faculty of Applied Sciences
Durban University of Technology
Durban 4000
KwaZulu-Natal
South Africa

Jane Misihairabgwi
Department of Human Biological and
Translational Medical Sciences
School of Medicine
Faculty of Health Sciences and Veterinary
Medicine
University of Namibia
Windhoek
Namibia

Jose Lorenzo
Facultad de Ciencias de Ourense
Área de Tecnología de los Alimentos
Universidad de Vigo
Ourense
Spain

Jumbale Mwarome
Department of Food Science
Nutrition and Technology
University of Nairobi
Nairobi
Kenya

Kehinde O. Oyeneye
Department of Economics
Crawford University
Faith City
Igbesa, Ogun State
Nigeria

Kehinde Oluyemi Ajayi
Department of Microbiology
Federal University of Technology
Akure, Ondo State
Nigeria
And
Department of Biological Sciences (Microbiology)
Precious Cornerstone University
Ibadan, Oyo State
Nigeria

Maame Akua Nyamekye
Department of Nutrition and Food Science
University of Ghana
Accra
Ghana

Maria Angula
Department of Human Biological and
Translational Medical Sciences
School of Medicine
Faculty of Health Sciences and Veterinary
Medicine
University of Namibia
Windhoek
Namibia

Mduduzi Mokoena
Department of Pathology
University of Limpopo
Mankweng
Polokwane
Limpopo
South Africa

Ohijeagbon O. Rebecca
Department of Food Sciences
Faculty of Food Science and Consumer Services
Ladoke Akintola University of Technology
Ogbomoso, Oyo State
Nigeria

Olabisi Omowumi Adebisi
Department of Molecular Biology and
Interdisciplinary Life Sciences
New Mexico State University
Las Cruces
New Mexico
USA

Olagboye O. Awosika
Department of Public Health Sciences
Faculty of Basic Medical Sciences
Adeleke University
Ede, Osun State
Nigeria

Olalekan J. Odukoya
Department of Food Science and Technology
College of Food Sciences and Human Ecology
Federal University of Agriculture
Abeokuta
Nigeria

Olufemi P. Sotayo
Department of Microbiology
Faculty of Science
University of Lagos
Akoka-Yaba
Lagos
Nigeria

Olusola A. Akinboboye
Department of Microbiology
Faculty of Pure and Applied Sciences
Southwestern University
Okun-Owa, Ogun State
Nigeria

Oluwafemi A. Adebo
Department of Biotechnology and Food
Technology
Faculty of Science
University of Johannesburg
Gauteng
South Africa

Oluwatosin A. Ijabadeniyi
Department of Biotechnology and Food
Technology
Durban University of Technology
Durban
KwaZulu-Natal
South Africa

Oluwatosin D. Ayannuga
Department of Microbiology
Faculty of Science
Adeleke University
Ede, Osun State
Nigeria

Opeyemi Lala
Department of Microbiology
Faculty of Science
Adeleke University
Ede, Osun State
Nigeria

Rebecca Olajumoke Oloniyo
Department of Food Nutrition and Home
Sciences
Prince Abubakar Audu University
Anyigba
Kogi State
Nigeria

Rukayat Abiola Abdulsalam
Department of Biotechnology and Food Science
Faculty of Applied Sciences
Durban University of Technology
Durban 4001
KwaZulu-Natal
South Africa

Saheed Sabiu
Department of Biotechnology and Food Science
Faculty of Applied Sciences
Durban University of Technology
Durban 4001
KwaZulu-Natal
South Africa

Tayo Funmilola Imafidon
Chemical Evaluation and Research
National Agency for Food and Drug
Administration and Control (NAFDAC)
Lagos State
Nigeria

Titilayo A. Ajayeoba
Department of Microbiology
Faculty of Science
Adeleke University
Ede, Osun State
Nigeria

Victor Ntuli
Department of Food Science and Technology
University of Venda
Thohoyandou
Limpopo
South Africa

Wisdom Selorm Kofi Agbemavor
Radiation Technology Centre
Biotechnology and Nuclear Agriculture Research
Institute
Ghana Atomic Energy Commission
Legon Accra
Ghana

Yemisi Adefunke Jeff-Agboola
Department of Science Laboratory Technology
University of Medical Sciences
Ondo City, Ondo State
Nigeria

Yetunde Mary Iranloye
Department of Food Science and Nutrition
College of Pure and Applied Sciences
Landmark University
Omu-Aran, Kwara State
Nigeria

Introduction

Every year, several million cases of foodborne diseases and deaths are reported in both developed and developing countries due to the consumption of contaminated food and water. The consumption of food contaminated by microbes has been associated with different foodborne diseases and forms an integral part of food safety concerns. Safety assurance of consumer goods has therefore become a global challenge to food processors and handlers, as well as to every stakeholder along the food value chain. Food safety assurance is further complicated by globalization and international food trade which exists among different countries. Failure to maintain good manufacturing and hygiene practices during food production, transportation, and storage; limitations of some food-processing methods in eliminating microbes and their toxins; adaptation and resistance of food microbes to changing climate conditions; and others are some of the challenges to food safety across the world.

A major constraint to the safety of consumer goods is the presence of food pathogens. Pathogens may be introduced into agricultural produce from irrigation water, insects, dusts, soil, fertilizers, and manure. In the United States, *Norovirus*, *Escherichia coli*, *Listeria monocytogenes*, and *Salmonella enterica* have been reportedly associated with contamination of fresh produce. For example, in 2011, *E. coli* O104:H4 outbreak spread from Germany to France, Denmark, and other countries, causing deaths and hemolytic uremic syndrome cases. Food pathogens are becoming more adaptive to their environment and have developed coping mechanisms to survive intolerable conditions, such as forming biofilms. With the development of these adaptive mechanisms, there is a need for interventions to mitigate pathogen contamination in food products. An assessment of the risks for biological hazards therefore becomes imperative in any food production and quality control system. Some other food contaminants such as pesticide residues, mycotoxins, antimicrobial residues, or antibiotic residues have been found in food products offered for sale. Even more challenging to food safety is food fraud, which involves the deliberate addition and/or substitution of food ingredients with non-food-grade ingredients or low-quality food ingredients, all aimed at profit maximization. In a bid to ensure food safety, the food retail sector faces numerous challenges, in terms of the inability to prevent microbial contamination, to prevent food fraud and maintain hygiene standards. Transmission of illness-causing microbes by food workers contributes significantly to all foodborne bacterial illnesses. This may be due to poor environmental and personnel hygiene practices, lack of adequate food safety training, the use of sick workers (while sick), no sick leave, and lack of good agricultural practices.

The sustainability of the food industry will largely depend on the ability to control and decisively eliminate compromise in food safety. This will consider all food threats and proffer measures to eliminate or reduce them in consumer foods, including the application of DNA-based techniques to food systems, predictive modeling, and food

https://doi.org/10.1515/9783110748345-205

safety culture. Good manufacturing practices will become pivotal to quality assurance during food production. Food sanitation programs, focusing on the design and construction of food plants, hazard analysis (HACCP), and basic hygiene practices of food handlers, will form the basis for industrial food manufacturing. Food legislation provides the boundaries for safe food manufacturing practices as well as guiding international food trade through the *Food Code* (*Codex Alimentarius*). The food laws will regulate the inclusion of substances as food additives and food ingredients, enhance the full disclosure of food ingredients to consumers, and guide against misrepresentation, mislabeling, and purposeful substitution of food products or ingredients. Furthermore, food regulators as well as relevant stakeholders need to frequently monitor both existing and emerging food safety risks and adopt appropriate smart food safety early warning systems or technologies.

This book provides in-depth information on food safety and toxicology, analyzing the present challenges to global food safety, and providing the pathway to safety and security in the food industry.

Funmilola Oluyemi Omoya*, Kehinde Oluyemi Ajayi
and Elizabeth Toyin Akinleye

Chapter 1
Hazard in food and global state of food safety

Abstract: Food is generally known as any substance consumed to provide nutritional support for an organism. It may be of plant or animal origin and contains essential nutrients needed for growth. A good food must be safe because it plays a major role in ensuring continuous advancement of a nation by enhancing the activities of trade and tourism. Food supply has been globalized, and this raises new challenges to ensuring the safety of the food supply. Food may be hazardous when it is contaminated by biological, chemical and physical hazards, these hazards could gain access to food during production processes or from the source during farm practices. The presence of hazardous agents in food substance rendered the food unsafe for human consumption, and when consumed becomes injurious to human health. Therefore, scientists may need to discover a faster way to detect hazardous agents in food and target certain allergy-causing proteins in foods. This may help decrease the risk of foodborne illnesses and aid in keeping food safe to eat.

1.1 Introduction

Food is generally known as any substance consumed for nutritional support of an organism [1]. Food may come from a plant or an animal and contains essential nutrients, such as carbohydrates, fats, proteins, vitamins, or minerals. Just as fuel is an important factor for vehicle mobility so also is food in its appropriate dietary proportion needed by mankind for their sustainability. Therefore, food is indispensable to humans and has significant importance in the survival and promotion of an excellent health in humans.

Safe food plays a vital role in ensuring continuous advancement of a nation by enhancing the trade and tourism activities [2]. This is mainly due to the global development which has resulted in the establishment of more complex and elongated global food chain. Although there have been recent campaigns for production of more

*Corresponding author: Funmilola Oluyemi Omoya, Department of Microbiology, The Federal University of Technology, Akure, PMB 704, Akure, Nigeria, e-mails: fomoya@yahoo.com; foomoya@futa.edu.ng
Kehinde Oluyemi Ajayi, Department of Microbiology, The Federal University of Technology, Akure, PMB 704, Akure, Nigeria; Department of Biological Sciences (Microbiology), Precious Cornerstone University, Ibadan, PMB 60, Agodi, Ibadan, Nigeria
Elizabeth Toyin Akinleye, Department of Microbiology, The Federal University of Technology, Akure, PMB 704, Akure, Nigeria

https://doi.org/10.1515/9783110748345-001

localized food, in reality, most people in developed countries prefer globalized food supply because consumers get some benefits [3, 28]. However, globalization in addition to diversities in local food production poses a new challenge for the safety of the food supply [4].

The advanced globalization of world's food supply will expose more population worldwide to food hazards. The threats posed by food safety remain today; hazards list that food safety practitioner manages increases on daily basis and the complexity is enormous. Also, climate changes have unimaginable impacts on food safety, both directly and indirectly, thereby predisposing the public to health risk [4].

Hazards in food occur as a result of introduction of any agent that has the ability to cause adverse health effects on consumers. Hazards could also occur in food due to exposure to hazardous agents that could cause contamination of the exposed food, and such agents could be of biological, chemical, allergenic, physical, nutritional, and/or biotechnology origin [5]. Hazards of different origin could be introduced to food any time during planting, harvesting, transportation, and storage of raw foods, as well as during formulation, processing, packaging, transportation, storage, preparation, and serving of processed foods.

1.2 Food safety

The availability of information and an increase in the number of data on diseases related to foodborne have boosted the awareness of general public on food safety [4]. Food safety can be pictured as, "protocols and steps targeted to keep foods from any or all hazards that could be present in different stages of food." Food safety has broad interpretation, which do not cover ways or protocols established to reduce food contamination of microbial origin but also include hazards introduced during other food processes. It may cause unavailability of nutrients in diet or formation of hazardous compounds [5].

The advantages of safe food supply include an increase in national economy, trade, and tourism, improvement of food and nutrition security, and underpin sustainable development [4]. The number of people buying and eating food prepared in public places has increased as a result of urbanization and changes in consumer habits which include travel [3]. Globalization has enhanced growing consumer demand for different foods, which has led to an increase in complex and longer global food chain. The continuous increase in world population, and mechanization and industrialization of agriculture and animal production to meet high demand for food have created both opportunities and threat for food safety [6]. Food safety is also predicted to be affected by climate change. Incidents of food safety at local level can quickly spread across the globe leading to international emergencies because of the fast speed of global food product movement. Disease outbreaks originated from food have been reported in every parts of the world in the last decade, and these were often spread through global trade. For example, due to the contamination of ready-to-eat meat with *Listeria monocytogenes* in South Africa, 1,060 cases of listeriosis and 216 deaths were reported in the year 2017/2018 [7]. In this case, contaminated ready-to-eat meat products were exported to 15 other African countries, and led to international response for risk management measures.

This showed the advantage of food safety in sustainable development goals' reality. Prioritization of food safety will enable a government to make public health-promoting policies and food safety systems [7].

1.3 Hazard in food

Continuous access to good amounts of safe and nutritious food is a major drive for life sustainability and good health promotion, and unsafe food containing harmful bacteria, viruses, parasites, or chemical residue is hallmark of over 200 infectious and noninfectious diseases [8]. In the world, almost 1 in 10 people have contacted various illnesses as they consume unsafe food and reported that 420,000 die each year, resulting in the loss of 33 million healthy life years lost [8]. According to the World Bank and World Health Organization, over US $100 billion is lost each year for pharmaceutical and medical due to unsafe food [8, 9]. The presence of hazards in food has created numerous cycles of disease, and malnutrition now affects infants, young children, elderly, and individuals with compromised immunity. Foodborne diseases affect the socioeconomic development by putting more burdens on healthcare systems, and nation's economies. Food supply now moves across different continents; collaboration between governments, producers, and consumers will help in eliminating food hazards and enforce food safety [8].

1.3.1 Biological hazards in food

The global impact of food hazard on health and quality assurance is a known fact, as well as the health effects, the various types of contaminated food most times have their economic costs underestimated, and also the outbreaks of foodborne diseases are mostly unnoticed, unreported, nor properly pass laboratory investigation. Globalization has resulted in widespread distribution of food and now serves as a medium whereby new and novel strains of pathogens that are known to be endemic in specific geographical area are being introduced [10]. Also, globalization has added to the control and eradication of both infectious and noninfectious foodborne hazards.

Global economic integration has helped raise incomes in many developing countries. However, rapid urbanization and other amenities are needed for safe food handling. Also, an increase in urbanization rate has resulted in commercial food production, therefore, an urgent need for training and retraining for food hygiene management practices [3, 28].

Biological hazards in food are defined as the presence of life organisms, or substances such as toxins produced by organisms that have deleterious effect on human health. The biological hazards are known to cause most foodborne illness outbreaks; hence, they are of paramount concern in food processing [25]. There are many risks associated with the biological hazards that have a tremendous impact on globalization. These organisms cause infection, intoxication, and sometimes even lead to death in humans [25].

There are different sources of biological hazards: environmental hazards such as bacteria in the soil, and agricultural runoff, poor sanitation practices, and cross-contamination associated with transportation, handling, processing, and storage. The microbial load and type of microorganisms in food is determined by the nature of food, packaging, storage environment, and conditions [11].

Biological hazards in food are microorganisms like bacteria, viruses, fungi, and parasites or their toxins that affect human health. Biological hazards are responsible for most of foodborne illness outbreaks which cause a great challenge to the food industry [5]. The factors associated with the emergence or outbreak of foodborne pathogens include climate changes and weather: for instance, hot weather conditions increase aflatoxin contamination in cheeses in 2013–2014; globalization is the major hallmark of hepatitis A virus outbreak in European Union and this is assumed to correlate with berry production in different Eastern European countries; modernization or change in manufacturing technologies was responsible for botulism outbreak in sauces or soups. Behavior and demographics of humans also played a major role in outbreak of foodborne illness, for example, higher consumption of meals away from home, genetic makeup, and infection susceptibility played a role in the emergence of *Clostridium difficile*. Furthermore, adaptation and modification of cells led to methicillin-resistant *Staphylococcus aureus* emergence in foods [12].

Foodborne bacteria are of different species; however, the major bacterial culprits in foodborne illnesses are *Bacillus cereus, Campylobacter jejuni, Clostridium botulinum, Clostridium perfringens, Escherichia coli* O157:H7, *Escherichia coli* O104:H4, *Listeria monocytogenes, Salmonella* species, *Shigella* species, *Yersinia enterocolitica, Staphylococcus aureus, Vibrio cholerae, Vibrio parahaemolyticus, Vibrio vulnificus,* and *Cronobacter sakazakii* [13]. Food contaminated with pathogenic and/or toxin-producing microorganisms could cause illness after consumption. Such illnesses could be infection or intoxication, or both forms. Pathogenic microorganisms cause adverse health effect on humans through mechanisms in which beneficial microorganisms are eliminated, use up the host nutrient, and destruction of the host tissue [5]. Food infection differs from food intoxication in the fact that it usually showed up on the host within hours of consuming the food suspected to be contaminated. Whereas food intoxications are caused by the presence of toxins that are already secreted by the microorganisms, either present in the food or in humans after ingestion.

Foodborne viruses are highly infectious. They penetrate the host cell, alter the function, and replicate their components as a means of reproduction. Following viruses, bacteriophage, hepatitis A virus, enteric virus (other than hepatitis A and noroviruses), norovirus, Norwalk, and rotavirus are to be considered in food safety. Contamination of food by people infected with virus, food ingredients such as water, and processing environment (i.e., equipment contamination) are various ways in which viruses gained entrance into food [13].

A parasite is any organism that is metabolically dependent on its host for growth and reproduction. Their presence in the host is usually detrimental to the host and does not supply the host with any nutrient. *Cryptosporidium parvum, Giardia duodenalis* or *intestinalis, Taenia* species, *Trichinella spiralis, Toxoplasma gondii, Entamoeba histolytica*, and *Entamoeba coli* are common parasites associated with foodborne illness. They gain entrance into food through similar routes in which viruses enter the food (i.e., poor personal hygiene practices and contaminations) [13].

Furthermore, in addition to other biological food safety hazards is prions, that is, proteinaceous infectious particles; they contain mainly protein [14]. They cause diseases such as bovine spongiform encephalopathy (BSE), which is called "mad cow disease," is a fatal disease of cattle's nervous system and is known as a transmissible spongiform encephalopathy (TSE). Other TSEs include scrapie, chronic wasting, and Creutzfeldt-Jakob disease in sheep, deer, and humans, respectively. Humans are infected by Creutzfeldt-Jakob disease through consumption of cattle infected with BSE [14]. Presently, treatment and vaccine are not available for the disease.

The biological hazards can originate from different sources such as fecal contamination, animal guts, soil and water contaminated by nontreated animal manure, and cross-contamination. There are different means by which cross-contamination can take place, and they are through cross-contamination of food products starting from processing environment as a result of unhygienic environment, and human contamination could be originated through poor personal hygiene, fecal contamination, and infection control failure which happens as a result of an unreported illness [26].

1.3.2 Physical hazards in food

Physical hazards or extraneous materials include those materials excluding biological hazards that could be present in a food which are originally not part of the food makeup, that is, foreign [15]. These hazards are normally nontoxic; however, they are associated with different protocols such as unsanitary production and processing environment, handling, storage, and food distribution. A few examples of physical hazards in food are insects, metal fragments, hair, pieces of plastic, wood chips, and glass [15]. These materials are termed hazardous because of sharpness, hardness, size, or shape which may result in health effects in humans, and such effects may include lacerations, perforations, and wounds or choking.

Physical hazards in food could be foreign materials (such as metal fragments in ground meat) unintentionally introduced to food products or objects that occurred naturally (e.g., bones in fish) could be injurious to the final consumer [16]. Food products may be contaminated by physical hazards during phases of production. The potential physical hazards in food could be classified as hard or sharp objects (these can cut the throat or mouth, damage the intestine, or even damage the teeth or gums); also food recall could occur because of the presence of physical hazards and this may affect the brand name and integrity of the company and products [29].

Physical hazards in food may have been originated from glass (e.g., glass containers), metal (e.g., fragments from equipment), plastics, stones incorporated in field crops during harvesting, wood, and natural components of food such as hard or sharp parts of a food (e.g., shells in nut products). The factors that determine a potential physical risk are sizes, types of consumer, types of product, and physical characteristics such as hardness, shape, and sharpness of food [29].

Extraneous materials in food are divided into unavoidable and avoidable. Unavoidable materials are by-products of the food processing or inherent materials in the product. For example, stems found in blueberries, dirt on potatoes, or insect fragments in figs' extraneous matter [15].

Avoidable materials are preventable in food, hence, are less tolerated than unavoidable. If proper good manufacturing practices (GMPs) are followed during food processing, the avoidable foreign materials will not be present. Avoidable extraneous materials may be any of the following: glass fragments, plastic and jewelry pieces, rubber chunks, barbules of feather, or any other foreign materials [15].

1.3.3 Allergenic hazards

Allergens are proteins that elicit abnormal immune response in a sensitive individual. Immunoglobulin E (IgE) is a major antibody involved in allergic reactions to food [17–19]. Some foods contain allergenic proteins as their natural constituents; this could cause a health risk to sensitive individuals. Symptoms of food allergy include but not limited to swelling of tongue and throat, nausea, rash, itchy skin, and death in serious cases [19]. Symptoms of a food allergy are usually sudden, triggered by small food quantity, and the occurrence always corroborates the time of food consumption. These associated symptoms result from immune system reaction due to the presence of allergen.

Consumption of foods that contained allergens must be avoided by allergic individuals to prevent immune reactions which may be life-threatening. Undeclared food allergens have their source from improper declaration of food allergen ingredients by the manufacturer or unintended due to allergen cross-contact [5].

Some foods that have been labeled to cause allergic reactions are peanuts, tree nuts such as almonds, Brazil nuts, cashews, and walnuts, sesame seeds, milk, eggs, some seafood, soy, and wheat [17].

It should be noted that not all adverse or severe reactions to food are mediated by IgE, for example, autoimmune disease called celiac disease; in this case, antibodies attack tissues in the presence of gluten which decreases the ability of the gut to absorb nutrients [5]. Individuals with celiac disease can make the individual to be nutrient deficient. Consumption of gluten-containing grains could increase the risk of lymphoma and osteoporosis [17].

Complete avoidance of the specific allergen by a sensitive individual is the only way to manage food allergies (and celiac disease) as there is no cure [5]. The consumer relies on food label information to manage food allergy. Inaccurate, undeclared, or hidden allergens on food labels could result in health hazard. Processing-associated contamination and other procedures can unintendedly produce allergenic by-products. Strict adherence to GMPs, hazard analysis and critical control points, and allergen prevention plans is very important [5].

1.3.4 Chemical hazards

Food products are contaminated when they come in contact with hazardous substances like chemicals at any stage during processing of food [4]. Chemicals are used during planting of food crops in farms; they are helpful, used as pesticides and for weed control. They are not usually hazardous if applied properly; however, they could be hazardous to consumers when used uncontrollably or the recommended treatment rates are exceeded.

Chemical hazards occur in food when chemicals are present above the permissible limit. Chemicals used as pesticides, animal treatment, manufacturing processes, and food additives could cause contamination.

Various hazardous chemicals in food industry include mycotoxins (e.g., aflatoxin, deoxynivalenol, ochratoxin, fumonisin, and patulin), natural toxins (e.g., glycoalkaloids from potatoes, hypoglycin from ackee fruit, cyanogenic glycoside from cassava root, bamboo shoots, and stone fruit), and marine toxins (marine toxins accumulated in fish and shellfish). Sources of marine toxins are decomposition and microscopic marine algae, environmental contaminants, food additives, for example, food colors (natural and synthetic), pH-adjusting agents, preservatives, bleaching agents, food enzymes, glazing and polishing agents, emulsifiers, and gelling agents, processing-induced chemicals (e.g., acrylamide, ethyl carbamate, and furan which are produced as a result of reactions during food processing), pesticides/agricultural products, and veterinary drug residues [5].

1.3.5 Nutritional hazards

Nutrients are essential to maintain good health. They include carbohydrates, proteins, fats, vitamins, and minerals. They are usually added as food fortification (e.g., addition of vitamins, minerals, and amino acids in foods) to ensure good health. Introduction of vitamins and minerals to food will help protect against nutritional deficiencies (e.g., milk fortified with vitamin D eliminate childhood rickets), and reduce the risk of diet-associated diseases (e.g., fortification of food with calcium and vitamin D help build strong bones and may reduce the risk of osteoporosis).

Despite the high health benefits derived from fortified food, overfortification can be hazardous, for instance, high consumption of vitamin A through food fortification by women could lead to birth defects during pregnancy [20].

1.3.6 Biotechnology-related hazards/novel foods

Biotechnology is the application of science and engineering to the direct or indirect use of living organisms or parts or products of living organisms in their natural or modified forms [27]. The genetically modified (GM) food could cause allergy, transfer of genes from GM food to either the body cells or bacteria that are present in the gastrointestinal tract of the consumer. In novel food, genetic modification is not used. They result from a new process not previously used for food or products that do not have a history of safe use as a food [21, 22].

1.3.7 Radiological hazards

Radiological hazards are very rare in food but their occurrence in food pose a higher health risk that occurs over a period of time [23]. Consumption of this in food could lead to the presence of radionuclides in humans and this depends on the radionuclide and the amount of radiation exposed, for example, radioactive iodine exposure could increase the risk of thyroid cancer [5].

1.3.8 Functional foods and hazards

In the twenty-first century, foods were not only consumed to stop hunger but also necessary for provision of nutrients to humans. Food serves as a means of improving the mental and physical states of consumers by preventing occurrences of disease. Hence, functional foods, also known as nutraceuticals, are the foods that provide nutritional and/or medical benefits which include prevention and treatment of diseases [24].

Functional foods are not exempted from the attack of these hazardous agents, especially the biological hazards. Microorganisms are ubiquitous and exhibit their activities in any favorable condition. Functional foods can be divided into foods originated from plant and animal origin (e.g., omega-3 fatty acids found in fish, probiotics; these are microbial food supplements having beneficial effects on consumers such as lactic acid-producing bacteria which are generally consumed in the form of yogurt) [29].

1.4 Conclusion

Food is an essential component of human's everyday life activities; sequel to this, food safety is of paramount concern. Food safety in the field of science generally includes all processes in food production cycle guided in ways that protect the health of consumers. This includes different routines and guidelines used to eliminate hazards in food. For food to be totally safe for consumption, the following are to be considered: the food origin, practices associated with labeling, additives, hygiene, control of hazards, and GMP. The prevention of different types of hazards in food is comprehensive; there is a need for food producers, national and international bodies to swift into action to curb risks associated with food safety in fast daily growing and more complex and globalized food chains.

References

[1]   Food and Agriculture Organization of the United Nations (FAO). (2015). Fishery Fact Sheets Collections. ASFIS List of Species for Fishery Statistics Purposes. In Fisheries and Aquaculture Department. Department of food safety and zoonoses (pp. 1–7).

[2]   Shrivastava, S. R., Shrivastava, P. S., & Ramasamy, J. (2016). Global food safety: challenges and recommended public health strategies. *International Journal of Preventive Medicine*, 7, 8.

[3]   Hoffmann, S., & Harder, W. (2010). Food Safety and Risk Governance in Globalized Markets. *Health Matrix: The Journal of Law-Medicine*, 20(1), 5–53.

[4]   Food and Agriculture Organization of the United Nations (FAO). (2018). There is no Food Security without Food Safety (pp. 8–13).

[5]   Al-Rub, F. A., Pittia, P., Shithab, P., Antonello, P., et al. (2020). Definition of food safety hazards (1st Edition) GAVIN eBooks, Lisle, IL 60532, USA pp. 5–8.

[6]   Food and Agriculture Organization of the United Nations (FAO). (2019). The World Health Organization (WHO). Codex Allimentarius International Food Standards. General standard for contaminants and toxins in food and feed. CODEX STAN, 193–1995. Available online: http://www.fao.org/fao-whocodexalimentarius/codex-texts/list-standards/en/ (accessed on 10 February 2020).

[7]   Child, K. (2018). Enterprise colony identified as source of *Listeria* outbreak. Times live. Retrieved July, 2021

[8]   World Health Organization. (2021). Health topics. Food safety. www.who.int/health-topics/food-safety (accessed on 2nd July, 2021).

[9]   World Bank. (2018). Food-borne illnesses cost US$ 110 Billion per year in low- and middle income countries. World bank Press release, October, 2018

[10]   Schirone, M., Visciano, P., Tofalo, R., & Suzzi, G. (2017). Editorial: Biological hazards in food. *Frontiers in Microbiology*, 7, 2154.

[11]   Sofos, J. N. (2014). Foods, materials, technologies and risks. *Encyclopaedia of Food Safety*, 3, 268–279.

[12]   Woolhouse, M. E. J., & Gowtage-Sequeria, S. (2005). Host range and emerging and remerging pathogens. *Emerging Infectious Diseases*, 11, 1842–1847.

[13] Bintsis, T. (2017). Foodborne pathogens. *AIMS Microbiology*, *3*(3), 529–563.

[14] Health Canada. (2013). Food related illness. Canadian food Inspection Agency (CFIA).

[15] Compendium of Analytical Methods (CAM). (2009). Extraneous materials overview. Guidelines for the general cleanliness of food – An overview.

[16] Nestor, P., & Theofilos, M. (2020). *Physical and mechanical hazards* (1st Edition, pp. 35–38).

[17] Gray, C. L., & Levin, M. E. (2014). Epidemiology of food allergy. *Current Allergy and Clinical Immunology*, *3*, 170–176.

[18] Sicherer, S. H. (2011). Epidemiology of food allergy. *Journal of Allergy Clinical Immunology: In Practice*, *127*, 594–602.

[19] Sicherer, S. H., & Sampson, H. A. (2014). Food allergy: Epidemiology, pathogenesis, diagnosis, and treatment. *Journal Allergy and Clinical Immunology*, *133*, 291–307.

[20] Eva, D. (2021). Vitamin A during pregnancy. Babycenter, reviewed by Erin, H June, 2021 Inspection Canada, (2014). *Modern Biotechnology*. A brief overview – Canadian food inspection agency.

[21] Bawa, A. S., & Anilakumar, K. R. (2013). Genetically modified foods: Safety, risks and public concerns – A review. *Journal of Food Science and Technology*, *50*, 1035–1046.

[22] Kuiper, H. A., Kleter, G. A., Noteborn, H. P. J. M., & Kok, E. J. (2001). Assessment of the food safety issues related to genetically modified foods. *The Plant Journal: For Cell and Molecular Biology*, *27*, 503–528.

[23] Food and Drug Administration (FDA). (2016a). Hazard analysis and risk based preventive controls for human food: Draft guidance of industry, http://www.FDA.gov/downloads/Food/GuidanceRegulation/GuidanceDocumentsRegulatoryInformation/ChemicalContaminantsMetalsNaturalToxinsPesticides/UCM374534, Retrieved, May, 2020.

[24] Saini, R. D. (2017). Chemistry of functional foods and their role in disease control. *International Journal of Biotechnology and Biochemistry*, *13*(2), 191–203.

[25] Al-Nabuisi, A., & Osaili, T. (2020a). *Microbiological and biological hazards* (1st Edition, pp. 9–24).

[26] Al-Nabulsi, A., & Osaili, T. (2020b). Food safety. Microbiological and Biological Hazards. Published by GAVIN eBooks (p. 9), 5911 Oak Ridge Way. Lisle. IL 60532, USA. ISBN978-1-951814-03-8.

[27] Maryam, B. M., Datsugwai, M. S., & Shehu, I. (2017). The role of biotechnology in food production and processing. *Engineering and Applied Sciences*, *2*(6), 113–124.

[28] World Health Organization. (2018a). Food Safety, Climate Change and the Role of WHO Department of food safety and zoonoses (pp. 1–7).

[29] World Health Organization (WHO). (2018b). Natural Toxins in food (Factsheet). Available online: https://www.who.int/news-

Elizabeth Sivhute, Jane Misihairabgwi and Ahmad Cheikhyoussef*

Chapter 2
Bacterial pathogens of food importance

Abstract: Bacterial species causing foodborne illnesses such as *Campylobacter jejuni*, *Clostridium botulinum*, *Listeria monocytogenes*, Shiga toxin-producing *Escherichia coli* O157, *Salmonella typhi*, *Shigella dysenteriae*, and *Vibrio cholerae* are the main representatives of bacterial pathogens of food importance, and they pose an increasing threat to the safety and health aspects of the public worldwide. To improve the global food safety and public health system, it is very important to develop novel and high technology-based solutions to fast detect and track record for the identification of these bacterial pathogens. Combination and comparative analysis of several advanced methods should be considered with promising potential for rapid and instant reporting. Food industry must ensure the safety of the produced foods to public and introduce faster, cost-effective, and safer processing steps to reduce/eliminate the presence of these biothreats in food production environment, and must advance toward efficient treatment for foodborne bacteria to save more lives and make sure that the world is safer for the next generations.

2.1 Introduction

There is a rising number of pathogenic bacteria outbreaks in food systems which result in an increasing number of infections worldwide, and the public health concerns are growing on their economic loss and impact in several developing [1–4, 165] and developed countries [5, 6, 169]. The presence of foodborne pathogens in food products and water is a big challenge for health authorities due to their rapid growth and dangerous impact on human health [7–9].

Several challenges have been reported during developing a rapid and sensitive method for foodborne pathogen detection in food environment, which is a very challenging process. These include sample preparation cost, food matrix complexity, required time frame for the detection of pathogens, and high cost associated with the advanced molecular assays and specialized resources required [10]. The World Health

*Corresponding author: Ahmad Cheikhyoussef, Science and Technology Division, Multidisciplinary Research Services (MRS), Centre for Research Services (CRS), University of Namibia, Private Bag 13301, Windhoek, Namibia, e-mail: acheikhyoussef@unam.na
Elizabeth Sivhute, Department of Food Science and System, School of Agriculture, University of Namibia, Windhoek, Namibia
Jane Misihairabgwi, Department of Human, Biological and Translational Medical Sciences, School of Medicine, Faculty of Health Sciences, University of Namibia, Private Bag 13301, Windhoek, Namibia

https://doi.org/10.1515/9783110748345-002

Organization (WHO) reported that in 2007 waterborne and foodborne infections are responsible for an approximate of 4 billion episodes of diarrhea every year, leading to 1–2 million deaths worldwide [11]. Kunadu et al. [12] indicated that developing countries are facing difficulties and challenges in reporting the estimates on food-borne illnesses and their causing factors and associated death statistics due to weak mentoring and surveillance systems. They further indicated that about 91 million in Africa encountered illnesses on annual basis because of foodborne diseases with eco-nomic losses for South Africa (2017–2018) and Tanzania (1997) caused by listeriosis and cholera outbreaks of US $260 million and US $36 million, respectively. In develop-ing countries, the main causes of foodborne diseases are lack of food safety surveil-lance systems and precision of regulatory instruments, lack of knowledge on food safety, and poor hygienic practices [12]. Therefore, the development of real-time and sensitive detection of pathogens will greatly monitor and control foodborne patho-gen-related fatalities [13]. This chapter summarizes and discusses the latest updates on the bacterial pathogens of food importance in terms of their distribution in differ-ent food systems, state-of-the-art recent advances of their reported detection methods, and their processing methods with the latest solutions and treatment.

2.2 Foodborne bacterial pathogens (FBPs) and foodborne illnesses

Foodborne diseases remain a major public health concern across the globe [9, 14–16, 157]. The Centers for Disease Control and Prevention (CDCP) in the United States reported that one in every six individuals acquires one type of foodborne illness every year [15]. Food-borne illnesses are caused by consuming food that has been contaminated by pathogens and/or their toxins [17], with their ability to be present in several types of foods within different environments [18]. Although foodborne illnesses are caused by different patho-genic organisms (bacteria, viruses, parasites, and fungi) [19], bacterial pathogens were pointed out and identified as the leading cause of foodborne illnesses [18], and have been implicated in many disease outbreaks [157]. Examples for the most common foodborne bacterial pathogens (FBPs) include *Bacillus cereus, Campylobacter jejuni, Clostridium botu-linum, Clostridium perfringens, Escherichia coli* O157:H7, Shiga toxin-producing *E. coli* (STEC), *Listeria monocytogenes, Salmonella enterica, Staphylococcus aureus*, and *Vibrio* species [18–22]. One of the FBPs is *S. aureus* which can secrete several virulence factors including enzymes (lipase, phosphatase, phospholipase, and hyaluronidase) and toxins (enterotoxins and exotoxins), which enable this bacterium in surviving and thriving its ecosystem due to their active cellular metabolism and high specificity [23, 24].

2.2.1 *Bacillus* spp.

The genus of *Bacillus* has rod-shaped, Gram-positive, aerobic, and/or sometimes anaerobic bacteria from the family Bacillaceae [18]. The genus has about 266 named species. Mostly, they can be found in several environments such as fresh and marine
water, and soils [14], and can form spores. Although most *Bacillus* spp. are not harmful to animals and humans, *Bacillus cereus* causes spoilage in canned foods and has
been linked to several foodborne outbreaks [14, 18, 20, 157]. Foodborne illness due to
B. cereus is caused by ingestion of contaminated food with more than 100,000 colony-
forming units per gram of the infected food. The common food vehicle is reheated
fried rice [14], although other food products, such as pasta salad, refried beans, spa-
ghetti, tomato sauce, and water, were also reported as sources of contamination [20].
B. cereus infection is characterized by abdominal pain, watery diarrhea, rectal tenes-
mus, and diarrhea, whereby symptoms can be developed within 6–15 h and continue
for 24 h [20].

2.2.2 *Campylobacter* spp.

Campylobacter is a Gram-negative, nonspore-forming bacterium that belongs to the
family Campylobacteriaceae [18]. Morphologically, they exist as either curved or spiral rods, and chemically, they are characterized as catalase- and oxidase-positive [19].
The same authors reported 25 species and 8 subspecies under this genus; *Bacteroides
ureolyticus* is also included under this genus. Among these species, *Campylobacter jejuni* and *Campylobacter coli* are the most important foodborne pathogens [19] that
have caused many diarrheal illnesses from several countries such as the United States
[18, 19, 25]. *Campylobacter* is mainly prevalent in intestinal tracts, oral cavity, and the
reproductive organs of humans and animals such as poultry, cattle, pigs, and sheep
[19]. Therefore, direct contact with infected animals or through equipment, water, or
during processing of carcasses in slaughtering area can lead to *Campylobacter* transmission [19]. Human infection with *Campylobacter* spp. results mainly from the consumption of contaminated products such as undercooked poultry [26]. *Campylobacter*
incubation period differs from 3 to 5 days [19]. Furthermore, human campylobacteriosis is characterized by abdominal pain, cramps, fever, malaise, vomiting, and watery/
bloody diarrhea [19].

2.2.3 *Clostridium* spp.

Clostridium genus is a Gram-positive bacterium that belongs to the family Bacillaceae
[18]. They include obligatory anaerobic or aerotolerant bacteria that are capable of
producing endospores [27]. The genus is comprised of around 250 species that include

important pathogens responsible for foodborne disease outbreaks [16, 27, 157]. *Clostridium botulinum* is a heterogeneous bacterium that is responsible for foodborne botulism [28]. Foodborne botulism is a severe form of food intoxication that results from the consumption of preformed botulinum neurotoxin [27]. Traditionally, eating non-well-processed sausages or home-canned foods can cause botulism [27]. Botulism is caused by eating contaminated foods such as garlic sauce, olives, potato salad, sauteed onions, and yoghurt [18]. When it comes to the detection of *C. botulinum* in food matrix, the focus is based on the identification of *C. botulinum* toxins such as botulism neurotoxin, instead of the bacteria itself [30]. The main leading infectious cause of antibiotic-associated diarrhea and colitis is *C. difficile* with 500,000.00 infection cases every year and approximately a recurrence risk of 20% after successful initial therapy [29]. *Clostridium perfringens*, previously known as *Clostridium welchii*, is one of the most common causes of bacterial foodborne diseases [18]. It is also classified among the primary causes of bacterial foodborne disease epidemics in many countries, for example, Australia and England [18, 27]. *C. perfringens* exists in intestines of humans and many animals and can, therefore, thrive in soil and areas polluted by human or animal feces [18]. Infections with *C. perfringens* are linked with consuming contaminated meat and poultry products [18]. Most likely, people suffering from *C. perfringens* infection will be exposed to diarrhea and stomach cramps after eating contaminated food with a timeline ranging from 6 to 24 h. Dieterle et al. [29] reported novel therapies and preventative strategies, to mention few: management of intestinal microbiome when antibiotic is being administered, newer antibiotics utilization, probiotics, and vaccinations for the primary and recurrent *C. difficile* infections.

2.2.4 *Escherichia coli*

According to Bintsis [18] and Abebe et al. [19], "*E. coli* is a Gram-negative, rod-shape, non-spore-forming bacterium that belong to Enterobacteriaceae." They are facultative anaerobes that can transform simple sugars such as glucose to form lactic, acetic, and formic acids via fermentation [19]. *E. coli* inhabits the gastrointestinal tract (GIT) of animals and humans [14]. While most *E. coli* strains are harmless to humans, other strains are pathogenic and cause serious human illnesses via toxin production [14, 18, 19]. According to pathogenic mechanisms for *E. coli*, this genus has been divided into six clusters, including enteropathogenic *E. coli*, enterohemorrhagic *E. coli*, also known as STEC, enterotoxigenic *E. coli*, enteroaggregative *E. coli* (EAggEC), enteroinvasive *E. coli*, and attaching and effacing *E. coli* [31]. According to Styles et al. [32], cattle are considered the natural reservoir of EHEC, and approximately three quarters of EHEC infections in humans are derived from bovine products. They further reported a potential quorum-sensing (QS)-based therapeutic approach in *E. coli* and identified a small molecule modulator of SdiA and provided the covalent inhibitor mechanism. SdiA from *E. coli* (SdiA$_{EC}$) is an important factor for EHEC colonization and to cause the disease [32]. "Antivirulence" approach is a strategy based on targeting the regulation, and the function of virulence factor has a potential therapeutic strategy to prevent or treat bacterial infections [33–35]. Among these *E. coli* pathotypes, STEC O157:H7 is the main cause for many foodborne disease outbreaks [21, 36, 158]. STEC, which has got the characteristics of both verotoxigenic *E. coli* and diarrheagenic EAggEC [21], is responsible for several human illnesses such as moderate diarrhea, hemolytic-uremic syndrome (HUS), and severe bloody diarrhea termed hemorrhagic colitis (HC) [19, 37, 38]. Pathogenic *E. coli* is among many pathogens which can get access to food of animal origin [19]; it can, therefore, be transmitted via consuming polluted water and foods, such as milk, raw, or non-well-cooked products, and vegetables [26, 163]. Sometimes, occupational exposure is another way for pathogenic *E. coli* to be transmitted from person to person with an incubation period ranging from 2 and 10 days [19]. After that, several symptoms start to develop such as abdominal pain, diarrhea, HC, HUS with acute kidney failure, thrombotic thrombocytopenic purpura, and vomiting [19, 37].

2.2.5 *Listeria* spp.

Listeria is a Gram-positive bacterium, psychotropic, motile, facultative anaerobic, nonspore-forming, and rod-shaped microbe [19, 157]. According to Abebe et al. [19], the genus is comprised of 10 species including *Listeria monocytogenes*, *L. ivanovii*, *L. innocua*, *L. grayi*, *L. murrayi*, *L. marthii*, *L. fleischmannii*, *L. seeligeri*, *L. welshimeri*, and *L. weihenstephanensis*. Two of these species *L. monocytogenes* and *L. ivanovii* are considered pathogenic [39]. *L. monocytogenes* has been responsible for most of the human cases of listeriosis [19], and it is the most dangerous in terms of the leading causes of death in humans in many countries [18, 39]. It is contributing to the highest proportion of ill people in hospitals with zoonosis in the European Union (EU) according to the investigation by the European Food Safety Authority and European Centre for Disease Prevention and Control [159].

The genus is widely distributed in nature [19] and is prevalent in different types of uncooked foods, for example, meats and vegetables [39]. Infection with *L. monocytogenes* occurs mostly after consuming polluted raw food [157] and contaminated cheeses which are either freshly prepared and surface-ripened soft and semisoft cheeses [40]. It can survive high salt concentrations, low pH, under refrigeration conditions, with incubation time ranging from 7 to 10 days, and symptoms including fever, fatigue, and gastrointestinal symptoms such as diarrhea, nausea, and vomiting [19]. These survival characteristics of *L. monocytogenes* make it an important risk in nonprocessed food such as ready-to-eat (RTE) products because heating step is not included to eliminate this pathogen before consumption [40]. Hossain et al. [41] reported on the molecular mechanism behind the postbiotic strains of *Lactobacillus curvatus* B.67 and *Lactobacillus plantarum* M.2 inhibition of *L. monocytogenes* could include: biofilm formation, expression levels of biofilm formation target genes, swimming motility, virulence, and quorum-sensing as they have been inhibited with increasing the postbiotics concentration suggesting their potential use as an efficient bio-preservatives to control for formation of *L. monocytogenes* biofilm in the food industry.

2.2.6 *Salmonella* spp.

Salmonella is rod-shaped (bacillus), Gram-negative bacteria that belong to Enterobacteriaceae [18, 19]. *Salmonella* genus has two pathogenic species: *S. enterica* and *Salmonella bongori* [18]. *S. enterica* is further divided into six subspecies [19]. Over 2,500 serotypes have been identified under the 2 species [18]. The serotypes are divided into typhoidal and nontyphoidal groups [22]. Certain serotypes are pathogenic and have been responsible for many foodborne disease outbreaks in both humans and animals across the globe [22, 42, 157]. The primary habitat of *Salmonella* species is the intestinal tract (IT) of farm animals, birds, insects, reptiles, and IT of humans [19]. Infections with *Salmonella* are mainly due to the consumption of food with fecal contamination [157]. Outbreaks caused by *Salmonella* are generally linked to contaminated eggs, meat, and meat products including poultry, with the ability of these bacteria to contaminate non-meat products like fruits and vegetables [26]. *Salmonella* incubation period ranges between 12 and 72 h [19], whereby clinical presentation of infection with this pathogen varies from enteric fever to gastroenteritis and bacteremia [18]. Abebe et al. [19] reported the symptoms for *Salmonella* that mainly focused on the GIT, such as "abdominal cramps, watery, greenish, and foul-smelling diarrhea or bloody diarrhea with mucous, nausea, and vomiting and non-GIT-based symptoms such as headache, fatigue, moderate fever, and prostration." Merino et al. [164] reported that several methodologies that rely on diverse calculations were utilized for the detection and evaluation of *Salmonella* spp. affixed to surfaces, such as confocal laser scanning microscopy, optical coherence tomography, and real-time polymerase chain reaction (RT-PCR). They further reported other methods for controlling the *Salmonella* biofilm formation, such as using lactic acid bacteria (LAB), bacteriocins, essential oils, QS inhibitors, nanomaterials, and phage therapy [164].

2.2.7 *Shigella* spp.

The genus of *Shigella* is a nonmotile, nonspore-forming, facultative anaerobic, rod-shaped, Gram-negative bacteria belonging to the family Enterobacteriaceae [18]. *Shigella*, which is genetically closely related to *E. coli*, is naturally found in humans and gorillas and is one of the prominent bacteria causing diarrhea in the world [18, 43]. Traditionally, *Shigella* sp. is made up of four serogroups that are considered as species: serogroup A as *S. dysenteriae*, serogroup B as *S. flexneri*, serogroup C as *S. boydii*, and serogroup D as *S. sonnei* [15]. The highest number of serotypes (38) is found under serogroups A, B, and C; meanwhile, only one serotype is found under serogroup D [18]. *Shigella* spp. are mainly found in poor sanitation and hygiene environments [15]. Person-to-person contact is considered the main source of transmission of shigellosis, as well as the ingestion of fecal-contaminated water and food [43]. Shigellosis outbreaks were reported from the following contaminated food products such as chicken, fresh produce, milk, salads, and shellfish [18].

2.2.8 *Staphylococcus* spp.

The genus *Staphylococcus* are catalase-positive, cocci-shaped Gram-positive bacteria belonging to the family Micrococcaceae [18, 19]. This genus comprises more than 40 species [19], with 16 species can be found in human beings as opportunistic microbes, considered as members of the skin's microbiota and of the mucous in mammals and birds [43]. Among the pathogenic species, *Staphylococcus aureus* is the leading cause of many staphylococcal food infections in both humans and animals [16, 19]. Kadariya et al. [16] and Sharma [43] attributed *Staphylococcus* genus to a wide spectrum of diseases, ranging from superficial skin infections to severe, and potentially lethal, aggressive diseases. Infection with *S. aureus* results from eating food contaminated with staphylococcal enterotoxins which are heat resistant [19, 44]. Food contamination occurs via direct contact, indirectly by skin fragments, or via respiratory tract droplets [28]. Most staphylococcal food poisoning cases have, however, been related to food contamination during preparation due to insufficient refrigeration, cooking or heating, and/or poor personal hygiene [18]. Examples on frequently affected foods with staphylococcal foodborne diseases are bakery products, dairy products, egg products, meat and meat products, milk, poultry, and salads, especially cream-filled pastries and cakes, and sandwich filling [16]. According to Kadariya et al. [16] and Bintsis [18], the time frame of 6–10 h was reported to be the incubation period with the following frequent symptoms : "nausea, abdominal cramps, headache, dizziness, chills, hypersalivation, perspiration, general weakness, muscular cramping and/or prostration, vomiting, and diarrhea" that may or may not contain blood.

2.2.9 *Yersinia* spp.

Yersinia are Gram-negative, nonspore-forming rods that belong to the family Enterobacteriaceae [18]. They considered as facultative anaerobes and proliferate under both aerobic and anaerobic conditions [18]. Taxonomically, the genus has been categorized into 6 biovars, 70 serotypes, and 10 species, with 3 species are the only ones considered as pathogenic to both humans and animals [43]. In humans, *Yersinia enterocolitica*, *Y. pestis*, and *Y. pseudotuberculosis* are responsible for foodborne infections [45]. Plague is caused by *Y. pestis*; meanwhile, *Y. pseudotuberculosis* is mainly an animal pathogen with the ability to infect humans if contaminated food or water being ingested, and *Y. enterocolitica* has raised as a cause of most foodborne gastroenteritis in humans [18, 46]. According to Leon-Velarde et al. [47], "*Y. enterocolitica* can cause mesenteric lymphadenitis, even sometimes sequelae such as reactive arthritis and erythema nodosum." Symptoms of yersiniosis include abdominal pain, mild fever, diarrhea, ileitis, and mesenteric lymphadenitis [18, 48]. Poor sanitation and improper sterilization techniques by food handlers are linked the most foodborne outbreaks. Yersiniosis outbreaks have been related to unpasteurized milk, oysters, and more commonly linked to eating undercooked dishes containing pork [46]. *Y. enterocolitica* can be transmitted via the fecal–oral route by the consumption of contaminated food or water with this pathogen. The detection of *Y. enterocolitica* in food is challenging because it shares similarities with other enteric bacteria necessitating further confirmation tests [48]. Several beneficial properties for *Yersinia* phages such as controlling contamination of food products have been reported , and their specificity could be utilized in the development of rapid and sensitive *Yersinia* identification and diagnostic tools in several foodstuffs and products [47].

2.2.10 *Vibrio* spp.

Bintsis [18] described *Vibrio* genus as "a motile, non-spore-forming, facultative anaerobic, Gram-negative rods that belongs to the Vibrionaceae." The genus contains more than 35 species, and is mainly found in typical environments of brackish or marine located in tropical areas [49]. Some species of *Vibrio* are pathogenic and can cause foodborne infection, usually associated with eating undercooked seafood [49, 50]. *V. cholerae*, *V. parahaemolyticus*, and *V. vulnificus* are considered the pathogenic *Vibrio* species [18]. *V. cholerae* is responsible for cholera and is transmitted by drinking contaminated water. Ingestion of *Vibrio* bacteria may lead mainly to watery diarrhea as well as other secondary symptoms such as vomiting, nausea, fever, renal failure, and circulatory collapse [14, 49].

2.3 FBP detection method

Rapid and cost-effective detection methods for FBP are the most effective strategies to control and prevent these public threats and ensure human health [17]. Enzyme-linked immunosorbent assay, high-performance liquid chromatography, biolumines-cence, chemiluminescent sensors, mass spectrometry, and PCR were reported to be the popular methods for the detection of pathogens or toxins [51–55, 161]. Conventional approaches to detect foodborne pathogens involve the identification of microorganism using morphological assessment via selective enrichment, biochemical analysis, and serological verification [56]. Although these methods have been used extensively in food safety with great achievements, they had some disadvantages which include inadequate sensitivity and reproducibility, lengthy steps, highly qualified staff, complexity, and expensive economic cost [56, 57]. Therefore, novel strategies with advantageous characteristics of cost-effective, rapid, and sensitive detection of FBP are of great importance for public health [57].

In the last decade, the use of nanotechnology and specifically functional nanomaterials with exceptional physiochemical properties has opened up new frontiers for the food sector with food safety inspection, food sensing and packaging, and foodborne pathogen detection methods with improved analytical capacity and smart applications to ensure fast and accurate detection and communication systems in food industry [58–66]; however, there are some concerns and challenges for the use of nanomaterials (nanosensors) within food system [67]. The detection methods for toxins from foodborne diseases caused by Gram-positive bacteria have been reported by Rajkovic et al. [28]. Although there were tremendous successes and development of fast and on-site assays for the detection of FBP, however, there are several limitations that these assays are facing in the field for an optimum and efficient implementation [17].

In the past two decades, RT-PCR and droplet digital PCR (ddPCR) have been used to perform both qualitative and quantitative detections of DNA in a wide range of samples successfully [68, 69] with the advantage of the later technique (ddPCR) to detect and quantify target DNA in complex food matrices in the presence of ample sequences over RT-PCR which has a disadvantage of negative sensitivity of the amplification to inhibitors result [70]. ddPCR is able to divide samples into high number (tens of thousands) of water-in-oil droplets before thermal cycling [160] with successful quantification of *Bacillus cereus* [71], *Salmonella typhimurium* [70], *Campylobacter* spp. [72], and *Yersinia enterocolitica* [73–75] in different food niches.

Considering the richness of generated data from the genome sequencing for several foodborne bacteria, the advancement of data mining, artificial intelligence, and machine learning techniques resulted in developing phage-derived bacterial detection systems of foodborne pathogens [76–81].

The discovery of nanoparticle-based surface-enhanced Raman spectroscopy technique has provided a greatest advance with regard to bacterial detection. It brought the signal generated by a single molecule on par with that of fluorescent dyes [82, 83] with several foodborne pathogens including *E. coli* [84] and *Salmonella* [85].

Liu et al. [86] developed a paper-based sensor system for a point-of-care (POC) nucleic acid amplification test with the ability to detect and quantify multiple genes from different pathogens such as *E. coli* and *Campylobacter jejuni*. They further provide the structure of the POC system which is composed of a paper sensor chip and a portable instrument, which is built on an internet of things (IoT) platform. According to Liu et al. [86], the advantageous characteristics of the newly developed system (paper-based sensor) include "reagent storage functionality, sample transportation, and nucleic acid amplification." They further added that, "via a WiFi network, IoT instrument uses an Arduino microcontroller to control temperature and collect fluorescence images, after that, it stores the data in a cloud storage for subsequent uses and future applications (POC detections of pathogens)." POC diagnostics are able to conduct tests in a closer distance to the patient care site with shorter responding time, minimize manual interference, and identify infectious agents and their antimicrobial resistance enabling clinical management to make fast decisions [87, 88]. This system enables and empowers the clinical practices in developed and developing countries, especially in settings where a centralized laboratory approach faces limitations and challenges [89]. A novel disposable all-printed electronic biosensor has been reported with the purpose of fast detection and classification of three well-known pathogens: *S. typhimurium* and *E. coli* strains JM109 and DH5-α [90].

Gootenberg et al. [91] reported a specific high-sensitivity enzymatic reporter unlocking (SHERLOCK) technology and developed it with a combination of recombinase polymerase amplification and CRISPR-Cas13 that has the ability to detect RNA or DNA from clinical samples through fluorescence and colorimetric readouts [91].

The new version of SHERLOCK called SHERLOCKv2 was developed by Kellner et al. [92], of which it performs multiplexed genotyping for an informed pharmacogenomic therapeutic development and application, field detection of genetically modified organisms, and to determine the presence of pathogens co-occurring [91]. It was also used for the bacterial genotyping detection and the presence of viral infectious agents, for example, confirmation of single-nucleotide variants and for the detection of antibiotic resistance genes [91].

Electrochemical DNA biosensors [93, 94, 168] and electrochemical methods [155] combined with nanostructured materials such as graphene and carbon nanotubes have been overutilized to design and fabricate electrochemical biosensors to detect food pathogens with high accuracy of sensitivity and selectivity [11, 95].

Although some progress has been reported in the nanomaterial-based electrochemical biosensor detection of pathogens in food matrices through electrochemical methodologies [11, 96–98, 155, 162], more attempts were reported on the newly developed graphene-based electrochemical and fluorescent biosensors that are considered as emerging methods for rapid detection of pathogenic *E. coli* O157:H7 [99–102], typhoidal *Salmonella* (*Salmonella typhi* and *Salmonella paratyphi* A) [103, 104, 167], *S. enterica* [105], and *S. aureus* [13]. Recently, gold nanoparticles (AuNPs) have been reported to possess multifunctional properties with potential medical applications and biological activities [106–108]. They are used for single and multiple detections of FBP [109]. They have been utilized to develop rapid, cheap, portable, and on-site biosensors for several food safety applications [110].

2.4 FBP prevention and processing method

There is a need to adopt new and efficient food processing methods to prevent the spread and cross-contamination of FBP in food industry. In the past two decades, this sector witnesses a tremendous development in terms of automation and application of smart materials and artificial intelligence for effective mitigation and prevention of these FBP incidences/outbreaks. Among the strategies to reduce/prevent bacterial pathogens from fresh produce such as fruits and vegetables is the use of surfactants [111], surfactants in combination with organic acids (peracetic acid) against *L. innocua* [112] and *L. monocytogenes* [166] on fresh apple, fresh blueberries, and derived products [113], and *Salmonella* spp. [114] on blueberry, tomato, and lettuce using water-assisted decontamination systems (WADS) of pulsed light and ultraviolet.

Recently, Bhilwadikar et al. [115] compared the popular household processes (simple household washing and chemical treatments) with recent techniques (cold plasma, ozone, and high hydrostatic pressure (HP)) to decontaminate microorganisms, for example, *Salmonella* spp., *E. coli,* and *L. monocytogenes,* and pesticides from fresh fruits and vegetables. Yoon and Lee [116] also compared the effectiveness of several strategies, including sanitizers such as chlorine-based sanitizers and aqueous disinfectants, technologies such as HP and ultraviolet light for the decontamination of fresh fruits and vegetables, and reported on their limitations, especially toward bacterial biofilm formation by *E. coli* O157:H7, *L. monocytogenes,* *S. aureus, S. Typhimurium,* and *Pseudomonas* spp.

An innovative hurdle method combined with standard techniques, such as "high hydrostatic pressure (HHP, 300 MPa), bacteriophage Listex™ P100, and a pediocin PA-1 producing *Pediococcus acidilactici* HA 6111-2," was proposed by Komora et al. [117] toward *L. monocytogenes* inactivation in a traditional fermented meat sausage from North of Portugal called Alheira. They achieved 5 log reduction, which is required according to the standard guidelines for RTE foods of the United States Department of Agriculture's Food Safety and Inspection Service (USDA-FSIS) [117]. Commercial starter cultures comprising bacteriocinogenic LAB such as Nisaplin® and Micro-GARD®, as well as bacteriocin solutions such as nisin and pediocin PA-1, are available for their well-documented and potential applications in fermented products [118]. Although bacteriophages can be exploited to ensure and increase food safety as they are considered as "generally recognized as safe," they have been applied and legally authorized in several food products for the biocontrol of *L. monocytogenes* in different countries, for example, the USA, Canada, Australia, and New Zealand [119]. Recently, the European Food Safety Agency (EFSA) declared a positive scientific opinion about bacteriophage P100 application in RTE food products in Europe [117]. The use of chemical treatment against FBPs has some negative effects; therefore, to evade such effects, enzyme-based techniques were given an attention [120]. Following the evaluation of the antagonistic effects of a commercial peptidase namely Flavourzyme, toward *S. enterica* and STEC, Nahar et al. [120] reported that the enzyme effectively inhibited the levels of relative expression of biofilm-forming, QS, and their virulence genes. Based on their findings, the researchers advocated that Flavourzyme can be used as a protective agent against FBPs, illustrating the role of enzymes' application as a novel defensive tool.

2.5 FBP treatments

The latest innovations in the detection and treatment of bacterial infections have been reported recently by Deusenbery et al. [121]. There are three categories for FBP treatments.

2.5.1 Non-pharmacological treatment

Globally, diarrheal foodborne infections are in the top listed frequently experienced symptoms [18, 122]. Although illnesses may be mild and self-limiting sometimes, offering a supportive therapy is crucial to prevent potentially deadly consequences [123].

2.5.1.1 Oral rehydration therapy

It has been shown to give better results in terms of severe diarrheal illness treatment, especially cholera [123], with proved prevention and treatment of dehydration in different age categories of patients [124]. Noteworthy to indicate that rehydration therapy must be guided by clinical assessment, especially in children susceptible to dehydration and require more assiduousness to determine the hydration status for those children [124].

2.5.1.2 Micronutrient

Micronutrient deficiency, particularly zinc deficiency, is very prevalent in developing countries where food supplement and dietary diversity choices are limited [123]. This can cause children to suffer from significant gastrointestinal losses due to persisting diarrheal incidents, which leads to deficit and hypersusceptibility to more complicated cases of infectious diarrhea. Supplementation with micronutrients such as zinc has been shown to improve the period of diarrhea in children with gastrointestinal losses because of recurrent diarrhea [123].

2.5.1.3 Probiotics

There is a growing interest and body of evidence on the use of probiotic bacteria and yeast supplementation in the prevention or during treatment of infectious diarrhea [123]. Probiotics colonize the host gut to minimize the environmental niche for the invading pathogens through their metabolite production in human GIT, especially bacteriocins [125, 126, 156], butyrate [127], short-chain fatty acids [128–132], and polyunsaturated fatty acids [133–136] and their interactions and communication pathways with human organs and physiological systems [137–140], resulting in improved diarrheal symptoms in a specific situation [123, 141]. LAB form the most popular strains of probiotics and have been used so far in food industry for their health benefits and function attributes [142]. Usually, they are found in fermented foods [143]. Once, probiotic-based food is ingested by the host, it has better survival ability in the acidic environment of the stomach, and later on, it inhabits the gut and starts converting prebiotics (nondigestible carbohydrates) into antimicrobial compounds such as lactic acid and other essential nutrients [137]. Recently, Hossain et al. [41] reported the ability of postbiotics, which are the metabolic by-products of *Lactobacillus curvatus* B.67 and *Lactobacillus plantarum* M.2 to suppress the *L. monocytogenes* biofilm growth on food contact surfaces, whereby the metabolic analysis revealed various organic acids, and well-known bacteriocin-encoding genes are the main factors responsible for the antilisterial effects. LABs possess antimicrobial properties, as the production of antilisterial gene, the coding gene to produce antimicrobial peptide, is considered as the main factor in listeriosis treatment that is linked to a wide array of potential functions and strategies of LABs [144].

2.5.2 Pharmacological treatment

Pharmacological treatments for foodborne infections involve the use of drugs to treat infections [123, 124]. Part of the drug classes used to address/treat the symptoms of foodborne illness is given in the following sections.

2.5.2.1 Antimotility agent

Diarrhea is the most common symptom of foodborne illness [18, 22]. Several agents are available to reduce stool frequency, volume, and urgency, which can reduce the burden of diarrhea and allow patients comfortably carrying out daily activities [124]. Antimotility-based drugs work very well during mild illness or when used together with antibiotics [123]. They effectively shorten the illness duration in travelers' diarrhea [124], but are ineffective in high-volume secretory diarrhea (e.g., cholera) or inflammatory colitis [123]. These antimotility agents include loperamide, diphenoxylate/

atropine, and opiates [123]. The use of these antimotility agents in children and STEC in suspected patients is, however, not recommended [124].

2.5.2.2 Antispasmodic

These are medications that are used to treat pain caused by abdominal cramping. These drugs include butylscopolamine, hyoscyamine, dicycloverine, scopolamine, and atropine [123].

2.5.2.3 Bismuth salt

Bismuth subsalicylate is a class of medication that is used for the treatment of gastrointestinal disorders worldwide [123]. It has antisecretory, anti-inflammatory, and antibacterial properties and is more effective in acute or traveler's diarrhea patients with uncomplicated symptoms [124].

2.5.2.4 Antiemetic

Nausea and vomiting are among the unpleasant symptoms of foodborne illness [18, 26]. They affect the use of oral rehydration, which leads to hospitalization for intravenous therapy [123]. For patients affected with clinically strong vomiting, antiemetic medication can improve symptoms and reduce the need for hospital admission and intravenous fluid administration [124]. The authors further added that cases of gastroenteritis-related vomiting in children can be treated through the use of a single dose of ondansetron (Zofran). Meanwhile, in adults, uses of antiemetics with gastroenteritis are reasonable; however, data regarding the effects are limited [124].

2.5.3 Antibiotic treatment

Ng et al. [145] reported that antibiotic treatment is the most successful strategy in the past decades and the most powerful antibacterial factor that has been used widely to comfort human discomfort resulted from infections caused by bacteria. Antibiotics are used strictly when a foodborne illness is suspected, especially when the ill person is feverish and has symptoms from the invading illness (e.g., gross hematochezia and leukocytes on fecal smear). According to Switaj et al. [124], "when the symptoms are persisting for more than one week or are severe (i.e., more than eight liquid stools per day), or in the case of hospitalization is required." Antibiotics are not used in toxin-mediated foodborne illness, including those caused by ingestion of bacteria-preformed toxins such as emetic toxin produced by *Staphylococcus aureus* and *Bacillus cereus* [123]. Treatment with antibiotics shortens the symptoms' duration in traveler's diarrhea patients. ETEC is the main agent responsible for the traveler's diarrhea worldwide [18, 22, 44]. Due to the elevated danger of HUS, ill persons administered with empiric antibiotic therapy should be controlled carefully if the suspected pathogen causing the infection is STEC [124]. Antibiotics used for treating foodborne illnesses include erythromycin, ciprofloxacin, azithromycin, fluoroquinolone, and ceftriaxone [123].

Bacterial detection has been achieved using the life cycle of lytic phages [146, 147]. Stone et al. [81] highlighted the importance of phages as antibiotic substitutes, considering the well-known rise of antibiotic-resistant bacterial infections that become a widespread phenomenon in clinics and hospitals. A chimeric phage nanoparticles method was established and introduced for fast, precise, and accurate in revealing the identity of pathogenies such as *Pseudomonas aeruginosa*, *V. cholerae*, and *E. coli* considering the high affinity and precision of phages for their bacterial hosts' advantages [148].

2.5.4 Carbohydrate vaccines: medicinal chemistry approach

Seeberger [149] reported an innovative and novel use of semi- and completely synthesized carbohydrate-based vaccines to address infections caused by bacteria using medicinal chemistry-based approach [150]. Currently, vaccines with carbohydrate (polysaccharides originated from the bacterial cell surface or conjugated) with a protein-structure-based carrier were successfully fabricated to shield and save millions of lives against fatal infections caused by *Streptococcus pneumoniae*, *Haemophilus influenzae* type b, and *Neisseria meningitides* [149]. Vaccine discovery through chemistry-based approach has been illustrated, and the efforts focused on synthesized glycan-based medicinal chemistry were summarized to improve the understanding around the antiglycan immune response in mammalian cells and define glycan epitopes for the novel synthesis of glycoconjugate vaccines [149, 151] against *S. pneumoniae* [170], *Clostridium difficile* [152], *Shigella* spp. [153, 154], and other foodborne bacteria.

References

[1]   Abed, A. H., Menshawy, A. M. S., Zeinhom, M. M. A., Hossain, D., Khalifa, E., Wareth, G., & Awad, M. F. (2021 May 29). Subclinical mastitis in selected bovine dairy herds in North Upper Egypt: Assessment of prevalence, causative bacterial pathogens, antimicrobial resistance and virulence-associated genes. *Microorganisms, 9*(6), 1175. doi:10.3390/microorganisms9061175.

[2]   Elmi, S. A., Simons, D., Elton, L., Haider, N., Abdel Hamid, M. M., Shuaib, Y. A., Khan, M. A., Othman, I., Kock, R., & Osman, A. Y. (2021 Jan 26). Identification of risk factors associated with resistant

Escherichia coli isolates from poultry farms in the East Coast of Peninsular Malaysia: A cross sectional study. *Antibiotics (Basel), 10*(2), 117. doi:10.3390/antibiotics10020117.

[3]   Jaja, I. F., Oguttu, J., Jaja, C. I., & Green, E. (2020 May 26). Prevalence and distribution of antimicrobial resistance determinants of *Escherichia coli* isolates obtained from meat in South Africa. *PLoS One, 15*(5), e0216914. doi:10.1371/journal.pone.0216914.

[4]   Tuem, K. B., Gebre, A. K., Atey, T. M., Bitew, H., Yimer, E. M., & Berhe, D. F. (2018). Drug resistance patterns of *Escherichia coli* in Ethiopia: A meta-analysis. *BioMed Research International, Hindawi*, 2018 May 6, 4536905. doi:10.1155/2018/4536905.

[5]   Kintz, E., Byrne, L., Jenkins, C., McCARTHY, N., Vivancos, R., & Hunter, P. (2019 Nov). Outbreaks of Shiga toxin-producing *Escherichia coli* linked to sprouted seeds, salad, and leafy greens: A systematic review. *Journal of Food Protection, 82*(11), 1950–1958. doi:10.4315/0362-028X.JFP-19-014.

[6]   Yang, K., Wang, A., Fu, M., Wang, A., Chen, K., Jia, Q., & Huang, Z. (2020 Jan 10). Investigation of incidents and trends of antimicrobial resistance in foodborne pathogens in eight countries from historical sample data. *International Journal of Environmental Research and Public Health, 17*(2), 472. doi:10.3390/ijerph17020472.

[7]   Alocilja, E. C., & Radke, S. M. (2003 May). Market analysis of biosensors for food safety. *Biosensors and Bioelectronics, 18*(5-6), 841–846. doi:10.1016/s0956-5663(0300009-5.

[8]   Dye, C. (2014 May 12). After 2015: Infectious diseases in a new era of health and development. *Philosophical Transactions of the Royal Society of London B: Biological Sciences, 369*(1645), 20130426. doi:10.1098/rstb.2013.0426.

[9]   Sheng, L., & Zhu, M. J. (2021). Practical in-storage interventions to control foodborne pathogens on fresh produce. *Comprehensive Review in Food Science Food Safety, 20*, 1–28. doi:https://doi.org/10.1111/1541-4337.12786.

[10]   Wisuthiphaet, N., Yang, X., Young, G. M., & Nitin, N. (2021). Application of engineered bacteriophage T7 in the detection of bacteria in food matrices. *Frontiers in Microbiology, 12*, 691003. doi:10.3389/fmicb.2021.691003.

[11]   Reta, N., Saint, C. P., Michelmore, A., Prieto-Simon, B., & Voelcker, N. H. (2018 Feb 21). Nanostructured electrochemical biosensors for label-free detection of water- and food-borne pathogens. *ACS Applied Materials and Interfaces, 10*(7), 6055–6072. doi:10.1021/acsami.7b13943.

[12]   Kunadu, A., Addo-Preko, E., & Asuming-Bediako, N. (2020). 8 Microbiological safety of foods. In O. Ijabadeniyi (Ed.), *Food science and technology: Trends and future prospects* (pp. 193–214). Berlin, Boston: De Gruyter. https://doi.org/10.1515/9783110667462-008

[13]   SK, K., Singh, E., Singh, P., Meyyappan, M., & HS, N. (2019). A review on graphene-based nanocomposites for electrochemical and fluorescent biosensors. *RSC Advances, 9*, 8778–8881. https://doi.org/10.1039/C8RA09577A

[14]   Bezirtzoglou, E., Maipa, V., Voidarou, C., Tsiotsias, A., & Papapetropoulou, M. (2009). Food-borne intestinal bacterial pathogens. Microbial Ecology in Health and Disease 2009 Jul; 12 (2): 96–104. https://doi.org/10.1080/089106000750060350

[15]   Debnath, F., Mukhopadhyay, A. K., Chowdhury, G., Saha, R. N., & Dutta, S. (2018 Mar 22). An outbreak of foodborne infection caused by Shigella sonnei in West Bengal, India. *Japanese Journal of Infectious Disease, 71*(2), 162–166. doi:10.7883/yoken.JJID.2017.304.

[16]   Kadariya, J., Smith, T. C., & Thapaliya, D. (2014). Staphylococcus aureus and staphylococcal food-borne disease: An ongoing challenge in public health. *BioMed Research International, 2014*, 827965. doi:10.1155/2014/827965. Epub 2014 Apr 1.

[17]   Petrucci, S., Costa, C., Broyles, D., Dikici, E., Daunert, S., & Deo, S. (2021). On-site detection of food and waterborne bacteria – Current technologies, challenges, and future directions. *Trends in Food Science and Technology, 115*, 409–421. doi:10.1016/j.tifs.2021.06.054.

[18]   Bintsis, T. (2017 Jun 29). Foodborne pathogens. *AIMS Microbiology, 3*(3), 529–563. doi:10.3934/microbiol.2017.3.529.

[19] Abebe, E., Gugsa, G., & Ahmed, M. (2020). Review on major food-borne zoonotic bacterial pathogens. *Journal of Tropical Medicine, 2020,* 4674235. 2020 Jun 29. doi:10.1155/2020/4674235.

[20] Choi, W., & Kim, S. S. (2020 Sep 1). Outbreaks, germination, and inactivation of *Bacillus cereus* in food products: A review. *Journal of Food Protection, 83*(9), 1480–1487. doi:10.4315/0362-028X.JFP-19-429.

[21] Priyanka, B., Patil, R. K., & Dwarakanath, S. (2016). A review on detection methods used for foodborne pathogens. *Indian Journal of Medical Research, 144,* 327–338.

[22] Schirone, M., Visciano, P., Tofalo, R., & Suzzi, G. (2019). Editorial: Foodborne pathogens: Hygiene and safety. *Frontiers in Microbiology, 10,* 1974. Published 2019 Aug 27. doi:10.3389/fmicb.2019.01974.

[23] Jiang, L., & Loo, S. C. J. (2021). Intelligent nanoparticle-based dressings for bacterial wound infections. *ACS Applied Bio Materials, 4*(5), 3849–3862. doi:10.1021/acsabm.0c01168.

[24] Sharma, A. K., Dhasmana, N., Dubey, N., Kumar, N., Gangwal, A., Gupta, M., & Singh, Y. (2017 Mar). Bacterial virulence factors: Secreted for survival. *Indian Journal of Microbiology, 57*(1), 1–10. doi:10.1007/s12088-016-0625-1.

[25] Leifert, C., Ball, K., Volakakis, N., & Cooper, J. M. (2008 Oct). Control of enteric pathogens in ready-to-eat vegetable crops in organic and 'low input' production systems: A HACCP-based approach. *Journal of Applied Microbiology, 105*(4), 931–950. doi:10.1111/j.1365-2672.2008.03794.x.

[26] Weam, B., Abraham, M., Doiphode, S., Peters, K., Ibrahim, E., Sultan, A., & Mohammed, H. O. (2016 Apr). Foodborne bacterial pathogens associated with the risk of gastroenteritis in the state of Qatar. *International Journal of Health Sciences (Qassim), 10*(2), 197–207.

[27] Long, S. C., & Tauscher, T. (2006 Sep). Watershed issues associated with Clostridium botulinum: A literature review. *Journal of water and health, 4*(3), 277–288. doi:10.2166/wh.2006.516.

[28] Rajkovic, A., Jovanovic, J., Monteiro, S., et al. (2020). Detection of toxins involved in foodborne diseases caused by Gram-positive bacteria. *Comprehensive Review in Food Science Food Safety, 19*(4), 1605–1657. doi:10.1111/1541-4337.12571.

[29] Dieterle, M. G., Rao, K., & Young, V. B. (2019 Jan). Novel therapies and preventative strategies for primary and recurrent *Clostridium difficile* infections. Annals of the New York Academy of Sciences, 1435(1), 110–138. doi:10.1111/nyas.13958Epub 2018 Sep 21.

[30] Walper SA, Lasarte Aragonés G, Sapsford KE, Brown CW 3rd, Rowland CE, Breger JC, Medintz IL. Detecting Biothreat Agents: From Current Diagnostics to Developing Sensor Technologies. ACS Sens. 2018 Oct 26;3(10):1894–2024.

[31] Traoré, A. N., Mulaudzi, K., Chari, G. J., Foord, S. H., Mudau, L. S., Barnard, T. G., & Potgieter, N. (2016 Aug 12). The impact of human activities on microbial quality of rivers in the Vhembe District, South Africa. *International Journal of Environmental Research and Public Health, 13*(8), 817. doi:10.3390/ijerph13080817.

[32] Styles, M. J., Early, S. A., Tucholski, T., West, K. H. J., Ge, Y., & Blackwell, H. E. (2020 Dec 11). Chemical control of quorum sensing in *E. coli*: Identification of small molecule modulators of SdiA and mechanistic characterization of a covalent inhibitor. *ACS Infectious Diseases, 6*(12), 3092–3103. doi:10.1021/acsinfecdis.0c00654.

[33] AE, C., Pierson, E., & Dt, H. (2007 Sep). Targeting virulence: A new paradigm for antimicrobial therapy. *Nature Chemical Biology, 3*(9), 541–548. doi:10.1038/nchembio.2007.24.

[34] Mühlen, S., & Dersch, P. (2016). Anti-virulence strategies to target bacterial infections. *Current Topics in Microbiology and Immunology, 398,* 147–183. doi:10.1007/82_2015_490.

[35] Rasko, D. A., & Sperandio, V. (2010 Feb). Anti-virulence strategies to combat bacteria-mediated disease. *Nature Reviews Drug Discovery, 9*(2), 117–128. doi:10.1038/nrd3013.

[36] Johnson, R. (2019). Foodborne illnesses and outbreaks from fresh produce. *In focus,* 1091–1092.

[37] Cooley, M. B., Jay-Russell, M., Atwill, E. R., Carychao, D., Nguyen, K., Quiñones, B., Patel, R., Walker, S., Swimley, M., Pierre-Jerome, E., Gordus, A. G., & Mandrell, R. E. (2013 Jun 6). Development of a robust method for isolation of Shiga toxin-positive *Escherichia coli* (STEC) from fecal, plant, soil and

water samples from a leafy greens production region in California. *PLoS One, 8*(6), e65716. doi:10.1371/journal.pone.0065716.

[38] Heijnen, L., & Medema, G. (2006 Dec). Quantitative detection of E. coli, E. coli O157 and other Shiga toxin producing E. coli in water samples using a culture method combined with real-time PCR. *Journal of Water and Health, 4*(4), 487–498.

[39] Orsi, R. H., & Wiedmann, M. (2016 Jun). Characteristics and distribution of Listeria spp., including Listeria species newly described since 2009. *Applied Microbiology and Biotechnology, 100*(12), 5273–5287. doi:10.1007/s00253-016-7552-2.

[40] Falardeau J, Trmčić A, Wang S. The occurrence, growth, and biocontrol of Listeria monocytogenes in fresh and surface-ripened soft and semisoft cheeses. Compr Rev Food Sci Food Saf. 2021 Jul;20(4): 4019-4048. doi: 10.1111/1541-4337.12768.

[41] Hossain, M. I., Mizan, M. F. R., Roy, P. K., Nahar, S., Toushik, S. H., Ashrafudoulla, M., Jahid, I. K., Lee, J., & Ha, S. D. (2021 Oct). *Listeria monocytogenes* biofilm inhibition on food contact surfaces by application of postbiotics from *Lactobacillus curvatus* B.67 and *Lactobacillus plantarum* M.2. *Food Research International, 148*, 110595. doi:10.1016/j.foodres.2021.110595.

[42] Lund, B. M., & O'Brien, S. J. (2011 Sep). The occurrence and prevention of foodborne disease in vulnerable people. *Foodborne Pathogens and Disease, 8*(9), 961–973. doi:10.1089/fpd.2011.0860.

[43] Sharma, A. (2018). The genus *Staphylococcus*: Harmful and beneficial microorganisms in the Pakistan. *Journal of Life and Social Sciences, 15*(2), 72–83.

[44] Pillsbury, A., Chiew, M., Bates, J., & Sheppeard, V. (2013 Jun 30). An outbreak of staphylococcal food poisoning in a commercially catered buffet. *Communicable Diseases Intelligence Quarterly Report, 37*(2), E144–8.

[45] Shoaib, M., Shehzad, A., Raza, H., & Niazi, S. (2019). A comprehensive review on the prevalence, pathogenesis and detection of *Yersinia enterocolitica*. *RSC Advances, 9*, 41010–41021.

[46] Sreedharan, A., Jones, C., & Schneider, K. (2015). Preventing Foodborne Illness : Yersiniosis FSHN12-09, 1–3, http://edis.ifas.ufl.edu.

[47] Leon-Velarde, C. G., Jun, J. W., & Skurnik, M. (2019 Nov 28). *Yersinia* phages and food safety. *Viruses, 11*(12), 1105. doi:10.3390/v11121105.

[48] Gupta, V., Gulati, P., Bhagat, N., Dhar, M. S., & Virdi, J. S. (2015 Apr). Detection of Yersinia enterocolitica in food: An overview. *European Journal of Clinical Microbiology & Infectious Diseases, 34*(4), 641–650. doi:10.1007/s10096-014-2276-7.

[49] Singh, A., & Lin, J. (2015). Microbiological, coliphages and physico-chemical assessments of the Umgeni River, South Africa. *International Journal of Environmental Health Research, 25*(1), 33–51. doi:10.1080/09603123.2014.893567.

[50] Mwabi, J. K., Mamba, B. B., & Momba, M. N. B. (2013). Removal of waterborne bacteria from surface water and groundwater by cost-effective household water treatment systems (HWTS): A sustainable solution for improving water quality in rural communities of Africa. *Water SA, 39*, 445–456.

[51] Koedrith, P., Thasiphu, T., Tuitemwong, K., Boonprasert, R., & Tuitemwong, P. (2014). Recent advances in potential nanoparticles and nanotechnology for sensing food-borne pathogens and their toxins in foods and crops: Current technologies and limitations. *Sensors and Materials, 26*, 711–736.

[52] Liu, F., Li, Y., Song, C., Dong, B., Liu, Z., Zhang, K., Li, H., Sun, Y., Wei, Y., Yang, A., Yang, K., & Jin, B. (2010 Sep 15). Highly sensitive microplate chemiluminescence enzyme immunoassay for the determination of staphylococcal enterotoxin B based on a pair of specific monoclonal antibodies and its application to various matrices. *Analytical Chemical, 82*(18), 7758–7765. doi:10.1021/ac101666y.

[53] Sauer, S., & Kliem, M. (2010 Jan). Mass spectrometry tools for the classification and identification of bacteria. *Nature Reviews Microbiology, 8*(1), 74–82. doi:10.1038/nrmicro2243.

[54]   Singh, A., Poshtiban, S., & Evoy, S. (2013). Recent advances in bacteriophage based biosensors for food-borne pathogen detection. *Sensors (Basel), 13*(2), 1763–1786. Published 2013 Jan 30. doi:10.3390/s130201763.

[55]   Zhu L, He J, Cao X, Huang K, Luo Y, Xu W. (2016 Mar 15). Development of a double-antibody sandwich ELISA for rapid detection of *Bacillus cereus* in food. *Scientific Reports, 6*:16092. doi: 10.1038/srep16092.

[56]   Liu, J. M., Wang, Z. H., Ma, H., & Wang, S. (2018). Probing and quantifying the food-borne pathogens and toxins: From in vitro to in vivo. *Journal of Agricultural and Food Chemistry, 66*(5), 1061–1066. doi:10.1021/acs.jafc.7b05225.

[57]   Li, D., Liu, L., Huang, Q., et al. (2021). Recent advances on aptamer-based biosensors for detection of pathogenic bacteria. *World Journal of Microbiology & Biotechnology, 37*, 45. doi:https://doi.org/10.1007/s11274-021-03002-9.

[58]   Cesewski, E., & Johnson, B. N. (2020 Jul 1). Electrochemical biosensors for pathogen detection. *Biosensors and Bioelectronics, 159*, 112214. doi:10.1016/j.bios.2020.112214. Epub 2020 Apr 12.

[59]   Cho, I.-H., & Ku, S. (2017). Current technical approaches for the early detection of foodborne pathogens: Challenges and opportunities. *International Journal of Molecular Sciences, 18*(10), 2078. doi:https://doi.org/10.3390/ijms18102078.

[60]   King, T., Osmond-Mcleod, M. J., & Duffy, L. L. (2018). Nanotechnology in the food sector and potential applications for the poultry industry. *Trends in Food Science and Technology, 72*, 62–73. doi: https://doi.org/10.1016/j.tifs.2017.11.015.

[61]   Kumar, H., Kuča, K., Bhatia, S. K., Saini, K., Kaushal, A., Verma, R., Bhalla, T. C., & Kumar, D. (2020). Applications of nanotechnology in sensor-based detection of foodborne pathogens. *Sensors, 20*(7), 1966. doi:https://doi.org/10.3390/s20071966.

[62]   Lugani, Y., Sooch, B. S., Singh, P., & Kumar, S. (2021). Nanobiotechnology applications in food sector and future innovations. *Microbial Biotechnology in Food and Health*, 197–225. doi:10.1016/B978-0-12-819813-1.00008-6.

[63]   Nile, S. H., Baskar, V., Selvaraj, D., Nile, A., Xiao, J., & Kai, G. (2020). Nanotechnologies in food science: Applications, recent trends, and future perspectives. *Nano-Micro Letters, 12*(1), 45. Published 2020 Feb 4. doi:10.1007/s40820-020-0383-9.

[64]   Rotariu, L., Lagarde, F., Jaffrezic-Renault, N., & Bala, C. (2016). Electrochemical biosensors for fast detection of food contaminants trends and perspective. *TrAC Trends in Analytical Chemistry, 79*, 80–87. https://doi.org/10.1016/j.trac.2015.12.017

[65]   Sahani, S., & Sharma, Y. C. (2021). Advancements in applications of nanotechnology in global food industry. *Food Chemistry, 342*, 128318. doi:10.1016/j.foodchem.2020.128318.

[66]   Zhang, R., Belwal, T., Li, L., Lin, X., Xu, Y., & Luo, Z. (2020). Nanomaterial-based biosensors for sensing key foodborne pathogens: Advances from recent decades. *Comprehensive Review in Food Science Food Safety, 19*(4), 1465–1487. doi:10.1111/1541-4337.12576.

[67]   Yang, T., & Duncan, T. V. (2021). Challenges and potential solutions for nanosensors intended for use with foods. *Nature Nanotechnology, 16*(3), 251–265. doi:10.1038/s41565-021-00867-7.

[68]   Law, J. W. F., Mutalib, N. S. A., Chan, K. G., & Lee, L. H. (2014). Rapid methods for the detection of foodborne bacterial pathogens: Principles, applications, advantages and limitations. *Frontiers in Microbiology, 5*, 1–19. doi:https://doi.org/10.3389/fmicb.2014.00770.

[69]   Persson, S., Eriksson, R., Lowther, J., Ellström, P., & Simonsson, M. (2018). Comparison between RT droplet digital PCR and RT real-time PCR for quantification of noroviruses in oysters. *International Journal of Food Microbiology, 284*, 73–83. doi:https://doi.org/10.1016/j.ijfoodmicro.2018.06.022.

[70]   Wang, M., Yang, J., Gai, Z., Huo, S., Zhu, J., Li, J., Wang, R., Xing, S., Shi, G., Shi, F., & Zhang, L. (2018). Comparison between digital PCR and real-time PCR in detection of *Salmonella typhimurium* in milk. *International journal of food microbiology, 266*, 251–256. https://doi.org/10.1016/j.ijfoodmicro.2017.12.011

[71]   Porcellato, D., Narvhus, J., & Skeie, S. B. (2016). Detection and quantification of Bacillus cereus group in milk by droplet digital PCR. *Journal of Microbiological Methods, 127*, 1–6. doi:https://doi.org/10.1016/j.mimet.2016.05.012.

[72]   Peruzy, M. F., Proroga, Y. T. R., Capuano, F., Corrado, F., Santonicola, S., De Medici, D., Delibato, E., & Murru, N. (2020b). Detection and quantification of Campylobacter in foods: New analytic approaches to detect and quantify campylobacter spp. in food samples. *Italian Journal of Food Safety, 9*, 88–92. doi:https://doi.org/10.4081/ijfs.2020.8591.

[73]   Cristiano, D., Peruzy, M. F., Aponte, M., et al. (2021). Comparison of droplet digital PCR vs real-time PCR for *Yersinia enterocolitica* detection in vegetables [published online ahead of print, 2021 Jun 29]. *International Journal of Food Microbiology, 354*, 109321. doi:10.1016/j.ijfoodmicro.2021.109321.

[74]   Peruzy, M. F., Murru, N., Perugini, A. G., Capuano, F., Delibato, E., Mercogliano, R., Korkeala, H., & Proroga, Y. T. R. (2017). Evaluation of virulence genes in *Yersinia enterocolitica* strains using SYBR Green real-time PCR. *Food Microbiology, 65*, 231–235. doi:https://doi.org/10.1016/j.fm.2017.03.004.

[75]   Peruzy, M. F., Aponte, M., Proroga, Y. T. R., Capuano, F., Cristiano, D., Delibato, E., Houf, K., & Murru, N. (2020a). *Yersinia enterocolitica* detection in pork products: Evaluation of isolation protocols. *Food Microbiology, 92*, 103593. https://doi.org/10.10

[76]   Alves-Barroco, C., Caço, J., Roma-Rodrigues, C., Fernandes, A. R., Bexiga, R., Oliveira, M., Chambel, L., Tenreiro, R., Mato, R., & Santos-Sanches, I. (2021 Jul 15). New Insights on *Streptococcus dysgalactiae* subsp. *dysgalactiae* Isolates. *Frontiers in Microbiology, 12*, 686413. doi:10.3389/fmicb.2021.686413.

[77]   Moye, Z. D., Woolston, J., & Sulakvelidze, A. (2018 Apr 19). Bacteriophage applications for food production and processing. *Viruses, 10*(4), 205. doi:10.3390/v10040205.

[78]   Paczesny, J., Ł, R., & Recent, H. R. (2020 Aug 3). Progress in the detection of bacteria using bacteriophages: A review. *Viruses, 12*(8), 845. doi:10.3390/v12080845.

[79]   Peters, T. L., Song, Y., Bryan, D. W., Hudson, L. K., & Denes, T. G. (2020 Oct 28). Mutant and recombinant phages selected from in vitro coevolution conditions overcome phage-resistant *Listeria monocytogenes*. *Applied and Environmental Microbiology, 86*(22), e02138–20. doi:10.1128/AEM.02138-20.

[80]   Philippe, C., Levesque, S., Dion, M. B., Tremblay, D. M., Horvath, P., Lüth, N., Cambillau, C., Franz, C., Neve, H., Fremaux, C., Heller, K. J., & Moineau, S. (2020 Jun 17). Novel genus of phages infecting *Streptococcus thermophilus*: Genomic and morphological characterization. *Applied and Environmental Microbiology, 86*(13), e00227–20. doi:10.1128/AEM.00227-20.

[81]   Stone, E., Campbell, K., Grant, I., & Understanding, M. O. (2019 Jun 18). Exploiting phage-host interactions. *Viruses, 11*(6), 567. doi:10.3390/v11060567.

[82]   Nie, S., & Emory, S. R. (1997 Feb 21). Probing single molecules and single nanoparticles by surface-enhanced Raman scattering. *Science, 275*(5303), 1102–1106. doi:10.1126/science.275.5303.1102.

[83]   Stiles, P. L., Dieringer, J. A., Shah, N. C., & Van Duyne, R. P. (2008). Surface-enhanced Raman spectroscopy. *Annual Review of Analytical Chemistry (Palo Alto Calif), 1*, 601–626. doi:10.1146/annurev.anchem.1.031207.112814.

[84]   Yang, D., Zhou, H., Haisch, C., Niessner, R., & Ying, Y. (2016). Reproducible E. coli detection based on label-free SERS and mapping. *Talanta, 146*, 457–463. doi:10.1016/j.talanta.2015.09.006.

[85]   Su, L., Zhang, P., D-w, Z., Y-j-q, W., & R-g, Z. (2015). Rapid detection of *Escherichia coli* and *Salmonella typhimurium* by surface-enhanced Raman scattering. *Optoelectronics Letters, 11*(2), 157–160. doi:10.1007/s11801-015-4216-x.

[86]   Liu, M., Zhao, Y., Monshat, H., Tang, Z., Wu, Z., Zhang, Q., & Lu, M. (2020 Dec 1). An IoT-enabled paper sensor platform for real-time analysis of isothermal nucleic acid amplification tests. *Biosensors and Bioelectronics, 169*, 112651. doi:10.1016/j.bios.2020.112651.

[87]   Chan, C. P., Mak, W. C., Cheung, K. Y., Sin, K. K., Yu, C. M., Rainer, T. H., & Renneberg, R. (2013). Evidence-based point-of-care diagnostics: Current status and emerging technologies. *Annual Review of Analytical Chemistry (Palo Alto Calif), 6*, 191–211. doi:10.1146/annurev-anchem-062012-092641.

[88] Holland, C. A., & Kiechle, F. L. (2005 Oct). Point-of-care molecular diagnostic systems – Past, present and future. Current Opinion in Microbiology, 8(5), 504–509. doi:10.1016/j.mib.2005.08.001.

[89] Niemz, A., Ferguson, T. M., & Boyle, D. S. (2011 May). Point-of-care nucleic acid testing for infectious diseases. Trends in Biotechnology, 29(5), 240–250. doi:10.1016/j.tibtech.2011.01.007.

[90] Ali, S., Hassan, A., Hassan, G., Eun, C. H., Bae, J., Lee, C. H., & Kim, I. J. (2018 Apr 12). Disposable all-printed electronic biosensor for instantaneous detection and classification of pathogens. Scientific Reports, 8(1), 5920. doi:10.1038/s41598-018-24208-2.

[91] Gootenberg, J. S., Abudayyeh, O. O., Kellner, M. J., Joung, J., Collins, J. J., & Zhang, F. (2018 Apr 27). Multiplexed and portable nucleic acid detection platform with Cas13, Cas12a, and Csm6. Science, 360 (6387), 439–444. doi:10.1126/science.aaq0179.

[92] Kellner, M. J., Koob, J. G., Gootenberg, J. S., Abudayyeh, O. O., & Zhang, F. (2019 Oct). SHERLOCK: Nucleic acid detection with CRISPR nucleases. Nature Protocols, 14(10), 2986–3012. doi:10.1038/s41596-019-0210-2.

[93] Wu, Q., Zhang, Y., Yang, Q., Yuan, N., & Zhang, W. (2019 Nov 12). Review of electrochemical DNA biosensors for detecting food borne pathogens. Sensors (Basel), 19(22), 4916. doi:10.3390/s19224916.

[94] Yang, L., & Bashir, R. (2008 Mar-Apr). Electrical/electrochemical impedance for rapid detection of foodborne pathogenic bacteria. Biotechnology Advance, 26(2), 135–150. doi:10.1016/j.biotechadv.2007.10.003.

[95] Muniandy, S., Dinshaw, I. J., Teh, S. J., Lai, C. W., Ibrahim, F., Thong, K. L., & Leo, B. F. (2017 Nov). Graphene-based label-free electrochemical aptasensor for rapid and sensitive detection of foodborne pathogen. Analytical and Bioanalytical Chemistry, 409(29), 6893–6905. doi:10.1007/s00216-017-0654-6.

[96] Curulli, A. (2021 May 15). Electrochemical biosensors in food safety: Challenges and perspectives. Molecules, 26(10), 2940. doi:10.3390/molecules26102940.

[97] Paniel, N., & Noguer, T. (2019 Sep 1). Detection of Salmonella in food matrices, from conventional methods to recent aptamer-sensing technologies. Foods, 8(9), 371. doi:10.3390/foods8090371.

[98] Sharifi, S., Vahed, S. Z., Ahmadian, E., Dizaj, S. M., Eftekhari, A., Khalilov, R., Ahmadi, M., Hamidi-Asl, E., & Labib, M. (2020 Feb 15). Detection of pathogenic bacteria via nanomaterials-modified aptasensors. Biosensors and Bioelectronics, 150, 111933. doi:10.1016/j.bios.2019.111933.

[99] Barreiros dos Santos, M., Agusil, J. P., Prieto-Simón, B., Sporer, C., Teixeira, V., & Samitier, J. (2013 Jul 15). Highly sensitive detection of pathogen Escherichia coli O157: H7by electrochemical impedance spectroscopy. Biosensors and Bioelectronics, 45, 174–180. doi:10.1016/j.bios.2013.01.009.

[100] Kaur, H., Shorie, M., Sharma, M., Ganguli, A. K., & Sabherwal, P. (2017 Dec 15). Bridged Rebar Graphene functionalized aptasensor for pathogenic E. coli O78: K80:H11detection. Biosensors and Bioelectronics, 98, 486–493. doi:10.1016/j.bios.2017.07.004.

[101] Pandey, A., Gurbuz, Y., Ozguz, V., Niazi, J. H., & Qureshi, A. (2017 May 15). Graphene-interfaced electrical biosensor for label-free and sensitive detection of foodborne pathogenic E. coli O157: H7. Biosensors and Bioelectronics, 91, 225–231. doi:10.1016/j.bios.2016.12.041.

[102] Xu, M., Wang, R., & Li, Y. (2017 Jan 1). Electrochemical biosensors for rapid detection of Escherichia coli O157:H7. Talanta, 162, 511–522. doi:10.1016/j.talanta.2016.10.050.

[103] Appaturi, J. N., Pulingam, T., Thong, K. L., Muniandy, S., Ahmad, N., & Leo, B. F. (2020 Jan 15). Rapid and sensitive detection of Salmonella with reduced graphene oxide-carbon nanotube based electrochemical aptasensor. Analytical Biochemistry, 589, 113489. doi:10.1016/j.ab.2019.113489.

[104] Dinshaw, I. J., Muniandy, S., Teh, S. J., Ibrahim, F., Leo, B. F., & Thong, K. L. (2017). Development of an aptasensor using reduced graphene oxide chitosan complex to detect Salmonella. Journal of Electroanalytical Chemistry, 806, 88–96. doi:https://doi.org/10.1016/j.jelechem.2017.10.054.

[105] Muniandy, S., Teh, S. J., Appaturi, J. N., Thong, K. L., Lai, C. W., Ibrahim, F., & Leo, B. F. (2019 Jun). A reduced graphene oxide-titanium dioxide nanocomposite based electrochemical aptasensor for

rapid and sensitive detection of *Salmonella enterica*. *Bioelectrochemistry, 127*, 136–144. doi:10.1016/j.bioelechem.2019.02.005.

[106] Hu, X., Zhang, Y., Ding, T., Liu, J., & Zhao, H. (2020 Aug 13). Multifunctional gold nanoparticles: A novel nanomaterial for various medical applications and biological activities. *Frontiers in Bioengineering and Biotechnology, 8*, 990. doi:10.3389/fbioe.2020.00990.

[107] Saha, K., Agasti, S. S., Kim, C., Li, X., & Rotello, V. M. (2012 May 9). Gold nanoparticles in chemical and biological sensing. *Chemistry Review, 112*(5), 2739–2779. doi:10.1021/cr2001178.

[108] Siddiqi, K. S., & Husen, A. (2017 Mar). Recent advances in plant-mediated engineered gold nanoparticles and their application in biological system. *Journal of trace elements in medicine and biology: Organ of the Society for Minerals and Trace Elements, 40*, 10–23. doi:10.1016/j.jtemb.2016.11.012.

[109] Pissuwan, D., Gazzana, C., Mongkolsuk, S., & Cortie, M. B. (2020 Jan). Single and multiple detections of foodborne pathogens by gold nanoparticle assays. *Wiley Interdisciplinary Reviews: Nanomedicine and Nanobiotechnology, 12*(1), e1584. doi:10.1002/wnan.1584.

[110] Hua, Z., Yu, T., Liu, D., & Xianyu, Y. (2021). Recent advances in gold nanoparticles-based biosensors for food safety detection. *Biosensors and Bioelectronics, 179*, 113076. doi:10.1016/j.bios.2021.113076.

[111] Falk, N. A. (2019 Sep). Surfactants as antimicrobials: A brief overview of microbial interfacial chemistry and surfactant antimicrobial activity. *Journal of Surfactants and Detergents, 22*(5), 1119–1127. doi:10.1002/jsde.12293.

[112] Pietrysiak, E., Kummer, J. M., Hanrahan, I., & Ganjyal, G. M. (2019 Nov). Efficacy of surfactant combined with peracetic acid in removing *Listeria innocua* from fresh apples. *Journal of Food Protection, 82*(11), 1965–1972. doi:10.4315/0362-028X.JFP-19-064.

[113] Pérez-Lavalle, L., Carrasco, E., & Valero, A. (2020 Oct 28). Strategies for microbial decontamination of fresh blueberries and derived products. *Foods, 9*(11), 1558. doi:10.3390/foods9111558.

[114] Huang, R., & Chen, H. (2019 May). Comparison of water-assisted decontamination systems of pulsed light and ultraviolet for Salmonella inactivation on blueberry, tomato, and lettuce. *Journal of Food Science, 84*(5), 1145–1150. doi:10.1111/1750-3841.14510.

[115] Bhilwadikar, T., Pounraj, S., Manivannan, S., Rastogi, N. K., & Negi, P. S. (2019 Jul). Decontamination of microorganisms and pesticides from fresh fruits and vegetables: A comprehensive review from common household processes to modern techniques. *Comprehensive Review in Food Science Food Safety, 18*(4), 1003–1038. doi:10.1111/1541-4337.12453.

[116] Yoon, J. H., & Lee, S. Y. (2018). Review: Comparison of the effectiveness of decontaminating strategies for fresh fruits and vegetables and related limitations. *Critical Reviews in Food Science and Nutrition, 58*(18), 3189–3208. doi:10.1080/10408398.2017.1354813.

[117] Komora, N., Maciel, C., Amaral, R. A., Fernandes, R., Castro, S. M., Saraiva, J. A., & Teixeira, P. (2021 Oct). Innovative hurdle system towards *Listeria monocytogenes* inactivation in a fermented meat sausage model – High pressure processing assisted by bacteriophage P100 and bacteriocinogenic Pediococcus acidilactici. *Food Research International, 148*, 110628. doi:10.1016/j.foodres.2021.110628.

[118] Silva, C. C. G., Silva, S. P. M., & Ribeiro, S. C. (2018 Apr 9). Application of bacteriocins and protective cultures in dairy food preservation. *Frontiers in Microbiology, 9*, 594. doi:10.3389/fmicb.2018.00594.

[119] Chibeu, A., Agius, L., Gao, A., PM, S., AM, K., & Balamurugan, S. (2013 Oct 15). Efficacy of bacteriophage LISTEX™P100 combined with chemical antimicrobials in reducing *Listeria monocytogenes* in cooked turkey and roast beef. *International Journal of Food Microbiology, 167*(2), 208–214. doi:10.1016/j.ijfoodmicro.2013.08.018.

[120] Nahar, S., Jeong, H., Kim, Y., et al. (2021). Inhibitory effects of Flavourzyme on biofilm formation, quorum sensing, and virulence genes of foodborne pathogens *Salmonella Typhimurium* and *Escherichia coli*. *Food Research International, 147*, 110461. ISSN 0963-9969, https://doi.org/10.1016/j.foodres.2021.110461.

[121] Deusenbery, C., Wang, Y., & Shukla, A. (2021). Recent innovations in bacterial infection detection and treatment. *ACS Infectious Diseases, 7*(4), 695–720. doi:10.1021/acsinfecdis.0c00890.

[122] Kendall, P. (2012). *Bacterial Foodborne Illness*. Food and Nutrition Series/ Health,Fact sheet no.9300, https://mountainscholar.org/handle/10217/182468 (accessed on 23-08-2021).

[123] Steiner, T. (2013 Sep). Treating foodborne illness. *Infectious Disease Clinics of North America, 27*(3), 555–576. doi:10.1016/j.idc.2013.05.006.

[124] Switaj, T. L., Winter, K. J., & Christensen, S. R. (2015 Sep 1). Diagnosis and management of foodborne illness. *American Family Physician, 92*(5), 358–365.

[125] Ahmad, V., Khan, M. S., Jamal, Q. M. S., Alzohairy, M. A., Al Karaawi, M. A., & Siddiqui, M. U. (2017 Jan). Antimicrobial potential of bacteriocins: In therapy, agriculture and food preservation. *International Journal of Antimicrobial Agents, 49*(1), 1–11. doi:10.1016/j.ijantimicag.2016.08.016.

[126] Fuochi, V., Emma, R., & Furneri, P. M. (2021). Bacteriocins, a natural weapon against bacterial contamination for greater safety and preservation of food: A review. *Current Pharmaceutical Biotechnology, 22*(2), 216–231. doi:10.2174/1389201021666200704145427.

[127] Salvi, P. S., & Cowles, R. A. (2021 Jul 14). Butyrate and the intestinal epithelium: Modulation of proliferation and inflammation in homeostasis and disease. *Cells, 10*(7), 1775. doi:10.3390/cells10071775.

[128] McLoughlin, R. F., Berthon, B. S., Jensen, M. E., Baines, K. J., & Wood, L. G. (2017 Sep 1). Short-chain fatty acids, prebiotics, synbiotics, and systemic inflammation: A systematic review and meta-analysis. *The American Journal of Clinical Nutrition, 106*(3), 930–945. doi:10.3945/ajcn.117.156265.

[129] Nogal, A., Valdes, A. M., & Menni, C. (2021Jan-Dec). The role of short-chain fatty acids in the interplay between gut microbiota and diet in cardio-metabolic health. *Gut microbes, 13*(1), 1–24. doi:10.1080/19490976.2021.1897212.

[130] Ranjbar, R., Vahdati, S. N., Tavakoli, S., Khodaie, R., & Behboudi, H. (2021 Sep). Immunomodulatory roles of microbiota-derived short-chain fatty acids in bacterial infections. *Biomedicine and Pharmacotherapy, 141*, 111817. doi:10.1016/j.biopha.2021.111817.

[131] Rauf, A., Khalil, A. A., Rahman, U. U., Khalid, A., Naz, S., Shariati, M. A., Rebezov, M., Urtecho, E. Z., Rddg, D. A., Anwar, S., Alamri, A., Saini, R. K., & Rengasamy, K. R. R. (2021 Mar). Recent advances in the therapeutic application of short-chain fatty acids (SCFAs): An updated review. *Critical Reviews in Food Science and Nutrition, 11*, 1–21. doi:10.1080/10408398.2021.1895064.

[132] van de Wouw, M., Boehme, M., Lyte, J. M., Wiley, N., Strain, C., O'Sullivan, O., Clarke, G., Stanton, C., Dinan, T. G., & Cryan, J. F. (2018 Oct). Short-chain fatty acids: Microbial metabolites that alleviate stress-induced brain-gut axis alterations. *The Journal of Physiology, 596*(20), 4923–4944. doi:10.1113/JP276431.

[133] Bartsch, B., Then, C. K., Harriss, E., Kartsonaki, C., & Kiltie, A. E. (2021 Apr 23). The role of dietary supplements, including biotics, glutamine, polyunsaturated fatty acids and polyphenols, in reducing gastrointestinal side effects in patients undergoing pelvic radiotherapy: A systematic review and meta-analysis. *Clinical and Translational Radiation Oncology, 29*, 11–19. doi:10.1016/j.ctro.2021.04.006.

[134] Gómez-Fernández, A. R., Faccinetto-Beltrán, P., Orozco-Sánchez, N. E., Pérez-Carrillo, E., Marín-Obispo, L. M., Hernández-Brenes, C., Santacruz, A., & Jacobo-Velázquez, D. A. (2021 Aug 12). Sugar-free milk chocolate as a carrier of omega-3 polyunsaturated fatty acids and probiotics: A potential functional food for the diabetic population. *Foods, 10*(8), 1866. doi:10.3390/foods10081866.

[135] Hastings, C. N., Sheridan, H., Pariante, C. M., & Mondelli, V. (2017). Does diet matter? The use of polyunsaturated fatty acids (PUFAs) and other dietary supplements in inflammation-associated depression. *Current Topics in Behavioral Neurosciences, 31*, 321–338. doi:10.1007/7854_2016_31.

[136] Mokkala, K., Pussinen, P., Houttu, N., Koivuniemi, E., Vahlberg, T., & Laitinen, K. (2018 Feb 27). The impact of probiotics and n-3 long-chain polyunsaturated fatty acids on intestinal permeability in pregnancy: A randomised clinical trial. *Beneficial Microbes, 9*(2), 199–208. doi:10.3920/BM2017.0072.

[137] Peng, M., Tabashsum, Z., Anderson, M., et al. (2020a). Effectiveness of probiotics, prebiotics, and prebiotic-like components in common functional foods. *Food Science and Food Safety, 19*, 1908–1933. doi:https://doi.org/10.1111/1541-4337.12565.

[138] Rajanala, K., Kumar, N., & Chamallamudi, M. R. (2021). Modulation of gut-brain axis by probiotics: A promising anti-depressant approach. *Current Neuropharmacology, 19*(7), 990–1006. doi:10.2174/1570159X19666201215142520.

[139] Jakubczyk, D., & Górska, S. (2021 Jun 21). Impact of probiotic bacteria on respiratory allergy disorders. *Frontiers in Microbiology, 12*, 688137. doi:10.3389/fmicb.2021.688137.

[140] Jamalkandi, S. A., Ahmadi, A., Ahrari, I., Salimian, J., Karimi, M., & Ghanei, M. (2021 Jun). Oral and nasal probiotic administration for the prevention and alleviation of allergic diseases, asthma and chronic obstructive pulmonary disease. *Nutrition research reviews, 34*(1), 1–16. doi:10.1017/S0954422420000116.

[141] Chen, K., Xin, J., Zhang, G., Xie, H., Luo, L., Yuan, S., Bu, Y., Yang, X., Ge, Y., & Liu, C. (2020 Aug 12). A combination of three probiotic strains for treatment of acute diarrhoea in hospitalised children: An open label, randomised controlled trial. *Beneficial Microbes, 11*(4), 339–346. doi:10.3920/BM2020.0046.

[142] Nichols, A. W. (2007). Probiotics and athletic performance: A systematic review. *Current Sports Medicine Reports, 6*, 269–273. doi:https://doi.org/10.1007/s11932-007-0044-5.

[143] Champagne, C. P., Gomes da cruz, A., & Daga, M. (2018). Strategies to improve the functionality of probiotics in supplements and foods. *Current Opinion in Food Science, 22*, 160–166. doi:https://doi.org/10.1016/j.cofs.2018.04.008.

[144] Yap, P.-C., MatRahim, N.-A., AbuBakar, S., & Lee, H. Y. (2021). Antilisterial potential of lactic acid bacteria in eliminating listeria monocytogenes in host and ready-to-eat food application. *Microbiological Research, 12*(1), 234–257. doi:https://doi.org/10.3390/microbiolres12010017.

[145] Ng, Z. J., Zarin, M. A., Lee, C. K., & Tan, J. S. (2020). Application of bacteriocins in food preservation and infectious disease treatment for humans and livestock: A review. *RSC Advances, 10*, 38937–38964. doi:https://doi.org/10.1039/d0ra06161a.

[146] Chen, J., Andler, S. M., Goddard, J. M., Nugen, S. R., & Rotello, V. M. (2017 Mar 6). Integrating recognition elements with nanomaterials for bacteria sensing. *Chemical Society Reviews, 46*(5), 1272–1283. doi:10.1039/c6cs00313c.

[147] Du Toit, A. (2019 Mar). Phage induction in different contexts. *Nature Reviews Microbiology, 17*(3), 126–127. doi:10.1038/s41579-019-0150-4.

[148] Peng, H., Borg, R. E., Nguyen, A. B. N., & Chen, I. A. (2020b May 22). Chimeric phage nanoparticles for rapid characterization of bacterial pathogens: Detection in complex biological samples and determination of antibiotic sensitivity. *ACS Sensors, 5*(5), 1491–1499. doi:10.1021/acssensors.0c00654.

[149] Seeberger, P. H. (2021 Apr 14). Discovery of semi- and fully-synthetic carbohydrate vaccines against bacterial infections using a medicinal chemistry approach. *Chemistry Review, 121*(7), 3598–3626. doi:10.1021/acs.chemrev.0c01210.

[150] Astronomo, R. D., & Burton, D. R. (2010 Apr). Carbohydrate vaccines: Developing sweet solutions to sticky situations? *Nature Reviews Drug Discovery, 9*(4), 308–324. doi:10.1038/nrd3012.

[151] Kay, E., Cuccui, J., & Wren, B. W. (2019 May 1). Recent advances in the production of recombinant glycoconjugate vaccines. *NPJ Vaccines, 4*, 16. doi:10.1038/s41541-019-0110-z.

[152] Monteiro, M. A., Ma, Z., Bertolo, L., Jiao, Y., Arroyo, L., Hodgins, D., Mallozzi, M., Vedantam, G., Sagermann, M., Sundsmo, J., & Chow, H. (2013 Apr). Carbohydrate-based Clostridium difficile vaccines. *Expert Review of Vaccine, 12*(4), 421–431. doi:10.1586/erv.13.9.

[153] Phalipon, A., Tanguy, M., Grandjean, C., Guerreiro, C., Bélot, F., Cohen, D., Sansonetti, P. J., & Mulard, L. A. (2009 Feb 15). A synthetic carbohydrate-protein conjugate vaccine candidate against *Shigella flexneri* 2a infection. *Journal of Immunology, 182*(4), 2241–2247. doi:10.4049/ jimmunol.0803141.

[154] Pozsgay, V., Chu, C., Pannell, L., Wolfe, J., Robbins, J. B., & Schneerson, R. (1999 Apr 27). Protein conjugates of synthetic saccharides elicit higher levels of serum IgG lipopolysaccharide antibodies in mice than do those of the O-specific polysaccharide from *Shigella dysenteriae* type 1. *Proceedings of the National Academy of Sciences of the United States of America, 96*(9), 5194–5197. doi:10.1073/ pnas.96.9.5194.

[155] Amiri, M., Bezaatpour, A., Jafari, H., Boukherroub, R., & Szunerits, S. (2018 Jun 22). Electrochemical methodologies for the detection of pathogens. *ACS Sensors, 3*(6), 1069–1086. doi:10.1021/ acssensors.8b00239.

[156] Barcenilla, C., Ducic, M., López, M., Prieto, M., & Álvarez-Ordóñez, A. (2022 Jan). Application of lactic acid bacteria for the biopreservation of meat products: A systematic review. *Meat Science, 183*, 108661. doi:10.1016/j.meatsci.2021.108661.

[157] Bianchi, F., & van den Bogaart, G. (2021). Vacuolar escape of foodborne bacterial pathogens. *Journal of Cell Science, 134*(5), jcs247221. 2020 Sep 1:10.1242/jcs.247221.

[158] Edelstein, M., Sundborger, C., Hergens, M. P., Ivarsson, S., Dryselius, R., Insulander, M., Jernberg, C., Hutin, Y., & Wallensten, A. (2014 Jun 6). Barriers to trace-back in a salad-associated EHEC outbreak, Sweden, June 2013. *PLOS Currents, 6*. doi:10.1371/currents. outbreaks.80bbab3af3232be0372ea0e904dcd1fe.

[159] European Food Safety Authority and European Centre for Disease Prevention and Control (EFSA and ECDC) (2019 Dec 11). The European Union one health 2018 zoonoses report. *EFSA Journal, 17*(12), e05926. doi:10.2903/j.efsa.2019.5926.

[160] Hindson BJ, Ness KD, Masquelier DA, Belgrader P, Heredia NJ, Makarewicz AJ, Bright IJ, Lucero MY, Hiddessen AL, Legler TC, Kitano TK, Hodel MR, Petersen JF, Wyatt PW, Steenblock ER, Shah PH, Bousse LJ, Troup CB, Mellen JC, Wittmann DK, Erndt NG, Cauley TH, Koehler RT, So AP, Dube S, Rose KA, Montesclaros L, Wang S, Stumbo DP, Hodges SP, Romine S, Milanovich FP, White HE, Regan JF, Karlin-Neumann GA, Hindson CM, Saxonov S, Colston BW. High-throughput droplet digital PCR system for absolute quantitation of DNA copy number. Anal Chem. 2011 Nov 15;83(22):8604–10.

[161] Gregor, C., Gwosch, K. C., Sahl, S. J., & Hell, S. W. (2018 Jan 30). Strongly enhanced bacterial bioluminescence with the ilux operon for single-cell imaging. *Proceedings of the National Academy of Sciences of the United States of America, 115*(5), 962–967. doi:10.1073/pnas.1715946115.

[162] Gupta, R., Raza, N., Bhardwaj, S. K., Vikrant, K., Kim, K. H., & Bhardwaj, N. (2021 Jan 5). Advances in nanomaterial-based electrochemical biosensors for the detection of microbial toxins, pathogenic bacteria in food matrices. *Journal of Hazardous Materials, 401*, 123379. doi:10.1016/j. jhazmat.2020.123379.

[163] McLarnan, S. M. (2017). *Escherichia coli* as a Water Quality Indicator Organism: A Case for Responsive, Science-Based Policy. All College Thesis Program, 2016–2019. 38. https://digitalcommons.csbsju.edu/honors_thesis/38

[164] Merino, L., Procura, F., Trejo, F. M., Bueno, D. J., & Golowczyc, M. A. (2019 May). Biofilm formation by *Salmonella* sp. in the poultry industry: Detection, control and eradication strategies. *Food Research International, 119*, 530–540. doi:10.1016/j.foodres.2017.11.024.

[165] Messele, Y. E., Abdi, R. D., Yalew, S. T., Tegegne, D. T., Emeru, B. A., & Werid, G. M. (2017 Aug 15). Molecular determination of antimicrobial resistance in *Escherichia coli* isolated from raw meat in Addis Ababa and Bishoftu, Ethiopia. *Annals of Clinical Microbiology and Antimicrobials, 16*(1), 55. doi:10.1186/s12941-017-0233-x.

[166] Shen, X., Cong, J., Mugendi, J., Hanrahan, I., & Zhu, M. J. (2021a Jun 18). Synergistic effects of lauric arginate and peracetic acid in reducing *Listeria monocytogenes* on fresh apples. *Frontiers in Microbiology, 12,* 641034. doi:10.3389/fmicb.2021.641034.

[167] Shen, Y., Xu, L., & Li, Y. (2021b). Biosensors for rapid detection of Salmonella in food: A review. *Comprehensive Reviews in Food Science and Food Safety, 20,* 149–197. doi:https://doi.org/10.1111/1541-4337.12662.

[168] Zhang, Z., Zhou, J., & Du, X. (2019a Mar 28). Electrochemical biosensors for detection of foodborne pathogens. *Micromachines (Basel), 10*(4), 222. doi:10.3390/mi10040222.

[169] Zhang, N., Liu, E., Tang, A., Ye, M. C., Wang, K., Jia, Q., & Huang, Z. (2019b May 22). Data-driven analysis of antimicrobial resistance in foodborne pathogens from six states within the US. *International Journal of Environmental Research and Public Health, 16*(10), 1811. doi:10.3390/ijerph16101811.

[170] Croxtall JD, Keating GM. Pneumococcal polysaccharide protein D-conjugate vaccine (Synflorix; PHiD-CV). Paediatr Drugs. 2009;11(5):349–57.

Rukayat Abiola Abdulsalam, Abidemi Ojo, Fatai Oladunni Balogun,
Feroz Mahomed Swalaha and Saheed Sabiu*

Chapter 3
Viruses and food safety

Abstract: Foodborne viruses, particularly norovirus and Hepatitis A, have emerged as significant contributors to high morbidity and mortality rates, presenting substantial challenges to both human health and the global economy. Their resilience to traditional preservation technologies has resulted in a profound economic and public health burden. Unlike bacteria, viruses do not replicate within food matrices, and their cell culture remains a complex and demanding task. This chapter present an overview of the major viruses associated with foodborne diseases, elucidating their health implications in humans, methods for detection in food, and strategies for control and prevention to ensure food safety. The findings from the review revealed that food and food products contaminated with viruses have negative impacts on the overall human health as well as agriculture vitality, economic prosperity, sustainable development, and food security. The prevalence of foodborne infections attributable to viruses makes microbiological quality the most crucial component of food quality and safety. Thus, food safety majorly focuses on controlling the contamination of foods by pathogens, including viruses. Also, while the COVID-19 pandemic significantly impacted the global food supply chain, current and emerging evidence suggest that it is not foodborne and poses no food safety risk. Consequently and to avoid the risk posed by foodborne viruses, Good Manufacturing Practices, Good Hygiene Practices, and Good Agricultural Practices should be embraced. Such practices will ensure food safety at every stage of the supply chain and thereby contributing to a robust prevention of possible future foodborne outbreaks of microbial origin including viruses.

Acknowledgments: We appreciate the Directorate of Research and Postgraduate Support, Durban University of Technology, and the National Research Foundation (NRF grant number 120433), South Africa, for the research supports to Dr. S. Sabiu. The role of Dr. D. Yadav is also acknowledged for her input on the initial outline of the manuscript.

*Corresponding author: Saheed Sabiu, Department of Biotechnology and Food Science, Faculty of Applied Sciences, Durban University of Technology, Durban 4001, KwaZulu-Natal, South Africa, e-mail: sabius@dut.ac.za

Rukayat Abiola Abdulsalam, Fatai Oladunni Balogun, Feroz Mahomed Swalaha, Department of Biotechnology and Food Science, Faculty of Applied Sciences, Durban University of Technology, Durban 4001, KwaZulu-Natal, South Africa

Abidemi Ojo, Centre for Applied Food Sustainability and Biotechnology, Central University of Technology, Bloemfontein, 9301, South Africa

https://doi.org/10.1515/9783110748345-003

3.1 Introduction

Viruses are microscopic and highly contagious organisms that can survive and remain infectious in environments and particularly in foods for long periods and may withstand unfavorable conditions [1]. While bacteria remain the most prevalent food-poisoning microbes, the role of viruses in food poisoning/contamination is emerging and posing a significant threat to food safety [2]. Foodborne viruses account for an estimated 18% of the United Kingdom's food poisoning incidents [174], whereas 9.4 million people are sickened, with 1,350 deaths attributable to common and clinically significant pathogens, including viruses that contaminate foods annually in the United States [3]. Of these 9.4 million cases in the United States, norovirus (NoV), a major foodborne pathogen, was reportedly accounted for 58% of the foodborne ailments [3]. In Africa, the scenario is not in any way better, where approximately 91 million persons consume pathogens such as virus-contaminated foods that make them sick, with about 137,000 attributable mortalities annually [4]. Food and food products containing pathogenic microbes, including viruses such as NoV, hepatitis A virus (HAV) and hepatitis E virus (HEV), cause diseases ranging from acute diarrhea and gastroenteritis to life-threatening ailments like cancers [5]. Unfortunately, virus-poisoned food impacts human health and well-being and grossly impedes agriculture vitality, economic prosperity, sustainable development, and food security. Generally, while every human is susceptible, infants, young children, pregnant women, aged, and immune-compromised persons are more vulnerable to virus-contaminated foods [5]. The relative number of foodborne illnesses due to viruses makes microbiological quality the most important aspect of food safety. Thus, food safety majorly focuses on controlling the contamination of foods by pathogens, including viruses. Here, we appraised the major foodborne viruses, the health implications of consuming virus-contaminated foods, common viral detection methods in food and food products, and approaches to viral control and prevention to ensure food safety and quality.

3.2 Foodborne viruses

The nonenveloped viruses are the most implicated foodborne viruses and are usually highly resistant to excessive pH, drying, and radiation, thus making them persistent in food samples [6]. Generally, the food and food products such as fresh produce (fruits and salads) as well as scollops, mussels, clams, and oysters, with little or no processing, are the most vulnerable to viral infections [2]. For instance, NoV and HAV have been found to contaminate shellfish and are regarded as the causative agents of most foodborne outbreaks in the developed countries [7]. Other viruses associated with foodborne infections are human rotavirus, HEV, astrovirus, and aichi virus [8]. Each virus group chooses its host and cell range as each has a unique infectious cycle.

Based on infection manifestation, foodborne viruses have been grouped as either those that infect the mucous membrane of the stomach and intestine (gastroenteritis) such as the NoV, human rotavirus, astrovirus, aichi virus, adenovirus, and sapovirus; those that are transmitted enterically (HAV and HEV), where they stay in the stomach and move to the liver to elicit their infectious effects; those infecting the respiratory system, or the enteroviruses that replicate in the intestine of humans but only cause sickness when they move to other sensitive parts of the body like the central nervous system (Table 3.1) [8, 7]. Irrespective of the group, these viruses are highly infectious when ingested through the oral route and are mostly shed in human feces [8, 7].

While zoonotic transmission through food has been reported in the HEV, NoV and HAV remain the two most infectious foodborne viruses and are normally transmitted from person to person [1]. Transmission may also occur through food handlers resulting in long duration of outbreak without knowing the source of infection. Thus, contamination of food by viruses could result from infectious food handlers who unwittingly transfer the virus to food. More so, the secondary circulation of a virus is very vast compared to bacteria. Of the over 10 families of viruses implicated in foodborne sicknesses ranging from diarrhea to intense liver diseases [9], the gastroenteritis group is the most prevalent [10].

Table 3.1: Foodborne viruses, site of infection in the humans, and food safety perspective.

Infection site	Virus	Food safety perspective
Intestinal system	Norovirus, human rotavirus, sapovirus, adenovirus	The occurrence is very high and extreme in infants and children less than 5 years. Of the viruses in this category, norovirus is the most prevalent, and viruses infecting the intestinal system remain the most prevalent foodborne viruses.
Liver	Hepatitis A and E viruses	The infections caused by these viruses are usually intense and fast becoming prominent in developed countries. Although most are foodborne infections, transmission could also be zoonotic, for example, from pigs.
Respiratory system	HPA-H5N1, SARS-CoV	HPA-H5N1 is an impinging public health infection that is fast becoming prominent in poultry. However, contact with infectious chickens may cause SARS-CoV.
Neural tissue and nervous system	Enterovirus, poliovirus, Nipah virus	Prominent infections in pigs are caused by the bat virus and may become foodborne when humans consume pigs.

Source: [8]. HPA-H5N1, highly pathogenic avian influenza H5 subtype; SARS-CoV, severe acute respiratory syndrome coronavirus-2.

3.2.1 Understanding the major foodborne viruses and their epidemiology

3.2.1.1 Norovirus (NoV)

Previously known as Norwalk virus, NoV is a single-stranded, positive-sense RNA virus belonging to the family Calciviridae. The NoV is the major genus of the five genera of the family Calciviridae with others being *Sapovirus, Lagovirus, Nebovirus,* and *Vesivirus* [11]. Of these genera, the NoV and *Sapovirus* are dominated by the clinically significant human enteric viruses [11]. Specifically, the NoV has a nonenveloped genome of 7.5 kb with a diameter of 23–40 nm. Its genome has three open reading frames (ORFs), designated as ORF-1, ORF-2, and ORF-3 encoding eight viral proteins (VPs). While ORF-2 and ORF-3 encode the structural components of the virions, VP1 and VP2, respectively, the ORF-1 encodes a polyprotein that is proteolytically processed into six nonstructural proteins, including the NoV protease and RNA-dependent polymerase [12].

There are at least 10 genomic constellations (G) of NoVs and those in GI, GII, GIV, GVIII, and GIX infect humans and cause illness [13]. For instance, in the United States, over 99% of NoV foodborne outbreaks are caused by the GI and GII viruses [14], while GVIII and GIX strains have been implicated in major foodborne outbreaks in Japan and the United States between 2004 and 2011 [13] and since 2013 [15], respectively. A variant of the GII (GII.4) has been recognized as a significant agent in global NoV foodborne outbreaks in long-term care facilities, with evidence of the emergence of new strains having a similar spread pattern with the influenza virus [169]. The GIV, on the other hand, occurs in environmental and clinical specimens, and one of its two subtypes (GIV.1) directly infects humans [16], while the other subtype (GIV.2) could indirectly infect humans through its primary hosts (felines and canines) [17].

Globally, NoV remains the major viral cause of endemic and epidemic acute gastroenteritis and is normally transmitted through person-to-person contact, ingestion of fecally contaminated water or food, or self-contamination following contact with contaminated environmental surfaces [165]. Specifically, while NoV foodborne outbreaks have occurred through infected handlers, several contaminated foods and food products such as fresh fruits and vegetables and marine *Bivalvia* have also been implicated [7]. The resulting infections are common in closed or semiclosed settings and crowded environments, including care and nursing homes, schools, hospitals, hotels, and cruise ships. In fact, most resort cases of food poisoning on cruise ships have been linked with NoV infection [2]. However, while data exist on the nature and degree of devastation inflicted by NoV infections of some parts of the world, the true contribution made by food to the rates of NoV infection in countries such as Scotland is currently unknown, and due to the significance of the virus to public health, the Scottish government is funding a large program of research to improve understanding in this area [174].

The prevalence of NoV is high in children less than 5 years, although adults are also infected. Following successful infection, NoV has an incubation period of 1–2 days in humans before the manifestation of symptoms like diarrhea and vomiting [18]. Subclinical symptoms last for less than a week, and infected individuals may shed the virus in feces for weeks. Both barf and feces contain the virus, and shedding may occur before symptoms appear [18].

Annually, at least 365 NoV foodborne outbreaks occurred in the United States between 2001 and 2008, while the occurrence slightly decreased to 252 between 2009 and 2012 [19, 20]. Many of these outbreaks were due to food handlers accounting for respective 53% and 70% increase in NoV prevalence in the period under review [20].

Besides handlers, NoV outbreaks have also been traced to irrigated water, although this is less common when compared to foodborne viral outbreaks. Occurrence of NoV has been reported in contaminated drinking and recreational water [21], with evidence of shellfish obtained from these water sources contaminated by feces [22].

3.2.1.2 Hepatitis A virus

HAV is a single-stranded, positive-sense RNA virus belonging to the family Picornaviridae. It has a genome size of about 7.5 kb with a diameter of 35 nm [6]. Serologically, the family Picornaviridae has only one genus, the *Hepevirus*. The genome of HAV is subdivided into six constellations (I–VI), of which types I, II, and III are human-specific and are further divided into classes A and B [23]. Infections involving HAV start in the intestine with 14–42 days' incubation period, from where they infect the liver [7]. The virus can spread easily via oral–fecal route causing fever, vomiting and nausea, jaundice, inflammation, and in extreme cases, liver failure [174]. While HAV infection could be discomforting, it is generally not severe, and most victims recover within months after stimulation of lifelong immunity [174].

Globally, poor hygiene culminating in contaminated foods and water as well as consumption of undercooked or raw shellfish from contaminated water and contact with HAV infected persons have been identified as major sources of HAV infection. Hence, regular maintenance of good hygiene measures and avoidance of contact with infected persons are practicable ways of controlling/preventing HAV [174].

HAV has been implicated in intense foodborne outbreaks globally, with varying degrees of prevalence correlating with the hygiene status of regions or countries [24]. For instance, foodborne HAV infections are most prevalent in underdeveloped countries, where most children infected are asymptomatic, making all adults immune [25]. However, in developed countries, HAV are not peculiar due to enhanced hygiene and living standards, though adults are more susceptible while only a few children are infected [26]. The potent outcome of HAV outbreaks is generally high in developing and underdeveloped countries [27, 164].

3.2.1.3 Hepatitis E virus

HEV, a member of the family Hepeviridae, is a single-stranded, positive-stranded RNA virus with a naked genome of size 7.2 kb and diameter of 35 nm [28]. The virus infects humans and animals such as chicken, fish, rabbit, camel, and rats [29]. The mammalian HEV is divided into four species (A–D). Group A is further divided into four constellations (1–IV), with 1 and II purely infecting humans, while III and IV are zoonotic and primarily through pigs [30, 31]. Zoonotic transmission of HEV has been detected in the hepatocytes and muscles of the infected animal and the environment [32, 33]. The presence of HEV in the environment suggests that enteric virus entered water bodies through sewage, thereby contaminating fresh produce [34] and seafood [35]. Transmission is generally via fecal contaminated water in endemic regions and is often considered a travel-related disease in developed countries. It has also been documented that HEV can be transmitted through the consumption of raw or under-processed meat [163]. The symptoms of HEV infections are generally mild and may not require treatment except in extreme cases with immunocompromised persons [174]. For instance, in 2015, approximately 20 million HEV infections including those of foodborne origin that culminated in 3.3 million symptomatic cases and 44,000 deaths were reported worldwide with compromised immunity being a contributing factor [155].

HEV is highly endemic in areas with poor hygienic conditions causing high mortality rate in pregnant women [37]. Table 3.2 summarizes some of the epidemiological data on selected viral foodborne outbreaks including those of NoV, HAV, and HEV.

Table 3.2: Epidemiological data of selected viral foodborne outbreaks.

Virus	Year	Country	Statistics and description of the viral foodborne outbreak	Reference(s)
NoV	2017	Japan	Consumption of shredded nori product contaminated during manufacturing resulted in outbreak that affected 2,094 persons between January and December 2017 in 4 remote regions (Wakayama 36.4%; Tokyo 57.0%; Fukuoka 1.9%; Osaka 4.7%) of the country	[162]
	2015	Taiwan	Consumption of pork liver in cold appetizers and lamb chops prepared by asymptomatic food handlers affected 169 persons with symptoms ranging from mild to severe ailments	[38]

Table 3.2 (continued)

Virus	Year	Country	Statistics and description of the viral foodborne outbreak	Reference(s)
	2012	South Korea	Consumption of seasoned green seaweed with radishes with 59 reported infections	[161]
	2006	Australia	Consumption of salads locally and unhygienically prepared by the food handlers	[39]
	2005	Spain	Three reported cases of infected persons follow consumption of an assorted shellfish dish containing raw and steamed bivalve mollusks Consumption of round of beef served by an asymptomatic food handler	[160] [40]
HAV	2016	United States	Consumption of smoothies containing strawberries that resulted in 134 active infections	[41]
	2013	Austria	A case of severe infection reported after consumption of frozen berry mix	[42]
	2011	United States	Pomegranate arils present in "Townsend Farms Organic Antioxidant Blend" were associated with HAV with 162 active cases in 10 (Arizona (23), California (79), Colorado (28), Hawaii (8), New Hampshire (1), New Jersey (1), New Mexico (11), Nevada (6), Utah (3), and Wisconsin (2)) states	http://www.cdc.gov/hepatitis/Outbreaks/2013/A1b-03-31/index.html
	2011	England	Consumption of semidried tomatoes linked to two HAV infections in 2011	[43]
	2011	Netherlands	Five positive cases of HAV reported after consumption of RTE salads containing semidried tomatoes	[44]
	2010	France	About 59 confirmed HAV infections linked to ingestion of semidried tomatoes in 2010	[45]
	2003	United States	Consumption of green onions resulted in 650 confirmed HAV infections in the United States in 2003	[46]

Table 3.2 (continued)

Virus	Year	Country	Statistics and description of the viral foodborne outbreak	Reference(s)
HEV	2015	Spain	Eight positive infections associated with HAV reported after consumption of wild boar meat	[47]
	2012	France	Consumption of uncooked pig liver sausage with one active case reported	[48]
	2011	France	Two HEV infections linked to likely consumption of raw figatelli	[49]
	2010	France	Consumption of raw figatelli with two reported cases	[50]
	2009	UK	Consumption of shellfish, lobster, crab, bacon, cured pork, pate, eggs, unpasteurized cheese, and venison were associated with 156 reported cases in 2009 in the UK	[51]
	2009–2012	Czech Republic	Consumption of home pig slaughter, pork meat, brawn, minced meat, liver sausage, and contact with precooked sausage, wild boar goulash resulted in 27 reported cases between 2009 and 2012	[159]
	2008	UK	Four vacationers' acute HEV infection was reported upon returning from a world cruise to the UK in 2008. Of 789 persons tested, 195 were seropositive for HEV and the infection was linked to shellfish consumption while on board	[51]
	2004	Hungary	One reported case of HEV infection following consumption of pork sausage prepared house-slaughtered meat	[52]
Sapovirus	2010	Japan	Largest Japanese foodborne outbreak of Sapovirus in the Aichi, Gifu, and Mie Prefectures. Illness was linked to eating a delivered box lunch. Of the 655 cases, 350 (53.4%) developed gastrointestinal symptoms within 48 h of eating	[53]

Table 3.2 (continued)

Virus	Year	Country	Statistics and description of the viral foodborne outbreak	Reference(s)
Rotavirus	2000	United States	An estimated 108 cases were associated with eating tuna or chicken salad sandwiches in Washington, DC, in 2000	[54]
Astrovirus	1991	Japan	A large outbreak of astrovirus that lasted for 5 days affected about 4,700 persons (students and teachers) from schools in Katano City, Osaka, Japan, in June 1991. It was linked to the contaminated food from a common supplier	[170]

3.3 Presence, stability, and control of viruses in fresh and ready-to-eat foods

3.3.1 Occurrence of viruses in food and food products

Viruses have been the main pathogens implicated in foodborne diseases related to submanufactured, subpacked foods, fresh produce, ready-to-eat (RTE) fruits and vegetables [55]. Several foodborne outbreaks have been linked with contaminated fresh produce and RTE food, especially by carrier food handlers [8, 8]. Fresh produce and RTE foods such as bivalve shellfish are frequently associated with foodborne viral outbreaks. For example, Schmidt et al. [39] reported an outbreak of foodborne NoV from Australia due to the locally prepared salad by the food handlers in 2006. Consequently, the WHO has established and identified NoV and HAV in fresh produce and RTE foods to be the major viruses in which adequate prevention measures should be taken seriously [8]. Several foodborne viruses like NoV, HAV, HEV, rotavirus, astroviruses, picornaviruses, parvoviruses, and enteroviruses have been identified, and they are transmitted through food, water, contact surfaces in the environment, and contact with infected persons [1, 157]. Due to being facultative obligate, viruses cannot propagate during food production, transportation, or while storing the food, and the organoleptic properties of the contaminated food will still be normal [1]. The contamination of foods and food products including fresh and RTE foods by viruses normally occurs when food/food product comes in touch with water that is contaminated by sewage or when an infectious food handler is in contact with food [7].

Fresh produce items (fruits and vegetables) are normally contaminated by viruses through the fecal route, especially during a preharvest stage. Sewage can contaminate water used for either irrigation, cleaning farm implements, or fertilizers and agrochemicals' applications [56]. Irrigation of vegetables such as lettuce, broccoli, cabbage, and cucumber with sewage-contaminated water can lead to foodborne outbreaks if these vegetables are taken directly by consumers without proper washing. Also, many farmers use feces as manure to spring up their produce without having prior knowledge of the havoc. Shellfish remains the main source of outbreaks via sewage contamination by a virus [56]. One explicit thing about sewage-related contamination is that food can become contaminated with several viruses; thus, infected individuals are prone to infections by multiple strains of viruses [172].

Food (fresh and RTE) can be contaminated at any time in the trophic level ranging from the producers to consumers in a manner that occurs ceaselessly. For instance, farm produce can be touched when harvesting, during packaging, and when supplying to retail stores or even doorstep [57]. In the same way, the retail outlet and cooks spread the virus in the process of handling prior to consumption. The food handlers majorly get contaminated from this virus (enteric) and subsequently shed the virus during defecation, baby napkin replacement, or washing of toilet areas and are then transmitted from the skin (hand) to foods and insensate surfaces such as knife grip and chopping board [57]. Another way of spreading viruses by food handlers is by vomiting. NoV has been associated with vomiting and diarrhea with a very abrupt onset, and outbreaks from virus-containing vomitus and inhaling of virus-containing aerosols have been documented [58]. Another way in which fresh and RTE food can be contaminated by a virus is through zoonotic transmission. There has been evidence that HEV is transmitted by raw meat, liver of dear, pork, organs, and feces of animals [59, 60].

3.3.2 Stability of viruses in food and food products

Stability or persistence is mostly used to describe the ability of a virus to establish an infection and still maintain infectivity within the food. This also includes how long it can stay and its concentration in the food [61]. For the enteric viruses, for instance, their stability in foods depends on the type of food, viral strain, temperature, pH, relative humidity, exposure to sunlight, and moisture content of the food [62]. Generally, it has been reported that the use of chilled storage temperature, either refrigeration or freezing, has little or no effect on the viruses in the stored products. For instance, storing fruits and vegetables at the temperature of 2–11 °C slows down respiration, browning, moisture loss, senescence, and microbial growth but immensely contribute to the spread of enteric viruses [62, 63]. More so, studies have shown that enteric viruses persist in food even after surpassing the food shelf life, and this was evident in several studies where enteric viruses such as HAV and NOV were shown to be infectious in mussel for 4 weeks without reduction in viral titer [64], or where HAV and rotavirus retained their infectivity following freezing for 90 days [65] as well as in murine norovirus (MNV-1) isolated in lettuce vegetable left at 25 °C for 4 days with 1-log reduction [66]. Furthermore, Shieh et al. [67] detected HAV on oysters stored at −20 °C for 12 and 42 days. These studies are attestations that food contaminated prior to freezing or chilling will still contain a high viral load with infectious tendency.

Contrary to the observation in fresh foods, the stability of viruses in dried food products depends on the nature of the inanimate object or formites present at the time of contamination. The formites are majorly contaminated through stool and vomit from patients with diarrhea and hepatitis [68]. While the enteric viruses are resistant to drying and survive on formites, though with reduced viability and persistency with time, other viruses transmitted via the oral–fecal route show little decay when drying up [69, 70]. The presence of HAV on sun-dried tomato fruits is another evidence supporting that contamination of food before drying can still contain ample viral particles [45]. Taken together, it could be seen that viruses can survive the food chain from production to consumers and persist on food, especially when the food is consumed raw.

3.3.3 Control of viruses in foods and food products

The perseverance of viruses in food processing and accompanying implications on human health and well-being over the years has prompted attention to the need to prevent contamination through effective control strategies. The strategy must start from the preharvest stage for fresh produce and at the postharvest stage for the RTE foods. While this seems to be a noble approach, it is worthwhile to know that viruses are unruly to most of the methods used for processing and preserving foods [156]. For instance, most of the viruses can survive at extreme pH conditions, while the enteric viruses can overcome the energy from ionizing radiation at a dose that affects the organoleptic properties of the product [156]. The ability viruses such as HAV and NoV to survive acidification or fermentation preservative method related to pH treatment have made them to prevail in viral outbreaks implicating acidic fruits like raspberries [71].

Similarly, the use of low temperature has little or no impinging effect in controlling viruses as freezing is seen as a method of preserving the virus [72].

Generally, viruses can remain active in food for days and even weeks in the environment without losing their infectivity. For a viral infection to be established from contaminated food, several factors such as the stability of the virus, amount of virion particles shed by the infectious individual, the processing of the food, the amount of virus needed to cause infection, and the immunity of the host are crucial for consideration [73]. More so, the inherent properties of food can also protect the virion from treatment [74]. Consequent upon these stringent hygienic measures such as supplying of clean and potable water to farmland, prevention of water bodies from sewage contamination, prevention of food from human feces, cleaning of food equipment and disinfection of surfaces where food is to be prepared, frequent washing of hands with soap and water, proper washing of fruits and vegetables before consumption, usage of portable water in the food processing plant, training and awareness of food handlers in hygienic practices and avoidance of sick handlers in food establishment must be in place to avoid possible viral foodborne outbreaks [1, 75–77, Mahendra and Yodit, 2020]. Equally crucial is constant monitoring of viruses in food since the organoleptic properties of food do not change if infected by a virus [1, 26, 155].

In addition, immunization and food processing technologies like thermal processing, high-pressure processing (HPP), UV and gamma irradiation, pulsed electric fields (PEFs), and chemical-based control (the use of sodium hypochlorite and hypochlorous acid from electrolyzed water) can control food contamination by enteric viruses [78]. Several reports have shown that thermal processing and HPP strategies are most effective in controlling food contamination by enteric viruses [74, 79, 80].

The use of high temperatures for various times has proved to be most effective in controlling or inactivating enteric viruses in foods. Temperatures at or over 90 °C for longer than 1.5 min are effective against varieties of enteric viruses [79, 80]. For instance, Butot et al. [81] reported the reduction of enteric viruses in vegetables after steam blanching at 95 °C for 2.5 min. Other studies have shown that treatment at 90 °C for 3 min was sufficient to inactivate most HAV in some seafood such as greenshell mussels and soft clams [82, 83]. Generally, inactivation of viruses by heat treatment or high temperature results from detrimental changes to the viral capsid, as demonstrated by Russell et al. [84], who reported rupturing and decomposition of the icosahedral capsid of adenovirus following treatment at high temperature. O'Brien and Newman [85] have also reported changes in the isoelectric point of viral capsid after treatment at 56 °C. Croci et al. [86] gave a related submission on the protein coat of HAV that coagulated and disintegrated at high heat treatment.

Unlike the temperature treatment, HPP inactivation of foodborne viruses compresses and releases pressure, usually from 100 to 700 MPa, on packaged foods that are submerged in a liquid such as potable water [74]. The mechanism of action of HPP in virus inactivation is by either dissociation or denaturation of the capsid for nonenveloped viruses, while for enveloped ones, the viral envelope is damaged [154] as against damage of nucleic acids [87]. It was also hypothesized that HPP could inactivate enteric viruses by detaching nucleic acid from the genome, preventing viral replication within the host cell. The HPP can be influenced by processing parameters such as pressure, temperature, and time, and nonprocessing parameters like water activity and pH foods [88], and studies have shown that foodborne enteric viruses can be inactivated at different pressure and treatment times [87, 89]. For instance, HAV was inactivated from strawberry puree and sliced green onions with the pressure of 375 MPa at 22 °C for 5 min [90], while Lou et al. [91] reported the treatment and inactivation of four human NoV strains at high pressures ranging from 200 to 600 MPa. Also, murine NoV treated at 350 MPa for 5 min at 30 °C was reduced by 1.15 log pfu/mL compared to 5.56 log pfu/mL reduction when treated at the same pressure and time at 5 °C [92]. Hence, it could be logically stated that viruses respond to the combination of pressure and temperature differently.

For the chemical-based approach, chlorine, chlorine oxide, and ozone remain the most widely used antimicrobials in food processing, and their efficacy depends on concentration, exposure time, nature of vegetable/fruit, the type of microorganism involved, number of soil particles in the processing water, temperature, and pH of water [93]. Generally, enteric viruses display a varying degree of susceptibility to chlorine inactivation, and it has been established that no single virus can be used as an indicator for disinfection. For instance, feline calicivirus (FCV) can only be inactivated at a dose of 5,000 mg/L with a 3 log reduction, a concentration that is four times higher than the FDA recommendation, making FCV resistant to all disinfectants used in washing fresh produce [94]. This could be the reason that most disinfectants are not sufficient for inactivating viruses when there is a foodborne outbreak. Other studies have also focused on NoV resistance to chlorine. A study by Keswick et al. [95], involving feeding eight volunteers with water inoculated with NoV treated with chlorine at a concentration of 3.75–6.25 mg/L, found that five participants tested positive for the virus. However, when the concentration of the inoculated water was increased to 10 mg/L, none of the volunteers tested positive. Shin and Sobsey [152] also investigated the effect of chlorine on NoV inactivation using the RT-PCR technique and concluded that the virus was not as resistant to chlorine as earlier reported in other studies. And attributed the observed differences to medium used to prepare the viral suspension and the cluster of virus presence.

Chlorine dioxide (ClO_2), a more soluble oxidizing agent in water than chlorine, is another chemical-based antiviral control alternative. Compared with Cl, ClO_2 limits water discoloration by oxidizing iron and manganese, unreactive with ammonia, and has a stable pH when compared to chlorine [96]. However, due to its instability and possibility of explosion when concentrated, it is normally prepared on-site for use, undermining its applications [151]. The efficacy of enteric virus inactivation by ClO_2 treatment has been speculated to target the viral capsid [97], while others have lent credence viral nucleic acid as its target [98]. Li et al. [99] observed the efficacy of inactivation of HAV by ClO_2 and proposed that the chemicals damaged the viral genome by preventing its replication while reacting with capsid protein, thus preventing attachment to the host cells. Unlike ClO_2, ozone (O_3) is sparingly soluble in water, and generally regarded as safe with an oxidative potential of 2.07 mV, making it 52% stronger than chlorine [150]. It is a broad-spectrum disinfectant that is generated on-site. It is used as a disinfectant for food treatment, water treatment, and also in food-processing industries. In solution, O_3 decomposes to hydroperoxyl (HO_2), hydroxyl (OH), and superoxide (O_2), and its activity depends on these radicals. Although its radicals oxidize the viral envelope [100], its mechanism of action on enteric viruses depends on the type of virus and strain type. Generally, most foodborne viruses are sensitive to O_3 because they are nonenveloped, allowing easy penetration of O_3 [101]. For the inactivation of nonenveloped viruses, O_3 targets either the viral capsid protein, point site of virus attachment to the host cell, or the nucleic acid. Its resultant effect on targets the nucleic acid is the prevention of viral replication. Opinions of targeting the capsid are divergent with the most prominent being capsid damage influenced by amino acid composition [102], and a culminative effect related to reduction in viral penetration after treatment of viral capsid with O_3 [168].

Ultraviolet (UV) light treatment at high doses has also been demonstrated against foodborne viruses, with the main targets being the viral nucleic acid and capsid proteins of enteric viruses, with effect manifesting as morphological changes of the target structures [149]. The efficacy of UV light can be influenced by the host cell involved, the strain of the virus, produce type, and experimental condition [103].

Irradiation and PEFs are other technological strategies employed to manage and control foodborne viral pathogens. The principle of irradiation is similar to that of pasteurization, hence, the name cold pasteurization [171]. The efficacy of irradiation on viral inactivation in foods is very low and this has been associated with the smaller size of most foodborne viruses such as enteric viruses, making them untreatable and resistant to ionizing irradiation. Besides, the nature of the food and temperature are other important factors of interest [104]. The PEF is a nonthermal technology that uses short electrical bursts to inactivate microorganisms in liquid or semiliquid food, with a minimal detrimental effect on the food [105]. Although the information on the use of PEFs on viral inactivation is scarce, however, the available evidence suggests that foodborne enteric viruses like rotavirus are resistant to PEF treatment at 20–29 kv/cm for 145.6 µs, and the reported effect has been attributed to the presence of viral capsid as compared to membrane found in bacterial cells [147].

Generally, it is noteworthy that identifying the major sources of viral contamination of foods is crucial as this will guide in adopting the most effective and adequate control measures in controlling virus-related foodborne issues.

3.4 Methods for detecting viruses in foods

It is a known fact that viruses do not replicate in food and, as such, virus-contaminated foods will not degenerate, and the sensory properties will remain essentially unaltered. This is one of the singular facts that has made suspecting the presence of viruses in food very difficult. Hence, efficient tools and strategies in detecting viruses in food samples are needed to understand transmission routes, control and preventative measures.

Over the years, considerable progress has been made in developing sensitive methods for detecting viruses in foods, and some of these include the molecular-based nonamplified (microarray [106–108]) and amplified (nucleic acid sequence-based amplification (NASBA) [109, 110] and the reverse transcription-real-time polymerase chain reaction (RT-qPCR) [111, 112]) methods. However, the use of RT-qPCR to detect enteric viruses is most dominating [112].

Regardless of the food sample, the general procedures for detecting food viruses require virus extraction from the food matrix, viral RNA purification, and molecular detection (Stals et al., 2012a). The outline of the strategies for molecular detection of enteric viruses such as NoV-I, NoV-II, HAV, and HEV is shown in Figure 3.1.

The methods used to extract viruses from food samples are dependent on the food compositions. Baert et al. [113] categorized foods into three: carbohydrates and water-based foods such as fruits and vegetables fall into the first category, while fat- and protein-based foods like RTE products are in the second category, and the third category includes the shellfish. Protocols for extracting viruses can be elution of viral particles (with or without the acid adsorption preceded step) or by proteinase K treatment or direct RNA extraction [114–116, 173].

Studies have used acid adsorption–elution to extract viruses from shellfish, water/carbohydrate-based foods, and fat/protein-based foods [117, Stals et al., 2011b]. In this process, the acid adsorption step precedes some elution steps; this involves the addition of an acid buffer to the oyster tissue, which allows the viral particles to adsorb to the oyster tissue, followed by the elution step. In the elution step, the supernatant was dis-

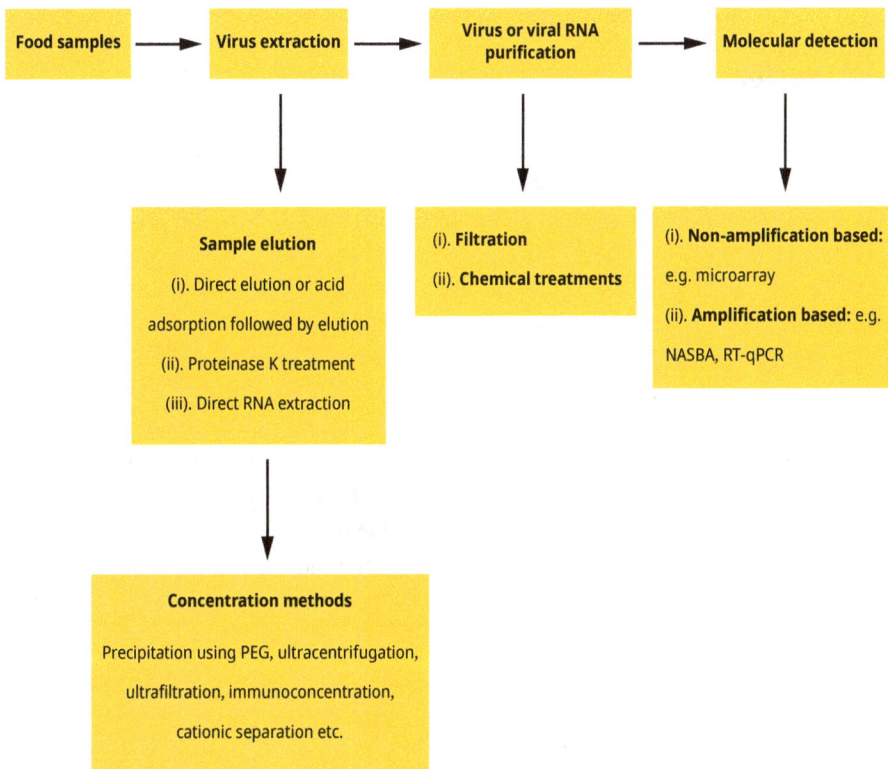

Figure 3.1: The overview of strategies for molecular detection of NoVs, HAV, and HEV in food samples (adapted from [153]).

carded, and the viruses were eluted using a more acidic or neutral buffered solution [118]. Other methods such as alkaline (pH 9 and 10.5) enrichment [119] and histo-blood group antigen-conjugated beads have also been used [120]. While viral extraction from shellfish has also been achieved by proteinase K treatment [114, 115], direct viral RNA extraction has been successfully applied to fats/protein foods and a variety of RTE foods [113, 116, 153].

The concentrations of viruses extracted depend on the initial number of viruses present in the food samples. The extraction step is quite challenging because it may result in low viral particles recovery; hence, it is inevitable to concentrate the eluted viral particles. Methods used in concentrating viral particles include viral precipitation with polyethylene glycol [121–123], ultrafiltration [124], ultracentrifugation [125, 126], cationic separation, and immunoconcentration [120, 127].

In the purification step, food debris and inhibitors possibly present in the extracted virus are removed by filtration through 0.45 and 0.20 μm filters [124] or by treatment using specific kits such as Vertrel® XF [128]. Chloroform:butanol system has also been employed as an efficient method to purify the extracted viral RNA [113, 153].

3.4.1 Nucleic acid sequence-based amplification (NASBA)

NASBA, also known as "self-sustained sequence replication" (3SR), is a sensitive transcription amplification system used in the replication of nucleic acids in vitro [129, 130]. NASBA is an isothermal method that produces multiple copies of single-stranded RNA. In NASBA, RNA amplification depends on the activities of three enzymes. These enzymes are ribonuclease H (RNase H), avian myeloblastosis virus RT, and T7 DNA-dependent RNA polymerase (DdRp) [109, 131]. The NASBA or 3SR reaction steps are shown in Figure 3.2. The reaction is initiated by the annealing of the first primer containing a T7 promoter sequence to sense RNA followed by the transcription of a strand of antisense DNA by RT; and the digestion of sense RNA by RNase H. The second primer with the T7 promoter sequence anneals to the single antisense DNA strand, and the complementary DNA (cDNA) strand is synthesized by RT. The resulting double-stranded DNA is used as a template by DdRp for the transcription of multiple copies of antisense RNA [109, 131]. The forward primer used for NASBA is gene-specific, complementary to the target RNA and contains a T7 promoter sequence at the 5'-end. The reverse primer may have a nonspecific sequence with a generic probe that can be used in electrochemiluminescence detection [109].

The NASBA amplicons can be detected by agarose gel electrophoresis and hybridization [109, 132]. The use of molecular beacons for real-time detection of the targeted product is another molecular technology that can be used in detecting NASBA amplicons [133].

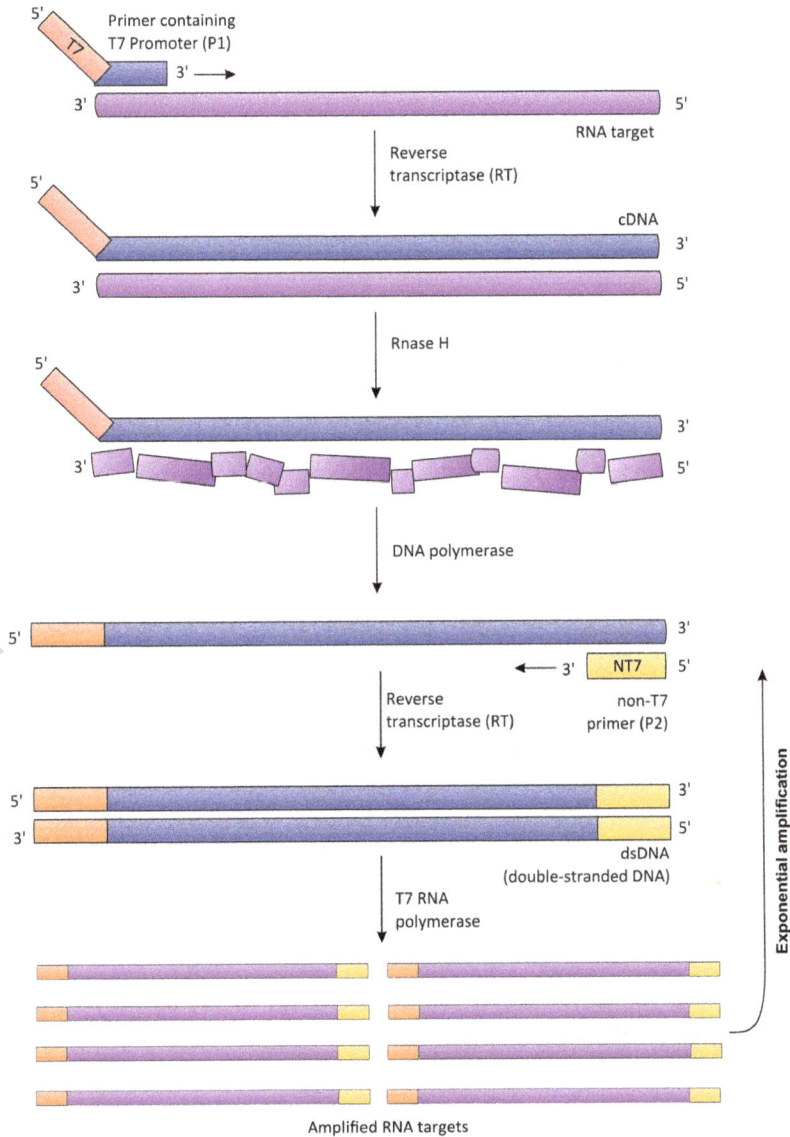

Figure 3.2: Nucleic acid sequence-based amplification.

3.4.2 The reverse transcription-real-time polymerase chain reaction (RT-qPCR)

The advent of RT-qPCR allows for the efficient detection of viruses. Until the introduction of RT-qPCR testing, the origin of 68% of foodborne outbreaks before 1997, including those later elucidated as NoV, was unknown [134], and this is contrary to the 67% increase in the known etiology of viral foodborne occurrences from 1998 to 2001 [135]. Hence, the introduction of RT-qPCR made the detection of NoV in suspected outbreaks detectable and stronger than those in the last 20 years. In RT-qPCR, RT catalyzes cDNA synthesis using RNA as the starting material [111, 136], and cDNA is used as a template in qPCR. The real-time PCR (qPCR) allows data collection as the PCR process occurs. The qPCR combines amplification and detection into one step, and the fluorescent chemistry such as SYBR® Green or TaqMan® used correlates the PCR product with the fluorescent intensity [137]. The RT-qPCR can be performed in a one-step assay, where RT and PCR are combined as a single reaction, or in a two-step assay, where RT is first performed, followed by the PCR step (Figure 3.3).

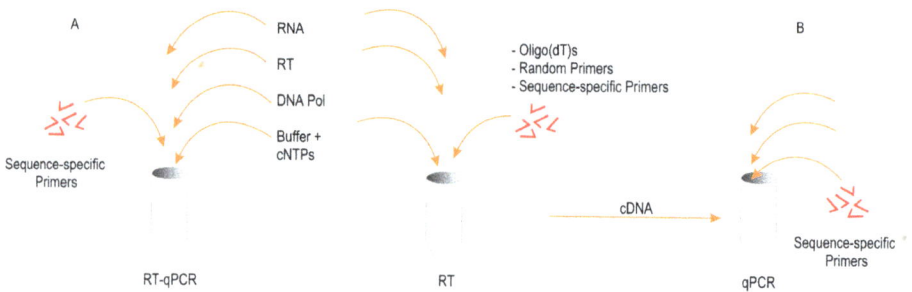

Figure 3.3: Reverse transcription-real-time polymerase chain reactions: (A) one-step reaction and (B) two-step reactions.

Most reported standardized methodologies (ISO) and several studies [121, 107, 138–140, 153, 166] have used RT-qPCR assays in detecting several enteric viruses like NoVI and NoVII, and HAV and HEV in foods, and the results are highly encouraging.

3.5 SARS-CoV-2 and food safety

The severe acute respiratory syndrome coronavirus 2 (SARS-CoV-2) is an enveloped, positive-sense, single-stranded RNA (29.8–29.9 kb) and belongs to the family Coronaviridae [141]. It was first discovered in Wuhan, China, in December 2019 and causes a highly transmittable and pathogenic disease (COVID-19) [142]. As of 20 September 2021, the global reported COVID-19 cases stood at 230,912,407 with 47,335,621 mortalities (https://coronavirus.jhu.edu/map.html). At the inception of the pandemic, data on SARS-CoV-2 and its survival in diverse formites, and more importantly, in food and food products were unavailable. This, coupled with the severity of the disease, prompted food scientists, food regulatory authorities, and food safety agencies to assess the risk of potentially transmitting SARS-CoV-2 from food packaging or contaminated food (Anelich et al., 2021). One of the concerns was to establish the possibility of fecal–oral route transmission of the virus as supportive evidences were emerging that the viral genetic material was detected in the systemic circulation and anal swabs of infected patients [143], and this constituted a crucial point, with diarrhea being a prominent symptom of COVID-19. However, reports on the fecal–oral transmission of the virus are divergent to date, with the most prominent scientific evidence premised on the main route of its transmission involving person-to-person via contact, aerosols, and respiratory droplets, establishing SARS-CoV-2 as a nonfoodborne virus [144]. Additionally, the diarrhea in COVID-19 victims has been opined to be unlikely from ingesting contaminated food but from the pathway of the virus through the respiratory system to the digestive tract [145].

Generally, while acknowledging that the current pandemic has significantly disrupted the global food supply chain, it however poses no food safety risk. This is premised on the convergent evidence-based scientific views that SARS-CoV-2 is neither foodborne nor transmitted by the food packaging material, but it could be spread through contacts with contaminated surfaces and subsequent introduction into the body via the external orifices linked to the respiratory tract [146]. Hence, upholding and maintaining good hygienic practices including proper cleaning and disinfection of food contact surfaces, and more importantly, utensils and high-touch surfaces in a food environment are highly recommended.

3.6 Conclusion and perspectives

NoV and HAV remain the most common foodborne viral pathogens, and the food type determines the degree of contamination and viral persistency; hence, some foods are more linked to a particular virus than others. Although every human is susceptible to virus-poisoned foods, the younger children, pregnant women, aged, and immunocompromised persons are more vulnerable [5]. Generally, most viruses implicated in foodborne outbreaks are nonenveloped and tend to be indefinitely persistent in the environment and are resistant to common food preservation technology. Besides resistance, viral detection in foods has been challenging over the years until recently where efficient tools and strategies including the molecular approaches such as NASBA [109, 110] and RT-qPCR [111] were developed and were proved to be efficient in appreciating viral transmission routes and approaches to prevent and/or control foodborne viral infections and outbreaks. More importantly, identification of the main sources of viral contamination of foods is germane and will provide guidance for the detection method(s) to be employed before the application of appropriate viral control measures.

Food and food products contaminated with viruses have negative impacts on the overall human health as well as agriculture vitality, economic prosperity, sustainable development, and food security. The prevalence of foodborne infections attributable to viruses makes microbiological quality the most crucial component of food quality and safety. Thus, food safety majorly focuses on controlling the contamination of foods by pathogens, including viruses. While the current COVID-19 pandemic has significantly impacted the global food supply chain, current and emerging evidence suggests that it is not foodborne and poses no food safety risk. Consequent upon the foregoing and to avoid the risk posed by foodborne viruses, good manufacturing practices, good hygiene practices, and good agricultural practices should be embraced. Such practices will ensure food safety at every stage of the supply chain, thereby contributing to a robust prevention of possible future foodborne outbreaks of microbial origin including viruses.

References

[1] Pal, M., & Ayele, Y. (2020). Emerging role of foodborne viruses in public health. *Biomed Research, 5*, 01–4.

[2] Carter, A. (2018). Viruses In Food: Things You Don't Want To Know. *Hygiene Food Safety*. https://hygienefoodsafety.org/viruses-in-food/

[3] Scallan, E., Hoekstra, R. M., Angulo, F. J., Tauxe, R. V., Widdowson, M. A., Roy, S. L., . . . Griffin, P. M. (2011). Foodborne illness acquired in the United States – major pathogens. *Emerging Infectious Diseases, 17*(1), 7.

[4] World Health Organization (WHO). (2019). First-ever world food safety day elevates attention to dangerous foodborne diseases in Africa. https://www.afro.who.int/news/first-ever-world-food-safety-day-elevates-attention-dangerous-foodborne-diseases-africa (Accessed on 7th July 2021).

[5] Centers for Disease Control and Prevention. (2016). Burden of foodborne illness: findings. *Centers for Disease Control and Prevention*.

[6] Coudray-Meunier, C., Fraisse, A., Mokhtari, C., Martin-Latil, S., Roque-Afonso, A. M., & Perelle, S. (2014). Hepatitis A virus subgenotyping based on RT-qPCR assays. *BMC Microbiology, 14*(1), 1–11.

[7] Maunula, L., & von Bonsdorff, C. H. (2016). Foodborne viruses in ready-to-eat foods. In *Food hygiene and toxicology in ready-to-eat foods* (pp. 51–68). Academic Press.

[8] FAO/WHO (Food and Agriculture Organization of the United Nations/World Health Organization) (2008). In *Microbiological hazards in fresh leafy vegetables and herbs: Meeting report* (Vol. 14). World Health Organization.

[9] Nørrung, B. (2013). Foodborne viruses: an introduction. *Food Borne Viruses and Prions and Their Significance for Public Health, 6*, 37.

[10] Acheson, D., Bresee, J. S., Widdowson, M. A., Monroe, S. S., & Glass, R. I. (2002). Foodborne viral gastroenteritis: challenges and opportunities. *Clinical Infectious Diseases, 35*(6), 748–753.

[11] Green, K. Y. (2013). Caliciviridae: The noroviruses. In *Fields virology* (6th ed., pp. 949–979). Philadelphia, PA: Wolters Kluwer Health/Lippincott Williams and Wilkins.

[12] Thorne, L. G., & Goodfellow, I. G. (2014). Norovirus gene expression and replication. *Journal of General Virology, 95*(2), 278–291.

[13] Chhabra, P., De graaf, M., Parra, G. I., Chan, M. C. W., Green, K., Martella, V., & Vinjé, J. (2019). Updated classification of norovirus genogroups and genotypes. *The Journal of General Virology, 100* (10), 1393.

[14] Cannon, J. L., Barclay, L., Collins, N. R., Wikswo, M. E., Castro, C. J., Magaña, L. C., . . . Vinjé, J. (2017). Genetic and epidemiologic trends of norovirus outbreaks in the United States from 2013 to 2016

demonstrated emergence of novel GII 4 recombinant viruses. *Journal of Clinical Microbiology, 55*(7), 2208–2221.

[15] Barclay, L., Davis, T., & Vinjé, J. (2021). Rare Norovirus GIV Foodborne Outbreak, Wisconsin USA. *Emerging Infectious Diseases, 27*(4), 1151.

[16] Eden, J. S., Lim, K. L., & White, P. A. (2012). Complete genome of the human norovirus GIV. 1 strain Lake Macquarie virus. *Journal of Virology 2012*, 86, 10251–10252.

[17] Martella, V., Campolo, M., Lorusso, E., Cavicchio, P., Camero, M., Bellacicco, A. L., & Buonavoglia, C. (2007). Norovirus in captive lion cub (Panthera leo). *Emerging Infectious Diseases, 13*(7), 1071.

[18] Atmar, R. L., Opekun, A. R., Gilger, M. A., Estes, M. K., Crawford, S. E., Neill, F. H., & Graham, D. Y. (2008). Norwalk virus shedding after experimental human infection. *Emerging Infectious Diseases, 14*(10), 1553.

[19] Hall, A. J., Eisenbart, V. G., Etingüe, A. L., Gould, L. H., Lopman, B. A., & Parashar, U. D. (2012). Epidemiology of foodborne norovirus outbreaks, United States, 2001–2008. *Emerging Infectious Diseases, 18*(10), 1566.

[20] Hall, A. J., Wikswo, M. E., Pringle, K., Gould, L. H., & Parashar, U. D. (2014). Vital signs: foodborne norovirus outbreaks – United States, 2009–2012. *MMWR: Morbidity and Mortality Weekly Report, 63*(22), 491.

[21] Gallay, A., De Valk, H., Cournot, M., Ladeuil, B., Hemery, C., Castor, C., & Desenclos, J. C. (2006). A large multi-pathogen waterborne community outbreak linked to faecal contamination of a groundwater system, France, 2000. *Clinical Microbiology & Infection, 12*(6), 561–570.

[22] Lopman, B., Gastanaduy, P., Park, G. W., Hall, A. J., Parashar, U. D., & Vinjé, J. (2012). Environmental transmission of norovirus gastroenteritis. *Current Opinion in Virology, 2*(1), 96–102.

[23] Costa-Mattioli, M., Cristina, J., Romero, H., Perez-Bercof, R., Casane, D., Colina, R., & Ferré, V. (2002). Molecular evolution of hepatitis A virus: a new classification based on the complete VP1 protein. *Virology Journal, 76*(18), 9516–9525.

[24] Mast, E. E., & Alter, M. J. (1993). October Epidemiology of viral hepatitis: an overview. *Seminars in Virology,* 4(5), 273–283. Academic Press.

[25] Hollinger, F. B., & Emerson, S. U. (2007). Hepatitis A virus. En libro: Fields Virology. *Knipe. DM y Howley, PM Philadelphia.*

[26] Franco, E., Meleleo, C., Serino, L., Sorbara, D., & Zaratti, L. (2012). Hepatitis A: Epidemiology and prevention in developing countries. *World Journal of Hepatology, 4*(3), 68.

[27] Koopmans, M., & Duizer, E. (2004). Foodborne viruses: an emerging problem. *International Journal of Food Microbiology, 90*(1), 23–41.

[28] Velebit, B., Djordjevic, V., Milojevic, L., Babic, M., Grkovic, N., Jankovic, V., & Yushina, Y. (2019, September). The common foodborne viruses: a review. In *IOP Conference series: Earth and environmental science* (Vol 333, No. 1, p. 012110). IOP Publishing.

[29] Sridhar, S., Lau, S. K., & Woo, P. C. (2015). Hepatitis E: A disease of reemerging importance. *Journal of the Formosan Medical Association, 114*(8), 681–690.

[30] Pavio, N., Meng, X. J., & Doceul, V. (2015). Zoonotic origin of hepatitis E. *Current Opinion in Virology, 10*, 34–41.

[31] Smith, D. B., & Simmonds, P. (2015). Hepatitis E virus and fulminant hepatitis–A virus or host-specific pathology? *Liver International, 35*(4), 1334–1340.

[32] Iaconelli, M., Purpari, G., Della Libera, S., Petricca, S., Guercio, A., Ciccaglione, A. R., & La Rosa, G. (2015). Hepatitis A and E viruses in wastewaters, in river waters, and in bivalve molluscs in Italy. *Food and Environmental Virology, 7*(4), 316–324.

[33] Kantala, T., Heinonen, M., Oristo, S., von Bonsdorff, C. H., & Maunula, L. (2015). Hepatitis E virus in young pigs in Finland and characterization of the isolated partial genomic sequences of genotype 3 HEV. *Foodborne Pathogens and Disease, 12*(3), 253–260.

[34] Maunula, L., Kaupke, A., Vasickova, P., Söderberg, K., Kozyra, I., Lazic, S., & Cook, N. (2013). Tracing enteric viruses in the European berry fruit supply chain. *International Journal of Food Microbiology, 167*(2), 177–185.

[35] Crossan, C., Baker, P. J., Craft, J., Takeuchi, Y., Dalton, H. R., & Scobie, L. (2012). Hepatitis E virus genotype 3 in shellfish, United Kingdom. Emergency. *Infection Disease,* 18, 2085–2087.

[36] Rutjes, S. A., Lodder, W. J., Bouwknegt, M., & de Roda Husman, A. M. (2007). Increased hepatitis E virus prevalence on Dutch pig farms from 33 to 55% by using appropriate internal quality controls for RT-PCR. *Journal of Virological Methods, 143*(1), 112–116.

[37] Aggarwal, R., & Naik, S. (2009). Epidemiology of hepatitis E: current status. *Journal of Gastroenterology and Hepatology, 24*(9), 1484–1493.

[38] Chen, M. Y., Chen, W. C., Chen, P. C., Hsu, S. W., & Lo, Y. C. (2016). An outbreak of norovirus gastroenteritis associated with asymptomatic food handlers in Kinmen, Taiwan. *BMC Public Health, 16*(1), 1–6. doi: https://doi.org/10.1186/s12889-016-3046-5.

[39] Schmid, D., Stüger, H. P., Lederer, I., Pichler, A. M., Kainz-Arnfelser, G., Schreier, E., & Allerberger, F. (2007). A foodborne norovirus outbreak due to manually prepared salad, Austria 2006. *Infection, 35*(4), 232–239.

[40] Barrabeig, I., Rovira, A., Buesa, J., Bartolomé, R., Pintó, R., Prellezo, H., & Domínguez, À. (2010). Foodborne norovirus outbreak: The role of an asymptomatic food handler. *BMC Infectious Diseases, 10*(September), 1–7. doi: https://doi.org/10.1186/1471-2334-10-269.

[41] FDA. (2016). *FDA Investigates Outbreak of Hepatitis A Illness Linked to Frozen Strawberries.* https://www.fda.gov/food/outbreaks-foodborne-illness/fda-investigates-outbreak-hepatitis-illnesses-linked-frozen-strawberries

[42] Wenzel, J. J., Schemmerer, M., Oberkofler, H., Kerschner, H., Sinha, P., Koidl, C., & Allerberger, F. (2014). Hepatitis a outbreak in Europe: Imported frozen berry mix suspected to be the source of at least one Infection in Austria in 2013. *Food and Environmental Virology, 6*(4), 297–300. doi: https://doi.org/10.1007/s12560-014-9165-1.

[43] Carvalho, C., Thomas, H. L., Balogun, K., Tedder, R., Pebody, R., Ramsay, M., & Ngui, S. L. (2012). A possible outbreak of hepatitis A associated with semidried tomatoes, England, July-November 2011. *Eurosurveillance, 17*(6), 1–4. doi: https://doi.org/10.2807/ese.17.06.20083-en. England.

[44] Fournet, N., Baas, D., Van pelt, W., Swaan, C., Ober, H., Isken, L., Cremer, J., Friesema, I., Vennema, H., Boxman, I., Koopmans, M., & Verhoef, L. (2012). Another possible food-borne outbreak of hepatitis a in the Netherlands indicated by two closely related molecular sequences, July to October 2011. *Eurosurveillance, 17*(6), 1–3. doi: https://doi.org/10.2807/ese.17.06.20079-en.

[45] Gallot, C., Grout, L., Roque-Afonso, A. M., Couturier, E., Carrillo-Santisteve, P., Pouey, J., & Vaillant, V. (2011). Hepatitis A associated with semidried tomatoes, France, 2010. *Emerging Infectious Diseases, 17*(3), 566.

[46] Flynn, D. (2009). Chi-Chi hepatitis A outbreak. *Food Safety News,* 1–2.

[47] Riveiro-Barciela, M., Sauleda, S., Quer, J., Salvador, F., Gregori, J., Pirón, M., Rodríguez-Frías, F., & Buti, M. (2017). Red blood cell transfusion-transmitted acute hepatitis E in an immunocompetent subject in Europe: a case report. *Transfusion,* 1–4. doi: https://doi.org/10.1111/trf.13876.

[48] Moal, V., Gerolami, R., & Colson, P. (2012). First human case of co-infection with two different subtypes of hepatitis E virus. *Intervirology, 55*(6), 484–487. doi: https://doi.org/10.1159/000335664.

[49] Anty, R., Ollier, L., Péron, J. M., Nicand, E., Cannavo, I., Bongain, A., Giordanengo, V., & Tran, A. (2012). First case report of an acute genotype 3 hepatitis E infected pregnant woman living in South-Eastern France. *Journal of Clinical Virology, 54*(1), 76–78. doi: https://doi.org/10.1016/j.jcv.2012.01.016.

[50] Renou, C., Pariente, A., Cadranel, J., Nicand, E., & Pavio, N. (2011). Clinical silent forms may partly explain the rarity of acute cases of autochthonous genotype 3c hepatitis E infection in France. *Journal of Clinical Virology,* 51, 139–141.

[51] Said, B., Ijaz, S., Kafatos, G., Booth, L., Thomas, H. L., Walsh, A., Ramsay, M., & Morgan, D. (2009). Hepatitis E outbreak on cruise ship. *Emerging Infectious Diseases, 15*(11), 1738–1744. doi: https://doi. org/10.3201/eid1511.091094.

[52] Reuter, G., Fodor, D., Kátai, A., & Szucs, G. (2006). Identification of a novel variant of human hepatitis E virus in Hungary. *Journal of Clinical Virology, 36*(2), 100–102. doi: https://doi.org/10.1016/j. jcv.2006.01.016.

[53] Kobayashi, S., Fujiwara, N., Yasui, Y., Yamashita, T., Hiramatsu, R., & Minagawa, H. (2012). A foodborne outbreak of sapovirus linked to catered box lunches in Japan. *Archives of Virology, 157*(10), 1995–1997.

[54] Bosh, A., Pinto, R. M., & Guix, S. (2014). Human astrovirus. *Clinical Microbiology Reviews, 27,* 1048–1074.

[55] Heaton, J. C., & Jones, K. (2008). Microbial contamination of fruit and vegetables and the behaviour of enteropathogens in the phyllosphere: a review. *Journal of Applied Microbiology, 104*(3), 613–626.

[56] Symes, S. J., Gunesekere, I. C., Marshall, J. A., & Wright, P. J. (2007). Norovirus mixed infection in an oyster-associated outbreak: an opportunity for recombination. *Archives of Virology, 152*(6), 1075–1086.

[57] Bidawid, S., Malik, N., Adegbunrin, O., Sattar, S. A., & Farber, J. M. (2004). Norovirus cross-contamination during food handling and interruption of virus transfer by hand antisepsis: experiments with feline calicivirus as a surrogate. *Journal of Food Protection, 67*(1), 103–109.

[58] Boone, S. A., & Gerba, C. P. (2007). Significance of fomites in the spread of respiratory and enteric viral disease. *Applied and Environmental Microbiology, 73*(6), 1687–1696.

[59] Matsuda, H., Okada, K., Takahashi, K., & Mishiro, S. (2003). Severe hepatitis E virus infection after ingestion of uncooked liver from a wild boar. *The Journal of Infectious Diseases, 188*(6), 944–944.

[60] Takahashi, K., Kitajima, N., Abe, N., & Mishiro, S. (2004). Complete or near-complete nucleotide sequences of hepatitis E virus genome recovered from a wild boar, a deer, and four patients who ate the deer. *Virology, 330*(2), 501–505.

[61] Guix, S., Caballero, S., Villena, C., Bartolomé, R., Latorre, C., Rabella, N., & Pintó, R. M. (2002). Molecular epidemiology of astrovirus infection in Barcelona Spain. *Journal of Clinical Microbiology, 40*(1), 133–139.

[62] Rzeżutka, A., & Cook, N. (2004). Survival of human enteric viruses in the environment and food. *FEMS Microbiology Reviews, 28*(4), 441–453.

[63] Seymour, I. J., & Appleton, H. (2001). Foodborne viruses and fresh produce. *Journal of Applied Microbiology, 91*(5), 759.

[64] Hewitt, J., & Greening, G. E. (2004). Survival and persistence of norovirus, hepatitis A virus, and feline calicivirus in marinated mussels. *Journal of Food Protection, 67*(8), 1743–1750.

[65] Butot, S., Putallaz, T., & Sanchez, G. (2008). Effects of sanitation, freezing and frozen storage on enteric viruses in berries and herbs. *International Journal of Food Microbiology, 126*(1–2), 30–35.

[66] Fallahi, S., & Mattison, K. (2011). Evaluation of murine norovirus persistence in environments relevant to food production and processing. *Journal of Food Protection, 74*(11), 1847–1851.

[67] Shieh, Y. C., Khudyakov, Y. E., Xia, G., Ganova-Raeva, L. M., Khambaty, F. M., Woods, J. W., & Fiore, A. E. (2007). Molecular confirmation of oysters as the vector for hepatitis A in a 2005 multistate outbreak. *Journal of Food Protection, 70*(1), 145–150.

[68] Cheesbrough, J. S., Barkess-Jones, L., & Brown, D. W. (1997). Possible prolonged environmental survival of small round structured viruses. *The Journal of Hospital Infection, 35*(4), 325–326.

[69] Abad, F. X., Pinto, R. M., & Bosch, A. (1994). Survival of enteric viruses on environmental fomites. *Applied and Environmental Microbiology, 60*(10), 3704–3710.

[70] Abad, F. X., Villena, C., Guix, S., Caballero, S., Pintó, R. M., & Bosch, A. (2001). Potential role of fomites in the vehicular transmission of human astroviruses. *Applied and Environmental Microbiology, 67*(9), 3904–3907.

[71]   Hjertqvist, M., Johansson, A., Svensson, N., Abom, P. E., Magnusson, C., Olsson, M., & Andersson, Y. (2006). Four outbreaks of norovirus gastroenteritis after consuming raspberries, Sweden, June-August 2006. *Weekly Releases(1997–2007)*, *11*(36), 3038.

[72]   Papafragkou, E., D'Souza, D. H., & Jaykus, L. A. (2006). Food-borne viruses: prevention and control. In *Viruses in foods* (pp. 289–330). Boston, MA: Springer.

[73]   Koopmans, M., von Bonsdorff, C. H., Vinjé, J., de Medici, D., & Monroe, S. (2002). Foodborne viruses. *FEMS Microbiology Reviews*, *26*(2), 187–205.

[74]   Grove, S. F., Lee, A., Lewis, T., Stewart, C. M., Chen, H., & Hoover, D. G. (2006). Inactivation of foodborne viruses of significance by high pressure and other processes. *Journal of Food Protection*, *69*(4), 957–968.

[75]   EFSA Panel on Biological Hazards (BIOHAZ), Ricci, A., Allende, A., Bolton, D., Chemaly, M., Davies, R., & Girones, R. (2017). Public health risks associated with hepatitis E virus (HEV) as a food-borne pathogen. *EFSA Journal*, *15*(7), e04886.

[76]   Ruggeri, F. M., Di Bartolo, I., Ponterio, E., Angeloni, G., Trevisani, M., & Ostanello, F. (2013). Zoonotic transmission of hepatitis E virus in industrialized countries. *New Microbiologica*, *36*(4), 331–344.

[77]   Tricco, A. C., Duval, B., De Serres, G., Gilca, V., Vrbova, L., Anonychuk, A., & Moher, D. (2006). A review of interventions triggered by hepatitis A infected food-handlers in Canada. *BMC Health Services Research*, *6*(1), 1–7.

[78]   Bosch, A., Gkogka, E., Le Guyader, F. S., Loisy-Hamon, F., Lee, A., Van Lieshout, L., & Phister, T. (2018). Foodborne viruses: Detection, risk assessment, and control options in food processing. *International Journal of Food Microbiology*, *285*, 110–128.

[79]   Bozkurt, H., D'Souza, D. H., & Davidson, P. M. (2014). Determination of thermal inactivation kinetics of hepatitis A virus in blue mussel (Mytilus edulis) homogenate. *Applied and Environmental Microbiology*, *80*(10), 3191–3197.

[80]   EFSA Panel on Biological Hazards (BIOHAZ) (2015). Evaluation of heat treatments, different from those currently established in the EU legislation, that could be applied to live bivalve molluscs from B and C production areas, that have not been submitted to purification or relaying, in order to eliminate pathogenic microorganisms. *EFSA Journal*, *13*(12), 4332.

[81]   Butot, S., Putallaz, T., Amoroso, R., & Sánchez, G. (2009). Enteric virus inactivation in minimally-processed berries and herbs. *Applied and Environmental Microbiology*, 4155–4161.

[82]   Hewitt, J., & Greening, G. E. (2006). Effect of heat treatment on hepatitis A virus and norovirus in New Zealand greenshell mussels (Perna canaliculus) by quantitative real-time reverse transcription PCR and cell culture. *Journal of Food Protection*, *69*(9), 2217–2223.

[83]   Sow, H., Desbiens, M., Morales-Rayas, R., Ngazoa, S. E., & Jean, J. (2011). Heat inactivation of hepatitis A virus and a norovirus surrogate in soft-shell clams (Mya arenaria). *Foodborne Pathogens and Disease*, *8*(3), 387–393.

[84]   Russell, W. C., Valentine, R. C., & Pereira, H. G. (1967). The effect of heat on the anatomy of the adenovirus. *Journal of General Virology*, *1*(4), 509–522.

[85]   O'Brien, R. T., & Newman, J. (1979). Structural and compositional changes associated with chlorine inactivation of polioviruses. *Applied and Environmental Microbiology*, *38*(6), 1034–1039.

[86]   Croci, L., Ciccozzi, M., De Medici, D., Di Pasquale, S., Fiore, A., Mele, A., & Toti, L. (1999). Inactivation of hepatitis A virus in heat-treated mussels. *Journal of Applied Microbiology*, *87*(6), 884–888.

[87]   Kingsley, D. H., Hoover, D. G., Papafragkou, E. F. I., & Richards, G. P. (2002). Inactivation of hepatitis A virus and a calicivirus by high hydrostatic pressure. *Journal of Food Protection*, *65*(10), 1605–1609.

[88]   Govaris, A., & Pexara, A. (2021). Inactivation of foodborne viruses by high-pressure processing (HPP). *Foods*, *10*(2), 215.

[89]   Li, X., Chen, H., & Kingsley, D. H. (2013). The influence of temperature, pH, and water immersion on the high hydrostatic pressure inactivation of GI 1 and GII. 4 human noroviruses. *International Journal of Food Microbiology*, *167*(2), 138–143.

[90] Kingsley, D. H., Guan, D., & Hoover, D. G. (2005). Pressure inactivation of hepatitis A virus in strawberry puree and sliced green onions. *Journal of Food Protection, 68*(8), 1748–1751.

[91] Lou, F., DiCaprio, E., Li, X., Dai, X., Ma, Y., Hughes, J., & Li, J. (2016). Variable high-pressure-processing sensitivities for genogroup II human noroviruses. *Applied and Environmental Microbiology, 82*(19), 6037–6045.

[92] Kingsley, D. H., Holliman, D. R., Calci, K. R., Chen, H., & Flick, G. J. (2007). Inactivation of a norovirus by high-pressure processing. *Applied and Environmental Microbiology, 73*(2), 581–585.

[93] Warf, C., Herdt, J., Foong-Cunningham, S., & Swanson, K. (2007). Critical factors that influence the effectiveness of aqueous antimicrobial interventions of fresh produce. In *IFT annual meeting* (pp. 042–01). *Chicago, IL*.

[94] Rutala, W. A., Barbee, S. L., Aguiar, N. C., Sobsey, M. D., & Weber, D. J. (2000). Antimicrobial activity of home disinfectants and natural products against potential human pathogens. *Infection Control & Hospital Epidemiology, 21*(1), 33–38.

[95] Keswick, B. H., Satterwhite, T. K., Johnson, P. C., DuPont, H. L., Secor, S. L., Bitsura, J. A., & Hoff, J. C. (1985). Inactivation of Norwalk virus in drinking water by chlorine. *Applied and Environmental Microbiology, 50*(2), 261–264.

[96] Aieta, E. M., & Berg, J. D. (1986). A review of chlorine dioxide in drinking water treatment. *Journal of American Water Works Association, 78*(6), 62–72.

[97] White, G. C. (1978). *Disinfection of wastewater and water for reuse*. Van Nostrand Reinhold.

[98] Taylor, G. R., & Butler, M. (1982). A comparison of the virucidal properties of chlorine, chlorine dioxide, bromine chloride and iodine. *Epidemiology and Infection, 89*(2), 321–328.

[99] Li, J. W., Xin, Z. T., Wang, X. W., Zheng, J. L., & Chao, F. H. (2004). Mechanisms of inactivation of hepatitis A virus in water by chlorine dioxide. *Water Research, 38*(6), 1514–1519.

[100] Langlais, B., Reckhow, D. A., & Brink, D. R. (1991). Practical application of ozone: Principle and case study. *Ozone in Water Treatment," Lewis Publishers. Chelsea, Mich*.

[101] Khadre, M. A., Yousef, A. E., & Kim, J. G. (2001). Microbiological aspects of ozone applications in food: a review. *Journal of Food Science, 66*(9), 1242–1252.

[102] Mudd, J. B., Leavitt, R., Ongun, A., & McManus, T. T. (1969). Reaction of ozone with amino acids and proteins. *Atmospheric Environment (1967), 3*(6), 669–681.

[103] Fino, V. R., & Kniel, K. E. (2008). UV light inactivation of hepatitis A virus, Aichi virus, and feline calicivirus on strawberries, green onions, and lettuce. *Journal of Food Protection, 71*(5), 908–913.

[104] Patterson, M. F. (1993). Food irradiation and food safety. *Reviews in Medical Microbiology, 4*(3), 151–158.

[105] Ramaswamy, R., Jin, T., Balasubramaniam, V. M., & Zhang, H. (2005). *Pulsed electric field processing: fact sheet for food processors* (Vol. 22). *Ohio State University Extension Factsheet*.

[106] Chen, H., Mammel, M., Kulka, M., Patel, I., Jackson, S., & Goswami, B. B. (2011). Detection and identification of common food-borne viruses with a tiling microarray. *The Open Virology Journal, 5*, 52.

[107] Derrick, J., Hollinghurst, P., O'Brien, S., Elviss, N., Allen, D. J., & Iturriza-Gómara, M. (2021). Measuring transfer of human norovirus during sandwich production: Simulating the role of food, food handlers and the environment. *International Journal of Food Microbiology, 348*, 109151.

[108] Won, Y. J., Lee, S. G., Paik, S. Y., & Lyoo, K. S. (2019). Development of an oligonucleotide-based microarray for the detection of foodborne viruses. *Journal of Consumer Protection and Food Safety, 14* (3), 287–291.

[109] Deiman, B., Van aarle, P., & Sillekens, P. (2002). Characteristics and applications of nucleic acid sequence-based amplification (NASBA). *Molecular Biotechnology, 20*(2), 163–179.

[110] Jean, J., Blais, B., Darveau, A., & Fliss, I. (2002). Simultaneous detection and identification of hepatitis A virus and rotavirus by multiplex nucleic acid sequence-based amplification (NASBA) and microtiter plate hybridization system. *Journal of Virological Methods, 105*(1), 123–132.

[111] Doak, S. H., & Zaïr, Z. M. (2012). Real-time reverse-transcription polymerase chain reaction: technical considerations for gene expression analysis. In *Genetic toxicology* (pp. 251–270). New York, NY: Springer.

[112] Rutjes, S. A., van den Berg, H. H., Lodder, W. J., & de Roda Husman, A. M. (2006). Real-time detection of noroviruses in surface water by use of a broadly reactive nucleic acid sequence-based amplification assay. *Applied and Environmental Microbiology, 72*(8), 5349–5358.

[113] Baert, L., Uyttendaele, M., & Debevere, J. (2008). Evaluation of viral extraction methods on a broad range of ready-to-eat foods with conventional and real-time RT-PCR for norovirus GII detection. *International Journal of Food Microbiology, 123*(1–2), 101–108.

[114] Plante, D., Bran Barrera, J. A., Lord, M., Iugovaz, I., & Nasheri, N. (2021). Development of an RNA Extraction Protocol for a Norovirus from Raw Oysters and Detection by qRT-PCR and Droplet-Digital RT-PCR. *Foods, 10*(8).

[115] Purpari, G., Macaluso, G., Di Bella, S., Gucciardi, F., Mira, F., Di Marco, P., . . . Guercio, A. (2019). Molecular characterization of human enteric viruses in food, water samples, and surface swabs in Sicily. *International Journal of Infectious Diseases, 80*, 66–72.

[116] Rajiuddin, S. M., Jensen, T., Hansen, T. B., & Schultz, A. C. (2020). An optimised direct lysis method for viral RNA extraction and detection of foodborne viruses on fruits and vegetables. *Food and Environmental Virology, 12*(3), 226–239.

[117] Dubois, E., Agier, C., Traoré, O., Hennechart, C., Merle, G., Cruciere, C., & Laveran, H. (2002). Modified concentration method for the detection of enteric viruses on fruits and vegetables by reverse transcriptase-polymerase chain reaction or cell culture. *Journal of Food Protection, 65*(12), 1962–1969.

[118] David, S. T., McIntyre, L., MacDougall, L., Kelly, D., Liem, S., Schallié, K., & Brassard, J. (2007). An outbreak of norovirus caused by consumption of oysters from geographically dispersed harvest sites, British Columbia, Canada, 2004. *Foodborne Pathogens and Disease, 4*(3), 349–358.

[119] Le Guyader, F. S., Mittelholzer, C., Haugarreau, L., Hedlund, K. O., Alsterlund, R., Pommepuy, M., & Svensson, L. (2004). Detection of noroviruses in raspberries associated with a gastroenteritis outbreak. *International Journal of Food Microbiology, 97*(2), 179–186.

[120] Morton, V., Jean, J., Farber, J., & Mattison, K. (2009). Detection of noroviruses in ready-to-eat foods by using carbohydrate-coated magnetic beads. *Applied and Environmental Microbiology, 75*(13), 4641–4643.

[121] Amroabadi, M. A., Rahimi, E., Shakerian, A., & Momtaz, H. (2021). Incidence of hepatitis A and hepatitis E viruses and norovirus and rotavirus in fish and shrimp samples caught from the Persian Gulf. *Arquivo Brasileiro de Medicina Veterinária e Zootecnia, 73*, 169–178.

[122] Cheong, S., Lee, C., Choi, W. C., Lee, C. H., & Kim, S. J. (2009). Concentration method for the detection of enteric viruses from large volumes of foods. *Journal of Food Protection, 72*(9), 2001–2005.

[123] Morales-Rayas, R., Wolffs, P. F., & Griffiths, M. W. (2010). Simultaneous separation and detection of hepatitis A virus and norovirus in produce. *International Journal of Food Microbiology, 139*(1–2), 48–55.

[124] Scherer, K., Johne, R., Schrader, C., Ellerbroek, L., Schulenburg, J., & Klein, G. (2010). Comparison of two extraction methods for viruses in food and application in a norovirus gastroenteritis outbreak. *Journal of Virological Methods, 169*(1), 22–27.

[125] Rzeżutka, A., Alotaibi, M., D'agostino, M., & Cook, N. (2005). A centrifugation-based method for extraction of norovirus from raspberries. *Journal of Food Protection, 68*(9), 1923–1925.

[126] Rzeżutka, A., D'Agostino, M., & Cook, N. (2006). An ultracentrifugation-based approach to the detection of hepatitis A virus in soft fruits. *International Journal of Food Microbiology, 108*(3), 315–320.

[127] Papafragkou, E., Plante, M., Mattison, K., Bidawid, S., Karthikeyan, K., Farber, J. M., & Jaykus, L. A. (2008). Rapid and sensitive detection of hepatitis A virus in representative food matrices. *Journal of Virological Methods, 147*(1), 177–187.

[128] Fumian, T. M., Leite, J. P. G., Marin, V. A., & Miagostovich, M. P. (2009). A rapid procedure for detecting noroviruses from cheese and fresh lettuce. *Journal of Virological Methods, 155*(1), 39–43.

[129] Fahy, E., Kwoh, D. Y., & Gingeras, T. R. (1991). Self-sustained sequence replication (3SR): an isothermal transcription-based amplification system alternative to PCR. *Genome Res, 1*(1), 25–33.

[130] Fakruddin, M. D., Mazumdar, R. M., Chowdhury, A., & Mannan, K. B. (2012). Nucleic acid sequence-based amplification (NASBA)-prospects and applications. *International Journal of Life Science and Pharma Research, 2*(1), 106–121.

[131] Mueller, J. D., Pütz, B., & Höfler, H. (1997). Self-sustained sequence replication (3SR): an alternative to PCR. *Histochemistry and Cell Biology, 108*(4), 431–437.

[132] Jean, J., D'souza, D. H., & Jaykus, L. A. (2004). Multiplex nucleic acid sequence-based amplification for simultaneous detection of several enteric viruses in model ready-to-eat foods. *Applied and Environmental Microbiology, 70*(11), 6603–6610.

[133] Polstra, A. M., Goudsmit, J., & Cornelissen, M. J. B. I. D. (2002). Development of real-time NASBA assays with molecular beacon detection to quantify mRNA coding for HHV-8 lytic and latent genes. *BMC Infectious Diseases, 2*(1), 1–10.

[134] Olsen, S. J., MacKinon, L. C., Goulding, J. S., Bean, N. H., & Slutsker, L. (2000). *Surveillance for foodborne-disease outbreaks* (pp. 1993–1997). United States.

[135] Gould, L. H., Walsh, K. A., Vieira, A. R., Herman, K., Williams, I. T., Hall, A. J., & Cole, D. (2013). Surveillance for foodborne disease outbreaks – United States, 1998–2008. *Morbidity and Mortality Weekly Report: Surveillance Summaries, 62*(2), 1–34.

[136] Farkas, D. H., & Holland, C. A. (2009). Overview of molecular diagnostic techniques and instrumentation. In *Cell and tissue based molecular pathology* (pp. 19–32). Churchill Livingstone.

[137] Wong, M. L., & Medrano, J. F. (2005). Real-time PCR for mRNA quantitation. One-step versus two-step real- time PCR. *Biotechniques, 39*(1), 75–85.

[138] Hennechart-Collette, C., Dehan, O., Laurentie, M., Fraisse, A., Martin-Latil, S., & Perelle, S. (2021). Detection of norovirus, hepatitis A and hepatitis E viruses in multicomponent foodstuffs. *International Journal of Food Microbiology, 337*, 108931.

[139] Macaluso, G., Guercio, A., Gucciardi, F., Di Bella, S., La Rosa, G., Suffredini, E., & Purpari, G. (2021). Occurrence of human enteric viruses in shellfish along the production and distribution chain in Sicily, Italy. *Foods, 10*(6), 1384.

[140] Schultz, A. C., & Myrmel, M. (2013). Methods for detection of viruses in food (Norovirus and hepatitis A virus). *Food Borne Viruses and Prions and Their Significance for Public Health, 6*, 79–111.

[141] Shereen, M. A., Khan, S., A., K., et al. (2020). COVID-19 infection: origin, transmission, and characteristics of human coronaviruses. *Journal of Advanced Research*, doi: https://doi.org/10.1016/j.jare.2020.03.005.

[142] Shode, F. O., Idowu, A. S. K., Uhomoibhi, O. J., & Sabiu, S. (2021). Repurposing drugs and identification of inhibitors of integral proteins (spike protein and main protease) of SARS-CoV-2. *Journal of Biomolecular Structure and Dynamics, 2021*, 1–16. doi: 10.1080/07391102.2021.1886993.

[143] Zhang, W., Du, R. H., Li, B., Zheng, X. S., Yang, X. L., Hu, B., Wang, Y. Y., Xiao, G. F., Yan, B., Shi, Z. L., & Zhou, P. (2020). Molecular and serological investigation of 2019-nCoV infected patients: implication of multiple shedding routes. *Emerging Microbes & Infections, 9*(1), 386–389.

[144] WHO (World Health Organization) *COVID-19 and Food Safety: Guidance for Food Businesses: Interim Guidance, 07 April 2020*. (2020). Available online at: https://www.who.int/publications/i/item/covid-19-and-food-safety-guidance-for-food-businesses (accessed September 20, 2021)

[145] Xiao, F., Tang, M., Zheng, X., Liu, Y., Li, X., & Shan, H. (2020). Evidence for gastrointestinal infection of SARS-CoV-2. *Gastroenterology, 158*(6), 1831–1833.

[146] Anelich, L., Lues, R., Farber, J. M., & Parreira, V. R. (2020). SARS-CoV-2 and risk to food safety. *Frontiers in Nutrition*, 7, 580551. doi: https://doi.org/10.3389/fnut.2020.580551.

[147] Khadre, M.A. and Yousef, A.E., 2002. Susceptibility of human rotavirus to ozone, high pressure, and pulsed electric field. Journal of Food Protection, 65(9), pp.1441–1446.

[148] Le Guyader, F. S., Le Saux, J. C., Ambert-Balay, K., Krol, J., Serais, O., Parnaudeau, S., & Atmar, R. L. (2008). Aichi virus, norovirus, astrovirus, enterovirus, and rotavirus involved in clinical cases from a French oyster-related gastroenteritis outbreak. *Journal of Clinical Microbiology*, 46(12), 4011–4017.

[149] Sena, J.D. and Jarvis, D.L., 1981. Modification of the poliovirus capsid by ultraviolet light. Canadian Journal of Microbiology, 27(11), pp.1185–1193.

[150] Castell-Perez, M.E. and Moreira, R.G., 2004. Decontamination systems. Preharvest and Postharvest Food Safety: Contemporary Issues and Future Directions, pp.337–348.

[151] Beuchat, L.R., 1996. Pathogenic microorganisms associated with fresh produce. Journal of food protection, 59(2), pp.204–216.

[152] Shin, G.A. and Sobsey, M.D., 2008. Inactivation of norovirus by chlorine disinfection of water. Water research, 42(17), pp.4562–4568.

[153] Stals, A., Baert, L., Van Coillie, E., & Uyttendaele, M. (2011). Evaluation of a norovirus detection methodology for soft red fruits. *Food Microbiology*, 28(1), 52–58.

[154] Grove, S.F., Lee, A., Lewis, T., Stewart, C.M., Chen, H. and Hoover, D.G., 2006. Inactivation of foodborne viruses of significance by high pressure and other processes. Journal of Food Protection, 69(4), pp.957–968.

[155] World Health Organization (2015). Hepatitis E vaccine: WHO position paper, May 2015. *Weekly Epidemiological Record= Relevé Épidémiologique Hebdomadaire*, 90(18), 185–200.

[156] Food and Agriculture Organization/World Health Organization (FAO/WHO), 2005. Food Safety Risk Analysis Part 1. An Overview and Framework Manual, Provisional Edition.

[157] Adams, M.R. and Moss, M.O., 2000. Food microbiology. Royal society of chemistry.

[158] Beuchat, L. R. (1998). Surface decontamination of fruits and vegetables eaten raw: a review. In *Food safety issues*. Geneva: World Health Organization, WHO/FSF/FOS/98.2.

[159] Chalupa, P., Vasickova, P., Pavlik, I. and Holub, M., 2014. Endemic hepatitis E in the Czech Republic. Clinical Infectious Diseases, 58(4), pp.509–516.

[160] Sala, M.R., Arias, C., Dominguez, A., Bartolomé, R. and Muntada, J.M., 2009. Foodborne outbreak of gastroenteritis due to Norovirus and Vibrio parahaemolyticus. Epidemiology & Infection, 137(5), pp.626–629.

[161] Park, J.H., Jeong, H.S., Lee, J.S., Lee, S.W., Choi, Y.H., Choi, S.J., Joo, I.S., Kim, Y.R., Park, Y.K. and Youn, S.K., 2015. First norovirus outbreaks associated with consumption of green seaweed (Enteromorpha spp.) in South Korea. Epidemiology & Infection, 143(3), pp.515–521.

[162] Sakon, N., Sadamasu, K., Shinkai, T., Hamajima, Y., Yoshitomi, H., Matsushima, Y., Takada, R., Terasoma, F., Nakamura, A., Komano, J. and Nagasawa, K., 2018. Foodborne outbreaks caused by human norovirus GII. P17-GII. 17–contaminated nori, Japan, 2017. Emerging infectious diseases, 24(5), p.920.

[163] Feagins, A.R., Opriessnig, T., Guenette, D.K., Halbur, P.G. and Meng, X.J., 2007. Detection and characterization of infectious Hepatitis E virus from commercial pig livers sold in local grocery stores in the USA. Journal of general virology, 88(3), pp.912–917.

[164] Pintó, R.M. and Saiz, J.C., 2007. Enteric hepatitis viruses. Perspectives in medical virology, 17, pp.39–67.

[165] Wikswo, M.E., Kambhampati, A., Shioda, K., Walsh, K.A., Bowen, A. and Hall, A.J., 2015. Outbreaks of acute gastroenteritis transmitted by person-to-person contact, environmental contamination, and unknown modes of transmission—United States, 2009–2013. Morbidity and Mortality Weekly Report: Surveillance Summaries, 64(12), pp.1–16.

[166] Lowther, J. A., Bosch, A., Butot, S., Ollivier, J., Mäde, D., Rutjes, S. A., & Leclercq, A. (2019). Validation of EN ISO method 15216-Part 1-Quantification of hepatitis A virus and norovirus in food matrices. *International Journal of Food Microbiology*, *288*, 82–90.

[167] Pal, M., Ayele, Y., & Durglishilvi, N. (2020). Emerging role of Aeromonas hydrophila as a foodborne pathogen of public health concern. *EC Microbiology*, *16*, 55–58.

[168] Riesser, V., Perrich, J., Silver, B., & McCammon, J. (1977). Possible mechanisms of poliovirus inactivation by ozone. In Forum on Ozone Disinfection. Proceedings of the International Ozone Institute (pp. 186–192). Syracuse, NY.

[169] Vega, E., Barclay, L., Gregoricus, N., Shirley, S.H., Lee, D. and Vinjé, J., 2014. Genotypic and epidemiologic trends of norovirus outbreaks in the United States, 2009 to 2013. Journal of clinical microbiology, 52(1), pp.147–155.

[170] Oishi, I., Yamazaki, K., Kimoto, T., Minekawa, Y., Utagawa, E., Yamazaki, S., . . . Glass, R. I. (1994). A large outbreak of acute gastroenteritis associated with astrovirus among students and teachers in Osaka Japan. *Journal of Infectious Diseases*, *170*(2), 439–443.

[171] Environmental Protection Agency (EPA). (2008). Food irradiation. Available from: http://www.epa.gov/radiation/sources/food_irrad.html. Accessed Dec 2008. 908–913

[172] Le Guyader, F. S., Bon, F., DeMedici, D., Parnaudeau, S., Bertone, A., Crudeli, S., Doyle, A., Zidane, M., Suffredini, E., Kohli, E., Maddalo, F., Monini, M., Gallay, A., Pommepuy, M., Pothier, P., & Ruggeri, F. M. (2006a). Detection of multiple Noroviruses associated with an international gastroenteritis outbreak linked to oyster consumption. *Journal of Clinical Microbiology*, 44, 3878–3882.

[173] Stals, A., Baert, L., Van Coillie, E., & Uyttendaele, M. (2012). Extraction of food-borne viruses from food samples: a review. *International Journal of Food Microbiology*, *153*(1–2), 1–9.

[174] Food Standards Scotland. 2021. *Foodborne Viruses*. Available: https://www.foodstandards.gov.scot/consumers/food-safety/foodborne-illness/foodborne-viruses20 (Accessed 15 August 2021).

Adeoye John Kayode*, Bartholomew Saanu Adeleke,
Ayodeji Charles Osunla, Folasade Banji-Onisile* and Aboi Igwaran

Chapter 4
Zoonoses: implications for food safety

Abstract: Foodborne zoonotic microorganisms are major pathogens that cause human diseases with significant implications for health and food safety. Infections due to these pathogens are acquired when foods contaminated with vegetative cells or their toxins are consumed, resulting in significant economic and public health consequences globally. Hence, this chapter highlights the background and significance of foodborne zoonotic agents belonging to the major classes of microbial species, including bacteria, fungi, viruses, protozoans, and infectious particles (prions). Generally, food animals tend to be the major reservoir of these pathogens, and foods or food products originating from animals are the major vehicle of transmission of the pathogens to humans. Toxin production and virulent factors are often responsible for the pathogenesis of these pathogens. Foodborne zoonotic pathogens cause human infections that are typically characterized by gastrointestinal symptoms, including vomiting, nausea, abdominal cramps, diarrhea, body pain, and more severe health issues such as neurological diseases. Prevention of these pathogens in foods could be achieved by the implementation of standardized hazard analysis critical control point (HACCP), good hygiene, good manufacturing practices, good operational sanitation practices, and pasteurization procedures to enhance food safety. The increasing spread of multidrug resistance among zoonotic foodborne pathogenic microbial species is worrisome as this could lead to the emergence of entirely new pathogenic strains in the food chain. Therefore, the implementation of coordinated monitoring and surveillance systems in the food chain (from farm to fork) is highly recommended.

*Corresponding author: Adeoye John Kayode, Department of Viticulture and Oenology, JH Neethling Building, Victoria Street, Stellenbosch, 7600, South Africa,
e-mail: johnkayus37@yahoo.com, johnkayus3749@gmail.com
*Corresponding author: Folasade Banji-Onisile, Department of Microbiology, School of Life Sciences, College of Agriculture, Engineering and Science, University of KwaZulu-Natal, Durban 4000, South Africa, e-mail: folasadeonisile@gmail.com
Bartholomew Saanu Adeleke, Food Security and Safety Focus Area, Faculty of Natural and Agricultural Sciences, North-West University, Private Bag X2046, Mmabatho 2735, South Africa; Department of Biological Sciences, Olusegun Agagu University of Sciences and Technology, PMB 353 Okitipupa, Ondo State, Nigeria
Ayodeji Charles Osunla, Department of Microbiology, Adekunle Ajasin University, PMB 001 Akungba Akoko, Ondo State, Nigeria
Aboi Igwaran, Department of Biochemistry and Microbiology, University of Fort Hare, Private Bag X1314, Alice 5700, South Africa

https://doi.org/10.1515/9783110748345-004

4.1 Introduction

Food is a substance composed of varieties of essential nutrients for human nutrition and sustenance. Food security is one of the critical issues around the globe, and the increase in the human population has led to a high demand for foods to meet human nutritional requirements [1, 2]. The safety of food is most often threatened by numerous pathogenic microbial species that are responsible for varieties of foodborne maladies [3]. For good reasons, food safety is a matter of intense public health concern globally. Millions of the annual cases of food foodborne diseases (infections and intoxications) raise concerns not only limited to food served in fast food outlets and restaurants but also include food bought in supermarkets. A school of thought earlier put forward that "food can never be entirely safe" [3]. Nonetheless, its safety is imperative and should be ensured as best as possible to limit microbial contamination, especially the zoonotic foodborne pathogens.

Zoonoses are infectious diseases caused by zoonotic pathogens (pathogens that are usually transmitted from animals to humans). Diseases originating from animals are of significant importance and they could be ranked based on their public health and socioeconomic impact [4]. As such, the spectrum of zoonotic diseases and their impact could be more felt in developing countries compared with modern industrialized countries [5]. Zoonotic diseases can be considered from various perspectives. In consideration of the global perspective, recognition of zoonotic infections has spanned through many centuries, and zoonotic infections constitute a greater number of emerging and re-emerging infectious diseases worldwide. Studies have shown that the progressive increase in the recognition of new zoonotic diseases is a consequence of increased human–animal interface due to development, industrialization, and encroachment into the natural habitats of wildlife, giving rise to an increased trajectory of zoonotic infections [6, 33]. In rural areas, foodborne zoonotic pathogens may be more frequently encountered because people share their home environment with several animal species under unhygienic conditions and poor sanitation. This also unavoidably contributes to human exposure to zoonotic pathogens in such environments where veterinary services and medical care may be inadequate or lacking.

Over 150 zoonotic diseases have been recognized from all known classes of infectious agents with different hosts and varying transmission modes [4]. Symptoms of foodborne zoonotic diseases may vary ranging from mild and transient (including nausea and vomiting) to life-threatening diseases such as paralysis, brain and nervous system dysfunction, kidney and liver failure [7]. Several zoonotic pathogens (about 60%) that cause diseases in humans have emerged from animals or products of animals [8, 9]. Animals especially vertebrates could either be reservoirs or act as mechanical vectors engaging in direct or indirect transmission between animals and humans. Wild animals like bats, rodents, drought animals (horses), and food animals like cattle, poultry, and fish are important reservoirs of foodborne zoonotic pathogens [7, 10]. Products of animals like egg, fish, milk, meat, and dairy products are at high risks of zoonotic pathogens [11]. In this background, high demand and consumption of such foods derived from animal sources have increased the risks of zoonotic foodborne diseases in humans.

Most of the pathogens responsible for zoonotic infection in humans have been involved in foodborne or waterborne outbreaks and include viruses, bacteria, protozoans, and fungi. These microorganisms primarily cause food spoilage. However, they also cause foodborne diseases in humans after consuming food contaminated with pathogens, thus constituting major food safety problems. Many foodborne zoonoses may not be accompanied by serious health issues or fatal consequences; however, in some instances, significant or long-term outcomes on an individual's overall health or psychological well-being and debilitating effects on work ability could be observed. Furthermore, loss of concentration and discomfort experienced daily may aggravate pains and medical expenses. This chapter, therefore, aims to provide insight into zoonotic foodborne pathogens and their implications in food safety.

4.2 Common transmission route of zoonotic foodborne pathogens

Transmission of foodborne zoonotic pathogens to humans is usually through the oral (ingesting contaminated food or water) route. This occurs when the transmissive stages or minimally processed food containing the infective stages of the pathogens are ingested. However, aerosols, direct animal–human contact (including contact with contaminated feces), herbage and soil, fomite, and vectors are also important transmission routes involved. Usually, humans can be the accidental, paratenic, or intermediate host. In this section, the oral route of transmission of zoonotic pathogens involving the water–food connection is of particular interest.

4.2.1 Water as a vehicle of zoonotic pathogens

The complexity of the water–food connection for zoonotic pathogens cannot be overemphasized with human and animal wastes (feces and urine) which being the key vehicle of the environmental transmissive stages of foodborne zoonotic pathogens. As such, human and animal waste disposal remains important issues of public health that are yet to be addressed in many countries. For instance, ova of *Schistosoma haematobium* and the spores of some microsporidia like *Encephalitozoon cuniculi* are discharged into the environment through urine. The transmissible stages of these pathogens can directly or indirectly contaminate food and water. Contaminated water is an important source either by direct consumption or as a channel for the dissemination of zoonotic pathogens in the food chain, especially during food production, preparation, or processing. Transmissible stages of the pathogens could be transported into drinking, recreational, fresh, marine, and irrigation waters and these, in turn, could contaminate food supply during irrigation or when used for manufacturing and ancillary purposes. Also, the preparation of foods particularly vegetables or salads and rinsing fruits in portable water contaminated with zoonotic pathogens in homes is possible. In addition, the likelihood of people contracting foodborne zoonotic diseases has risen due to increased consumer vogue and beliefs that suggest eating raw or undercooked vegetables retains heat-labile nutrients or preserves the plants' natural flavour [12].

4.2.2 Food as a vehicle of zoonotic pathogens

Foodborne zoonoses occur when contaminated foods (especially food animal or animal products) are consumed. Usually, zoonotic pathogens contaminate food at different stages (during production, harvesting or slaughtering, processing, distribution, preparation, storage, and consumption) along the food chain. Food contamination risks may be dependent on the state of health of food handlers, food safety laws and regulatory system, financial strength for safer processing facilities, food hygiene practices, personal hygiene, and education or knowledge of food handlers [5, 13]. In recent times, the impact of foodborne pathogens on public health, morbidity and mortality rates, and the global economy is receiving significant recognition [5, 14]. A large proportion of humans worldwide are affected by foodborne diseases yearly and about 600 million become ill each year from consuming contaminated food. This includes about 220 million children under 5 years [15, 16]. About 1 in 10 people fall ill and about 33 million healthy life years are lost [15]. In developed countries, about 30% of the population suffers from foodborne illnesses and an estimate of up to 2 million deaths are recorded yearly in developing countries. Syndrome of foodborne infection or food poisoning ensues after ingesting food contaminated with pathogenic microorganisms or their toxins, adulterants, and other relevant contaminants. Foodborne illnesses could cause debilitating infections including meningitis or severe diarrhea, long-lasting disability, or death. Foods derived from animals including raw or undercooked meat and dairy products are commonly regarded as a source of foodborne diseases because they provide a more suitable environment for the proliferation of various microbial species due to the high protein content. The fight against foodborne diseases is challenging due to a rapidly changing pattern of human food consumption, globalization of the food market, and climate change. Also, the emergence and re-emergence of new pathogens and the increasing spread of multidrug resistance among pathogenic microorganisms highlight the implications of zoonotic pathogens in food safety.

4.3 Common foodborne zoonotic pathogens

The major etiological agents of zoonotic foodborne infections are viruses, bacteria, parasites, and fungi and are usually acquired by consuming foods contaminated with these pathogenic microbial species. Some of these pathogens include viruses (noroviruses, hepatitis A or E viruses (HAV/HEV), and adenoviruses), enteric bacteria (*Escherichia coli*, *Salmonella*, *Campylobacter*, *Listeria*, or the Enterobacteriaceae family), protozoans (*Entamoeba histolytica*, *Cryptosporidium parvum*, *Leishmania* spp., and *Sarcocystis* spp.), and fungi (*Mucor circinelloides* and *Basidiobolus ranarum*) [5, 17–20].

4.3.1 Bacterial zoonotic foodborne pathogens

Bacteria are prominent pathogens responsible for most of the health issues affecting humans. They are responsible for almost two-thirds of about 250 different foodborne diseases recognized and constitute serious concerns to food safety globally [21]. The World Health Organization (WHO) named *Salmonella*, *Campylobacter*, *Escherichia coli*, *Listeria monocytogenes*, and *Vibrio cholerae* among the 31 pathogens responsible for common foodborne illnesses annually and usually characterized by severe and fatal outcomes [22]. More importantly, these bacterial pathogens share a zoonotic origin, and Gram-negative bacteria were more frequently implicated in zoonotic foodborne infections than Gram-positive bacteria. These bacteria pathogens pose health threats to millions of people annually. Symptoms such as vomiting, nausea, diarrhea, abdominal pain, fever, headache, and severe and fatal outcomes are usually observed in associated cases.

4.3.1.1 *Salmonella*

This is an enteric rod-shaped Gram-negative bacterium and important pathogen of medical importance responsible for one of the most common foodborne illnesses in both developed and developing countries. They are rated among the four key global causes of diarrheal diseases, though the incidence rates may vary within regions. They are naturally found in the environment, but animals (domestic and wild) are the popular carriers of the pathogen. *Salmonella* species are zoonotic pathogens, and they primarily inhabit the intestinal tract of birds, reptiles, insects, animals, and humans. Within the food chain, *Salmonella* can adapt to different environmental conditions including growth at a high temperature (54 °C) and low temperature 2–4 °C (optimum 37 °C), pH 4.0–9.5 (optimum 6.5–7.0) and low water activity of 0.94. Food such as eggs, poultry, and other products from animals are usually involved in the transmission of the pathogen [5, 22]. Most cases of salmonellosis present mild outcomes. Notwithstanding, outcomes could be life-threatening [15]. In humans, salmonellosis broadly presents two outcomes that are either invasive systemic typhoid fever or self-limiting gastroenteritis, which are predominantly dependent on the serovar involved in infections. As such, serovars of *Salmonella* are categorized as typhoidal serovars (*Salmonella typhi* and *Salmonella paratyphi* A, B, and C) causing typhoid and nontyphoidal *Salmonella* (NTS; *Salmonella typhimurium*) causing gastroenteritis. *Salmonella enterica* serotype *enteritidis* and *Salmonella* serotype *typhimurium* are the most common NTS and broad host range serovars causing gastroenteritis transmitted from animals to humans [23, 15]. Recent reports identified that variants of NTS were found associated with invasive systemic infections and high rates of mortality among people with compromised immunity in sub-Saharan Africa [24]. *Salmonella typhi* crosses the intestinal epithelium after ingestion of contaminated food and is then spread to systemic sites like the spleen, liver, gall bladder, and bone marrow. Symptoms including diarrhea, headache, fever, stomach pain, muscle aches, nausea, constipation, and sometimes vomiting may be observed [15, 23]. Symptoms of nontyphoidal salmonellosis are mild and, in most cases, patients could recover without definite treatment. In some instances, particularly in elderly patients and children, dehydration could be severe and life-threatening. Foods such as eggs, poultry, meat, and other products from animals have been involved in outbreaks of salmonellosis. Also, the fecal–oral route is a popular transmission route and persons having contact with infected and often asymptomatic animals such as pets can be infected [15].

4.3.1.2 *Escherichia coli*

A Gram-negative anaerobic, rod-shaped, coliform bacterium commonly found adapted to the gut of warm-blooded animals and humans. Cattle are the main reservoir of the pathogen. Other ruminants including deer, goats, and sheep are significant reservoirs. *E. coli* can infect mammals including cats, dogs, rabbits, horses, pigs, and birds (chickens and turkeys). Most strains of this pathogen are harmless; however, some strains can cause serious foodborne infection [25]. Based on the clinical symptoms that manifest during infection and the virulence factors harbored, the pathogenic strains were classified into intestinal pathogenic *E. coli* (InPEC) and extraintestinal pathogenic *E. coli* (ExPEC). The InPEC includes the Shiga toxin-producing *E. coli* also known as the enterohemorrhagic *E. coli* (*E. coli* O157:H7) or verocytotoxin-producing *E. coli*, enteroinvasive *E. coli*, enteroaggregative *E. coli*, enteropathogenic *E. coli*, diffusely adherent *E. coli*, and the enterotoxigenic *E. coli*. The ExPEC includes neonatal meningitis *E. coli* and uropathogenic *E. coli* [26]. Foodborne infections involving *E. coli* are commonly transmitted to humans through the consumption of contaminated foods such as undercooked meat, meat products or hamburgers, dried cured salami, unpasteurized milk/fresh-pressed apple cider, yoghurt, cheese, fresh fruits, and vegetables including lettuce, spinach, coleslaw, salad, and sprouts [22]. Also, cross-contamination during food preparation with beef and other meat products, contaminated surfaces, kitchen utensils, and fecal contamination of food and water could lead to infection [25]. Symptoms of *E. coli* infection include diarrhea, abdominal cramps or bloody diarrhea (hemorrhagic colitis), vomiting, and fever. Infection may persist between 3 and 8 days, and recovery of most patients occurs within 10 days of infections. However, infection in the elderly and young children may result in life-threatening diseases (hemolytic uremic syndrome) [25].

4.3.1.3 *Campylobacter*

This is a Gram-negative, rod-shaped, nonsporulating, spirally curved bacterial species known to colonize the gastrointestinal tracts of many animals. About 32 *Campylobacter* species and 13 subspecies are known, and the pathogenic ones include *Campylobacter jejuni, C. rectus, C. insulaenigrae, C. helveticus, C. fetus, C. concisus, C. lari, C. hyointestinalis, C. mucosalis, C. upsaliensis, C. coli, C. ureolyticus,* and *C. sputorum* have been identified in over 400–500 million cases of human infections [27]. *Campylobacter jejuni* subspecies *jejuni* and *Campylobacter coli* are responsible for about 95% and 5% of zoonoses infections, respectively. These pathogens have the unique ability to thrive in a microaerophilic atmosphere that contains 5% oxygen (O_2) and 10% carbon dioxide (CO_2) within the temperature range of 30–46 °C (40–42 °C optimum). They do not proliferate outside the alimentary tract of warm-blooded animals, but they can survive several weeks in foods preserved at refrigeration temperatures. This unique attribute makes them pathogens of interest in the food chain. Campylobacteriosis is usually acquired when food contaminated by the pathogen is consumed especially among certain groups including children, elderly, and immunocompromised persons, patients suffering from inflammatory bowels and hemoglobinopathy patients. The period of incubation varies in patients depending on the infective dose exposure but takes 1–7 days to develop symptoms, and it may be longer for patients exposed to a lower infective dose. Symptoms such as watery diarrhea, fever with bloody stool, vomiting, abdominal pain, and dehydration have been reported. Although symptoms may disappear within 5–7 days, in some instances, complications may occur in about 1% of cases. Complications such as peripheral neuropathies (Guillain-Barré syndrome – a neurological disorder characterized by anemia, sensory loss, possible involvement of respiratory muscles, and weakness), irritable bowel syndrome, functional intestinal disorder, reactive arthritis involving the ankles and knees, which occurs about a month postinfection and may develop for as long as 5 years. *Campylobacter* is not usually involved in large-scale outbreaks but rather sporadic outbreaks involving small groups. Contaminated pork and beef are known to cause about 20–30% of campylobacteriosis cases, whereas poultry meats such as meats from turkeys, ducks, ostriches, laying hens and broilers, and poultry products have been identified as the major causes of about 50–80% cases of campylobacteriosis [27]. Ruminants (cattle, goats, and sheep) are known reservoirs of the pathogen. Raw milk and drinking water are usually involved in foodborne cases of *Campylobacter*.

4.3.1.4 *Listeria*

This is a Gram-positive bacterium made up of 26 species including *Listeria monocytogenes, L. innocua, L. seeligeri, L. ivanovii, L. welshimeri, L. marthii,* and *L. grayi* [28–30]. Two species (*L. monocytogenes* mainly infects animals and humans, while *L. ivanovii* infects ruminant animals) cause listeriosis [31]. *L. monocytogenes* was further classified into 14 serotypes including the recently described serotype 4 h (a novel hypervirulent hybrid belonging to the sub-lineage II) [32]. These pathogens have a competitive advantage to survive harsh environmental conditions including a broad pH range (4.4–9.4), a low water activity of 0.9, high salt concentrations, low temperatures (at −4 °C), and at a relatively high temperature (45 °C maximum). *Listeria*, an intracellular pathogen, penetrates the intestinal cells and host cells including the liver, spleen, placenta, brain, and heart. The actin-based motility mechanism enhances bacteria proliferation to other cells and escapes the host's immunological responses. *Listeria* is an environmental pathogen and has been isolated from environmental waters [34], soil, feces, agricultural environment, feed, processing facilities, and various types of foods and ready-to-eat foods (RTE) or meats [33]. *Listeria* may colonize animals including cattle, horses, sheep, goats, pigs, wild birds and poultry, fishes, and shellfish, and these animals may also disseminate the pathogen into the environment, especially into pastures. Foods including unpasteurized dairy products, seafood, and various RTE foods have been implicated in listeriosis outbreaks. Symptoms set in after contaminated foods are consumed and the incubation period could be long, usually between 1 and 70 days depending on the state of health and infection caused in individuals. Listeriosis infection is often accompanied by mild symptoms (such as fever, headache, diarrhea, myalgia, miscarriages, and death in neonates) in noninvasive cases; and severe and fatal health consequences including meningitis, septicemia, abortion, pneumonia, and encephalitis in cases of invasive listeriosis particularly, among immunocompromised individuals. Listeriosis is often classified among the most serious foodborne infections due to the severe health outcomes, high hospitalization, and high death rate (20–30%) [22]. The severity of the infection could be strain-dependent and the virulence determinants harbored [34].

4.3.1.5 *Vibrio*

This is a Gram-negative, rod-shaped, motile bacteria with polar flagellation that is naturally adapted to a wide range of aquatic habitats including freshwater, marine, and estuarine environments. Their abundance in an environment is usually reflective of the environmental temperatures as they tend to prefer warm, slightly salty (brackish) water. More than 100 *Vibrio* spp. have been described, about 12 were known to infect both humans and animals including *Vibrio cholerae, V. parahaemolyticus, V. furnissii, V. vulnificus, V. metschnikovii, V. alginolyticus, V. fluvialis, V. mimicus, V. damse-*

lae subsp. *damselae* (*Photobacterium*), *V. harveyi* (*V. carchariae*), and *V. hollisae* (*Grimontia*). However, *Vibrio anguillarum* and *V. tapetis* are majorly known to cause serious infections among vertebrate and invertebrate animals [35]. *V. cholerae* is responsible for a severe diarrheal (cholera) disease transmitted through contaminated water or contact with an infected person, and the infection could lead to a fatal outcome if untreated. On the other hand, noncholera *Vibrio* spp. can cause vibriosis, an infection acquired through the consumption of raw or undercooked contaminated seafood or exposure to seawater. Millet gruel, rice, vegetables, and various types of seafood have been implicated in cholera outbreaks. Several clinical manifestations are usually observed in noncholera *Vibrio* disease and could be mild or self-limiting gastroenteritis except in the case of wound infection that can lead to septicemia rapidly and often characterized by a high mortality rate caused by the opportunistic *V. vulnificus* [36, 37].

4.3.1.6 *Staphylococcus*

This is a genus of Gram-positive nonspore-forming cocci bacteria currently made up of more than 50 species [38]. Among these, *S. aureus*, *S. saprophyticus*, *S. epidermidis*, *S. hominis*, and *S. haemolyticus* are considered potential human pathogens but the first three are the most common isolates. These bacteria naturally inhabit the skin, nose, and mucous membranes in humans and many animal species and are also abundant in the environment. The prevalence of foodborne infection due to *S. aureus* is high and ranked second to salmonellosis. The ability of *S. aureus* to coagulate plasma distinguished it from the other two pathogens. Despite being regarded as the normal flora, *S. aureus* is also an important opportunistic human and animal pathogen that causes foodborne infection and sepsis in hospitals and nurseries. Dermal infections like superficial pyogenic (pus-forming) and underlying tissues, as well as serious systemic infections, are caused by *S. aureus*. It can produce a range of toxins including cytotoxins (general systemic toxins) and toxic shock superantigens and enterotoxins (food poisoning). Infections such as mastitis in dairy animals, bone, and joint lesions (bumblefoot) in poultry occasionally cause skin infections in livestock. Animals such as cats, dogs, and horses are vulnerable to *S. aureus* and may play important roles in transmission. *S. aureus* can cause food spoilage, food intoxication, and food poisoning thereby reducing food quality.

4.3.1.7 *Yersinia*

This is a Gram-negative, cocci or rods, nonsporulating bacteria. *Yersinia*, a member of the family Enterobacteriaceae, is comprised of 17 species. Three of the species are pathogenic to humans including *Y. pseudotuberculosis* causing rodentiosis but rarely reported, *Y. pestis* causing plague, and *Y. enterocolitica* causing yersiniosis, and gastrointestinal tract infection in humans. *Y. enterocolitica*, a zoonotic and psychrophilic enteric pathogen, can grow at a temperature ranging between 0 and 45 °C and optimally between 25 and 32 °C. It can survive and proliferate in the viable-but-nonculturable state and can also produce thermostable toxins in cooling under refrigeration. This characteristic highlights its relevance in the food chain. It can grow at a pH lower than 9 and water activity not lower than 0.96. *Y. enterocolitica* biotypes have been reported including 1A, 1B, 2, 3, 4, 5 (six biotypes) and classified based on their biochemical reactions such as hydrolysis of esculin and salicin, decarboxylation of *O*-nitrophenyl-β-D-galactopyranoside and ornithine and nitrate reduction. Further classification of *Y. enterocolitica* into serotypes (most pathogenic serotypes 1B/O:8, 2/O:5,27, 2/O:9, 3/O:3, and 4/O:3) was based on the structure of the lipopolysaccharide O-antigen, a composition of the superficial lipopolysaccharide of the cell layer. The incubation period could range between 1 and 11 days, and symptoms may be mild but severe enteritis and gastritis which could disappear within 1–3 weeks may develop in some cases. All humans are susceptible to yersiniosis but children <5 years, immunocompromised persons, and the elderly are more vulnerable. Fever, diarrhea (usually bloody), and stomachache symptoms in adults could be likened to appendicitis. Infections due to *Y. enterocolitica* are usually characterized by nonspecific symptoms sharing symptoms with other microbial infection, and this has led to the misdiagnosis of many cases of yersiniosis infections. However, complications involving osteoarthritis, purulent hepatitis, endocarditis, erythema, bacteremia, nodosum, myocarditis, splenitis, or nephritis, and less often sepsis may be observed [7]. Other bacterial foodborne zoonotic pathogens were described in Table 4.1.

4.3.2 Parasitic zoonotic foodborne pathogens

The dependence of parasites on the host cell for survival has been a major concern to human health with pronounced pathological disorders. In the world today, more than 50 identifiable parasitic diseases are known with emergence and re-emergence in nature [55, 56]; and pathogens of animal origin are of major concern to health and food safety. Food of animal origin rich in proteins, minerals, vitamins, and energy is mainly required for human nutrition. Other animal food products from goats, cattle, and sheep, among others play an important role in meat production for human use [57]. Nevertheless, in the dairy industry, milking in an unhygienic environment can result in the transmission of parasitic zoonotic diseases. Also, ingestion of contaminated aquatic foods, raw foods, and poultry products can be a vehicle for the transmission of parasitic zoonotic diseases.

Table 4.1: Some zoonotic pathogens that can be transmitted to humans from the consumption of contaminated food/food animals.

Pathogen type	Pathogen (disease)	Intermediate/ definitive host(s)	Infection route/ probable means of transmission to humans	Human disease caused/clinical signs and symptoms	Human-to-human/animal-to-human transmission	Possible outbreak/ epidemic/ pandemic	Risk factors
Bacteria	*Bacillus anthracis* (anthrax)	Large mammals with hooves (wild ungulates), e.g., moose, white-tailed and mule deer, and reindeer	Inhalation/ ingestion or wound contamination by spores of the bacterium, consumption of raw or undercooked meat. Animals get infected by spores in soil, vegetation, water contaminated from carcasses, or contaminated hay or livestock feed. The pathogen is found mostly in soil with high moisture, organic, calcium, and alkaline pH	Human and animal infections Inhalation: short breath, chest pain, nausea, cough, stomach pains, vomiting, sweats, headache, body ache, and fatigue Cutaneous: skin sore Gastrointestinal infection: sore throats, swelling of neck glands, fever, nausea, vomiting, headache, fainting, abdominal swelling, and diarrhea	Rare human-to-human or animal-to-animal transmission, the cutaneous form is considered noncontagious. People can get infected after having contact with infected animals or their products such as wool, hide, hair, or meat.	Shared contaminated food/water could lead to an outbreak	Laboratory, animal husbandry/butchery workers

(continued)

Table 4.1 (continued)

Pathogen type	Pathogen (disease)	Intermediate/definitive host(s)	Infection route/probable means of transmission to humans	Human disease caused/clinical signs and symptoms	Human-to-human/animal-to-human transmission	Possible outbreak/epidemic/pandemic	Risk factors
	Clostridium botulinum (botulism) [39]	Mammals (marine mammals – seals and whales) and fish (especially salmon)	Consuming parboiled or raw fish, seal oil, seal meat, or other wild meats that were faultily aged or fermented, or contaminated by the pathogen or toxin	Numerous symptoms including nausea, blurred vision, constipation, vomiting, abdominal pain, blurred vision, fatigue, motor nerves paralysis, and fatal cases may result in respiratory paralysis	No	Consuming shared contaminated products could lead to an outbreak	Children <1 fed with honey, consuming, lightly preserved foods, inadequately processed home-canned foods, or bottled foods
	Brucella spp. (brucellosis)	Mammals (wild) including elk, reindeer/caribou, muskoxen, white-tailed and mule deer, hares, bison, goats, moose, sheep, cattle, fox, wolf, mink, rodents, and marine mammals	Handling newborn calves, carcasses, and fetuses of infective animals. Raw, frozen, or dried meat/milk and marrow consumption. Dogs: uncooked infected tissue consumption	Acute or insidious (systemic) bacterial disease including intermittent fever, headache, chills, joint pain, weight loss, weakness, sweating, and sometimes fatal cerebral infection	Transmission of the pathogen from animal to human usually by direct contact. Human-to-human transmission is rare.	Outbreak is possible when shared contaminated product is consumed	Human brucellosis risk factors include consumption of raw meat, unpasteurized cheese and milk from infected animals, and contact with placentas from fetal abortion

Mycobacterium tuberculosis, M. pinnipedii, M. bovis, and others (mycotuberculosis and tuberculosis) [40]	Terrestrial mammals such as bison, deer, elk, mule deer and wolves, white-tailed deer, moose, and marine mammals (commonly wild seals and crustaceans on rare occasions)	Through inhalation, ingesting undercooked/raw/ unpasteurized milk or meat products, direct contact with broken skin	Localized skin infection, cough, shortness of breath, abdominal pain, constipation vomiting, nausea, blurred vision, and fatigue	Transmission of the pathogen from animal to human usually by direct contact. Direct human–to–human transmission.	Outbreak is possible when a shared contaminated product is consumed. Direct human-to-human transmission	Raw or soured milk, aerosol inhalation, urine, and feces from infected animals, sharing home environment with animals
Francisella tularensis (tularemia)	Squirrels, beavers, hares, wolves, muskrats, bears, voles, and other wildlife	Humans' infection is acquired through ingestion of contaminated food/ water, raw/ insufficiently cooked meat, inhalation, bite (from arthropod), contact with infected tissues or fluid from animals	Abdominal pain, vomiting, ulceration, diarrhea, pneumonia, conjunctivitis, septicemia, skin lesions/ulcerations, hepatosplenomegaly, lymph adenomegaly	Transmission of the pathogen from animal to human is usually by direct contact. No human-to-human transmission.	Outbreak rare but possible when shared contaminated product is consumed, source of contaminated water, or abundance of vector	Proximity of individuals close to animals/disease reservoirs, especially farmers and veterinarians

(continued)

Table 4.1 (continued)

Pathogen type	Pathogen (disease)	Intermediate/ definitive host(s)	Infection route/ probable means of transmission to humans	Human disease caused/clinical signs and symptoms	Human-to-human/animal-to-human transmission	Possible outbreak/ epidemic/ pandemic	Risk factors
	Erysipelothrix rhusiopathiae (erysipelas)	Aquatic and terrestrial mammals including white-tailed and mule deer, swine, turkeys, ducks, sheep, cats, chickens, muskoxen, caribou, fish, birds, and arthropods	Ingestion of contaminated meat/fish and their products, exposure to infected animals via skin wounds. Environmental sources of infection are possible	Severe cases of infection involve diffuse cutaneous or systemic disease, endocarditis, septicemia, abscesses, arthritis, meningitis, endocarditis, pneumonia	Transmission of the pathogen from animal to human usually by direct contact. No human-to-human transmission.	No	Occupational, including butchers, slaughterhouse workers, fishermen, housewives, and veterinarians
	Coxiella burnetii (Q fever)	Goat, sheep, cattle, northern seals, and seabirds	Dust inhalation contaminated by animal urine, feces, milk, or birth materials, drinking/ ingesting contaminated water, food, or animal products (cheese and milk)	Mild symptoms include muscle aches, fever, cough, nausea, weight loss, fatigue, headache, vomiting, cough, diarrhea, chest, or stomach pain Severe symptoms include hepatitis, pneumonia, miscarriage, pre-term delivery, low infant birth weight, and stillbirth	Rare but highly transmissible from animal to human usually by direct contact. Human-to-human transmission is not high.	Outbreak is occasional but shared contaminated water or food source could lead to an outbreak	Occupational exposure includes farming, veterinary medicine, and animal research

Leptospira interrogans (leptospirosis)	Wild and domestic animals including pet dogs, pigs, coyotes, foxes, otters, white-tailed and mule deer, skunks, raccoons, northern fur seals, opossums, beavers, otters, and rodents	Contact with body fluids, urine of infected animals, water and soil contaminated by fluids and urine from infected animals	Headache, high fever, muscle pain, chills, bleeding, vomiting, nausea, diarrhea, headache, skin rash, sore throat, kidney, kidney damage, liver failure, meningitis, respiratory distress, and death	Animal-to-human and human-to-human transmissions are possible.	Contaminated water source	Occupational exposure of slaughterhouse personnel
Borrelia burgdorferi (Lyme disease) [41]	Songbirds, seabirds, ungulates, ticks, white-footed mice, white-tailed deer, cattle, and dairy animals	Oral transmission, food, and their products (e.g., raw milk/milk) infected with the pathogen and tick's bite that initially fed on an infected animal	Flulike symptoms, rash, often in a bull's-eye pattern, joint pain and weakness in the limbs, fever, arthritis, and paralysis. Infection may involve the cardiac, skin, muscular, and nervous systems	Animal-to-human transmission.	No	Skin exposure, spending time in a wooded or grassy environment, prompt, and proper tick

(continued)

Table 4.1 (continued)

Pathogen type	Pathogen (disease)	Intermediate/ definitive host(s)	Infection route/ probable means of transmission to humans	Human disease caused/clinical signs and symptoms	Human-to-human/animal-to-human transmission	Possible outbreak/ epidemic/ pandemic	Risk factors
	Bacillus cereus (food poisoning) [42]	Sheep, cattle, goat, pig, rabbit, chicken, soil, and marine sponges	Infant formula, cooked rice, animal products, inadequately cooked foods, meat products, soups, sauces, vegetables, pudding, fried rice, beef, pork, chicken, fish, raw/ pasteurized milk, dairy products, pasta, flour and raw rice, potato, pasta, pastries, salads	Diarrhea-type food poisoning symptoms may mimic *Clostridium perfringes* food poisoning including moderate nausea, diarrhea, abdominal pain, and rectal tenesmus. Emetic food poisoning includes nausea, abdominal cramps, and vomiting. Symptoms may mimic those of *Staphylococcus aureus* poisoning	No	Shared contaminated food sources could lead to an outbreak	Inadequately prepared/refrigerated foods due to the ability to form spores, surviving longer at extreme temperatures

	Reservoir	Transmission	Disease/Symptoms	Transmissibility	Outbreak potential	Risk factors
Proteus spp.	Water, soil, rats, feces, urine, hair remnants, and intestinal tracts of mammals including humans	Contaminated food, dairy milk (from goat, sheep, and cattle), chicken, and food animals	Urinary tract infection (UTI), bacteremia and intra-abdominal infections, opportunistic infections within the respiratory tract, wounds, and burns Symptoms of the UTI include cloudy urine, pain or burning during urination, fever, chills, and abdominal pain. Other symptoms include pyuria, frequent urination, cystitis, hematuria, and pyelonephritis	Animal-to-human and human-to-human transmissions are possible.	Shared contaminated food sources could lead to an outbreak	Unprotected anal sex in men, sexual intercourse between men and women, immunodeficiency, improper cleaning of catheter, prolonged catheterization, immunodeficiency (i.e., <200 µL CD4, underlying illness)
Viruses						
Adenovirus (Adenoviridae) dsDNA nonenveloped virus	Humans, feces, environment, birds, lizards, and mammals	The fecal–oral route (ingestion of contaminated food or water) and contaminated surfaces	Gastroenteritis, respiratory infections, hepatitis, nephritis or encephalitis, pancreatitis, hemorrhagic colitis	Animal-to-human and human-to-human transmissions are possible.	Shared contaminated water/food sources could lead to an outbreak	Immunocompromised persons and young children <2 years

(continued)

Table 4.1 (continued)

Pathogen type	Pathogen (disease)	Intermediate/definitive host(s)	Infection route/probable means of transmission to humans	Human disease caused/clinical signs and symptoms	Human-to-human/animal-to-human transmission	Possible outbreak/epidemic/pandemic	Risk factors
	Astrovirus (Astroviridae) +ssRNA nonenveloped virus	Cattle, swine, sheep, felines, cervids, canines, rodents, rabbits, bats, and mink	Fecal oral route, contaminated food and water, or sewage, vegetables or fruits irrigated with contaminated water, contaminated surfaces, or fomites	Gastroenteritis in children, watery diarrhea, fever, vomiting, anorexia, abdominal pain, and headache, usually mild and self-limiting	Animal-to-human and human-to-human transmissions are possible.	Shared contaminated water/food sources could lead to an outbreak	Immunocompromised persons and young children
	SARS-CoV (Coronaviridae) +ssRNA enveloped virus	Bats, racoon dogs, or palm civets	Fecal food contamination, respiratory droplets, body fluids, feces, contaminated surfaces, live or exotic animals	Respiratory and intestinal disease or epidemic gastrointestinal disease, chills, fever, headache, diarrhea	Animal-to-human and human-to-human transmissions are possible.	Shared contaminated water/food sources could lead to an outbreak	Immunocompromised persons

Nipah virus (Paramyxoviridae) – ssRNA enveloped virus	Bats or pigs	Fruits partially consumed by bat or infected persons, contaminated palm saps, contact with infected domestic animals, body fluids, e.g., urine, blood, nasal secretions, and feces	Respiratory disease, acute encephalitis, cough, fever, headache, shortness of breath	Animal-to-human and human-to-human transmissions are possible.	–	Hospital workers and caretakers, close contact with infected pigs
Avian influenza/ influenza A virus (H5N1 influenza virus, Orthomyxoviridae) – ssRNA enveloped virus	Wild birds, e.g., waterfowl, poultry, chicken, ducks, and geese	Feces, contaminated food and water, secretions from infected animals (poultry), inhalation of aerosolized droplets, contaminated fomites, and contaminated meats (duck meat)	Acute respiratory tract infection, sore throat, fever, malaise, cough, vomiting, abdominal pain	Animal-to-human and human-to-human transmssions are possible.	Shared contaminated water/food sources could lead to an outbreak	Contact to infected animals, immunocompromised persons

(continued)

Table 4.1 (continued)

Pathogen type	Pathogen (disease)	Intermediate/ definitive host(s)	Infection route/ probable means of transmission to humans	Human disease caused/clinical signs and symptoms	Human-to-human/animal-to-human transmission	Possible outbreak/ epidemic/ pandemic	Risk factors
	Rotavirus (Reoviridae) dsRNA nonenveloped virus [43]	Infect a wide range of animals including pigs, cattle, horses, cats, dogs, chickens, and humans	The fecal–oral route, contaminated food, and water (the respiratory route is possible)	Gastroenteritis, diarrhea, abdominal discomfort, vomiting, fever, dehydration	Animal-to-human and human-to-human transmissions are possible.	Shared contaminated facilities and water/food sources could lead to an outbreak	Children <5 years
	Aichivirus (Picoviridae) +ssRNA nonenveloped virus [44]	Oysters, shellfish, wild boars, sewage, river, and groundwater	Fecal oral route, consumption of contaminated food and water, oysters, and seafood	Fever, gastroenteritis, nausea, vomiting, diarrhea, and abdominal pain	–	Shared contaminated facilities, water/ food sources could lead to an outbreak	Children <5 years
	Calicivirus (Caliciviridae) +ssRNA virus	Antarctic marine mammals including whales, elephant seals, fur seals, walruses, domestic and wild cats, rabbits, reptiles, livestock, amphibians, and birds	Infected animal/ secretion's contact with the broken skin	Fluid-filled blisters on the extremities	Human-to-human transmission is rare but possible through contact with broken skin/ contact with blister fluid.	Rare but possible through contact/ handling of the animals	Infect humans of all ages

				Animal-to-human transmission.		
Sealpox/ Parapoxvirus	Harbor and gray seals	Direct contact through pox lesions on infected mammals	Painful nodular lesions	Animal-to-human transmission.	Rare but possible through contact with broken skin/contact with blister fluid	Usually infect seal handlers
Rabies (Rhabdoviridae) –ssRNA enveloped virus	Principal reservoir host: bats, wolves, arctic foxes, and red foxes	Humans and domestic dog bites from infected wildlife	Affects the central nervous system of mammals; discomfort/general weakness, headache, fever, anxiety, prickling at the site of the bite, delirium, agitation, confusion, hallucinations, insomnia, and hydrophobia (fear of water). Could be fatal if untreated	Human-to-human is extremely rare; through bite or organ transplant.	No	40% of people bitten by rabid animals are children <15 years
Prion Chronic wasting disease (CWD)	Wild cervids: Moose, white-tailed and mule deer, elk, and reindeer in Fennoscandia (potentially caribou)	Consumption of infected offal, so public health officials still advise caution	N/A	There was experimental evidence of CWD transmission to nonhuman primates. Other TSEs have spread from animals to humans.	Potentially possible through consumption of shared contaminated product	Meat consumption

(continued)

Table 4.1 (continued)

Pathogen type	Pathogen (disease)	Intermediate/definitive host(s)	Infection route/probable means of transmission to humans	Human disease caused/clinical signs and symptoms	Human-to-human/animal-to-human transmission	Possible outbreak/epidemic/pandemic	Risk factors
Protozoan	*Cryptosporidium* spp. (cryptosporidiosis)	Small ruminants (sheep, cattle, goat, ovine, cervine, and caprine), calves, equines, rodents, and primates	Consumption of raw or uncooked meat. Accidental ingestion of fecal-oocyte-contaminated water. Cross-contamination of meat and meat products, drinking water, and fomites (clothes and footwear) used in livestock farms or wildlife parks that have been exposed to the feces of an infected human or animal	Watery diarrhea, stomach cramps or pain, dehydration, nausea, vomiting, fever, and weight loss	–	Consumption of contaminated foods and sharing of fomites exposure to feces of animals	Livestock husbandry, veterinary, immunocompromised individual

Giardia duodenalis (giardiasis) [45, 46]	Cattle, goats, sheep, pigs, dogs, cats, rodents, equines, rabbits, and shellfish	Consumption of raw or undercooked contaminated meat or filter-feeding shellfish	Diarrhea, bloating, malnutrition, and weight loss	Direct contact with the infected animals.	Consumption of raw meat and seafood	Animal contact or consumption of water and food contaminated by animals, immunocompromised individual
Toxoplasma gondii (toxoplasmosis)	Goat, sheep, pig, cat, bird, and fish	Consumption of cyst contaminated raw meat or organs of an infected intermediate host. Ingestion of sporulated oocysts' contaminated water. Congenital transmission from mother to fetus from an infected person	Flu, schizophrenia, epilepsy, Alzheimer's, and Parkinson's disease	Shedding and prevalence of oocytes in the animal host.	Ingestion of oocytes through contaminated water	Proximity to infected domesticated or wild reservoir hosts
Sarcocystis spp. (sarcocystosis) [47, 48]	Cattle, pig, and sheep	Ingestion of sporocyst contaminated water by animal feces. Consumption of raw or undercooked meat	Muscular sarcocystosis, febrile myositis, severe myalgia, fever, fatigue, vasculitis, bronchospasm, pruritic eruptions, headache, and arthralgia	Consumption of meat infected with sporocysts of infected animals.	Ingestion of cysts of the parasite from the infected animals	Occupational risk, consumption of raw foods, proximity to a wide range of animals and wildlife

(continued)

Table 4.1 (continued)

Pathogen type	Pathogen (disease)	Intermediate/definitive host(s)	Infection route/probable means of transmission to humans	Human disease caused/clinical signs and symptoms	Human-to-human/animal-to-human transmission	Possible outbreak/epidemic/pandemic	Risk factors
Trematode	*Paragonimus* spp. (paragonimiasis) [49, 50]	Crustaceans, water snails, cats, dogs, and other carnivores	Consumption of raw animal food products. Drinking inadequately treated water contaminated by procercoid-infected crustaceans, eating the raw or insufficiently heated meat of the second intermediate or paratenic hosts, such as frogs and snakes, that are infected with plerocercoids	–	Placing poultices of frog or snake flesh on open wounds or other lesions, in particular on the eyes.	Consumption of infected or contaminated seafood	Immunocompromised individuals, packaged seafood, raw or undercooked food, poor hygiene water environment
	Clonorchis sinensis (clonorchiasis)	Fish, snails, shrimps, and mammals	Eating raw or partially cooked fish	Cholelithiasis, cholangitis, cholecystitis, and cholangiocarcinoma	Animal-to-human transmission	–	Preserved food, and immunocompromised individuals

	Fasciola spp. (fasciolosis) [51]	Sheep, cattle, and man	Ingestion of raw infected meat or consumption of raw or undercooked meat	Malaise, fever, abdominal pain, nausea, urticaria, hepatomegaly, vomiting, and weight loss due to liver damage	Animal-to-human transmission by direct contact with infected animals.	–	Veterinary professionals, immunocompromised individuals, diseased animals
Cestode	*Taenia* spp. (taeniosis)	Pig, cattle, and man	Ingestion of raw or undercooked contaminated food/meat, or via exposure by hand–mouth contact to eggs derived from the contaminated environment	Weight loss, abdominal pain, upset stomach, and loss of appetite	Animal-to-human transmission.	–	Direct contact with the infected animals, stray or roaming animals in the urban or peri-urban areas
	Dibothriocephalus spp. (diphyllobothriosis)	Fresh or marine fishes (perch and pike)	Consumption of unfrozen fish products or raw fish	Weight loss, diarrhea, abdominal discomfort, and vomiting	Animal-to-human transmission.	–	Seafood traders/ suppliers, immunocompromised individual, diseased animals

(continued)

Table 4.1 (continued)

Pathogen type	Pathogen (disease)	Intermediate/ definitive host(s)	Infection route/ probable means of transmission to humans	Human disease caused/clinical signs and symptoms	Human-to-human/animal-to-human transmission	Possible outbreak/ epidemic/ pandemic	Risk factors
Nematode	*Trichinella* spp. (trichinosis) [52]	Rodents and pigs	Ingestion of raw or undercooked meat (bear and pork)	Nausea, diarrhea, vomiting, fever, fatigue, abdominal discomfort, constipation, headaches, cough, chills, muscle pain, itchy skin	Animal-to-animal transmission is possible.	Shared contaminated facilities and water/food sources could lead to an outbreak	Improper food preparation
	Gnathostoma spinigerum (gnathostomiasis) [53, 54]	Pigs, dogs, fish, and frogs	Ingestion of raw or insufficiently cooked meat (poultry and freshwater fish)	Fever, lack of appetite, excessive tiredness, nausea, vomiting, abdominal pain, diarrhea	Animal-to-human transmission is possible.	Shared contaminated facilities and water/food sources could lead to an outbreak	Consumption of raw undercooked freshwater fish and geographical exposure

+ssRNA, single-stranded positive-sense RNA virus; −ssRNA, single-stranded negative-sense RNA virus; dsDNA, double-stranded DNA virus; dsRNA, double-stranded RNA virus.

The transmission of parasitic zoonoses in foods has been a major concern to human health and safety.

The prevalence of foodborne parasitic pathogens, protozoans, and helminths among the world populace can be dependent on the prevailing environmental conditions which influence their survival with water, soil, and food being of particular significance [12]. Some other factors that influence the transmission of zoonotic diseases in developing countries include poverty, illiteracy, and global warming due to climate change. Factors such as high population density, culinary habits, poverty, open space defecation, lack of potable water, personal hygiene, and abundance of stray animals are known to contribute to the frequent spread of parasitic zoonotic diseases. The impact of global warming on the spread of zoonotic disease influences both human and animal lifestyles such that the vectors become more prevalent in the susceptible host. Giving updated information on the classes, occupation, animal-related, food-borne, and soil-related parasitic zoonoses will help avert their implications on human health and food production.

The importance of parasitic zoonoses has been understudied, which may likely be due to the lack of awareness by a health personnel. Nevertheless, high demand and consumption of raw food and food products from livestock, use of polluted water, inadequate food processing, and poor hygiene practices have contributed to the spread of zoonotic parasites in the food chain. The current status of some parasitic zoonoses directly or indirectly poses health risks to humans and animals in many parts of the world [55]. To address this menace, there is a need to employ appropriate measurable surveillance and control strategies for parasitic zoonoses through collaborative research in animals and infected humans. Some major examples of parasitic zoonotic pathogens affecting humans include *Leishmania* spp., *Toxoplasma gondii, Cryptosporidium parvum, Entamoeba histolytica, Sarcocystis* spp., *Giardia duodenalis, Babesia microti, Balantidium coli, Gastrodiscoides hominis,* and *Echinococcus granulosus* [19, 58–61]. Notably, parasitic zoonoses, such as cryptosporidiosis, leishmaniosis, toxoplasmosis, taeniasis, cysticercosis, echinococcosis, trichinellosis, amoebiasis, giardiasis, sarcocystosis, clonorchiasis, fasciolopsiasis, hydatidosis, sparganosis, angiostrongyliasis, gnathostomiasis, trichostrongylosis, toxocariasis, and ancylostomiasis, infect humans through the consumption of contaminated foods [62, 49]. The presence of these parasites plays an important role in food safety, and human and animal health. Parasitic zoonoses can be classified based on etiological agent, transmission cycle, reservoir host, and principal host as presented in Table 4.1.

4.3.3 Viral zoonotic foodborne pathogens

Only a few foodborne zoonotic viral diseases have been identified in contrast to the abundance of bacterial and parasitic zoonoses. Viral zoonotic foodborne pathogens are usually nonenveloped viruses that do not possess a lipid envelope. This class of zoonotic viruses (nonenveloped viruses) are generally resistant to environmental stress including heat, desiccation, organic solvent, and extreme pH. As such, typical inactivation treatments active against the vegetative stages of bacterial foodborne pathogens or nonenveloped viruses (such as influenza virus) are not effective against them. These features enhance them to resist common sanitation processes and food processing technologies, hence, their prolonged and persistent stability in foods and the environment. Major viral foodborne pathogens are caliciviruses (norovirus), adenoviruses, rotaviruses, astroviruses, sapoviruses, hepatitis A virus (HAV), and hepatitis B virus (HEV). The increasing prevalence of HEV and sapovirus was reported in the European Union (EU) and Asia [63, 44]. Viral foodborne zoonosis is usually characterized by symptoms ranging from mild (usually non-bloody) diarrhea, pains, abdominal cramps, nausea or vomiting, acute or severe neurological diseases, and even death. Globally, zoonotic foodborne viruses are the leading cause of gastroenteritis and may also cause illness and enterically transmitted hepatitis after migrating to other human organs from the intestine. From a global perspective, the burden of foodborne illnesses is more pronounced for viral zoonotic pathogens transmitted due to unhygienic practices by food handlers during food production, infected food handlers shedding viruses, or ingesting contaminated food. This is also applicable to viruses transmitted through the fecal–oral route, after ingestion, causing infection in their host followed by gut's epithelial cell lining invasion and subsequently replicating in the same site or other body parts.

4.3.3.1 Hepatitis A virus

HAV are nonenveloped positive single-stranded RNA viruses, belonging to the family Picornaviridae and genus *Hepatovirus*. HAV were classified into eight lineages based on genetic diversity, of which genotypes I–III infect humans. Genotypes I to III were further classified into subgenotypes IA, IB, IIA, IIB, IIIA, and IIIB, while genotypes IV and V are found primarily in nonhuman primates [20]. There is no risk of introduction from a reservoir as hepatoviruses have only been found in humans and primates. To support source tracking of foodborne outbreaks in regions known for the endemic circulation of HAV, further segregation into geographically defined clusters is usually employed [64]. The transmissibility of HAV is lesser with lower incidences in regions with good hygiene and proper sanitation compared to noroviruses (regions with high socioeconomic status). About 1.5 million people are infected with the virus each year. HAV is one of the childhood infections in highly endemic regions that runs an asymptomatic course, triggering lifelong immune responses [65]. HAV can persist in contaminated food, water, and the environment for several weeks. They retain infectivity after 92 days at 25 °C in seawater due to their stability outside a host. Thus, the virus is tolerant to freezing and desiccation and will favorably survive on vegetables throughout production until consumption [20]. Water and foodborne outbreaks due to HAV have been documented. The most challenging risk of foodborne HAV is usually the introduction of the pathogen through food into regions where population immunity is relatively limited. Foods susceptible to contamination during the production phase include produce irrigated with contaminated water, for example, soft fruits such as strawberries, raspberries, green onions, and lettuce; or bivalve-feeding mollusks like clams, oysters, and mussels. HAV is the second virus listed and considered a significant cause of severe diseases by the Centre for Disease Control assessment of foodborne pathogens and could be related to the high severity of the infection. Symptoms include loss of appetite, fatigue, fever, nausea, vomiting, dark urine, jaundice (yellowing of eyes and skin), and joint pain and can lead to more severe illnesses like liver failure and finally death.

4.3.3.2 Hepatitis E virus

HEV is a small non-enveloped virus and belongs to the genus *Orthohepevirus*, family Hepeviridae. The genus *Orthohepevirus* is made up of four species, including *Orthohepevirus* A, B, C, and D, of which HEVs infecting humans belong to *Orthohepevirus* A. Species of *Orthohepevirus* A virus contains eight genotypes ranging from HEV1-8, determined by comparison of the amino acid sequences of the concatenated open reading frames 1 and 2 (ORF1 and ORF 2). Among these, the HEV1 and HEV2 infect humans causing waterborne outbreaks. HEV3 and HEV4 are zoonotic, infecting humans and swine, and found to be responsible for HEV foodborne infection in industrialized countries. HEV5 and HEV6 have been found in wild boar while HEV7 was detected in camels. This group of viruses infects humans, primates, wild boar, pigs, rabbits, and deer [20, 63, 66], while rabbits, wild boar, and pigs are known reservoirs [67]. Globally, human exposure to HEV is estimated at 2.3 billion people, with 20 million infections, 3.3 million asymptomatic cases, and about 44,000 deaths in 2015 according to the WHO report [20]. HEV1 and HEV2 are endemic in developing countries causing waterborne infections, especially in places where sanitation is poor. Foodborne outbreaks are mostly linked to HEV3 and HEV4 strains and are usually transmitted via the fecal–oral route, that is, contamination of drinking water by feces on infected persons. Direct contact with reservoir animals and consumption of contaminated water, foods (raw or undercooked foods from infected animals or shellfish), and vertical transmission from pregnant women or blood transfusion are also associated with the transmission of HEV. Foods such as filter-feeding bivalve shellfish, raw or undercooked meat, and produce are high-risk foods [20]. HEV is primarily a liver disease and has been associated with nonhepatic diseases like monophasic neurologic disorders of the peripheral nervous system, glomerulonephritis, acute pancreatitis, hemolytic anemia, mixed cryoglobulinemia, severe thrombocytopenia. Foodborne HEV infection presents self-limiting clinical symptoms ranging from asymptomatic to acute liver failure. Symptomatic infections are usually observed in pregnant women, and they are at a higher risk of acute liver failure compared to other infected individuals. Chronic HEV has also been reported among other groups of immunocompromised individuals including hematological cancer and human immuno-deficiency virus patients. There has been evidence of neurological cases of acute and chronic HEV infection after the viral RNA was detected in the cerebral spinal fluid of peripheral neuropathy patients which resolved upon clearing [63].

4.3.3.3 Norovirus

This is a small nonenveloped, positive-sense RNA virus belonging to the family Calicivir-idae and is made up of five genera including the *Norovirus* and *Sapovirus* causing human infections. Additionally, one other potential genus detected in rhesus macaques was described. *Norovirus* has been detected in cats, cattle, dogs, mice, pigs, and sheep while *Sapovirus* was detected in pigs. *Norovirus* causes gastroenteritis, abdominal cramps, vomiting, nausea, malaise, muscle pain, and low-grade fever in humans while those infecting animals could cause different clinical syndromes including systemic dis-eases with hemorrhagic syndromes, oral lesions, upper respiratory tract infections, and others [68, 20]. Predominantly, noroviruses and sapoviruses are the only known *Calici-virus* causing human diseases excluding vesiviruses. For noroviruses, seven genogroups (GI–GVII) were recognized including GI, GII, and GIV were known to infect humans. GII and GIV viruses have been detected in pigs and carnivores (lion cub and dog), respec-tively, while GIII viruses infect cattle and sheep and GV viruses infect mice [20, 65, 69]. Noroviruses are an important etiological agent of diarrhea documented worldwide. However, few studies have compared the standardized comparisons of an international and true estimate of diseases. Evidence has emerged from community studies, identify-ing these viruses as the number one cause of community-acquired gastroenteritis in-fecting at least one out of four or five persons per year, especially among young children and the elderly. The propensity of noroviruses to cause an outbreak is the best-described feature of the virus with a very low infectious dose (1–10 particles), and a huge amount of the viruses are shed by infected persons (up to 1,010 million per gram of stool). Most common noroviruses evolve through the selection of filter variants that escape the blocking activities of the receptors from the antibodies triggered prior to in-fection and accumulation of mutations. This has resulted in the increased emergence of new variants with the incidence of *Norovirus* outbreaks. It is challenging to attribute the number of diseases caused by the viruses to a particular mode of transmission, par-ticularly foodborne spread. Sewage contamination could be a source of transmission early in the food chain and lack of good personal hygiene, particularly when RTE items are involved. Also, environmental contamination, person-to-person transmission via contact, and food–waterborne transmission are possible. In the EU in 2008, shellfish, crustacean, mollusks, and their products were the most frequently implicated food items in *Norovirus* outbreaks. Also, fruits and vegetables irrigated with contaminated water could cause outbreaks. Outbreaks usually occur frequently when food handled and prepared by another person is consumed in restaurants, hospitals, nursing homes, schools, cruise ships, summer camps, and even within family dinners [20]. Summarily, noroviruses have been recognized as a common cause of foodborne gastroenteritis *Campylobacter* and more commonly compared to *Salmonella*. Other relevant viral foodborne zoonotic pathogens were described in Table 4.1.

4.3.4 Prion zoonotic foodborne agents

Prion is a proteinaceous infectious particle that can trigger abnormal folding of the normal protein of the brain. It triggers the neurodegenerative disorder that typically causes about 1–2 deaths per million population annually. The pathological misfolding and aggregation of a cellular glycoprotein, the prion protein (PrP, encoded by PRNP), was identified as the cause of prion disease. Prion's diseases are transmissible between individuals because the misfolding of the normal PrP is promoted by the pathologic form of PrP (PrP^{Sc}) [70]. Prion infections comprise several disease conditions affecting both animals and humans and are sometimes spread to humans by infected meat. The diseases occur when normal protein found on the surface of many cells becomes abnormal and clumps in the brain resulting in brain damage. The accumulation of abnormal protein in the brain can cause personality changes, memory impairment, and difficulty with movement. Unfortunately, experts' knowledge about prion disease is limited despite that prion's disorders are generally fatal. There are different types of prion disease including the Creutzfeldt-Jakob disease (CJD), variant CJD (vCJD), variably protease-sensitive prionopathy, Gerstmann-Straussler-Scheinker disease, Kuru, and fatal insomnia. The CJD prion disease is usually the most common and may be inherited. It could develop suddenly without any known risk factors. Most cases are sporadic and tend to strike people around age 60. The vCJD is related to the "mad cow disease" and may be acquired by the consumption of infected animal products. The emergence was earlier linked to an epidemic of bovine spongiform encephalopathy in the United Kingdom through the consumption of contaminated beef. Several other cases were reported in France, the USA, and Canada, and there are possibilities of future cases considering the protracted incubation period of the disease [71]. The disease usually affects younger people and could be acquired through meat consumption which may likely cause normal human prion protein to develop abnormally. Other relevant prion's foodborne zoonotic disease was described in Table 4.1.

4.3.5 Fungal zoonotic foodborne pathogens

Fungi constitute a fundamental part of the natural environment and play several roles in food production. Some themselves serve as sources of food while others are food spoilage agents. They are nonphotosynthetic saprophytic microorganisms found in soil and dead organic matter or as plant, animal, or human parasites. Of about

1.5 million species of fungi known, only 300 were known to cause disease in humans [17]. Apart from being food contaminants, some fungi can be harmful to humans. Molds and yeasts cause serious food spoilage at storage with enormous economic losses, but they can cause several acute and chronic diseases in humans associated with the mycotoxins produced. Various mycotoxin producers including *Alternaria* spp., *Aspergillus* spp., *Fusarium* spp., and *Penicillium* spp. could contaminate produce including vegetables in the field or during storage [72, 73]. The non-infectious but harmful effects of mycotoxins have been well described; however, the risk and extent of invasive infections posed by fungi in food are unknown. Cases of invasive infections resulting from the ingestion of food, dietary supplements, or beverages appear to be rare [74].

Some clinically significant fungal zoonotic pathogens including *Basidiobolus ranarum, Cryptococcus neoformans, Histoplasma capsulatum, Malassezia* spp., *Microsporum* spp., *Paracoccidioides brasiliensis, Penicillium marneffei, Sporothrix schenckii, Trichophyton verrucosum, Arthroderma vanbreuseghemii, Arthroderma benhamiae, Histoplasma capsulatum* var. *capsulatum*, and *H. capsulatum* var. *duboisii* have been identified [17, 75]. However, limited studies had evaluated fungi as zoonotic foodborne pathogens. A study in Texas, USA, reported a foodborne outbreak linked to the consumption of mold (*Mucor circinelloides*)-contaminated yoghurt involving more than 200 individuals that suffered from nausea, vomiting, and diarrhea [76]. The pathogen can cause fatal infections in immunocompromised persons and is not known to produce mycotoxin [77]. *M. circinelloides* has been isolated from various foods, including meat, mung beans, soybeans, hazelnuts, and walnuts. The pathogen is considered an emerging pathogen causing human illness, particularly cutaneous infection, and is occasionally isolated from birds, cattle, swine, and humans.

Basidiobolomycosis is a rare but emerging fungal infection caused by *Basidiobolus ranarum* [78] found worldwide and is usually isolated from foodstuffs, leaves of deciduous trees, decaying plant materials, and soil. *B. ranarum* is occasionally present in the gastrointestinal tract of reptiles (geckos and garden lizards), amphibians (toads and frogs), fish, mammals (including insectivorous bats, dogs, horses, and humans), and the feces of wallabies and kangaroos. Infections due to *B. ranarum* is usually subcutaneous or gastrointestinal. Bloody diarrhea, nonbilious vomiting, abdominal pain, and low-grade fever have been identified as the symptoms of ingestion of food contaminated with *B. ranarum*. Infections due to the traumatic implantation or inhalation of fungus present in plant debris have been reported in tropical countries, including Africa, Asia, the USA, South America, and Europe. The fungal spores could cause hard enlarged lumps under the skin of the arms or legs by growing slowly after entering the human tissue. Also, ingestion of soil or animal feces through contaminated food is another infection route [17, 78].

4.4 Conclusion and recommendation

Conclusively, the progressive increase in the recognition of new zoonotic foodborne agents, immunocompromised human populations that are practically susceptible to the pathogens, human interaction with the environment, and the rate of antibiotic resistance among pathogens may likely lead to the emergence of more new zoonotic foodborne diseases, in addition to existing ones. The burden of foodborne diseases is enormous and remains disproportionate to public health and the scientific attention accorded. Hence, the need for public awareness and the orientation of the risk factors associated with unhygienic food preparation procedures and handling. Consumption of raw animal products and unstandardized slaughtering processes may increase the burden of foodborne illnesses globally. As such, effective implementation of the standardized hazard analysis critical control point, good manufacturing practices, good operational sanitation practices, and pasteurization procedures must be enforced to enhance food safety. Therefore, the implementation of coordinated monitoring and surveillance systems within the food chain (from farm to fork) is highly recommended.

References

[1] Adeleke, B. S., & Babalola, O. O. (2020). Oilseed crop sunflower (Helianthus annuus) as a source of food: Nutritional and health benefits. *Food Science & Nutrition*. https://doi.org/10.1002/fsn3.1783.

[2] Salami, S. A., Luciano, G., O'Grady, M. N., Biondi, L., Newbold, C. J., Kerry, J. P., & Priolo, A. (2019). Sustainability of feeding plant by-products: A review of the implications for ruminant meat production. *Animal Feed Science Technology*. https://doi.org/10.1016/j.anifeedsci.2019.02.006.

[3] Borchers, A., Teuber, S. S., Keen, C. L., & Gershwin, M. E. (2010). Food safety. *Clinical Reviews in Allergy & Immunology*. https://doi.org/10.1007/s12016-009-8176-4.

[4] Schantz, P. M. (1991). Parasitic zoonoses in perspective. *International Journal for Parasitology*. https://doi.org/10.1016/0020-7519(91)90006-s.

[5] Abebe, E., Gugsa, G., & Ahmed, M. (2020). Review on Major Food-Borne Zoonotic Bacterial Pathogens. *Journal of Tropical Medicine*. https://doi.org/10.1155/2020/4674235.

[6] Cutler, S. J., Fooks, A. R., & Van Der Poel, W. H. M. (2010). Public health threat of new, reemerging, and neglected zoonoses in the industrialized world. *Emerging Infectious Diseases*. https://doi.org/10.3201/eid1601.081467.

[7] Chlebicz, A., & Śliżewska, K. (2018). Campylobacteriosis, Salmonellosis, Yersiniosis, and Listeriosis as Zoonotic Foodborne Diseases: A Review. *International Journal of Environmental Research and Public Health*. https://doi.org/10.3390/ijerph15050863.

[8] Bidaisee, S., & Macpherson, C. N. L. (2014). Zoonoses and one health: A review of the literature. *Journal of Parasitology Research*. https://doi.org/10.1155/2014/874345.

[9] Ejo, M., Garedew, L., Alebachew, Z., & Worku, W. (2016). Prevalence and Antimicrobial Resistance of Salmonella Isolated from Animal-Origin Food Items in Gondar, Ethiopia. *BioMedicine Research International*. https://doi.org/10.1155/2016/4290506.

[10] Heredia, N., & García, S. (2018). Animals as sources of food-borne pathogens: A review. *Animal Nutrition*. https://doi.org/10.1016/j.aninu.2018.04.006.

[11] Haileselassie, M., Taddele, H., Adhana, K., & Kalayou, S. (2013). Food safety knowledge and practices of abattoir and butchery shops and the microbial profile of meat in Mekelle City, Ethiopia. *Asian Pacific Journal of Tropical Biomedicine*. https://doi.org/10.1016/S2221-1691(13)60085-4.

[12] Slifko, T. R., Smith, H. V., & Rose, J. B. (2000). Emerging parasite zoonoses associated with water and food. *International Journal for Parasitology*. https://doi.org/10.1016/S0020-7519(00)00128-4.

[13] Aklilu, A., Kahase, D., Dessalegn, M., Tarekegn, N., Gebremichael, S., Zenebe, S., Desta, K., Mulugeta, G., Mamuye, Y., & Mama, M. (2015). Prevalence of intestinal parasites, Salmonella and Shigella among apparently health food handlers of Addis Ababa University student's cafeteria, Addis Ababa, Ethiopia. *BMC Research Notes*. https://doi.org/10.1186/s13104-014-0967-x.

[14] Ayana, Z., Yohannis, M., Abera, Z., & Ababa, A. (2015). Food-Borne Bacterial Diseases in Ethiopia The Federal Democratic Republic of Ethiopia Sugar Corporation, 1 Diversified Products and Agro-processing. *Academy of Journal Nutrition*.

[15] WHO. (2018a). Salmonella (non-typhoidal). World Heal. Organ.

[16] WHO. (2015). WHO estimates of the global burden of foodborne diseases: foodborne disease burden epidemiology reference group 2007-2015. World Health Organization. foodborne disease burden epidemiology reference group 2007-2015.

[17] Adebiyi, A. I., & Oluwayelu, D. O. (2018). Zoonotic fungal diseases and animal ownership in Nigeria. *Alexandria Journal of Medicine*. https://doi.org/10.1016/j.ajme.2017.11.007.

[18] Guo, Y., Li, N., Ryan, U., Feng, Y., & Xiao, L. (2021). Small ruminants and zoonotic cryptosporidiosis. *Parasitology Research*. https://doi.org/10.1007/s00436-021-07116-9.

[19] Karanis, P. (2006). A review of an emerging waterborne medical important parasitic protozoan. *The Japanese Journal of Protozoology*, *39*, 5–19. https://doi.org/10.18980/jjprotozool.39.1_5.

[20] O'Shea, H., Blacklaws, B. A., Collins, P. J., McKillen, J., & Fitzgerald, R. (2019). Viruses Associated With Foodborne Infections. *Reference Module in Life Sciences*. https://doi.org/10.1016/b978-0-12-809633-8.90273-5.

[21] Argaw, S., Addis, M., Umeda, K., Ono, H. K., Wada, T., Motooka, D., Nakamura, S., Nakamura, H., Hu, D.-L., Gupta, P. K., Tchounwou, P. B., Yedjou, C. G., Patlolla, A. K., Sutton, D. J., Kumar, N., Singh, A., Sharma, D. K., Kishore, K., Bhalla, T. C., Monika, S., & Savitri. (2019) *A review on staphylococcal food poisoning*. Food Safety and Human Health.

[22] WHO. (2020). Food Safety: Major foodborne illnesses and causes. World Heal. Organ.

[23] Johnson, R., Mylona, E., & Frankel, G. (2018). Typhoidal Salmonella: Distinctive virulence factors and pathogenesis. *Cellular Microbiology*. https://doi.org/10.1111/cmi.12939.

[24] Feasey, N. A., Dougan, G., Kingsley, R. A., Heyderman, R. S., & Gordon, M. A. (2012). Invasive non-typhoidal salmonella disease: An emerging and neglected tropical disease in Africa. *Lancet*. https://doi.org/10.1016/S0140-6736(11)61752-2.

[25] WHO (2018b). E. coli. World Heal. Organ.

[26] Iwu, C. D., Kayode, A. J., & Igere, B. E. (2022). High levels of multi drug resistant Escherichia coli pathovars in preharvest environmental samples: a ticking time bomb for fresh produce related disease outbreak. 10, 1–15. https://doi.org/10.3389/fenvs.2022.858964.

[27] Igwaran, A., & Okoh, A. I. (2019). Human campylobacteriosis: A public health concern of global importance. *Heliyon*. https://doi.org/10.1016/j.heliyon.2019.e02814.

[28] Carlin, C. R., Liao, J., Weller, D., Guo, X., Orsi, R., & Wiedmann, M. (2021). Listeria cossartiae sp. Nov., listeria immobilis sp. nov., listeria portnoyi sp. nov. and listeria rustica sp. nov., isolated from agricultural water and natural environments. *International Journal of Systematic and Evolutionary Microbiology*. https://doi.org/10.1099/ijsem.0.004795.

[29] Carlin, C. R., Liao, J., Weller, D. L., Guo, X., Orsi, R., & Wiedmann, M. (2021). Corrigendum to: Listeria cossartiae sp. nov., Listeria farberi sp. nov., Listeria immobilis sp. nov., Listeria portnoyi sp. nov. and Listeria rustica sp. nov., isolated from agricultural water and natural environments *International Journal of Systematic and Evolutionary Microbiology*, 71. https://doi.org/10.1099/ijsem.0.004885.

[30] Quereda, J. J., Leclercq, A., Moura, A., Vales, G., Gómez-Martín, Á., García-Muñoz, Á., Thouvenot, P., Tessaud-Rita, N., Bracq-Dieye, H., & Lecuit, M. (2020). Listeria valentina sp. nov., isolated from a water trough and the faeces of healthy sheep. *International Journal of Systematic and Evolutionary Microbiology*. https://doi.org/10.1099/ijsem.0.004494.

[31] Kayode, A. J., Igbinosa, E. O., & Okoh, A. I. (2020). Overview of listeriosis in the Southern African Hemisphere – Review. *Journal of Food Safety*, 40, e12732. https://doi.org/10.1111/jfs.12732.

[32] Yin, Y., Yao, H., Doijad, S., Kong, S., Shen, Y., Cai, X., Tan, W., Wang, Y., Feng, Y., Ling, Z., Wang, G., Hu, Y., Lian, K., Sun, X., Liu, Y., Wang, C., Jiao, K., Liu, G., Song, R., Chen, X., Pan, Z., Loessner, M. J., Chakraborty, T., & Jiao, X. (2019). A hybrid sub-lineage of Listeria monocytogenes comprising hypervirulent isolates. *Nature Communications*. https://doi.org/10.1038/s41467-019-12072-1.

[33] Kayode, A. J., & Okoh, A. I. (2022a). Assessment of the molecular epidemiology and genetic multiplicity of Listeria monocytogenes recovered from ready-to-eat foods following the South African listeriosis outbreak. *Scientific Reports*, 12, 20129. https://doi.org/10.1038/s41598-022-20175-x.

[34] Kayode, A. J., & Okoh, A. I. (2022b). Incidence and genetic diversity of multi-drug resistant Listeria monocytogenes isolates recovered from fruits and vegetables in the Eastern Cape Province, South Africa. *International Journal of Food Microbiology*. https://doi.org/10.1016/j.ijfoodmicro.2021.109513.

[35] Austin, B. (2010). Vibrios as causal agents of zoonoses. *Veterinary Microbiology*. https://doi.org/10.1016/j.vetmic.2009.03.015.

[36] Baker-Austin, C., Oliver, J. D., Alam, M., Ali, A., Waldor, M. K., Qadri, F., & Martinez-Urtaza, J. (2018). Vibrio spp. infections. *Nature Reviews Disease Primers*. https://doi.org/10.1038/s41572-018-0005-8.

[37] Osunla, C. A., & Okoh, A. I. (2017). Vibrio pathogens: A public health concern in rural water resources in sub-Saharan Africa. *International Journal of Environmental Research and Public Health*. https://doi.org/10.3390/ijerph14101188.

[38] Grace, D., & Fetsch, A. (2018). *Staphylococcus aureus – a foodborne pathogen: epidemiology, detection, characterization, prevention, and control: an overview*. Staphylococcus Aureus.

[39] Keatts, L. O., Robards, M., Olson, S. H., Hueffer, K., Insley, S. J., Joly, D. O., Kutz, S., Lee, D. S., Chetkiewicz, C. L. B., Lair, S., Preston, N. D., Pruvot, M., Ray, J. C., Reid, D., Sleeman, J. M., Stimmelmayr, R., Stephen, C., & Walzer, C. (2021). Implications of zoonoses from hunting and use of wildlife in North American Arctic and Boreal biomes: pandemic potential, monitoring, and mitigation. *Front Public Health*. https://doi.org/10.3389/fpubh.2021.627654.

[40] Ibrahim, S., Cadmus, S. I. B., Umoh, J. U., Ajogi, I., Farouk, U. M., Abubakar, U. B., & Kudi, A. C. (2012). Tuberculosis in humans and cattle in Jigawa State, Nigeria: Risk factors analysis. *Veterinary Medicine International*. https://doi.org/10.1155/2012/865924.

[41] Farrell, G. M., & Marth, E. H. (1991). Borrelia burgdorferi: another cause of foodborne illness?. *International Journal of Food Microbiology*. https://doi.org/10.1016/0168-1605(91)90117-8.

[42] ECDC, E. (2018). The European Union summary report on trends and sources of zoonoses, zoonotic agents and food-borne outbreaks in 2017. *EFSA Journal*, 16, e05500. https://doi.org/10.2903/j.efsa.2018.5500.

[43] Ghosh, S., & Kobayashi, N. (2014). Exotic rotaviruses in animals and rotaviruses in exotic animals. *VirusDisease*. https://doi.org/10.1007/s13337-014-0194-z.

[44] Rivadulla, E., & Romalde, J. L. (2020). A comprehensive review on human aichi virus. *Virologica Sinica*. https://doi.org/10.1007/s12250-020-00222-5.

[45] Cai, W., Ryan, U., Xiao, L., & Feng, Y. (2021). Zoonotic giardiasis: an update. *Parasitology Research*. https://doi.org/10.1007/s00436-021-07325-2.

[46] Dixon, B. R. (2021). Giardia duodenalis in humans and animals – Transmission and disease. *Research of Veterinary Sciences*. https://doi.org/10.1016/j.rvsc.2020.09.034.

[47] Prakas, P., Balčiauskas, L., Juozaitytė-Ngugu, E., & Butkauskas, D. (2021). The role of mustelids in the transmission of Sarcocystis spp. Using cattle as intermediate hosts. *Animals*. https://doi.org/10.3390/ani11030822.

[48] Rosenthal, B. M. (2021). Zoonotic sarcocystis. *Research in Veterinary Science*. https://doi.org/10.1016/j.rvsc.2021.02.008.

[49] Rabone, M., Wiethase, J., Clark, P. F., Rollinson, D., Cumberlidge, N., & Emery, A. M. (2021). Endemicity of paragonimus and paragonimiasis in sub-Saharan Africa: A systematic review and mapping reveals stability of transmission in endemic foci for a multi-host parasite system. *PLoS Neglected Tropical Diseases*. https://doi.org/10.1371/journal.pntd.0009120.

[50] Zhou, P., Chen, N., Zhang, R. L., Lin, R. Q., & Zhu, X. Q. (2008). Food-borne parasitic zoonoses in China: perspective for control. Trends in Parasitology. https://doi.org/10.1016/j.pt.2008.01.001.

[51] Sabourin, E., Alda, P., Vázquez, A., Hurtrez-Boussès, S., & Vittecoq, M. (2018). Impact of human activities on fasciolosis transmission. *Trends in Parasitology*. https://doi.org/10.1016/j.pt.2018.08.004.

[52] Malov, V. A., Volchkova, E. V., Konnova, Y. A., Tsvetkova, N. A., Konnov, V. V., Pischasov, S. V., & Malova, A. V. (2021). Trichinosis: outbreak description and brief literature review. *Epidemiology of Infectious Disease*. https://doi.org/10.17816/eid52976.

[53] Chai, J. Y., Han, E. T., Shin, E. H., Park, J. H., Chu, J. P., Hirota, M., Nakamura-Uchiyama, F., & Nawa, Y. (2003). An outbreak of gnathostomiasis among Korean emigrants in Myanmar. *The American Journal of Tropical Medicine and Hygiene*. https://doi.org/10.4269/ajtmh.2003.69.67.

[54] Liu, G. H., Sun, M. M., Elsheikha, H. M., Fu, Y. T., Sugiyama, H., Ando, K., Sohn, W. M., Zhu, X. Q., & Yao, C. (2020). Human gnathostomiasis: a neglected food-borne zoonosis. *Parasites & Vectors*. https://doi.org/10.1186/s13071-020-04494-4.

[55] Rahman, M. T., Sobur, M. A., Islam, M. S., Ievy, S., Hossain, M. J., Zowalaty, M. E. E., Rahman, A. M. M. T., & Ashour, H. M. (2020). Zoonotic diseases: Etiology, impact, and control. Microorganisms. https://doi.org/10.3390/microorganisms8091405.

[56] Short, E. E., Caminade, C., & Thomas, B. N. (2017). Climate Change Contribution to the Emergence or Re-Emergence of Parasitic Diseases. *Infectious Diseases: Research and Treatment*. https://doi.org/10.1177/1178633617732296.

[57] Morales, F. D. A. R., Genís, J. M. C., & Guerrero, Y. M. (2019). Current status, challenges and the way forward for dairy goat production in Europe. *Asian-Australasian Journal of Animal Sciences*. https://doi.org/10.5713/ajas.19.0327.

[58] Alvarez Rojas, C. A., Mathis, A., & Deplazes, P. (2018). Assessing the contamination of food and the environment with Taenia and Echinococcus eggs and their zoonotic transmission. *Current Clinical Microbiology Reports*. https://doi.org/10.1007/s40588-018-0091-0.

[59] Bednarska, M., Bajer, A., Drozdowska, A., Mierzejewska, E. J., Tolkacz, K., & Welc-Falęciak, R. (2015). Vertical transmission of Babesia microti in BALB/c mice: Preliminary report. *PLoS One*. https://doi.org/10.1371/journal.pone.0137731.

[60] Giarratana, F., Nalbone, L., Napoli, E., Lanzo, V., & Panebianco, A. (2021). Prevalence of Balantidium coli (Malmsten, 1857) infection in swine reared in South Italy: A widespread neglected zoonosis. *Veterinary World*. https://doi.org/10.14202/vetworld.2021.1044-1049.

[61] Sah, R., Acosta, L., & Toledo, R. (2019). A case report of human gastrodiscoidiasis in Nepal. *Parasitology International*. https://doi.org/10.1016/j.parint.2019.03.014.

[62] Pinto-Ferreira, F., Caldart, E. T., Pasquali, A. K. S., Mitsuka-Breganó, R., Freire, R. L., & Navarro, I. T. (2019). Patterns of transmission and sources of infection in outbreaks of human toxoplasmosis. *Emerging Infectious Diseases*. https://doi.org/10.3201/eid2512.181565.

[63] Harrison, L., & DiCaprio, E. (2018). Hepatitis E Virus: An Emerging Foodborne Pathogen. *Frontiers in Sustainable Food Systems*. https://doi.org/10.3389/fsufs.2018.00014.

[64] Robertson, B. H., Jansen, R. W., Khanna, B., Totsuka, A., Nainan, O. V., Siegl, G., Widell, A., Margolis, H. S., Isomura, S., Ito, K., Ishizu, T., Moritsugu, Y., & Lemon, S. M. (1992). Genetic relatedness of hepatitis A virus strains recovered from different geographical regions. *Journal of General Virology*. https://doi.org/10.1099/0022-1317-73-6-1365.

[65] Kroneman, A., Vennema, H., Deforche, K., Avoort, H., Peñaranda, S., Oberste, M. S., Vinjé, J., & Koopmans, M. (2011). An automated genotyping tool for enteroviruses and noroviruses. *Journal of Clinical Virology*. https://doi.org/10.1016/j.jcv.2011.03.006.

[66] Smith, D. B., Simmonds, P., Izopet, J., Oliveira-Filho, E. F., Ulrich, R. G., Johne, R., Koenig, M., Jameel, S., Harrison, T. J., Meng, X. J., Okamoto, H., Van Der Poel, W. H. M., & Purdy, M. A. (2016). Proposed reference sequences for hepatitis E Virus subtypes. *Journal of General Virology*. https://doi.org/10.1099/jgv.0.000393.

[67] Doceul, V., Bagdassarian, E., Demange, A., & Pavio, N. (2016). Zoonotic hepatitis E virus: Classification, animal reservoirs and transmission routes. *Viruses*. https://doi.org/10.3390/v8100270.

[68] Farkas, T., Sestak, K., Wei, C., & Jiang, X. (2008). Characterization of a rhesus monkey Calicivirus representing a new genus of Caliciviridae . *Virology Jounral*. https://doi.org/10.1128/jvi.00070-08.

[69] Zheng, D. P., Ando, T., Fankhauser, R. L., Beard, R. S., Glass, R. I., & Monroe, S. S. (2006). Norovirus classification and proposed strain nomenclature. *Virology*. https://doi.org/10.1016/j.virol.2005.11.015.

[70] Coulthart, M. B., Geschwind, M. D., Qureshi, S., Phielipp, N., Demarsh, A., Abrams, J. Y., Belay, E., Gambetti, P., Jansen, G. H., Lang, A. E., & Schonberger, L. B. (2016). A case cluster of variant Creutzfeldt-Jakob disease linked to the Kingdom of Saudi Arabia. *Brain*. https://doi.org/10.1093/brain/aww206.

[71] Maheshwari, A., Fischer, M., Gambetti, P., Parker, A., Ram, A., Soto, C., Concha-Marambio, L., Cohen, Y., Belay, E. D., Maddox, R. A., Mead, S., Goodman, C., Kass, J. S., Schonberger, L. B., & Hussein, H. M. (2015). Recent us case of variant Creutzfeldt-Jakob disease – global implications. *Emerging Infectious Diseases*. https://doi.org/10.3201/eid2105.142017.

[72] Johannessen, G. S., & Torp, M. (2005). Improving the safety of organic vegetables. *Improving the Safety of Fresh Fruit and Vegetable*s. https://doi.org/10.1533/9781845690243.2.331.

[73] Sharifzadeh, A., Jebeli Javan, A., Shokri, H., Abbaszadeh, S., & Keykhosravy, K. (2016). Evaluation of antioxidant and antifungal properties of the traditional plants against foodborne fungal pathogens. *Journal Mycology Medical*. https://doi.org/10.1016/j.mycmed.2015.11.002.

[74] Benedict, K., Chiller, T. M., & Mody, R. K. (2016). Invasive Fungal Infections Acquired from Contaminated Food or Nutritional Supplements: A Review of the Literature. *Foodborne Pathogens and Disease*. https://doi.org/10.1089/fpd.2015.2108.

[75] Seyedmousavi, S., Guillot, J., Tolooe, A., Verweij, P. E., & de Hoog, G. S. (2015). Neglected fungal zoonoses: Hidden threats to man and animals. *Clinical Microbiology & Infection*, https://doi.org/10.1016/j.cmi.2015.02.031.

[76] Lee, S. C., Blake Billmyre, R., Li, A., Carson, S., Sykes, S. M., Huh, E. Y., Mieczkowski, P., Ko, D. C., Cuomo, C. A., & Heitman, J. (2014). Analysis of a food-borne fungal pathogen outbreak: Virulence and genome of a Mucor circinelloides isolate from yogurt. *mBio*, https://doi.org/10.1128/mBio.01390-14.

[77] Vellanki, S., Navarro-Mendoza, M. I., Garcia, A., Murcia, L., Perez-Arques, C., Garre, V., Nicolas, F. E., & Lee, S. C. (2018). Mucor circinelloides: Growth, Maintenance, and Genetic Manipulation. *Current Protocols in Microbiology*, https://doi.org/10.1002/cpmc.53.

[78] Shreef, K., Saleem, M., Saeedd, M. A., & Eissa, M. (2018). Gastrointestinal basidiobolomycosis: an emerging, and a confusing, disease in children (a multicenter experience). *European Journal of Pediatric Surgery*. https://doi.org/10.1055/s-0037-1598104.

[79] Moratal, S., Dea-Ayuela, M. A., Cardells, J., Marco-Hirs, N. M., Puigcercós, S., Lizana, V., & López-Ramon, J. (2020). Potential risk of three zoonotic protozoa (Cryptosporidium spp., giardia duodenalis, and toxoplasma gondii) transmission from fish consumption. *Foods*, https://doi.org/10.3390/foods9121913.

[80] Tornese, M., Rossi, M. L., Coca, F., Cricelli, C., & Troncoso, A. (2008). [Epidemiology and risk factors associated to foodborne and infant botulism: where and when? *Revista Chilena de Infectologia*. https://doi.org//S0716-10182008000100004.

[81] Kayode, A. J., Banji-Onisile, F. O., Olaniran, A. O., & Okoh, A. I. (2021). An overview of the pathogenesis, transmission, diagnosis, and management of endemic human coronaviruses: A reflection on the past and present episodes and possible future outbreaks. *Pathogens*, https://doi.org/10.3390/pathogens10091108.

[82] Kayode, A. J., Semerjian, L., Osaili, T., Olapade, O., & Okoh, A. I. (2021). Occurrence of Multidrug-Resistant Listeria monocytogenes in Environmental Waters: A Menace of Environmental and Public Health Concern. *Frontiers of Environmental Science*, *9*, 1–16. https://doi.org/10.3389/fenvs.2021.737435.

[83] Spika, J. S., Shaffer, N., Hargrett Bean, N., Collin, S., Macdonald, K. L., & Blake, P. A. (1989). Risk Factors for Infant Botulism in the United States. *American Journal of Diseases of Children*. https://doi.org/10.1001/archpedi.1989.02150190078026.

Fakoya Soji* and Abisoye Solomon Fiyinfoluwa

Chapter 5
Biofilm production by pathogens and control

Abstract: A biofilm is a composite made up of microbial colonies, typically those of bacterial origin, which live in an exopolysaccharide medium and adhere to other surfaces. Food borne illness and food poisoning/intoxication are frequently caused by improper handling of biofilm and ineffective controls in the food sector. This study made significant reference to the theory surrounding the composition, production, and control of biofilm in the food business. The prevention of food borne diseases can be greatly aided by the use of creative techniques to deal with the problems presented by biofilm development. Hence, a better understanding of microbial biofilm is required for the development of ingenious and effective control strategies in the management of biofilm hiccups associated with food industries.

5.1 Historical background of biofilm

A single type of cell or multiple bacterial colonies come together to form a complex microbiome structure known as a biofilm that adheres to surfaces [1]. These cells are enclosed in an extracellular polymeric matrix that frequently consists of eDNA, proteins, and polysaccharides and has a high level of antibiotic resistance [2]. Bacteria can multiply into two different ways: as planktonic cells and as sessile collections of bacteria known as biofilms. According to Muhammad et al. [1], a biofilm is a collection of germs whose cells adhere to one another on a surface and are shielded by an extracellular polymeric substance (EPS) made by the bacteria. Using a primitive microscope, a Dutch scientist Antonie van Leeuwenhoek discovered microbial biofilms when he first saw "animalcule" on the surfaces of teeth [3]. Later in 1973, Characklis discovered that in addition to being more resilient, biofilms are more resistant to disinfectants like chlorine. In 1978, Costerton coined the phrase "biofilm" and outlined its relevance. Biofilms can be found on hard surfaces that have been exposed to or submerged in an aqueous solution in industrial settings, lodging facilities, wastewater channels, restrooms, labs, and healthcare institutions. It can also resemble carpets floating on a liquid's surface. Both living and dead surfaces can support its development [4].

---

*__Corresponding author: Fakoya Soji__, Department of Biological Sciences, Olusegun Agagu University of Science and Technology, Okitipupa, Ondo State, Nigeria, e-mail: so.fakoya@oaustech.edu.ng
__Abisoye Solomon Fiyinfoluwa__, Department of Biological Sciences, Olusegun Agagu University of Science and Technology, Okitipupa, Ondo State, Nigeria

https://doi.org/10.1515/9783110748345-005

5.2 Introduction to biofilm

A biofilm is a grouping of microbes that may contain one or more microbial species. Numerous microbe species reside in close proximity and participate in intricate social interactions both within and between species in the densely populated microbial colonies found in biofilms [5]. When planktonic bacteria cling to biotic or abiotic surfaces, biofilms are created. Biofilms can persist for a very long time in a range of environmental niches because they are multicellular. Higher rates of gene transfer, collaboration, and stratification among microorganisms are possible in biofilms due to their "multicellular lifestyle" [4]. Numerous regulatory networks translate information into changes in gene expression, which in turn affect the spatial and temporal reorganization of the bacterial cell, during the transition of microorganisms from planktonic growth mode to biofilm [6]. The ability of many bacteria to naturally generate biofilms, which protect them from potentially damaging environmental factors such pH fluctuations, oxygen radicals, biocides, food deprivation, and antimicrobial agents [7], allows them to survive in a wide range of environments. The creation of biofilm was regarded as undesirable in the healthcare, marine, and industrial sectors, notably in the food business. Biofilms, however, are of enormous ecological and environmental value. According to Arunasri and Mohan [8], various bioelectrochemical systems rely on enhanced biofilms on the electrode surface. The formation and maintenance of microbial biofilms on the electrode surface have a major impact on the final products. In order to increase product yields, it is essential to comprehend the mechanisms underpinning biofilm formation, the changes in gene expression that take place during biofilm formation in microorganisms, and cutting-edge techniques for avoiding biofilm formation. The idea that bacteria exist as distinct organisms in a "planktonic state" is one of the most widespread misconceptions concerning microbial life. Instead, it has been demonstrated that sessile, stationary microbial communities develop on a range of surfaces on their own. Examples of these surfaces include domestic and commercial pipes, biomaterials like contact lenses, medical equipment like urinary catheters and implants, as well as plant and animal tissues. These microbe aggregations, also known as biofilms, can be made up of different bacterial and fungal populations and are classified as either monomicrobial or polymicrobial aggregates. Microbes can interchange substrates due to their close proximity, disseminate metabolic wastes, and remove hazardous end products, all of which are necessary for the coexistence of many species. Because of their distinctive form, biofilm communities can shield the bacteria they contain from the immune system, shear forces, and antimicrobials. A biofilm formed from the bacteria *Pseudomonas aeruginosa* and *Staphylococcus aureus* is depicted in Plate 5.1.

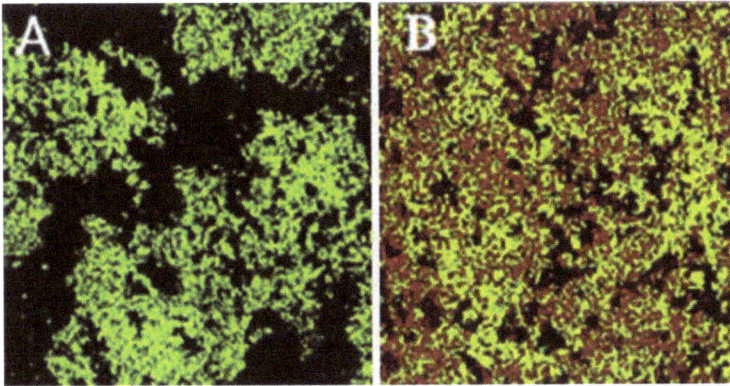

Plate 5.1: *P. aeruginosa* (green) and *S. aureus* combined to produce a polymicrobial biofilm (red). Confocal laser scanning microscope from Zeiss, ortho visualization (Image courtesy of Dr Elena Jordan-Lluch, University of Nottingham, uploaded by Liang Yang).
Source: [9].

5.3 Composition of biofilms

A group of bacteria known as a biofilm produces EPSs, such as proteins (1.2%), enzymes, DNA (1%), polysaccharides (1–2%), and RNA (1%). Water (up to 97%) makes up the majority of the biofilm together with these components and is in charge of allowing the passage of nutrients throughout the biofilm matrix (Table 5.1). A zone of tightly packed cells without any evident holes and a water channel for nutrient transport make up the two basic parts of the architecture of biofilms [10]. In biofilms, microbial cells are arranged in ways that considerably vary from one another in terms of their physiology and physical characteristics. Typically, neither the human immune system nor medications can penetrate bacterial biofilms. Biofilm-producing microorganisms have a greater ability to withstand and destroy antimicrobial medications, which can lead to protracted therapy. The bacteria that form biofilms activate particular genes that encourage the synthesis of stress genes in response to many condi-

Table 5.1: Chemical makeup of biofilms.

S. no.	Components	Proportion of the matrix
1	Bacterial cell	2.5%
2	DNA and RNA	<1–2%
3	Polysaccharides	1–2%
4	Protein	<1–2% (including enzymes)
5	Water	Up to 97%

Source: [12].

tions, such as cell density, nutritional requirements, temperature, pH, and osmolarity, to turn on resistant phenotypes [11]. Based on comparisons between the biofilm's circulation system and its water channels, biofilms are thought to be the first multicellular organisms [67]. The biofilm's many components, which are depicted in Table 5.1, reveal both its integrity and resistance to a variety of environmental conditions [68].

5.4 How biofilms are formed?

According to Okada et al. [13], the process by which microbial cells move from a planktonic to a sessile phase of development is known as biofilm production. Furthermore, it has been proposed that the expression of particular genes that regulate the biofilm's growth is necessary for biofilm formation. A number of variables that encourage adaptation to various dietary and environmental settings result in the formation of biofilms [14]. After attaching to a surface, the bacteria move through a number of phases during which they undergo various alterations (Figure 5.1). There is evidence that certain processes are triggered by the microorganisms that produce biofilms. These crucial steps are necessary for biofilm formation: first contact with a surface, development of a microcolony, building of three-dimensional structures, maturation, and spread of biofilms.

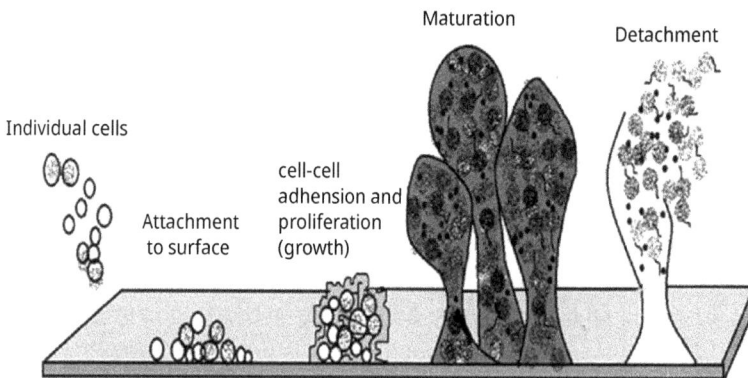

Figure 5.1: Three stages of the biofilm life cycle include attachment, colony growth (including the development of microclusters and three-dimensional structures), and clump-based separation [4].

5.4.1 Attachment

A bacterial cell creates a reversible link with the surface or another microbe that is already clinging when it is sufficiently close to a support or surface that it travels very slowly. The optimal environment for bacteria to attach to and flourish while creating biofilms can be found in a solid–liquid interface system (such as blood or water)

[4]. Rough, hydrophilic, and coated surfaces will promote the most frequent adhesion and biofilm growth. Increased flow rate, hotter water, or higher nutrient contents may also lead to more attachment if they do not go above critical levels. When there is a diverse community of organisms, the presence of locomotor structures on the cell surface, such as flagella, pili, fimbriae, proteins, or polysaccharides, is significant and may perhaps be helpful for biofilm growth [15].

5.4.2 Microcolony formation

When the binding between the two components becomes stable, the bacterial attachment to the biological tissue or physical surface causes the formation of a microcolony [16]. Chemical cues set off the earliest stages of bacterial development in the biofilm. When the signal intensity exceeds a specific threshold, the genetic mechanism for generating exopolysaccharides is triggered. In the implanted exopolysaccharide matrix, the bacteria divide as a result of this chemical signal, which eventually leads to the development of microcolonies.

5.4.3 Three-dimensional structure formation and maturation

A few biofilm-related genes start to express after the microcolony stage of biofilm development. These gene products are necessary for the EPS, which serves as the biofilm's main structural component [1]. It has been asserted that the development of extracellular matrix can be brought on solely by bacterial adhesion. After the matrix develops, water-filled tubes are created to carry nutrients throughout the biofilm. These water channels, in accordance with studies, serve as the circulatory system of a community, feeding it with different nutrients and removing waste from the microcolonies of the biofilm [17].

5.4.4 Detachment

Researchers have frequently noted that after biofilms have established, bacteria frequently eject themselves from them. The bacteria can swiftly grow and spread by doing this. Bacterial cells that are planktonic separate from the biofilm in a planned natural way. Mechanical stress can occasionally cause bacteria to become separated from the colony and invade the environment. Most of the time, some bacteria quit making EPS and disperse into the environment. Biofilm cells can separate from growing cells, scatter into aggregates due to flowing effects, or sense the presence of a quorum, for example [18]. An enzyme breaks down alginate, eliminating the cells from the biofilm. The phenotypic traits of organisms appear to be influenced by the dispersal of biofilms. Dispersed cells from the biofilm can preserve some traits of the biofilm, such as antibiotic resistance. The biofilm-forming, growth-disseminated cells can swiftly revert to their typical phenotype as planktonic organisms. The various stages of the biofilm life cycle are shown in Figure 5.2.

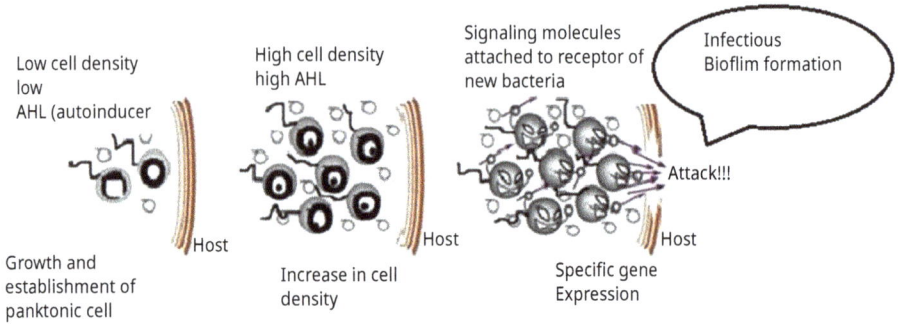

Figure 5.2: Cell density-dependent gene expression in quorum sensing [66].

5.4.5 Quorum sensing

Numerous bacterial species can communicate with one another during biofilm development, thanks to a method called quorum sensing (QS) [19]. Other cells require a system of cues to coordinate their gene expression and adjust to the population density in their immediate environment. Signaling molecules help in gene transcription both within and between bacterial species when they bind to the receptors of newly formed bacteria during QS. In terms of biofilm formation, food scarcity, and environmental stress conditions like disinfectants, antibiotics, bacterial colonization, the identification of bothersome species, the establishment of normal intestinal flora, as well as the prevention of harmful intestinal flora, the QS system enables communication between intraspecies and interspecies. Many bacteria with therapeutic value employ QS to control the total synthesis of virulence factors. QS is a product of several processes, including AI development, detection, and reaction, in Gram-positive bacteria. Numerous Gram-positive bacteria contain oligopeptide autoinducing peptides (AIPs), which are recognized by membrane-bound two-component signal transduction systems [70]. These pro-AIPs are encoded as precursors and exhibit sequence diversity.

5.5 Biofilm-forming bacteria

Nearly all microorganisms (99.9%) have the potential to build biofilms on a variety of surfaces, including biological and inert surfaces [20]. Biofilms are collections of EPS-producing microorganisms that adhere to surfaces. Because of its antibiotic resistance and disorders linked to indwelling medical devices, biofilm poses a serious threat to public health. According to the research by Sekhar et al. [20], *H. influenzae* can form a biofilm inside a human body while eluding the immune system. In addition to *P. aeruginosa, S. epidermidis, E. coli* spp., *S. aureus, Enterobacter cloacae*, and *K. pneumoniae*, numerous other bacterial species have also been found to be capable of producing biofilms (Table 5.2). Below are a few descriptions of the bacterial species that create biofilms.

Table 5.2: List of common bacterial species that produce biofilms [19].

S. no.	Common biofilm-forming bacterial species
1	*E. coli*
2	*P. aeruginosa*
3	*S. epidermidis*
4	*S. aureus*
5	*E. cloacae*
6	*K. pneumoniae*
7	*Actinomyces israelii*
8	*Haemophilus influenzae*
9	*Burkholderia cepacia*
10	*Staphylococcus epidermidis*
11	*Listeria monocytogenes*

5.5.1 Epstein-Barr virus

Numerous nosocomial and public illnesses, including prostatitis and urinary tract infections, are brought on by the rod-shaped Gram-negative bacterium *E. coli*. It can produce biofilm in addition to secreting polysaccharides and toxins. It can also create biofilm in a lab [19]. High-molecular-weight *E. coli* capsules are affixed to the cell surface. By preventing bacterial surface attachment, *E. coli* capsules indirectly promote the growth of biofilm. The ability of *E. coli* to generate biofilms is influenced by a variety of environmental factors [72]. Due to the presence of exopolymers, the *E. coli* biofilm can be hundreds of microns thick, making antibiotic treatment difficult.

5.5.1.1 *Pseudomonas aeruginosa*

Along with other *Pseudomonas species, P. aeruginosa* is a well-known Gram-negative opportunistic pathogen that is found naturally on the human skin [21]. *P. aeruginosa*, a common human pathogenic bacterium, can be isolated from a variety of sources, including people, plants, and animals. Because of its propensity to produce biofilm, it is known that this type of biofilm contributes to persistent infections. Silver has the ability to eliminate *P. aeruginosa* biofilm. It has been discovered that the bactericidal concentration of antibacterial silver is 10–100 times higher for the destruction of bacterial biofilms than for the elimination of planktonic microorganisms. *P. aeruginosa* produces biofilms, which are organized collections of several cell types wrapped in an extracellular polysaccharide matrix. Antibiotic resistance is not at all present in them. As a result of their initial adherence to a solid surface, which starts the creation of biofilms, microcolonies are formed. These microcolonies develop into mature biofilms that are exopolysaccharide-coated. *P. aeruginosa* is a multidrug-resistant bacterium when it comes to the antibiotic ciprofloxacin, which is frequently used to treat lung infections [65].

5.5.1.2 *Staphylococcus aureus*

The bacterium *Staphylococcus aureus*, which is multidrug resistant, is the cause of a number of nosocomial illnesses. A biofilm develops on catheters and open wounds [22]. To put together, the extracellular matrix in the cytoplasm, *S. aureus* recycles proteins. When forming biofilms in polluted environments, *S. aureus* is more adaptable and flexible because of the cytoplasmic proteins' dual roles as matrix proteins and cytoplasmic proteins. In chronic wounds, they might also promote the development of mixed-species biofilms [23].

5.5.1.3 *Streptococcus epidermidis*

Streptococcus epidermidis, a well-known opportunistic pathogen, is more likely to infect immunocompromised persons, intravenous drug users, people with AIDS, those taking immunosuppressive medications, and premature infants [24]. The propensity of *S. epidermidis* to produce biofilms increases the risk of contamination after surgical implantation of polymeric devices. The pathophysiology of *S. epidermidis* infections associated with medical devices is ultimately mediated by biofilm formation. The primary mechanism by which staphylococci to promote attachment is polysaccharide intercellular adhesion (PIA). The most crucial element of the extracellular matrix in *Staphylococcus* cells is PIA, which serves as a barrier and holds the cells together. Rohde et al. [25] found that 27% of *S. epidermidis* strains that generated biofilms when they were isolated from infections of prosthetic joints did so without the assistance of PIA.

5.5.1.4 *Enterobacter cloacae*

Numerous nosocomial infections in humans are brought on by the Gram-positive bacterium *E. cloacae*, including infections of the lower respiratory tract, bacteremia, urinary tract, endocarditis, intra-abdominal infections, septic arthritis, skin and soft tissue infections, osteomyelitis, and ocular infections. The most often isolated species that causes nosocomial infections is *Enterobacter*, which has recently become a significant pathogenic bacterium [26]. *E. cloacae* causes bloodstream infections, which are a major cause of disease and mortality in both industrialized and poor nations. Additionally, it leads to diseases like biliary and urinary tract infections that are linked to biofilms. Additionally, both ampicillin and narrow-spectrum cephalosporins inherently inhibit the growth of *Enterobacter*. Cephalosporin mutations with extended and broad spectrums can also happen often. Along with having β-lactamase, it also exhibits resistance to third-generation cephalosporins.

5.5.1.5 *Klebsiella pneumoniae*

Nosocomial infections are frequently brought on by *K. pneumoniae*, a member of the *Klebsiella* genus that is Gram-negative. A sizeable part of nosocomial infections, including pneumonia, septicemias, and soft tissue infections, is caused by *K. pneumoniae*, a highly significant species within the genus *Klebsiella* [27]. *K. pneumoniae* biofilm obtained from clinical strains was examined using field emission scanning electron microscopy and confocal laser scanning microscopy. Over 40% of the *K. pneumoniae* strains studied in this study were found to be able to form biofilms when isolated from human waste products such as urine, blood sputum, and wound swabs. However, as the temperature rose from 35 to 40 °C, biofilm continued to develop on inanimate surfaces. Additionally, many *K. pneumoniae* strains can successfully produce biofilms as opposed to just one strain [28].

5.6 Biofilms and antibiotic resistance

Mechanisms of antibiotics and biocides' efflux mechanisms, direct active molecule inactivation, changing the body's sensitivity to the target of action, and lowering the drug concentration before reaching the target site are the four different types of biofilm resistance. Antibiotic resistance in biofilms can change over time in both its severity and its underlying causes. Despite the fact that this evidence cannot be discounted in terms of resistance in the production of adherent cells, the main data points to the inability of conventional processes to account for the high resistance to antibacterial agents associated with biofilms. As a result, it has been hypothesized that the resistance provided by biofilms or adherent bacteria may possess some intrinsic processes and serve as the foundation of traditional antibiotic resistance [69]. The high resilience of biofilms has been attributed to a number of causes (Figure 5.3). These include (a) limited diffusion, (b) enzyme neutralizations, (c) heterogeneous functions, (d) sluggish growth rate, (e) the existence of persistent (nondividing) cells, (f) and the biofilm phenotype, which includes adaptive mechanisms including efflux pumps and membrane modification.

ANTIBIOTICS RESISTANCE ASSOCIATED TO BIOFILMS

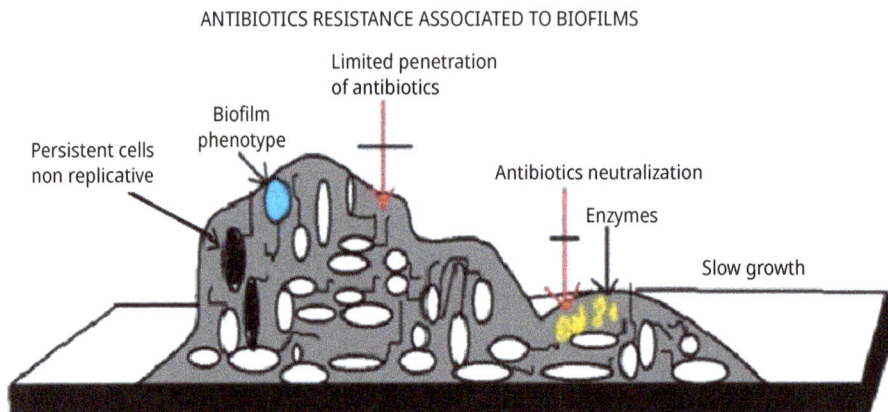

Figure 5.3: Biofilm is linked to antibiotic resistance, a description of the main mechanisms underlying antibiotic resistance, including enzyme neutralizations, the presence of persistent (nondividing) cells, and the biofilm phenotype [29].

5.6.1 Low penetration of antibiotics

Antibiotics may diffuse through the matrix of the biofilm. Exopolysaccharide functions as a physical barrier, which has an impact on how well antibiotics diffuse or penetrate into the deeper layers of biofilm. When molecules come into direct touch with this matrix, which slows down their movement to the biofilm's center, antibiotic resistance develops. This may also hamper high-molecular-weight substances like complement system proteins and lysozyme, and bacteria in liquid culture are more susceptible to medication exposure than those in biofilms with compact structures. Bacteria that are attached to biofilms but do not produce polysaccharides and are easily targeted by immune system cells can get away. When antibiotics bind to the biofilm matrix, they lose their effectiveness. *P. aeruginosa* contains the anionic exopolysaccharide known as alginate. The fact that fluoroquinolones and aminoglycosides only partially permeate the skin is explained by the existence of this matrix. Antibiotic penetration is insufficient to fully explain biofilm resistance; other mechanisms are probably at work. Also, it has recently been suggested that patients have plenty of time to build a defensive reaction to stress due to the slow dispersion of antibiotics [30].

5.6.2 Neutralization by enzymes

Antibiotic resistance in biofilm may be brought on by the presence of neutralizing enzymes that destroy or render antibiotics inactive. These enzymes are proteins that boost resistance by modifying and hydrolyzing antimicrobials biochemically. These enzymes from the biofilm surface build up in the glycocalyx due to antibiotics. Delayed antibiotic absorption and antibiotic degradation in the biofilm both heighten enzyme neutralization. The reason of antibiotic resistance in cystic fibrosis, which is brought on by *P. aeruginosa*, is excessive production of cephalosporinase AmpC enzymes. This enzyme confers β-lactam resistance when there are higher quantities of carbapenems present [18]. Growth was observed in a study utilizing filters impregnated with antibiotics on *K. pneumoniae* biofilm (mutant cells – lactamases), indicating that there may be another route of resistance that needs to be investigated despite efficient drug diffusion.

5.6.3 Heterogeneous nature

A microelectrode with probes was used to directly measure the oxygen concentration in distinct biofilm locations in studies to evaluate the rate of microbial proliferation in biofilms. Both aerobic and anaerobic processes occur simultaneously in the biofilms due to their metabolic and structural variability. As a result, antibiotic resistance may differ depending on where in the biofilms it is discovered. On the surface of biofilms, antibiotics are very active, but as the biofilm grows slowly or not at all, the sensitivity of the cells to antimicrobials declines [31]. We can better understand how differently the microbial communities in the various layers of the biofilm are responsive to various antibiotic treatments by looking at their aerobic or facultative anaerobic microbial populations. Antibiotics affect planktonic forms differently than they do adherent cells. Aminoglycosides are affected by oxygen limitations, and oxygen availability and pH gradients have an impact on microbe development in anaerobic environments.

5.6.4 Cell's slow rate of growth

Because there are not enough resources for microorganisms to grow quickly, they become resistant to antibiotics. In the case of a biofilm, a gradient of nutrients results in metabolically active cells at the exterior or surface layer and dormant cells at the core [32]. Only developing bacteria are impacted by penicillin and ampicillin. Other antibiotics that target cells in the stationary phase include fluoroquinolones, aminoglycosides, aminoglycosides, and cephalosporin. Hence, due to delayed development, resistance to various antibiotics has been found in several bacterial strains, including *E. coli* resistance to cetrimide, *P. aeruginosa* resistance to piperacillin and tobramycin, and *S. epidermidis* resistance to ciprofloxacin. It has been determined that delayed development was the cause of this resistance. During the host's innate immune response, specific naturally occurring peptides are produced that have antibacterial properties and defend the body [33]. The peptide polymyxin E can successfully treat both cancer patient biofilms and *P. aeruginosa*-induced cystic fibrosis. Patients with cystic fibrosis can use ciprofloxacin and tetracycline to remove active developing cells, and it is hypothesized that combining colistin with the other two drugs will be very effective at removing *P. aeruginosa* [34].

5.6.5 Existence of persistent cells

Only a small portion of the bacteria, referred to as persistent cells, remain after the biofilm has been treated with antibiotics. These cells may or may not transmit their resistance to their progeny and revert to their initial state after the stress or pressure has been relieved. The persistent cells temporarily stop replicating for a short time to ensure the survival of the community. Their coping mechanism is different from how stressed-out cells behave (environmental damage). Persistent cells can tolerate many antibiotic doses as they struggle to survive. Persistent cells increase in number when the density of bacterial cells in the stationary phase is at its highest, indicating that their principal function is survival. There are several signs that persistent cells are present in biofilms, such as the following: (a) the presence of a biphasic dimension in biofilms, which means that while a sizable portion of the cell population is attacked, the remaining portion is not attacked (resistant) even after receiving a strong dose of antibiotics; (b) the role of persistence genes as regulatory circuits; and (c) the capacity of bacteriostatic antibiotics to encourage the development of persistent cells [73].

5.6.6 Biofilm phenotype

By creating biofilms, bacteria produce a range of compounds referred to as secondary metabolites. These components are not necessary for the cell to grow. As signaling molecules, these metabolites aid in the formation of biofilms. The biofilm phenotype is defined as community cells that do not react to antibiotic therapy. These characteristics have raised the possibility that specific genes exist. When comparing the gene expression of *B. subtilis* biofilm and their planktonic culture cells, DNA microarrays only found a 6% difference, whereas *P. aeruginosa* showed a 1% difference. Unfortunately, no helpful data for characterizing this process has yet been produced by this differential gene expression [11].

5.6.7 Efflux pumps

Efflux pumps are protein structures that can either express constantly or seldom. These pumps might have substrate-specific characteristics. Similar molecules can be moved by these pumps, which may contribute to multidrug resistance [35]. Efflux pumps found in the periplasm of bacteria are in charge of the antagonistic accumulation of antibiotics. They demonstrate resistance to a number of antibiotics, such as tetracycline, macrolides, fluoroquinolones, and β-lactam, bringing their concentration to a dangerously low level. It has been discovered that prokaryotes contain efflux transporters from five different families. Excessive expression of the efflux pumps in *P. aeruginosa* biofilms has been linked to antibiotic resistance [36].

5.6.8 Modifications to membrane proteins

The permeability of the outer membrane is essential for the diffusion of antibiotics through several channels. The outer membrane channel proteins (porins) of Gram-negative bacteria are essential for this process because they move hydrophilic molecules from the external environment to the periplasm. Mutations in the genes that create porins can lead to the production of altered or dysfunctional proteins. Compared to normal porins, these mutant porins are less permeable to hydrophobic substances [37]. *P. aeruginosa* contains a specific porin termed OprD that enhances the absorption of imipenem and basic amino acids. OprD loss, which results in the breakdown of the three-dimensional imipenem molecule, is the root cause of imipenem resistance in *P. aeruginosa*. The uneven expression of porin-coding genes in biofilm leads to antibiotic resistance. When the expression of the ompC gene and three other (osmotically regulated) genes is increased, the bacterial cells form biofilms.

5.6.9 Phase change

The varied phenotype present in biofilms is crucial, which gives rise to infections with resistance. Only a handful of the numerous genera and species for which writers have documented this behavior include *Pseudomonas*, *Staphylococcus*, and specific Enterobacteriaceae species. Small colony variants, or SCVs, are little colonies formed by bacteria in biofilms that go inactive. Antibiotics that are dependent on the growth phase have a significantly lower chance of killing these mutant cells, and likewise suffer from a catalase deficit, impairing oxidative metabolism [71]. Higher levels of adhesion, autoaggregation, hydrophobicity, and low-level motility are observed in biofilms where identifiable colony morphological alterations caused by SCVs are present. It is an essential component of biofilms' survival mechanisms since it can tolerate a variety of extremely stressful environmental conditions. Phase variation was once believed to be the result of cellular internal rearrangement, but it is now believed to be the result of the interaction of genetic elements.

5.7 Food and food biofilm portals of contact

Surfaces of equipment used to prepare and handle food and beverages are commonly contaminated by germs (perhaps including hazardous pathogens) even after washing and disinfection procedures [38]. In the vicinity of the food industry, biofilms are quickly created by treating the material's surface and the reversible binding of cells to that surface. As a result, the binding solidifies, and microcolonies begin to grow. Normally, bacteria cling to surfaces and form biofilms, which are EPSs that include spatially structured populations [39]. In the food industry, biofilms, which are microorganisms that live on surfaces and increase cross-contamination of food, may alter the dynamics of cleaning and disinfection. A biofilm is a grouping of bacteria that is kept in an extracellular polymeric material matrix and is firmly attached to a surface. It poses a serious threat to the food sector. To prevent biofilm formation, remove bacteria from irreversible attachment stages, and stop them from increasing surface adhesion, bacteria attachments must be eliminated with a potent disinfectant [40].

Biofilms have become more resilient to disinfectants in a wide variety of food industries, including the processing of seafood, beverages, brewing, dairy products, meat, and poultry. Strong evidence suggests that their biofilm lifestyle, especially when compared to planktonic cells, makes them more resistant to antimicrobial treatments (Figure 5.4).

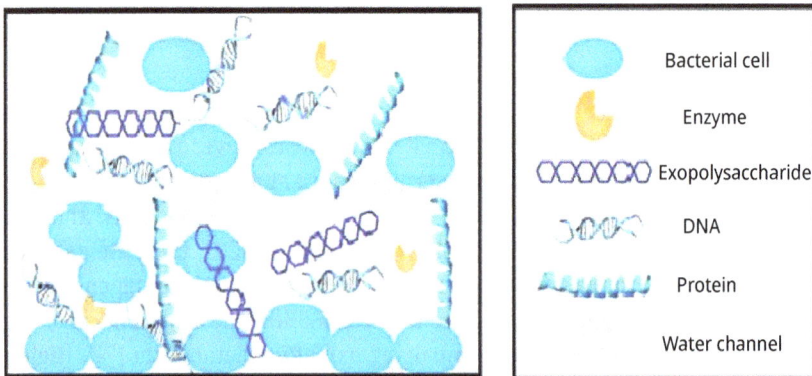

Figure 5.4: Structure of biofilms during growth and maturation phases [41].

Species that produce biofilm structures can be harmful to humans and are seen in manufacturing settings in the food business. These pathogens use the artificial substrates used in food preparation, including wood, glass, stainless steel, polyethylene, rubber, and polypropylene [42]. The characteristics of the bacterial growth form on food in a processing environment involve several behaviors when cleaning and disinfection processes are taken into account. In order to properly manage biofilm growth in the food industry, the proper tactic must be used. A list of pertinent biofilm-forming microorganisms in the food industry is given in Table 5.3.

Table 5.3: Biofilm-forming pathogens in the food industry.

Pathogen	Characteristics	Contaminated food	Examples of harmful spoilage effects	References
Listeria monocytogenes	Gram-positive, rod-shaped, facultative anaerobic	Meat, poultry, candied apples, frozen vegetables, soft cheeses, fruit, ready-to-eat foods, and unpasteurized milk	Aged, expectant, and immune-compromised individuals are at risk for listeriosis	[43]
Bacillus cereus	Facultatively anaerobic, spore-forming, and Gram-positive	Dairy products, rice, vegetables, and meat	Diarrhea and vomiting symptoms	[44]
Campylobacter jejuni	Aerobic, anaerobic, and Gram-negative	Poultry, unpasteurized milk, and animals	Fever, nausea, vomiting, stomach cramps, and bloody diarrhea	[45]
Escherichia coli	Gram-negative, rod-shaped	Fresh meat, fruits, and veggies, and raw milk	Diarrhea outbreaks and hemolytic uremic syndrome	[46]
Salmonella enterica	Gram-negative, rod-shaped, flagellate, facultative aerobic	Poultry meat, bovine, ovine, porcine, and fish	Can cause gastroenteritis or septicemia	[47]
Staphylococcus aureus	Gram-positive, nonspore-forming, nonmotile, facultative anaerobic	Meat, poultry, egg, dairy, salads, bakery items, particularly cream-filled pastries and cakes, and sandwich fillings	Methicillin resistance, can cause vomiting and diarrhea	[5]
Pseudomonas spp.	Psychrotrophic, motile, Gram-negative rod-shaped	Veggies, fruits, cuts of meat, and low-acid dairy products	Produces blue discoloration on fresh cheese	[48]
Pectinatus spp.	Gram-negative, nonspore-forming, anaerobic	Beer and brewery environment	Beer becomes murky and smells like rotten eggs due to rapid cell growth, which produces sulfur compounds	[49]

5.8 Control of biofilms in food industry

Today, many physical and chemical techniques are utilized to restrict the growth of biofilm in the food industries. The physical methods involve ultrasonication and hot steam in contrast to the chemical treatments, which make use of chemicals like sodium hypochlorite, peracetic acid, hydrogen peroxide, and sodium hydroxide solutions. These control approaches are also appropriate for use on working surfaces and pipes. Studies have shown that sodium hypochlorite, peracetic acid, and iodine are among the most effective. In order to effectively dissolve the biofilm, the exposure duration must be increased for at least 5 min. Thereafter, the equipment must be thoroughly cleaned with a brush and water to remove all traces of germs. Together with effective tepid water cleaning, bacteria at pipeline dead ends are also eradicated utilizing chemical techniques. Many methods for regulating biofilms have been documented in the literature; some of them are described here.

5.8.1 Pilicides

Pili are extracellular fibers made by bacteria that facilitate surface colonization and adhesion [50]. The chaperone-usher approach is a specific method used to assemble these pili [51]. Disrupting this pili-construction mechanism is a novel and workable method due to its potential application in the removal of biofilms. Moreover, pilicides, which are tiny synthetic compounds, have been developed by scientists to stop the growth of pili.

5.8.2 Enzymes

One more effective technique for eliminating biofilm is the use of enzymes. Given that the biofilm is composed of EPSs, these enzymes may be able to break down EPSs. The bulk of EPS and bacteria are found in biofilm. Planktonic cells are released when the biofilm is broken down by enzymes [52].

5.8.3 Blocking of quorum sensing

Several signaling pathways are required for bacterial biofilm formation and human drug resistance. One of them is the exchange of proteins or tiny organic molecules, as well as electrical signals [53]. Using QS inhibitors is the unique strategy for managing biofilm that has received the most investigation. For the past 10 years, numerous researchers have been searching for compounds that could suppress the QS system [54].

5.8.4 Currents in electricity

Many bacterial species have been shown to be susceptible to electrical current's bactericidal effects [55]. A study utilizing low-intensity current revealed a considerable reduction in the amount of both live *Staphylococcus* and *Pseudomonas* biofilm bacteria. Using electrical currents, electromagnetic fields, and ultrasound has enhanced biofilm removal results in both in vivo and in vitro research [56].

5.8.5 Exterior coatings

One of the most effective ways to remove or prevent bacterial biofilm from surfaces is to coat objects with metals, antiseptics, or antimicrobials [57].

Different antibacterial and antibiofilm substances are required since biofilms cannot be removed using conventional management methods. One approach focuses on nanoscale agents, which has the potential. Due to their unique properties, nanoparticles differ from their bulk chemical counterparts [46]. Due to their potent antibacterial properties, silver compounds have been used to prevent microbial infections since the ancient Egyptians, Greeks, and Romans, for example [58].

5.8.6 Bacteriophages

Use of bacteriophages is a novel and efficient technique for removing or eradicating biofilm [59]. The ability of bacteriophages to stop or slow down in vivo biofilm formation has also been proven. A genetically altered lytic phage containing a biofilm-degrading enzyme outperformed wild-type phages at eradicating biofilms in a research [12]. Similar to this, phage cocktails (a collection of several phages) can be used to totally eradicate bacterial biofilms [60].

Nonthermal plasma is a low-temperature, partly ionized gas with intriguing antibacterial properties. It is made by mixing UV light with oxygen, nitrogen, ozone, water, and helium at atmospheric pressure and under an electrical discharge. Although it can remove Gram-positive (*Bacillus* spp.) or Gram-negative (*Pseudomonas* spp.) bacteria biofilms in about 10 min, its use is now restricted to certain laboratory applications due to its high cost [61].

5.8.7 Using cold plasma in the food industry

With a focus on microbe control, enzyme inactivation, pesticide breakdown, allergen mitigation, and so on, the use of cold plasma in the food business is growing. This is because a wide range of reactive components are employed in the plasma-generating process and designs for configurations that create plasma are developed quickly. Being the primary path of cold plasma application in food-related industries, it has been amply demonstrated to be a successful technique to obtain food decontamination caused by numerous diseases and spoilage microorganisms. Several cold plasma-generation systems with diverse configuration designs have been studied and optimized recently for surface microbial decontamination of food goods [62]. Cold plasma is increasingly being used in the food sector for purposes other than antibacterial control, such as inactivating enzymes on food. The primary, secondary, and tertiary structures of different enzymes determine how they function, and interactions between protein molecules and reactive cold plasma are frequently what cause changes to different protein structures [63].

5.9 Conclusion

Understanding how bacteria behave during biofilm development and why they could be resistant to control measures requires a close examination of this bacterial biofilm. Food sector biofilms pose a significant financial and health risk. Understanding the diverse gene expression and regulation during biofilm growth will require a thorough molecular investigation. Biofilm growth can be regulated or avoided by focusing on these several molecular processes. In order to penetrate biofilms and reach bacteria that are deeply embedded in them, new application tactics are needed because biofilms exhibit a variety of resistance due to chemicals' restricted penetration. The major defenses against these are QS inhibitors, which are used to limit the development of bacterial biofilms and prevent the synthesis of polysaccharides [64]. Numerous researches have cited bacteriophages as a viable strategy for controlling biofilm disruption. Simply said, a deeper comprehension of bacterial biofilm is necessary to build creative, efficient management strategies that eventually lead to sanitary food sector products.

References

[1]   Muhammad, M. H., Idris, A. L., Fan, X., Guo, Y., Yu, Y., Jin, X., Qiu, J., Guan, X., & Huang, T. (2020). Beyond risk: bacterial biofilms and their regulating approaches. *Frontiers in Microbiology*, 11, 928. doi: 10.3389/fmicb.2020.00928.

[2]   Reygaert, W. C. (2018). An overview of the antimicrobial resistance mechanisms of bacteria. *AIMS Microbiology*, 4(3), 482–501. doi: 10.3934/microbiol.2018.3.482. PMID: 31294229; PMCID: PMC6604941.

[3]   Bisht, K., & Wakeman, C. A. (2019). Discovery and therapeutic targeting of differentiated biofilm subpopulations. *Frontiers in Microbiology*, 10, 1908. doi: 10.3389/fmicb.2019.01908.

[4]   Costerton, J. W. (1999). Introduction to biofilm. *International Journal of Antimicrobial Agents, 11*(3–4), 217–221.

[5]   Giaouris, E., Heir, E., Desvaux, M., Hébraud, M., Møretrø, T., Langsrud, S., Doulgeraki, A., Nychas, G. J., Kačániová, M., Czaczyk, K., Ölmez, H., & Simões, M. (2015). Intra- and inter-species interactions within biofilms of important foodborne bacterial pathogens. *Frontiers in Microbiology, 20*(6), 841, doi: 10.3389/fmicb.2015.00841, PMID: 26347727; PMCID: PMC4542319.

[6]   Liu, C., Sun, D., Zhu, J., Liu, J., & Liu, W. (2020). The regulation of bacterial biofilm formation by cAMP-CRP: a mini-review. *Frontiers in Microbiology, 11*, 802. doi: 10.3389/fmicb.2020.00802.

[7]   Hentzer, Á., Eberl, Á., & Givskov, Á. (2005). Transcriptome analysis of *Pseudomonas aeruginosa* biofilm development: anaerobic respiration and iron limitation. *Biofilms, 2*(1), 37–61.

[8]   Arunasri, K., & Mohan, S. V. (2019). Biofilms: microbial life on the electrode surface. In *Microbial electrochemical technology* (pp. 295–313). Elsevier.

[9]   Su, C. C., Binu, K., Thomas, S., Van der maarel, J. R. C., Yang, L., Rice, S. A., Doyle, P., & Kjelleberg, S. (2014). Dynamic remodeling of microbial biofilms by functionally distinct exopolysaccharides. *mBio, 5*(4), e01536–14. doi: https://doi.org/10.1128/mBio.01536–14.

[10] Quan, K., Hou, J., Zhang, Z., Ren, Y., Peterson, B. W., Flemming, H.-C., Christian, M., Busscher, H. J., & Van der mei, H. C. (2021). Water in bacterial biofilms: pores and channels, storage and transport functions. *Critical Reviews in Microbiology*. doi: 10.1080/1040841X.2021.1962802.

[11] Fux, C. A., Costerton, J. W., Stewart, P. S., & Stoodley, P. (2005). Survival strategies of infectious biofilms. *Trends in microbiology, 13*(1), 34–40.

[12] Lu, T. K., & Collins, J. J. (2007). Dispersing biofilms with engineered enzymatic bacteriophage. *Proceedings of the National Academy of Sciences. 104*(27), 11197–11202.

[13] Okada, M., Sato, I., Cho, S. J., Iwata, H., Nishio, T., Dubnau, D., & Sakagami, Y. (2005). Structure of the *Bacillus subtilis* quorum-sensing peptide pheromone ComX. *Nature. Chemistry & Biology, 1*(1), 23–24.

[14] Rivera, E. J., Marcy, G. W., Butler, R. P., Fischer, D., Vogt, S. S., & Lissauer, J. J. (2001). A pair of resonant planets orbiting GJ 876. *The Astrophysical Journal, 556*(1), 296.

[15] Donlan, R. M., & Costerton, J. W. Biofilms: survival mechanisms of clinically relevant microorganisms. *Clinical Microbiology Reviews, 15*(2), 167–193. doi: 10.1128/CMR.15.2.167-193.2002. PMID: 11932229; PMCID: PMC118068.

[16] Trunk, T., Khalil, H. S., & Leo, J. C. (2018). Bacterial autoaggregation. *AIMS Microbiology, 4*(1), 140–164. doi: 10.3934/microbiol.2018.1.140. PMID: 31294207; PMCID: PMC6605025.

[17] Davey, M. E., & O'toole, G. A. (2000). Microbial biofilms: from ecology to molecular genetics. *Microbiology and Molecular Biology Reviews, 64*(4), 847–867. doi: 10.1128/MMBR.64.4.847-867.2000. PMID: 11104821; PMCID: PMC99016.

[18] Pandey, N., & Cascella, M. (2022). Beta lactam antibiotics. In *StatPearls*, Treasure Island (FL): StatPearls Publishing; PMID: 31424895.

[19] Naves, P., Del Prado, G., Huelves, L., Rodriguez-Cerrato, V., Ruiz, V., Ponte, M, C., & Soriano, F. (2010). Effects of human serum albumin, ibuprofen and N-acetyl-L-cysteine against biofilm formation by pathogenic Escherichia coli strains. *Journal of Hospital Infection, 76*(2), 165–170.

[20] Sekhar, S., Kumar, R., & Chakraborti, A. (2009). Role of biofilm formation in the persistent colonization of *Haemophilus influenzae* in children from northern India. *Journal of. Medical Microbiology, 58*(11), 1428–1432.

[21] Pressler, T. E, Jensen, P. Ø., Bjarnsholt, T., Phipps, R., Rasmussen, T. B., Calum, H., Christoffersen, L., & Høiby, N. (2007). Rapid necrotic killing of polymorphonuclear leukocytes is caused by quorum-sensing-controlled production of rhamnolipid by Pseudomonas aeruginosa. *Microbiology, 153*(5), 1329–1338.

[22] Vuong, C., & Otto, M. (2002). *Staphylococcus epidermidis* infections. *Microbes and Infection, 4*(4), 481–489.

[23] Lucy, M. C., Butler, S. T., & Garverick, H. A. (2014). Endocrine and metabolic mechanisms linking postpartum glucose with early embryonic and foetal development in dairy cows. *Animal, 8*(1), 82–90.

[24] Otto, M., Süßmuth, R., Jung, G., & Götz, F. (1998). Structure of the pheromone peptide of the *Staphylococcus epidermidis* agr system. *FEBS Letters, 424*(1–2), 89–94.

[25] Rohde, H., Burandt, E. C., Siemssen, N., Frommelt, L., Burdelski, C., Wurster, S., & Mack, D. (2007). Polysaccharide intercellular adhesin or protein factors in biofilm accumulation of *Staphylococcus epidermidis* and *Staphylococcus aureus* isolated from prosthetic hip and knee joint infections. *Biomaterials, 28*(9), 1711–1720.

[26] Ma, L., Conover, M., Lu, H., Parsek, M. R., Bayles, K., & Wozniak, D. J. (2009). Assembly and development of the *Pseudomonas aeruginosa* biofilm matrix. *PLoS Pathogens, 5*(3), e1000354.

[27] Ellis, M. E. (1998). *Gram-negative bacillary pneumonia* (pp. 136–139). Cambridge: Cambridge University Press.

[28] Chung, P. Y. (2016). The emerging problems of Klebsiella pneumoniae infections: carbapenem resistance and biofilm formation. *FEMS Microbiology Letters, 363*(20), fnw219.

[29] Hogan, D., & Kolter, R. (2002). Why are bacteria refractory to antimicrobials?. *Current Opinion in Microbiology, 5*(5), 472–477.

[30] Mah, T. F. C., & O'toole, G. A. (2001). Mechanisms of biofilm resistance to antimicrobial agents. *Trends in Microbiology. 9*(1), 34–39.

[31] Stewart, P. S., & Franklin, M. J. (2008). Physiological heterogeneity in biofilms. *Nature Reviews Microbiology, 6*(3), 199–210.

[32] Munita, J. M., & Arias, C. A. (2016). Mechanisms of antibiotic resistance. *Microbiology Spectrum, 4*(2), 10.1128/microbiolspec.VMBF-0016-2015, doi: 10.1128/microbiolspec.VMBF-0016-2015. PMID: 27227291; PMCID: PMC4888801.

[33] Zharkova, M. S., Orlov, D. S., Golubeva, O. Y., Chakchir, O. B., Eliseev, I. E., Grinchuk, T. M., & Shamova, O. V. (2019). Application of Antimicrobial Peptides of the Innate Immune System in Combination With Conventional Antibiotics – A Novel Way to Combat Antibiotic Resistance? *Front Cellular and Infection Microbiology, 9*, 128. doi: 10.3389/fcimb.2019.00128.

[34] Pamp, S. J., Gjermansen, M., Johansen, H. K., & Tolker-Nielsen, T. (2008). Tolerance to the antimicrobial peptide colistin in *Pseudomonas aeruginosa* biofilms is linked to metabolically active cells, and depends on the pmr and mexAB-oprM genes. *Molecular microbiology, 68*(1), 223–240.

[35] Abdi, S. N., Ghotaslou, R., Ganbarov, K., Mobed, A., Tanomand, A., Yousefi, M., Asgharzadeh, M., & Kafil, H. S. (2020). *Acinetobacter baumannii* efflux pumps and antibiotic resistance. Infections and Drug Resistance, 13, 423–434. doi: 10.2147/IDR.S228089. PMID: 32104014; PMCID: PMC7024869.

[36] Zheng, P., Renee, R., Bernard, R. G., Tong-junlin, & Zhenyu, C. (2019). Antibiotic resistance in *Pseudomonas aeruginosa*: mechanisms and alternative therapeutic strategies. *Biotechnology Advances, 7*(1), 177–192.

[37] Hancock, R. E. (1997). The bacterial outer membrane as a drug barrier. *Trends in Microbiology, 5*(1), 37–42.

[38] Marouani-Gadri, N., Augier, G., & Carpentier, B. (2009). Characterization of bacterial strains isolated from a beef-processing plant following cleaning and disinfection-Influence of isolated strains on biofilm formation by Sakai and EDL 933 *E. coli* O157: H7. *International Journal of Food Microbiology, 133*(1–2), 62–67.

[39] Satpathy, S., Sen, S. K., Pattanaik, S., & Raut, S. (2016). Review on bacterial biofilm: an universal cause of contamination. Biocatalysis & Agricultural Biotechnology, 7, 56–66. doi: 10.1016/j. bcab.2016.05.002.

[40] Carrascosa, C., Raheem, D., Ramos, F., Saraiva, A., & Raposo, A. (2021). Microbial biofilms in the food industry – a comprehensive review. *International Journal of Environmental Research and Public Health, 18*(4), 2014. doi: 10.3390/ijerph18042014. PMID: 33669645; PMCID: PMC7922197.

[41]   Rabin, N., Zheng, Y., Opoku-Temeng, C., Du, Y., Bonsu, E., & Sintim, H. O. (2015). Biofilm formation mechanisms and targets for developing antibiofilm agents. *Future Medicinal Chemistry*, *7*, 493–512. doi: 10.4155/fmc.15.6.

[42]   Colagiorgi, A., Bruini, I., Di Ciccio, P. A., Zanardi, E., Ghidini, S., & Ianieri, A. (2017). Listeria monocytogenes biofilms in the wonderland of food industry. *Pathogens*, *6*, 41. doi: 10.3390/pathogens6030041.

[43]   Rothrock, M. J., Davis, M. L., Locatelli, A., Bodie, A., McIntosh, T. G., Donaldson, J. R., & Ricke S, C. (2017). Listeria Occurrence in Poultry Flocks: Detection and Potential Implications. *Frontiers in Veterinary Science*, *4*. doi: 10.3389/fvets.2017.00125.

[44]   Grigore-Gurgu, L., Bucur, F. I., Borda, D., Alexa, E. A., Neagu, C., & Nicolau, A. I. (2019). Bacterial Biofilms. In *IntechOpen*. London, UK: Biofilms Formed by Pathogens in Food and Food Processing Environments.

[45]   Chlebicz, A., & Śliżewska, K. (2018). Campylobacteriosis, salmonellosis, yersiniosis, and listeriosis as zoonotic foodborne diseases: a review. *International Journal of Environmental Research and Public Health*, 15, 863. doi: 10.3390/ijerph15050863.

[46]   Galié, S., García-Gutiérrez, C., Miguélez, E. M., Villar, C. J., & Lombó, F. (2018). Biofilms in the food industry: health aspects and control methods. *Frontiers in Microbiology*, *9*, 898. doi: 10.3389/fmicb.2018.00898.

[47]   Nguyen, H. D. N., Yang, Y. S., & Yuk, H. G. (2014). Biofilm formation of *Salmonella Typhimurium* on stainless steel and acrylic surfaces as affected by temperature and pH level. *LWT Food Science and Technology*, *55*, 383–388. doi: 10.1016/j.lwt.2013.09.022.

[48]   González-Rivas, F., Ripolles-Avila, C., Fontecha-Umaña, F., Ríos-Castillo, A. G., & Rodríguez-Jerez, J. J. (2018). Biofilms in the spotlight: detection, quantification, and removal methods. *Comprehensive Reviews in Food Science and Food Safety*, *17*, 1261–1276. doi: 10.1111/1541-4337.12378.

[49]   Van Houdt, R., & Michiels, C. W. (2010). Biofilm formation and the food industry, a focus on the bacterial outer surface. *Journal of Applied Microbiology*, *109*, 1117–1131. doi: 10.1111/j.1365-2672.2010.04756.x.

[50]   Roberts, J. J., Zhang, X., Perricaudet, M., Mallet, J., & Le Gal La Salle, G. (1994). The use of adenovirus vectors for intracerebral grafting of transfected nervous cells. *Neuroreport*. 5(7), 801–804.

[51]   Surette, M. G., & Bassler, B. L. (1998). Quorum sensing in *Escherichia coli* and *Salmonella typhimurium*. *Proceedings of the National Academy of Sciences*, *95*(12), 7046–7050.

[52]   Xavier, J. B., Picioreanu, C., Rani, S. A., Van Loosdrecht, M., & Stewart, P. S. (2005). Biofilm control strategies based on enzymic disruption of the extracellular polymeric substance matrix – a modeling study. *Microbiology*, *151*(12), 3817–3832.

[53]   Kim, M. K., Ingremeau, F., Zhao, A., Bassler, B. L., & Stone, H. A. (2016). Local and global consequences of flow on bacterial quorum sensing. *Nature Microbiology*, *1*, 15005. doi: 10.1038/nmicrobiol.2015.5.

[54]   Bjarnsholt, T., Kristiansen, S., Phipps, R., Nielsen, A. K., Jensen, P. Ø., & Givskov, M. (2007). Silver against *Pseudomonas aeruginosa* biofilms. *Acta Pathologica, Microbiologica, et Immunologica Scandinavica*, *115*(8), 921–928.

[55]   Del Pozo, J. L., Rouse, M. S., & Patel, R. (2008). Bioelectric effect and bacterial biofilms. A systematic review. *The International Journal of Artificial Organs*, *31*(9), 786–795.

[56]   Caubet, R., Pedarros-Caubet, F., Chu, M., Freye, E., De Belem Rodrigues, M., Moreau, J. M., & Ellison, W. J. (2004). A radio frequency electric current enhances antibiotic efficacy against bacterial biofilms. *Antimicrobial Agents and Chemotherapy*, *48*(12), 4662–4664.

[57]   O'Grady, N. P., Alexander, M., Dellinger, E. P., Gerberding, J. L., Heard, S. O., Maki, D. G., & Weinstein, R. A. (2002). Guidelines for the prevention of intravascular catheter–related infections. Centers for Disease Control and Prevention. *MMWR. Recommendations Reports: Morbidity and Mortality Weekly Report. Recommendations and Reports*, *51*(10), 1–29.

[58] Ebrahiminezhad, A., Raee, M. J., Manafi, Z., Sotoodeh Jahromi, A., & Ghasemi, Y. (2016). Ancient and novel forms of silver in medicine and biomedicine. *Journal of Advanced Medical Sciences and Applied Technologies, 2*, 122–128. doi: 10.18869/nrip.jamsat.2.1.122.

[59] Merril, C. R., Scholl, D., & Adhya, S. L. (2003). The prospect for bacteriophage therapy in Western medicine. *Nature Reviews Drug Discovery. 2*(6), 489–497.

[60] Jamal, M., Hussain, T., Das, C. R., & Andleeb, S. (2015). Characterization of Siphoviridae phage Z and studying its efficacy against multidrug-resistant Klebsiella pneumoniae planktonic cells and biofilm. *Journal of Medical Microbiology, 64*(4), 454–462.

[61] Scholtz, V., Pazlarova, J., Souskova, H., Khun, J., & Julak, J. (2015). Nonthermal plasma – A tool for decontamination and disinfection. *Biotechnology Advances, 33*, 1108–1119. doi: 10.1016/j.biotechadv.2015.01.002.

[62] Dasan, B. G., Mutlu, M., & Boyaci, I. H. (2016). Decontamination of Aspergillus flavus and *Aspergillus parasiticus* spores on hazelnuts via atmospheric pressure fluidized bed plasma reactor. *International Journal of Food Microbiology, 216*, 50–59.

[63] Misra, N. N., Pankaj, S. K., Segat, A., & Ishikawa, K. (2016). Cold plasma interactions with enzymes in foods and model systems. *Trends in Food Science and Technology, 55*, 39–47.

[64] Ng, W. L., & Bassler, B. L. (2009). Bacterial quorum-sensing network architectures. *Annual. Review of Genetics, 43*, 197–222.

[65] Lyczak, J. B., Cannon, C. L., & Pier, G. B. (2002). Lung infections associated with cystic fibrosis. *Clinical Microbiology Reviews, 15*(2), 194–222.

[66] Jamal, M., Ufaq Tasneem., Hussain, T. and Andleeb, S. (2015). Bacterial Biofilm: Its Composition, Formation and Role in Human Infections Research and Reviews: Journal of Microbiology and Biotechnology, 4(3) https://api.semanticscholar.org/CorpusID:37657223.

[67] Gilbert, P., Das, J., & Foley, I. (1997). Biofilm susceptibility to antimicrobials. *Advances in Dental Research, 11*(1), 160–167.

[68] Vinodkumar, C. S., Kalsurmath, S., & Neelagund, Y. F. (2008). Utility of lytic bacteriophage in the treatment of multidrug-resistant *Pseudomonas aeruginosa* septicemia in mice. *Indian Journal of Pathology and Microbiology, 51*(3), 360.

[69] Nickel, J. C., & Costerton, J. W. (1993). Bacterial localization in antibiotic-refractory chronic bacterial prostatitis. *The Prostate. 23*(2), 107–114.

[70] Håvarstein, L. S., Coomaraswamy, G., & Morrison, D. A. (1995). An unmodified heptadecapeptide pheromone induces competence for genetic transformation in Streptococcus pneumoniae. *Proceedings of the National Academy of Sciences. 92*(24), 11140–11144.

[71] Rusthoven, J. J., Davies, T. A., & Lerner, S. A. (1979). Clinical isolation and characterization of aminoglycoside-resistant small colony variants of Enterobacter aerogenes. *The American Journal of Medicine, 67*(4), 702–706.

[72] Larson, E. L., Gomez-Duarte, C., Lee, L. V., Della–Latta, P., Kain, D. J., & Keswick, B. H. (2003). Microbial flora of hands of homemakers. *American Journal of Infection Control, 31*(2), 72–79.

[73] Mendoza, M. T. H., & Esp, B. (2004). El papel del biofilm en el proceso infeccioso y la resistencia. *Nova, 2*(2), 71–80.

Angela Parry-Hanson Kunadu, Maame Akua Nyamekye
and Celestina Gosu-Attapkah

Chapter 6
Bacteria stress adaptation: implication and control

Abstract: Bacteria have an incredible ability to adapt to diverse environments, most of which are hostile. It is an important feature that enable them to survive and grow. Along the food value chain, right from the farm to plate, bacteria must overcome sublethal and lethal ranges of temperature, moisture, pH, solutes, desiccation and several oxidizing agents applied to food to preserve, prepare, store and improve safety of food. Recently, bacterial antimicrobial resistance has increased public health risk to zoonotic bacterial agents with increasing resistance to second- and third-line antibiotics used to treat infections from resistant bacterial agents. Antimicrobial resistance is another adaptation system that is occurring faster than anticipated and new technologies are exploring means of inhibiting bacteria without enabling opportunities for resistance or adaptation. This chapter discusses updates on molecular mechanisms by which bacteria respond and overcome stresses encountered in food environments. Sigma factors are particularly important in driving targeted response to stressful stimuli and are also important in providing cross protection to multiple and sometimes unrelated stresses that bacteria face. The chapter also discuss novel technologies to control bacteria in food environments.

6.1 Bacterial stress adaptation

Bacteria, by their nature, may encounter diverse environmental conditions including their natural habitat, and transition through food and food processing environments, physical, chemical, and biological conditions of the food, and possible access to human, animal, and/or plant host. Bacteria have thus developed an incredible adaptability to a variety of environmental stresses. These environmental conditions may be favorable, such as unlimited access to available moisture, simple nutrients, and where other optimal intrinsic and extrinsic factors, for instance, pH and temperature, respectively, prevail. However, bacteria in nature usually exist in suboptimal or hostile physical, chemical, and biological conditions that injure cell components and thus exert a stressful state to the cell. The ability of bacteria as single cells to adapt to these

Angela Parry-Hanson Kunadu, University of Ghana and Texas A&M University
Maame Akua Nyamekye, University of Ghana
Celestina Gosu-Attapkah, University of Ghana and Ghana Standards Authority

https://doi.org/10.1515/9783110748345-006

stressful conditions dictates their survival or growth outcomes in their environmental niches [1]. Several bacteria have adaptive mechanisms that enable their survival during stress. These adaptations have vast implications on food systems, in food processing, preservation, and packaging, in food contact surface cleaning and sanitization, and in the protection of consumer health and safety. It is necessary to also appreciate that the intrinsic and extrinsic features of food ecosystems, and the implicit characteristics of diverse microorganisms in it, have interconnecting effects on the adaptive capabilities of the microbial populations.

Foodborne pathogen responses to stress have been underestimated in the past. The food industry is constantly evolving to meet consumer expectations of less processed foods of premier nutritional and safety status. As a result of this development, foodborne pathogens are often exposed to sublethal stressors, possibly jeopardizing food safety by activating resistance mechanisms in bacterial systems. Bacteria may be exposed to stress conditions in foods, particularly those that have been minimally processed or that have been processed using hurdle technology. Their responses to stressful conditions applied in minimal food processing technologies may allow them to survive under lethal conditions, improve resilience to future processing and preservation conditions, and boost pathogenicity [2].

Most pathogenic bacteria linked to foodborne infections, including Gram-negative toxigenic *E. coli* and *Salmonella enterica* serovars, and Gram-positive *Listeria monocytogenes* can survive under diverse environments which include limited nutrients or starvation, varying pH, thermal or cold temperatures, osmolarity, desiccation or reduced water activity, exposure to oxidizing agents, bile salts, and antimicrobial peptides in raw and processed foods, and survive extreme conditions in the human gut [3]. It is essential that food preservation techniques meet specific prerequisites, such as extending shelf life of perishable products and maintaining the safety and wholesomeness of food, by inactivating spoilage and pathogenic microorganisms.

Bacteria have inherent abilities to survive and sometimes grow under suboptimal environmental conditions. For example, spore-forming bacteria, such as *Bacillus cereus*, form endospores that allow survival over prolonged periods in hostile environments. Other viable but nonreplicating bacteria called persisters exhibit temporary multidrug resistance phenotypes that are distinguishable from antibiotic resistance strains that have antibiotic resistance genotypes. It is believed that the inherent toxin–antitoxin systems may encourage a dormant bacterial cell state that produces temporary antibiotic resistance [4–6]. Viable-but-nonculturable (VBNC) cells are formed upon long periods of exposure to stressful conditions [7]. VBNC cells undergo limited metabolism, including gene expression, are unable to form colonies on solid media, but maintain cell membrane integrity [8]. VBNC cells can recover under suitable resuscitation conditions to become culturable with no increase in cell concentration because of regrowth [9–11].

6.2 Mechanisms of protection against stress

6.2.1 The significance of sigma factors in transcription

Regulation of bacterial gene expression often occurs at the transcription through reactions between RNA polymerase and sigma factors [12]. RNA polymerase is the enzyme required for copying specific genes on DNA molecules into RNA sequences by synthesizing the messenger RNA (mRNA) transcript.

Recognizing appropriate genes in certain environmental situations as well as producing mRNA transcripts that can be translated into new proteins require sigma factors. Sigma factors are cellular components that attach to the RNA core enzyme to produce the RNA polymerase holoenzyme. Sigma factors have the key task of directing RNA polymerase to the promoter sites to initiate synthesis of mRNA transcripts. Sigma factors are important actors in the regulation of mRNA synthesis.

Sigma S (σ^S) and sigma B (σ^B) are known to modulate general stress response in Gram-negative and Gram-positive bacteria, respectively. σ^S is an alternative sigma factor that guides activation of several genes responsible for cell protection, repair of injuries incurred by environmental stresses and neutralization of toxic effects caused by stressful environmental conditions. σ^S activity is common to Gram-negative *E. coli*, *Salmonella enterica*, and *Shigella flexneri* [13]. σ^S-mediated stress response occurs in glucose-rich media for protection against specific stresses such as acid stress response. However, in a minimal media, σ^S activates multiple protective genes that enable cell protection and viability under conditions of exposure to nontargeted and unrelated stresses as found in the stationary phase [14]. In Gram-positive bacteria, σ^B mediates general stress response. It is activated upon exposure to low pH, ethanol, thermal stress, salt stress, glucose, oxygen or phosphate starvation, and entry into stationary phase.

6.3 Survival mechanism under stressful environmental conditions

6.3.1 Thermal stress

One of the most popular and successful strategies for inhibiting bacterial contamination in food and food-processing environment is thermal treatment of food. Heat treatments are applied in food pasteurizations, blanching, cooking, cleaning-in-place (CIP) systems and for cleaning and sanitizing reusable food packaging materials. Pathogenic bacteria such as *E. coli* and *Salmonella* are mesophiles and naturally heat labile. However, they possess the ability to adapt to sublethal heat which activates resistance to potential lethal heat doses used in food applications. Their resistance to heat treatment, which is targeted at the control of bacterial pathogens, have implications on food safety [15].

Heat shock response (HSR) is common in bacterial systems [16]. High temperatures cause protein denaturation and aggregation of misfolded proteins. Proteins are essential constituents of bacterial cells; they occur on the cellular surface, across and within cell wall and cell membranes, and in the cytoplasm. Protein conformation is essential to their functionality; therefore, denatured proteins compromise cell integrity and may cause cell death. HSR involves coordinated responses and highly regulated formation of heat shock proteins (HSPs) upon upshift in temperature. The HSPs are composed of proteases and chaperones that have the responsibility of protecting cellular proteins by aiding in protein folding of newly synthesized polypeptides, preventing accumulation of misfolded proteins, and repairing proteins that have been partly misfolded due to exposure to heat [16]. They also assist in degradation of misfolded or denatured proteins that cannot be rescued or are not needed by the cell. Some molecular chaperones have been discovered in *E. coli*, including DNA (from the HSP70 family of chaperones), DnaJ, GrpE, GroEL (from the HSP60 family of chaperones), and GroES, and the majority of them are heat sensitive [17]. Proteases remove damaged proteins. Some proteases identified as part of HSP in *E. coli* cells include multicomponent proteases (ClpP and HslV). These proteases associate with co-chaperones that function as recognition subunits (ClpA or ClpX for ClpP and HslU for HslV) [16].

The alternative sigma factor σ^{32}, which is transcribed from the *rpoH* gene, mediates HSR [18]. Sigma 32 (σ^{32}) is transcribed at basal concentrations under normal growth temperature. Under stressful conditions or at elevated temperatures, σ^{32} concentration increases rapidly and transiently. It associates with RNA polymerase to the direct synthesis of HSP by recognizing promoter sites in the genome. Several proteases, including FtsH, HslVU, and ClpAP, degrade σ^{32} growth at 30 °C. However, when σ^{32} is bound to RNAP, σ^{32} is protected from damage. At low temperature, DnaK-DnaJ-GrpE chaperone sequesters σ^{32} from RNA polymerase. Upon heat shock, there is a rise in the concentration of denatured or unfolded proteins that have a higher binding affinity to DnaK or DnaJ. This lessens the concentration of free DnaK/DnaJ molecules accessible to bind σ^{32}. As cells attain thermal adaptation, the concentrations of unbound DnaK and DnaJ increase and become available to bind σ^{32} redirecting it toward degradation [17, 19].

6.3.2 Cold shock

L. monocytogenes is a versatile bacterium that can grow at psychrophilic to mesophilic temperatures (2–45 °C), with optimum growth occurring at 30 and 37 °C [20]. Growth and survival of *L. monocytogenes* at chilled temperatures is a characteristic feature of this pathogen that makes its control a challenge in the food industry [21]. Since refrigeration is a common technology for food preservation in industrial and domestic settings, understanding mechanisms of *L. monocytogenes* survival at low temperatures can inform the use of appropriate measures for efficient pathogen control.

L. monocytogenes synthesizes cold shock proteins (Csps) and cold acclimation proteins (Caps) during growth at low temperatures in response to a temperature down shock [22]. In the study by Bayles et al. [22], *L. monocytogenes* was subjected to a cold temperature shock from 37 to 5 °C, after which cold-inducible proteins were identified. About 12 Csps were induced in cold-shocked cultures from the study, whereas four Caps were formed during balanced growth at 5 °C compared to 37 °C. A pathogen's cold acclimation is often aligned with changes in microbial gene expression. Cui et al. [23] also reported increased mRNA expression of chaperone proteases like GroEL, ClpP, and ClpB which suggests that these enzymes may be associated with destruction of misfolded or denatured proteins that accrue in the cell due to growth at low temperatures [23].

σ^B has been found in Gram-positive bacterial pathogens. It is activated in response to sharp reduction in temperature. As σ^B concentration increases, cryoprotectant solutes such as betaine and carnitine accumulate in *L. monocytogenes* [24]. The cryoprotectants play a significant role in enhancing survival and growth at close to freezing temperatures. The capability of *L. monocytogenes* to survive and grow at refrigeration temperatures is of keen concern in refrigerated foods, such as soft cheeses, refrigerated meat spreads, and refrigerated smoked seafood, which are consumed without any additional processing. Foods like deli meats and hot dogs are also a risk, unless completely reheated. For that reason, testing of *L. monocytogenes* is one of the regulated requirements for refrigerated ready-to-eat foods in many jurisdictions.

6.3.2.1 Modification of cell membranes

During conditions of stress, bacteria modify their cell membranes to restrict movement of toxic materials into the cell or loss of cellular constituents. The bacterial cell membrane is made up of a phospholipid bilayer, which is maintained in a fluid state during optimal conditions, to enhance enzyme activity and enable solute transport. Under conditions of cold stress, most bacteria condense their cell wall components to form a crystalline structure. *Listeria* has high concentrations of iso- and anteiso-branched-chain fatty acids in their cell membranes [25]. Beales [26] reported increased proportion of C15:0 with a corresponding reduction in C17:0 in *L. monocytogenes* at chilled temperatures <7 °C. This shortening of fatty acids, and additional characteristics such as increase in unsaturated fatty acids, and a switch to anteiso-branching from iso-branching to reduce carbon–carbon interaction between neighboring chains are necessary to maintain membrane fluidity at chilled temperatures for cell growth [25, 26].

6.3.3 Osmotic stress

Salt and sugar are common solutes used to decrease water activity of foods, and consequently, increase its shelf life by exerting osmotic pressure on microbial cells. However, some microbes such as *E. coli* and *L. monocytogenes* can adapt and survive in high solute concentration environments in food for extended periods. *L. monocytogenes* can survive in 3 M NaCl concentrations [27]. Responses of bacteria to osmotic stress are termed osmoadaptation. Osmoadaptation comprises a sequence of activities resulting in the expression of relevant resistance genes and a resultant change in physiological status [28]. Osmotic pressure, usually referred to as osmolality, is a collective term referring to total concentrations of solutes in solution at a particular temperature. Bacterial cell membranes are elastic and semipermeable. Thus, water molecules can move into and out of the cell when solute concentration in the bacterial environment decreases or increases, respectively. In environments that have higher solute concentrations compared to the cell cytoplasm (hypertonic condition), there is water outflow from the cell, which causes plasmolysis. Plasmolysis refers to detachment and shrinking of the cytoplasmic membrane [29]. When solute concentrations in the cell environment is lower than the cell cytoplasm (hypotonic conditions), there is influx of water into the cell. This influx of water causes the cell membrane to expand, exerting pressure on the cell wall, and may lead to plasmoptysis, and eventually cause cell lysis. Plasmoptysis refers to overstretching of the cell wall leading to cell rupture. The pressure exerted on the cell wall is called turgor pressure. Osmoregulatory mechanisms generally occur upon osmotic up shock or down shock. The response mechanism depends on multiple factors including available proteins that support cell protection within seconds of osmotic stress. Responses that require protein synthesis may take a while to take effect [29].

One key mechanism *Listeria* uses in coping with osmotic stress is regulation of protein synthesis to increase or decrease the production of multiple proteins. When exposed to high salt environment, *L. monocytogenes* upregulates a group of genes collectively known as salt shock proteins (SSPs) transiently and stress acclimatization proteins over a longer period, even after the normalization of environmental conditions. Among SSPs, Ctc, a general stress response protein, protects the cell against salt inhibition when osmoprotectants are absent [30]. The *ctc* gene is under the control of σ^B. Osmoprotection is also achieved by active transport of potassium ions or glutamate into the cell followed by compatible solutes such as betaine, trehalose, carnitine, glycine, and proline [31]. In the absence of compatible solutes in food substrates, microbes can synthesize them from precursors in their environment [31]. Other studies have shown cross-protection of osmoadapted cells to acid stress [32].

6.3.4 Acid stress adaptation

Bacteria often encounter acid stress in acidic foods such as fermented foods and acidi-
fied foods and naturally acidic foods such as orange juice. Foodborne bacterial pathogens also encounter acidic conditions in the gastrointestinal tract and in the phagosome
of macrophages [33]. Movement of organic acids or protons from acidic environments
to the cells interior results in acidification of the cytoplasm causing damage to cellular
components and disrupting key biochemical processes. Bacteria can tolerate and adapt
to sublethal environments using a variety of stress response mechanisms. Most bacteria
have inherent pH homeostasis systems that protect cells against shifts in environmental
pH (e.g., pH 4.5–8.0). Under severe acidic conditions (usually <pH 4.5), protons leak into
the cell's cytoplasm faster than the cells that can maintain pH homeostasis. In such situations, several acid adaptation proteins are involved in cell protection. These include
activation of protective mechanisms to directly reduce the impact of acid-based cell
damage and activation of general stress response which is a broad-based cellular response to multiple stresses. Acid tolerance response (ATR) and acid resistance (AR)
have been reported in several neutrophilic bacterial systems [34, 103, 104]. For example,
exposing *L. monocytogenes* to a slightly acidic pH of 5.5 produces ATR, where the cells
become resistant to subsequent harsh low pH environment [35].

ATR depends on growth phase, while AR systems depend on the glucose concentration and the amino acids, arginine decarboxylase, glutamate decarboxylase, and/or
ornithine decarboxylase systems. Acid adaptation of exponential phase cells at sublethal doses confers protection to subsequent exposure to lethal acid loses of pH 3.0.
There are about 50 acid shock proteins that are expressed during the log-phase ATR
[34]. The acid shock proteins aid in maintaining a near-neutral internal pH while protecting critical cellular components for denaturation. Log-phase ATR also confers
cross-protection to other environmental stresses not yet encountered. These include
oxidative stress, high or low osmotic stress, thermal stress, and DNA damage. Cross-
protection is σ^s-dependent. On the other hand, stationary phase cells are transiently
tolerant to acidic conditions. However, acid-adapted stationary phase cells can survive in lethal acid conditions for prolonged periods. Stationary-phase acid tolerance is
not only σ^s-dependent but also dependent on PhoP and Fur proteins [36] and thus also
elicit cross-protection characteristics to other lethal stresses. In glucose-rich medium,
Gram-negatives such as *E. coli* activate glutamate or arginine decarboxylase systems
to aid in efflux of protons as they leak into the cytoplasm (Figure 6.1). However, in
glucose-limited medium, the oxidative system is activated alongside the glutamate or
arginine decarboxylase system. The oxidative system is composed of RpoS (σ^s), also
known to regulate several stress response genes in limited medium. The σ^s directs
RNA polymerase to transcribe several acid shock genes that play several protective
and injury repair roles in the cell. In addition, outer membrane porins, ompC and
ompF, help to restrict the entry of toxic compounds into the cell and activate selective
movement of substances into and out of the cell. Finally, the outer membrane fatty

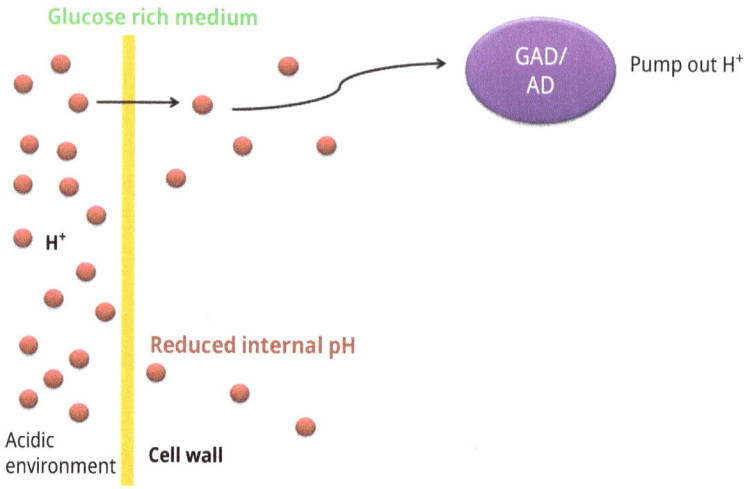

Figure 6.1: Acid resistance mechanism in glucose-rich medium.

acids become tightly packed through the synthesis of cyclopropane fatty acids to restrict diffusion of protons into the cell (Figure 6.2).

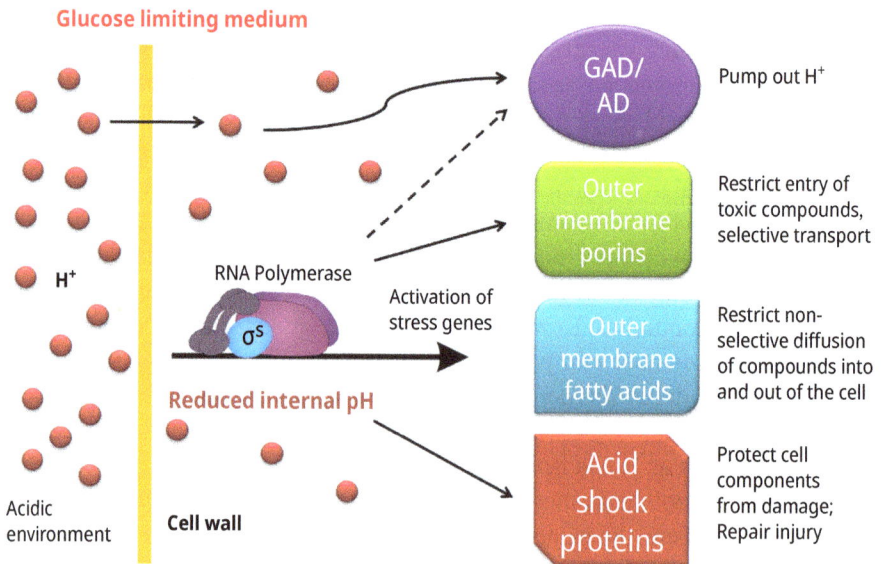

Figure 6.2: Acid resistance mechanism in glucose-limiting medium. GAD, glutamate decarboxylase system; AD, arginine decarboxylase system.

Protein synthesis is required for the acid stress response [37]. Many chaperones have been discovered as important acid-tolerant molecules involved in cell protection and repair of damages to cellular components [38].

Since the Gram-negative bacteria's periplasm lacks the inner membrane's protection, enzymes, transporters, and transmembrane antiporters, they are highly sensitive to acid stress. This causes denaturation and aggregation of the periplasmic proteins [39]. HdeA and HdeB are periplasmic chaperones that protect gut-associated bacteria against stomach acid damage and acid stress induced by organic acids [40]. HdeA protects proteins from acid-induced damage by attaching to them at low pH, causing activation of the chaperones [41]. HdeA has been identified to pay a significant role in the reconstruction and solubilization of denatured insoluble proteins [101]. Proteins like DegP and SurA help HdeA shield protein damage at low pH by acting as chaperones [39]. They contribute in the recovery of protein by promoting refolding during protein renaturation. HdeB functions similar to HdeA but at a much more stringent acidic pH [43].

6.3.5 Desiccation

One of the primary methods used to limit bacterial growth in food matrices is to reduce water activity (A_w), because water is required for metabolism [44]. Drying is a traditional method of preserving food based on the reduction of moisture in food. Low-moisture foods continue to play a significant role in our diet. As a result, several drying techniques have been developed for food preservation. Such foods are called low-moisture foods or intermediate-moisture foods. Some of these drying techniques include spray drying, supercritical fluid drying, and freeze drying with spray and freeze drying [45]. These foods have extended shelf life, and some are stable for years.

Serovars of S. enterica possess several defenses to protect them against the damaging effects of desiccation. A major component influencing Salmonella and other bacteria's survival and persistence on contact surfaces, surfaces of horticultural products, in dried and low moisture foods, and in relatively dry environments like soil, is water loss by evaporative drying or through matric water stress. Bacterial extracellular defenses against desiccation include multifunctional glycocalyces that protect bacteria against full desiccation. The extracellular matrix created by glycocalyx surface layers, which are often composed of exopolysaccharides and associated proteins, has a gel-like extracellular structure and frequently contains significant amounts of bound water. This water is slowly lost due to evaporation and other matric pressures, resulting in reduction of A_w in the surrounding environment [46]. Components that protect Salmonella typhimurium from desiccation damage include extracellular cellulose and curli, which is made up of aggregated fimbriae [47]. CsgD, a regulatory protein that regulates extracellular cellulose and curli production, is necessary for the formation of desiccation-resistant Salmonella biofilms. Furthermore, the O-antigen polysaccharide chain of lipopolysaccharide seems to contribute substantially to S. typhimurium's desiccation resistance [48].

6.3.5.1 Intracellular defenses against desiccation stress

Trehalose is a disaccharide that functions as a suitable solute to prevent loss of water from the cell and assists in preserving the structure and functionality of proteins and membrane lipids throughout drying events [49–52]. With extreme desiccation, trehalose may effectively substitute water, preventing protein denaturation and stabilizing membrane phospholipids. Trehalose also cross-protects *Salmonella* from other types of stresses linked to desiccation like heat stress and oxidative stress. The general stress regulator, σ^S, regulates trehalose production.

6.3.6 Antibiotic resistance: agricultural use of antibiotics

Antibiotic use in agriculture plays a significant role in antimicrobial resistance (AMR). For example, in the United States, approximately 80% of marketed antibiotics are used in animal feed. In 2010, 63,200 tons of antibiotics were used in cattle rearing around the world, far more than was used by humans. Antibiotics are usually added to healthy animal feed and drinking water to treat sick animals and prevent disease (prophylaxis) in healthy animals as well as to rearing subtherapeutic herds to improve weight gain and increase feed efficiency. Colistin, a critical antibiotic used as a last resort to treat serious human infections, is one of the most extensively used antibiotics in animal production around the world to enhance the growth of livestock, particularly pigs [53]. Several studies have reported multidrug resistance of bacteria found in food of both plant and animal origin and food-processing environments. These include bacteria that are resilient to critically important antimicrobics, solely for bacterial infections treatment in humans, such as fluoroquinolones, third- and fourth-generation cephalosporins, macrolides, and carbapenems [54].

6.4 Implications of bacteria stress adaptation

Bacterial stress adaptation characterizes the pathogen's ability to respond to stressful situations [55]. This increase in bacterial survival and endurance in severe settings is thought to be the result of a variety of physiological responses [56]. Regulation of expression, morphological changes in cell shape, and conformational changes in cell membrane structure are a few of the physiological changes that occur simultaneously in the cell. Furthermore, these physiological alterations in response to stressful situations can result in cross-resistance to additional stressful environmental conditions, as well as changes in colonization or virulence [57]. Bacterial stress adaptive responses have significant impact on the entire food chain. This could have serious ramifications in the assurance of food safety. It has potential impact on food security, public health, and the economy.

6.4.1 Biofilm formation

In spite of disinfection methods in use today, bacterial biofilms can form on surfaces in food and beverage manufacturing environments [58]. A biofilm is made up of heterogeneous bacterial population that adheres to surfaces and is covered in hardened matrix of extracellular polymeric substances [59, 60]. Diffusion of substances into the complex matrix is constrained such that bacteria in the inner strata of the biofilm are less susceptible to disinfectants applied on the surface and may respond to sublethal disinfectant doses by adapting to stress.

Bacteria, including foodborne pathogens like *Salmonella enterica, L. monocytogenes,* and *E. coli,* as well as spoilage bacteria like *Pseudomonas* spp., *Brochothrix thermosphacta,* and *Lactobacillus* spp. associated with meat, commonly grow as biofilms, rather than free living cells on food contact surfaces [61]. In meat industries, bacterial biofilms have been found on conveyor belts, tables, and cutting blades. These biofilms are extremely difficult to eliminate with conventional cleaning and sanitization protocols. The biofilms cause sporadic or persistent contamination of products, particularly in the meat industry resulting in a myriad of issues related to food safety, regulatory compliance, spoilage, food waste, company reputation, and financial loss [45, 62, 63]. Moreover, contamination events due to biofilms are difficult to identify.

6.4.2 Increased food waste

Food spoilage is considered as any shift in sensory attributes, including tactile, visual, olfactory, or flavor attributes, which is undesirable to the consumer [64]. Food spoilage, which leads to wastage, is a complex process and can happen at any point in the food chain. Large volumes of food are wasted each year due to microbial deterioration regardless of the current advances in preservation methods [65]. Although there are many different raw materials and processing setups, the microflora that grows during storage and in rotten foods may be explained using information about the food's origin, the substrate base, and their intrinsic and extrinsic characteristics including temperature, atmosphere, A_w, and pH. As inherent microflora adapt to these stresses imposed by preservation methods, they are able to grow, metabolize food substrates, and cause food spoilage.

Studies have shown that less intense preservation temperatures and conditions may have a variety of effects on the microbial community in the preserved food: some microbes may be unaffected, others may be injured, and some may be killed [66, 67], while some will become VBNC [67, 68]. To prevent false-positive or false-negative outcomes, it is critical to detect injured and VBNC microorganisms in industrial food manufacturing [69]. Underestimating the product's safety status could have serious implications; as the existence and possible regrowth of bacteria in a stress-adaptive state could cause decrease in shelf life or foodborne disease outbreak, and so could pose a food quality or food safety risk.

Spore-forming bacteria are generally Gram-positive bacteria and are linked with deterioration of heat-treated foods as their spores can endure high processing temperatures. These bacteria could be facultative anaerobes (able to grow with or without oxygen) or strict anaerobes (thrive in the absence of oxygen). Certain spore-formers are thermophilic, meaning their optimum growth occurs at elevated temperatures (up to 55 °C) and may therefore survive in foods that receive sublethal heat treatment [64]. *Leuconostoc, Carnobacterium, Lactobacillus, Lactococcus,* and *Weissella* are a few psychotropic bacteria that have been implicated in the spoilage of a variety of refrigerated modified-atmosphere packaged meat-based foods [70].

Gram-negatives, on the other hand, tend to enter the VBNC state when injured during or following adaptation. The VBNC state enables bacteria to escape monitoring mechanisms such as growth in selective media, and thus escape stringent food control protocols [71]. They can, however, repair and repopulate in foods while maintaining their metabolic activity and pathogenicity [72]. The presence of these organisms in food may cause food contamination and/or food spoilage, which may lead to increased food insecurity.

6.4.3 Antimicrobial resistance and their public health implications

AMR has emerged as an important public health risk, posing serious challenges to the prevention and treatment of chronic diseases. Despite several steps taken in past 10 years to address this problem, worldwide AMR trends show no indications of slowing down. AMR develops when bacteria, viruses, fungi, and parasites are able to adapt and thrive in the presence of drugs that previously harmed them [73]. The fact that some antibiotics are no longer effective in treating infectious diseases indicates an uncertain future in healthcare. Resistant microbial infection causes extended hospitalizations, as well as increased healthcare expenses, higher alternative drug costs, and treatment failures [73]. AMR has been linked to a total cost of more than €9 billion annually in Europe alone [73]. Because antibiotics' effectiveness will be diminished as AMR trends remain, doctors turn to last-option classes of drugs such as carbapenems and polymyxins, which are not always widely available in impoverished countries, are expensive, and have a variety of side effects.

6.5 Controls

Naturally, many food products are perishable; thus, they must be protected throughout the food value chain to maintain their wholesomeness and safety until the end of their intended shelf life. Food's raw materials, equipment parts, and contact surfaces are typical sources of contamination in the food industry. Cross-contamination from bacteria-infected surfaces can lead to food contamination. Furthermore, biofilms are associated with higher tolerance to biocides due to the arrangement of bacterial cells inside a polymer matrix, which decreases the penetration of the biocidal agent [74]. There are certain drawbacks to current hygiene practices for food facilities, such as the potential toxicity of sanitizers or the development of resistance to these sanitizers over time.

6.5.1 Novel inactivation technologies

Nonthermal preservation technologies are becoming more widely used because of a growing emphasis on minimal processing with the goal of maintaining or improving food safety while reducing the adverse effect of processing on quality characteristics of food products. These methods include the use of bacteriophage to inhibit specific bacterial pathogens, pulsed field and ultraviolet (UV) radiation, cold plasma (CP), and hydrostatic pressure.

6.5.1.1 Bacteriophage

Bacteriophages are viruses that attack bacteria. Phage applications as biocontrol agents against pathogenic bacteria are being employed in the food chain. Even though certain intrinsic characteristics can influence phage performance, innovations are needed to enhance bacteria-phage therapies, particularly *Salmonella*-phage therapy in the dairy industry. Using phages in several hurdle technology, that is, combining several technologies to ensure food safety, has already proven to be promising [75]. With regard to agri-food business, combination of biocontrol approaches has become an appealing choice for the control of harmful bacteria, because consumers consider biological methods as more suitable than chemical treatments. As a result, the ecological influence of phages on microbial communities in the entire ecosystem, including foods, must be assessed, to develop validated phage control strategies for applications throughout the food value chain [76].

6.5.1.2 Pulsed light and ultraviolet radiation

In the packaging business, the use of pulsed UV light decontamination method has been well demonstrated. Also, because UV is efficient against a wide range of pathogens, as well as bacterial endospores, and spoilage microorganisms, it could be successfully used for food and food packaging surface sanitization [77]. Pulsed light can inactivate a variety of microorganisms on a variety of matrices in a matter of seconds. Reports suggests that, depending on the type of microorganism, as high as 6 log units can be inhibited with this technology, although smaller colony count reductions of up to 3 log units have been discovered on coarse surfaces, for example, meat products [77–79].

6.5.1.3 Pulsed electric field

Pulsed electric fields are employed in a variety of food-processing applications, where they are used to induce stressful conditions in microorganisms, augment mass transfer in food items, and enhance microbial inactivation in food settings. In comparison to conventional heat treatments, this approach has a lot of potential for microbial inhibition, because it allows for effective inactivation of vegetative cells at comparatively low temperatures [80]. It has applications in cold pasteurization for heat-labile liquid food products [81].

6.5.1.4 Cold plasma

The use of CP has become a promising technology to decontaminate food contact surfaces. On the other hand, the evaluations of the plasma process are challenging since the process is unstandardized in that it uses various plasma sources, operating gases, and processing conditions leading to a variety of outcomes, and each outcome necessitates a new assessment [82]. The technology is widely employed in a variety of industries, including illumination, material design, medicine, and the food sector for material decontamination activities. Direct applications in food processing is, however, yet to materialize industrially, although there are studies to assess its efficacy in aflatoxin decontamination in grains.

6.5.1.5 High hydrostatic pressure

One of the most recognized nonthermal methods used in industrial food production is high hydrostatic pressure treatment (HHP). It can effectively reduce vegetative bacteria while preserving sensitive food constituents [83]. Currently, HHP equipment are only applied in batch production for foods packaged in flexible packaging materials.

To do this, the packaged objects are put in a pressure container connected to a high-pressure pump and a pressure intensifier [84].

6.5.2 Choice of cleaning agents

6.5.2.1 Disinfectants

Disinfectants are chemical agents that are applied to surfaces to kill all pathogenic bacteria. To control surface biocontamination, they must be utilized regularly in food processing environments and have regulatory approval [85]. The food sector uses a variety of disinfectants with distinct mechanisms of action, such as aldehydes, peroxygens, quaternary ammonium compounds, and halogen-realizing agents [86, 87]. In the food sector, many ways to limit biofilm growth have been utilized for many years. Most efforts have been to use physical and chemical methods to prevent bacterial contamination. Nevertheless, questions about the effectiveness and safety of these methods have prompted the discovery, development, and use of new methods for eliminating and/or limiting bacterial growth on surfaces and biofilm development. New biocides used as disinfectants in the food industry need to be user-friendly and safe for the planet. Various essential oils (EOs) with bacteriostatic and bactericidal properties are derived from herbs and spices. EO components constitute an intriguing opportunity to expand the current range of disinfectants [85, 88].

6.5.2.2 Enzymatic cleaning

Enzymatic cleaning makes use of beneficial bacteria to break down soil by producing enzymes that specifically degrade molecules found in industrial soil. Enzymatic cleaning has been suggested as an efficacious alternative for preventing biofilms in the food sector. A comparative study by Dalehalle et al. [89] reported a much higher efficacy of enzymatic cleaning compared to conventional cleaning of CIP systems. During traditional cleaning, several surfaces were heavily polluted with spoilage bacteria, but enzymatic cleaning reduced the total microflora concentration. Enzymatic cleaning has the potential to enhance the hygiene of food-processing facilities.

6.5.2.3 Ultrasound

In ultrasonic cleaning, an aqueous substance is used to transmit sound waves to remove adherence from a surface. Detergents added to the liquid aid in the removal process. Ultrasound cleaning uses a mechanism of ultrasonic cavitation which occurs when the ultrasonic waves are powerful enough to disintegrate the molecular bonds that hold the contact surface and insoluble impurities together. Cavitation occurs in both transient and stable states [102]. Transient cavitation occurs when unstable bubbles form in the immediate vicinity of an ultrasonicating instrument for less than one sound cycle and then disintegrate into smaller bubbles that explode. The bubble explosions create pressure shock waves that carry significant shear and drag forces to cause erosion, emulsification, and molecular and microbiological damage, primarily due to the impact of microjets [90–92]. Stable cavitation, on the other hand, produces bubbles with a significantly higher pressure intensity due to the gas contained in the aqueous medium. The continuous release of acoustic energy causes the bubbles produced to oscillate for a number of cycles while growing by gas and vapor diffusion until they reach an equilibrium size which is inversely correlated to the frequency of the sound field [92, 93].

6.5.3 Ozone

One of the most potent oxidants utilized in food is ozone, a gas made up of three oxygen atoms (O_3). Ozone has a wide range of uses in the treatment of water since it is effective at disinfection and at degrading both organic and inorganic contaminants. From the inactivation of bacteria in vegetables, to the sterilization of water in treatment systems, the use of ozonation technology has gained many applications today in food production systems [94].

Many studies have reported the efficacy of ozone for equipment and surface disinfection in food industries [95, 96]. The use of ozonated water in CIP systems has also proven to be effective in line sanitizing [97, 98]. Additionally, ozone is chosen over conventional disinfectants like chlorine due to chlorine's comparatively slow rate of inactivation at regulated quantities and consumer fears about chemical residues in food products and damaging effects on the environment.

The use and application of ozone, in the gaseous form or in its aqueous phase, have been studied in food applications to understand its antimicrobial efficacy and shelf life extension potential. The application of ozone in food may range from various washing processes using ozonated water, to storage in atmospheres saturated with ozone and the direct introduction of ozone in fluid foods [99]. All these applications have been shown to have efficacy in the control of pathogenic microorganisms including *L. monocytogenes*, *Staphylococcus aureus*, *Salmonella* sp., and *Pseudomonas* in foods [100].

One of the main advantages of employing ozone in any place in a food-processing plant is the decrease of microbial load on food surfaces. Having this advantage enhances the shelf life of some foods. The strong oxidative capability of ozone, on the other hand, may oxidize some food components present, which may affect nutrient levels. As a result, prolonged exposure to ozone, especially at high concentrations, increases the chance that a food will interact with ozone and change its basic chemical composition. It is crucial that users employ precisely the right amount of ozone to achieve microbial inhibition in a hurdle technology concept while reducing any unintended oxidative food degradation by excessive ozone exposure.

6.6 Conclusion

Bacterial stress adaptation causes significant challenges to food quality, food safety, and sanitization of food processing and preparation environments. While bacteria have developed sophisticated adaptive mechanisms that confer protection to stresses applied in traditional processing operations and sanitization, advancements in technology provide viable solutions to controlling persistent bacteria in food environments due to the activation of adaptive systems and general stress response regulators. Some of the solutions discussed include bacteriophages, ozone, pulsed field, ultrasound, and high hydrostatic pressure. As these technologies become more accessible, the novel biocontrol strategies will also become more feasible.

References

[1]   NicAogáin, K., & O'Byrne, C. P. (2016). The role of stress and stress adaptations in determining the fate of the bacterial pathogen Listeria monocytogenes in the food chain. *Frontiers in Microbiology, 7,* 1865.

[2]   Chung, H. J., Bang, W., & Drake, M. A. (2006). Stress response of Escherichia coli. *Comprehensive Reviews in Food Science and Food Safety, 5*(3), 52–64.

[3]   Spector, M. P., & Kenyon, W. J. (2012). Resistance and survival strategies of *Salmonella* enterica to environmental stresses. *Food Research International, 45*(2), 455–481.

[4]   Harms, A., Brodersen, D. E., Mitarai, N., & Gerdes, K. (2018). Toxins, targets, and triggers: an overview of toxin-antitoxin biology. *Molecular Cell, 70*(5), 768–784.

[5]   Schuster, C. F., & Bertram, R. (2013). Toxin–antitoxin systems are ubiquitous and versatile modulators of prokaryotic cell fate. *FEMS Microbiology Letters, 340*(2), 73–85.

[6]   Wang, X., Lord, D. M., Hong, S. H., Peti, W., Benedik, M. J., Page, R., & Wood, T. K. (2013). Type II toxin/antitoxin MqsR/MqsA controls type V toxin/antitoxin GhoT/GhoS. *Environmental Microbiology, 15*(6), 1734–1744.

[7]   Pinto, D., Santos, M. A., & Chambel, L. (2015). Thirty years of viable but nonculturable state research: unsolved molecular mechanisms. *Critical Reviews in Microbiology, 41*(1), 61–76.

[8]   Ayrapetyan, M., & Oliver, J. D. (2016). The viable but non-culturable state and its relevance in food safety. *Current Opinion in Food Science, 8,* 127–133.

[9]   Bogosian, G., & Bourneuf, E. V. (2001). A matter of bacterial life and death. *EMBO Report, 2*(9), 770–774.

[10] Mukamolova, G. V., Yanopolskaya, N. D., Kell, D. B., & Kaprelyants, A. S. (1998). On resuscitation from the dormant state of Micrococcus luteus. *Antonie Van Leeuwenhoek, 73*(3), 237–243.

[11] Ramamurthy, T., Ghosh, A., Pazhani, G. P., & Shinoda, S. (2014). Current perspectives on viable but non-culturable (VBNC) pathogenic bacteria. *Frontiers in Public Health, 2,* 103.

[12] Borukhov, S., & Nudler, E. (2003). RNA polymerase holoenzyme: structure, function and biological implications. *Current Opinion in Microbiology, 6*(2), 93–100.

[13] Boor, K. J. (2006). Bacterial Stress Responses: What Doesn't kill them can make them stronger. *PLOS in Biology*, 4(1), e23.

[14] Hecker, M., & Völker, U. (2001). General stress response of *Bacillus subtilis* and other bacteria. *Advances in Microbial Physiology*, 44, 35–91. doi: 10.1016/.s00652911(01)44011-2. PMID:11407115.

[15] Dawoud T, M., Davis M, L., Park S, H., Kim S, A., Kwon Y, M., Jarvis, N., O'Bryan, C, A, S. Z., Crandall P, G., & Ricke, S. C. (2017). The potential link between thermal resistance and virulence in *Salmonella*: A review. *Frontiers in veterinary science*, 4, 93. doi: 10.3389/fvets.2017.00093.

[16] Roncarati, D., & Scarlato, V. (2017). Regulation of heat-shock genes in bacteria: from signal sensing to gene expression output. *FEMS Microbiology Reviews*, *41*(4), 549–574.

[17] Missiakas, D., Schwager, F., Betton, J. M., Georgopoulos, C., & Raina, S. (1996). Identification and characterization of HsIV HsIU (ClpQ ClpY) proteins involved in overall proteolysis of misfolded proteins in Escherichia coli. *The EMBO Journal*, *15*(24), 6899–6909.

[18] Schumann, W. (2016). Regulation of bacterial heat shock stimulons. *Cell Stress and Chaperones*, 21, 959–968. doi: 10.1007/s12192-016-0727-z.

[19] Linke, K. (2005). *Characterization of the DnaK-DnaJ-GrpE system under oxidative heat stress* (Doctoral dissertation, Technische Universität München).

[20] Dortet, L., Radoshevich, L., Veiga, E., & Cossart, P. (2019). Listeria monocytogenes. In *Encyclopaedia of microbiology* (Fourth Edition, pp. 803–818). Academic Press, ISBN 9780128117378. https://doi.org/10.1016/B978-0-12-801238-3.02297-2.

[21] Rocourt, J., & Cossart, P. (1997). Food Microbiology: fundamentals and frontiers. In *American Society for Microbiology*. *Washington*.

[22] Bayles, D. O., Annous, B. A., & Wilkinson, B. J. (1996). Cold stress proteins induced in Listeria monocytogenes in response to temperature downshock and growth at low temperatures. *Applied and Environmental Microbiology*, *62*(3), 1116–1119.

[23] Cui, S., Huang, F., Wang, J., Ma, X., Cheng, Y., & Liu, J. (2005). A proteomic analysis of cold stress responses in rice seedlings. *Proteomics*, 5(12), 3162–3172.

[24] Becker, E., Huynh-Do, U., Holland, S., Pawson, T., Daniel, T. O., & Skolnik, E. Y. (2000). Nck-interacting Ste20 kinase couples Eph receptors to c-Jun N-terminal kinase and integrin activation. *Molecular & Cellular Biology*, *20*(5), 1537–1545.

[25] Annous, B. A., Becker, L. A., Bayles, D. O., Labeda, D. P., & Wilkinson, B. J. (1997). Critical role of anteiso-C15: 0 fatty acid in the growth of Listeria monocytogenes at low temperatures. *Applied and Environmental Microbiology*, *63*(10), 3887–3894.

[26] Beales, N. (2004). Adaptation of microorganisms to cold temperatures, weak acid preservatives, low pH, and osmotic stress: a review. *Comprehensive Reviews in Food Science and Food Safety*, *3*(1), 1–20.

[27] Cole, M. B., Jones, M. V., & Holyoak, C. (1990). The effect of pH, salt concentration and temperature on the survival and growth of *Listeria monocytogenes*. *Journal of Applied Bacteriology*, 69, 63–72.

[28] Papafili, A., Hill, M. R., Brull, D. J., McAnulty, R. J., Marshall, R. P., Humphries, S. E., & Laurent, G. J. (2002). Common promoter variant in cyclooxygenase-2 represses gene expression: evidence of role in acute-phase inflammatory response. *Arteriosclerosis, Thrombosis & Vascular Biology*, *22*(10), 1631–1636.

[29] Wang, X., & Wood, T. K. (2011). Toxin-antitoxin systems influence biofilm and persister cell formation and the general stress response. *Applied and Environmental Microbiology*, *77*(16), 5577–5583.

[30] Gardan, R., Duché, O., Leroy-Sétrin, S., & Labadie, J. (2003). Role of ctc from Listeria monocytogenes in osmotolerance. *Applied and Environmental Microbiology*, *69*(1), 154–161.

[31] Pichereau, V., Hartke, A., & Auffray, Y. (2000). Starvation and osmotic stress induced multi resistances: influence of extracellular compounds. *International Journal of Food Microbiology*, *55*(1-3), 19–25.

[32] Faleiro, M. L., Andrew, P. W., & Power, D. (2003). Stress response of Listeria monocytogenes isolated from cheese and other foods. *International Journal of Food Microbiology*, *84*(2), 207–216.

[33] Cotter, P. D., & Hill, C. (2003). Surviving the acid test: responses of gram-positive bacteria to low pH. *Microbiology and Molecular Biology Reviews, 67*(3), 429–453.

[34] Audia, J. P., Webb, C. C., & Foster, J. W. (2001). Breaking through the acid barrier: an orchestrated response to proton stress by enteric bacteria. *International Journal of Medical Microbiology, 291*(2), 97–106.

[35] O'Driscoll, B., Gahan, C. G., & Hill, C. (1996). Adaptive acid tolerance response in Listeria monocytogenes: isolation of an acid-tolerant mutant which demonstrates increased virulence. *Applied and Environmental Microbiology, 62*(5), 1693–1698.

[36] Lee, I. S., Slonczewski, J. L., & Foster, J. W. (1994). A low-pH-inducible, stationary-phase acid tolerance response in Salmonella typhimurium. *Journal of Bacteriology, 176*(5), 1422–1426.

[37] Liu, Y., Tang, H., Lin, Z., & Xu, P. (2015). Mechanisms of acid tolerance in bacteria and prospects in biotechnology and bioremediation. *Biotechnology Advances, 33*(7), 1484–1492.

[38] Nicolaou, S. A., Gaida, S. M., & Papoutsakis, E. T. (2010). A comparative view of metabolite and substrate stress and tolerance in microbial bioprocessing: from biofuels and chemicals, to biocatalysis and bioremediation. *Metabolic Engineering, 12*(4), 307–331.

[39] Hong, S. H., Wang, X., O'Connor, H. F., Benedik, M. J., & Wood, T. K. (2012). Bacterial persistence increases as environmental fitness decreases. *Microbial Biotechnology, 5*(4), 509–522.

[40] Mates, A. K., Sayed, A. K., & Foster, J. W. (2007). Products of the Escherichia coli acid fitness island attenuate metabolite stress at extremely low pH and mediate a cell density-dependent acid resistance. *Journal of Bacteriology, 189*(7), 2759–2768.

[41] Tapley, T. L., Körner, J. L., Barge, M. T., Hupfeld, J., Schauerte, J. A., Gafni, A., . . . Bardwell, J. C. (2009). Structural plasticity of an acid-activated chaperone allows promiscuous substrate binding. *Proceedings of the National Academy of Sciences, 106*(14), 5557–5562.

[42] Tapley, T. L., Franzmann, T. M., Chakraborty, S., Jakob, U., & Bardwell, J. C. (2010). Protein refolding by pH-triggered chaperone binding and release. *Proceedings of the National Academy of Sciences, 107*(3), 1071–1076.

[43] Kern, R., Malki, A., Abdallah, J., Tagourti, J., & Richarme, G. (2007). Escherichia coli HdeB is an acid stress chaperone. *Journal of Bacteriology, 189*(2), 603–610.

[44] Maserati, A., Fink, R. C., Lourenco, A., Julius, M. L., & Diez-Gonzalez, F. (2017). General response of Salmonella enterica serovar Typhimurium to desiccation: A new role for the virulence factors sopD and sseD in survival. *PLoS ONE, 12*(11), e0187692. https://doi.org/10.1371/journal.pone.0187692.

[45] Emami, F., Vatanara, A., Park, E. J., & Na, D. H. (2018). Drying technologies for the stability and bioavailability of biopharmaceuticals. *Pharmaceutics, 10*(3), 131.

[46] Ophir, T., & Gutnick, D. L. (1994). A role for exopolysaccharides in the protection of microorganisms from desiccation. *Applied and Environmental Microbiology, 60*(2), 740–745.

[47] White, A. P., Gibson, D. L., Kim, W., Kay, W. W., & Surette, M. G. (2006). Thin aggregative fimbriae and cellulose enhance long-term survival and persistence of Salmonella. *Journal of Bacteriology, 188*(9), 3219–3227.

[48] Garmiri, P., Coles, K. E., Humphrey, T. J., & Cogan, T. A. (2008). Role of outer membrane lipopolysaccharides in the protection of Salmonella enterica serovar Typhimurium from desiccation damage. *FEMS Microbiology Letters, 281*(2), 155–159.

[49] Crowe, J. H., Hoekstra, F. A., & C, L. M. (1992). Anhydrobiosis. *Annual Review of Physiology, 54*, 579–599.

[50] Elbein, A. D., Pan, Y. T., Pastuszak, I., & Carroll, D. (2003). Review – New insights on trehalose: a multifunctional molecule. *Glycobiology, 13*(4), 17R–27R. doi: 200310.1093/glycob/cwg047.

[51] Furuki, T., Oku, K., & Sakurai, M. (2009). Thermodynamic, hydration and structural characteristics of alpha, alpha-trehalose. *Frontiers in Bioscience, 14*, 3523–3535.

[52] Potts, M. (1994). Desiccation tolerance of prokaryotes. *Microbiology Reviews, 58*(4), 755–805.

[53] WHO. (2017). *Global priority list of antibiotic-resistant bacteria to guide research, discovery, and development of new antibiotics. World Health Organization, Geneva, Switzerland.* http://www.who.int/medicines/publications/WHO-PPL-Short_Summary_25Feb-ET_NM_WHO.pdf?ua//1. Retrieved from Geneva, Switzerland: http://www.who.int/medicines/publications/

[54] Mukerji, S., O'Dea, M., Barton, M., Kirkwood, R., Lee, T., & Abraham, S. (2017). Development and transmission of antimicrobial resistance among Gram-negative bacteria in animals and their public health impact. *Essays in Biochemistry, 61*(1), 23–35.

[55] Yousef, A. E., & Courtney, P. D. (2003). Basics of stress adaptation and implications in new-generation foods. *Microbial Stress Adaptation and Food Safety, 1*, 1–30.

[56] Wesche, A. M., Gurtler, J. B., Marks, B. P., & Ryser, E. T. (2009). Stress, sublethal injury, resuscitation, and virulence of bacterial foodborne pathogens. *Journal of Food Protection, 72*(5), 1121–1138.

[57] Dubois-Brissonnet, F. (2012). *Adaptation of Salmonella to antimicrobials in food-processing environments* (pp. 123–146). Annous BA, Gurtler J. Rijeka: Intech.

[58] Vestby, L. K., Møretrø, T., Langsrud, S., Heir, E., & Nesse, L. L. (2009). Biofilm forming abilities of Salmonella are correlated with persistence in fish meal-and feed factories. *BMC Veterinary Research, 5*(1), 1–6.

[59] Costerton, J. W., Lewandowski, Z., Caldwell, D. E., Korber, D. R., & Lappin-Scott, H. M. (1995). Microbial biofilms. *Annual Review of Microbiology, 49*(1), 711–745.

[60] Høiby, N., Bjarnsholt, T., Givskov, M., Molin, S., & Ciofu, O. (2010). Antibiotic resistance of bacterial biofilms. *International Journal of Antimicrobial Agents, 35*(4), 322–332.

[61] Giaouris, E., Heir, E., Hébraud, M., Chorianopoulos, N., Langsrud, S., Møretrø, T., . . . Nychas, G. J. (2014). Attachment and biofilm formation by foodborne bacteria in meat processing environments: causes, implications, role of bacterial interactions and control by alternative novel methods. *Meat Science, 97*(3), 298–309.

[62] Petruzzi, L., Corbo, M. R., Sinigaglia, M., & Bevilacqua, A. (2017). Microbial spoilage of foods: Fundamentals. In *The microbiological quality of food* (pp. 1–21). Woodhead Publishing.

[63] Shi, X., & Zhu, X. (2009). Biofilm formation and food safety in food industries. *Trends in Food Science and Technology, 20*(9), 407–413.

[64] Rawat, S. (2015). Food Spoilage: Microorganisms and their prevention. *Asian Journal of Plant Science and Research, 5*(4), 47–56.

[65] Gram, L., Ravn, L., Rasch, M., Bruhn, J. B., Christensen, A. B., & Givskov, M. (2002). Food spoilage – interactions between food spoilage bacteria. *International Journal of Food Microbiology, 78*(1-2), 79–97.

[66] Khadgi, N., Upreti, A. R., & Li, Y. (2017). Simultaneous bacterial inactivation and degradation of an emerging pollutant under visible light by $ZnFe_2O_4$ co-modified with Ag and rGO. *RSC Advances, 7*(43), 27007–27016.

[67] Schottroff, F., Fröhling, A., Zunabovic-Pichler, M., Krottenthaler, A., Schlüter, O., & Jäger, H. (2018). Sublethal injury and viable but non-culturable (VBNC) state in microorganisms during preservation of food and biological materials by non-thermal processes. *Frontiers in Microbiology, 9*, 2773.

[68] Siderakou, D., Zilelidou, E., Poimenidou, S., Tsipra, I., Ouranou, E., Papadimitriou, K., . D., Zilelidou, E., Poimenidou, S., Tsipra, I., Ouranou, E., Papadimitriou, K., & Skandamis, P. (2021). Assessing the survival and sublethal injury kinetics of Listeria monocytogenes under different food processing-related stresses. *International Journal of Food Microbiology, 346*, 109159.

[69] Foddai, A. C., & Grant, I. R. (2020). Methods for detection of viable foodborne pathogens: current state-of-art and future prospects. *Applied Microbiology and Biotechnology, 104*(10), 4281–4288.

[70] Andreevskaya, M., Jääskeläinen, E., Johansson, P., Ylinen, A., Paulin, L., Björkroth, J., & Auvinen, P. (2018). Food spoilage-associated *Leuconostoc, Lactococcus,* and *Lactobacillus* species display different survival strategies in response to competition. *Applied and Environmental Microbiology, 84*(13), e00554–18.

[71] Kumariya, R., Garsa, A. K., Rajput, Y. S., Sood, S. K., Akhtar, N., & Patel, S. (2019). Bacteriocins: Classification, synthesis, mechanism of action and resistance development in food spoilage causing bacteria. *Microbial Pathogenesis*, *128*, 171–177.

[72] Pradhan, D., & Negi, V. D., (2019), Stress-induced adaptations in Salmonella: A ground for shaping its pathogenesis. *Microbiological Research*, *229*, 12631. https://doi.org/10.1016/j.micres.2019.126311

[73] Dadgostar, P. (2019). Antimicrobial resistance: implications and costs. Infection and drug resistance, 12, 3903.

[74] Chen, B., Han, J., Dai, H., & Jia, P. (2021). Biocide-tolerance and antibiotic-resistance in community environments and risk of direct transfers to humans: Unintended consequences of community-wide surface disinfecting during COVID-19? *Environmental Pollution*, 117074.

[75] Singh, S., & Shalini, R. (2016). Effect of Hurdle Technology in Food Preservation: A Review. *Critical Reviews in Food Science and Nutrition*, 56, 641–649.

[76] LeLièvre, V., Besnard, A., Schlusselhuber, M., Desmasures, N., & Dalmasso, M. (2019). Phages for biocontrol in foods: What opportunities for Salmonella sp. control along the dairy food chain? *Food Microbiology*, *78*, 89–98.

[77] Van Impe, J., Smet, C., Tiwari, B., Greiner, R., Ojha, S., Stuli´c, V., et al. (2018). State of the art of nonthermal and thermal processing for inactivation of micro-organisms. *Journal of Applied Microbiology*, *125*, 16–35. doi: 10.1111/jam.13751.

[78] Heinrich, V., Zunabovic, M., Varzakas, T., Bergmair, J., & Kneifel, W. (2016). Pulsed light treatment of different food types with a special focus on meat: a critical review. *Critical Reviews in Food Science and Nutrition*, *56*(4), 591–613.

[79] Zunabovic, M., Heinrich, V., & Jaeger, H. (2017). Application of pulsed light for the decontamination of foods. In R. Rai & J. B. Aswathanarayan (Eds.), *Food safety and protection* (pp. 379–390). London: Taylor & Francis.

[80] Schottroff, F., Krottenthaler, A., & Jaeger, H. (2017). Stress induction and response, inactivation, and recovery of vegetative microorganisms by pulsed electric fields. In D. Miklavcic (Ed.), *Handbook of electroporation* (pp. 1–19). Cham: Springer International Publishing.

[81] Lasekan, O., Ng, S., Azeez, S., Shittu, R., Teoh, L., & Gholivand, S. (2017). Effect of pulsed electric field processing on flavor and color of liquid foods. *Journal Food Processing and Preservation*, *41*, e12940. doi: 10.1111/jfpp.12940.

[82] Schlüter, O., Ehlbeck, J., Hertel, C., Habermeyer, M., Roth, A., Engel, K.-H., et al. (2013). Opinion on the use of plasma processes for treatment of foods. *Molecular Nutrition & Food Research*, 57, 920–927. doi: 10.1002/mnfr.201300039.

[83] Gayan, E., Govers, S. K., & Aertsen, A. (2017). Impact of high hydrostatic pressure on bacterial proteostasis. *Biophysical Chemistry*, 231, 3–9. doi: 10.1016/j.bpc.2017.03.005.

[84] Elamin, W. M., Endan, J. B., Yosuf, Y. A., Shamsudin, R., & Ahmedov, A. (2015). High pressure processing technology and equipment evolution: a review. *Journal of Engineering Science and Technology Reviews*, *8*, 75–83. doi: 10.25103/jestr.085.11.

[85] Chinedu, M., Stephen, O., & Uchenna, U. (2014). Comparative studies of the efficacy of some disinfectants on human pathogens. *Global Journal of Medicine Researches and Studies*, 1(4), 103–110.

[86] Gaulin, C., Lê, M. L., Shum, M., & Fong, D. (2011). Disinfectants and sanitizers for use on food contact surfaces. *National Centre for Environmental Health Canada, Available online at*: http://www.ncceh.ca/sites/default/files/Food_Contact_Surface_Sanitizers_Aug_2011.pdf.

[87] Pfuntner, A. (2011). Sanitizers and disinfectants: the chemicals of prevention. *Food Safety Magazine*, *16*, 18–19.

[88] Exner, M., Vacata, V., Hornei, B., Dietlein, E., & Gebel, J. (2004). Household cleaning and surface disinfection: new insights and strategies. *Journal of Hospital Infection*, *56*, 70–75.

[89]   Dalehalle, L., Taminiau, B., Fastrez, S., Fall, A., Ballesteros, M., Burteau, S., & Daube, G. (2020). Evaluation of enzymatic cleaning on food processing installations and food products bacterial microflora. *Frontiers in Microbiology*, (11), 1827. doi: 10.3389/fmicb.2020.01827.

[90]   Krefting, D., Mettin, R., & Lauterborn, W. (2004). High-speed observation of acoustic cavitation erosion in multibubble systems. *Ultrasonics Sonochemistry*, *11*, 119–123.

[91]   Maisonhaute, E., Prado, C., White, P. C., & Compton, R. G. (2002) Surface acoustic cavitation understood via nanosecond electrochemistry. Part III: shear stress in ultrasonic cleaning. *Ultrasonics Sonochemistry*, *9*, 297–303.

[92]   Tanimura, Y., Yoshida, K., & Watanabe, Y. (2010). A study on cleaning ability of oscillating bubbles driven by low-frequency ultrasound. *Japanese Journal of Applied Physics*, *49*, 07HE20.

[93]   Holzfuss, J., Rüggeberg, M., & Billo, A. (1998). Shock wave emissions of a sonoluminescing bubble. *Physical Review Letters*, *81*, 5434–5437.

[94]   O'Donnell, C., Tiwari, B. K., Cullen, P. J., & Rice, R. G. (2012). *Ozone in food processing* (First Edition). Blackwell Publishing Ltd.

[95]   Moore, G., Griffith, C., & Peters, A. (2000). Bactericidal properties of ozone and its potential application as a terminal disinfectant. *Journal of Food Protection*, *63*(8), 1100–1106.

[96]   Zucker, I., Lester, Y., Alter, J., Werbner, M., Yecheskel, Y., Gal-Tanamy, M., & Dessau, M. (2021). Pseudoviruses for the assessment of coronavirus disinfection by ozone. *Environmental Chemistry Letters*, *19*(2), 1779–1785.

[97]   Avila-Sierra, A., Vicaria, J. M., Lechuga, M., Martínez-Gallegos, J. F., Olivares-Arias, V., Medina-Rodríguez, A. C., . . . Jurado-Alameda, E. (2021). Insights into the optimisation of the Clean-In-Place technique: Cleaning, disinfection, and reduced environmental impact using ozone-based formulations. *Food & Bioproducts Processing*, *129*, 124–133.

[98]   Englezos, V., Rantsiou, K., Cravero, F., Torchio, F., Giacosa, S., Segade, S. R., & Rolle, L. (2019). Minimizing the environmental impact of cleaning in winemaking industry by using ozone for cleaning-in-place (CIP) of wine bottling machine. *Journal of Cleaner Production*, *233*, 582–589.

[99]   Brodowska, A. J., Nowak, A., Kondratiuk-Janyska, A., Piątkowski, M., & Śmigielski, K. (2017). Modelling the ozone-based treatments for inactivation of microorganisms. *International Journal of Environmental Research and Public Health*, *14*(10), 1196.

[100] Patil, S., *et al.* (2011) Assessing the mechanism of microbial inactivation during ozone processing. In *ICEF 11 International Conference on Engineering and Food*, Athens, Greece, May 2011. doi: 10.21427/ D7N62X

[101] Malki, A., Le, H-T., Milles, S., Kern, R., Caldas, T., Abdallah, J. & Richarme, G. (2008). Solubilization of protein aggregates by acid stress chaperones HdeA and HdeB. Journal of Biological Chemistry, 283(20):13679–87. doi: 10.1074/jbc.M800869200.

[102] Otto, C., Zahn, S., Rost, F., Jaros, D., & Rohm, H. (2011). Physical Methods for Cleaning and Disinfection of Surfaces. *Food Engineering Reviews*, *3*, 171–188. doi: 10.1007/s12393-011-9038-4.

[103] Álvarez-Ordóñez, A., Fernández, A., Bernardo, A., & López, M. (2010b). Arginine and lysine decarboxylases and the acid tolerance response of Salmonella Typhimurium. *International Journal of Food Microbiology*, *136*(3), 278–282.

[104] Álvarez-Ordóñez, A., Fernández, A., Bernardo, A., & López, M. (2010a). Acid tolerance in Salmonella typhimurium induced by culturing in the presence of organic acids at different growth temperatures. *Food Microbiology*, *27*(1), 44–49.

Adebola O. Oladunjoye and Elizabeth U. Awani-Aguma

Chapter 7
Foodborne illnesses: prevention and control

Abstract: Food is necessary for the well-being of individuals. However, it has to be wholesome and nutritious in order to carry out its function of promoting health and providing nourishment to the body as well as prevent foodborne illnesses. Foodborne illnesses are defined as illnesses that emanate from the ingestion of contaminated food products. Contamination of food could inadvertently take place within any point along the food value chain – from farm (during production) to the table (during consumption). They are generally classified as foodborne infections and foodborne poisoning/intoxication. The former refers to diseases that occur as a result of ingestion of food containing pathogens that proliferate in the gastrointestinal tracts and tissues of man, whereas the latter refers to illnesses that result from the ingestion of toxins that could be biological or chemical sources or ingredients and additives utilized in food production intentionally or unintentionally to enhance some quality attributes in the food product. Foodborne illnesses have been identified to have caused an estimated 48 million diseases, 128,000 hospitalizations, and an average of 3,000 recorded death cases in the United States. This impact has been caused by several agents of biological (pathogens and biotoxins), chemical (chemicals applied in cultivation, production, processing, etc., of food), and physical (metal shavings from can, plastics) origin. Foodborne illnesses are dangerous and could result in an outbreak that could lead to loss of lives; hence, there is the necessity to ameliorate or arrest the spread of the outbreak via methods of surveillance and investigations that aid to determine the cause, trend, and possibly proffer solutions that may either reduce the effect of the outbreak or put an end to it and also create the means to prevent future occurrences. The study of illness occurrences is referred to as epidemiology, and it can be achieved through descriptive, cohort, and molecular methods. The major way to prevent foodborne illnesses is to ascertain the safety of food for human consumption in a certain number of steps, including cleaning, separation, cooking, and chilling. The use of bacteriophages and bacteriocins has been successful in the biological control of foodborne illnesses.

Adebola O. Oladunjoye, Elizabeth U. Awani-Aguma, Department of Food Technology, University of Ibadan, Ibadan, Nigeria

https://doi.org/10.1515/9783110748345-007

7.1 Introduction

Food has been described as any substance that is designed for human consumption and is necessary for life's nourishment and sustenance, whether it is processed, semi-processed, or uncooked [1]. Food quality and safety are extremely important and have drawn the attention of all stakeholders. The agro-industrial market and consumer awareness have increasingly been liberalized, and the food supply chain is becoming cohesive [2]. As a result, food safety and quality have become key problems. It is not enough to just yield food; it must be safe to ingest as well as confer health and nutritional benefits. The food supply chain has to be nutrient-rich from manufacturing to consumption [1]. Food quality refers to the characteristics/attributes that influence consumers' choice and satisfaction of food products [3]. The concept of food quality involves sensory attributes, which can easily be observed by the human sense organs, and hidden attributes, such as nutrition and safety, which will entail instrumentation to determine their measure [1]. These attributes may be positive or negative; positive attributes include flavor, color, texture, appearance, and shape, while negative attributes include contamination, spoilage, adulteration, contamination, and food safety hazard [4]. These quality attributes serve as factors that determine the safety of food products consumed.

Foodborne illnesses include a wide range of challenges that exist from farm to fork. Growing, harvesting, transporting, and preparing food in unsanitary conditions and without sufficient temperature and environmental controls are dangerous as these factors provide a means for animals and humans to become infected [5]. The term "zoonosis" refers to the spread of infections between animals and humans, either directly or indirectly. The consequences of zoonosis in the food processing and production industry include loss in work days and economic losses. Some individuals ranging from children under the age of five, for example, are more vulnerable to zoonotic diseases and are more likely to die from them. Pregnant women, anyone over the age of 65, and the immune-compromised, such as cancer patients receiving chemotherapy as well as HIV-positive patients, are all at risk [6].

Foodborne illnesses are ailments occasioned by eating contaminated food. Infectious microorganisms such as bacteria, viruses, and parasites as well as non-infectious chemicals represent part of over 250 foodborne-recognized dangers [7]. Even though infections from these pathogenic organism contaminations in food often result in diarrhea and vomiting, no conventional ailing symptom occurs for general foodborne illnesses. These substances sometimes exist in food as an intrinsic component or added inadvertently (or intentionally) during food manufacturing, processing, or preparation. Furthermore, a great number of foodborne agents can be transmitted via other route besides food, which include water or contact with contaminated farm animals, birds, or humans [8]. Only a minimal number of foodborne diseases are fully contaminated via foods, and the severity of diseases associated with foodborne infections varies with the causative agent and food management strategies. *Listeria monocytogenes,* for example, is assumed to be entirely foodborne, but *Escherichia coli* O157:H7 infection can be obtained by consuming infected food or water and also via direct contact with ailing animals or people.

7.2 Foodborne illnesses or diseases

Foodborne diseases was defined by [9] as the illness that occurs as a result of ingesting spoiled or poisoned food that has been contaminated by toxicants or microorganisms at any point along the food value chain (from farm to the table). These diseases have been classified broadly into two groups: foodborne infection and foodborne intoxication [4, 9–12]. Foodborne infection occurs when an individual ingests food containing pathogenic microorganisms, which invade and proliferate in the gastrointestinal tracts or other tissues. Examples of such pathogens include *Campylobacter, Salmonella, Yersinia enterocolitica,* and *Vibrio* [12]. Foodborne intoxication is a term that is used to describe diseases that emanate from the consumption of food containing toxins of microbial sources or biotoxicants or other toxic substances that were included in food as additives intentionally or unintentionally. Foodborne intoxication can be characterized by short incubation periods and can manifest without fever [9].

According to the FAO [13], foodborne diseases relate to both acute and chronic diseases that are commonly transmitted through food and take into account morbidity, disability, long-term implications as well as mortality. Typically, the meal is tainted by pathogens or toxicants. Such food holds enough pathogenic organisms or toxic substances to cause illness [14].

In the United States, foodborne diseases have orchestrated an approximated 48 million illnesses (approximately one from every six Americans), 128,000 clinical issues, and 3,000 recorded deaths annually. Unknown substances are responsible for 38.4 million diseases, 71,878 clinical cases, and 1,686 fatalities among them [6]. Over 250 distinct foodborne ailments have reportedly been identified, with microorganism of bacterial, viral, and parasitic origin observed to be the most common causes. Furthermore, bacteria such as *Campylobacter* spp., *Salmonella* spp., *E. coli* O157:H7, and *Norovirus* represent the highest known causative agents of common causes of foodborne illness. Toxins or poisons found in food are responsible for other ailments. Many diseases that are disseminated through food are also spread through water or via direct contact with people (www.cdc.gov/foodborneburden). Determining whether an infection is foodborne is challenging since symptoms could be hidden for several days or weeks following exposure or contamination to a pathogenic contaminant, and the most common pathogens are not foodborne [8].

7.3 Classification of foodborne diseases

Foodborne intoxication and infections are the two major groups of foodborne diseases, depending on the causative agent, as expressed in Figure 7.1.

Foodborne infections are diseases resulting from causative agents that exist in their viable state. These pathogens are consumed in foods, and this can result in an infection. These pathogens react with the tissues of the body and result in infection [15]. The origin of these pathogens is fungal, bacterial, viral, or parasitic. Foodborne infections often do not manifest immediately after the consumption of contaminated food as the pathogens tend to have long incubation periods that are mostly depicted by fever [16].

Foodborne poisoning/intoxication occurs as a result of ingestion of preformed toxins from microorganisms, intoxication released by toxic plants or animal issues, or ingestion of food tainted by toxin of chemical origin [15]. Foodborne intoxicated foods may contain biotoxicants, toxins of metabolic products, or poisonous substances. Tissues of some plants and animals are the sources of biotoxicants. Toxins from metabolic products, on the other hand, are formed and discharged by microorganisms such as bacteria, viruses, and parasites. These agents replicate in the gastrointestinal tracts of their host. Poisonous substances may gain entrance into food either intentionally by food processors or unintentionally anywhere along the food supply chain (processing, transportation, and storage) [6].

Table 7.1: Differences between foodborne infection and foodborne intoxication.

	Foodborne infections	Foodborne intoxications
Cause	Bacteria, parasites, and viruses	Toxins
Mechanism of action	Invade and/or multiply within the gastrointestinal tract of host	No invasion or multiplication within the intestinal lining of the host
Incubation period	Few hours to few days	Minutes to hours
Symptoms	Diarrhea, nauseating feeling, vomiting, and stomach cramps	Blur vision, fatigue, breathing failure, numbness, sensory and stimulus dysfunction, vomiting, nausea, and diarrhea
Mode of transfer	Spreads by contact via fecal-oral means	Not communicable
Risk factors to food contamination	Poor cooking, cross-contamination, unsanitized hand contact, and poor personal hygiene	Inadequate cooking and improper holding temperatures

Source: [11].

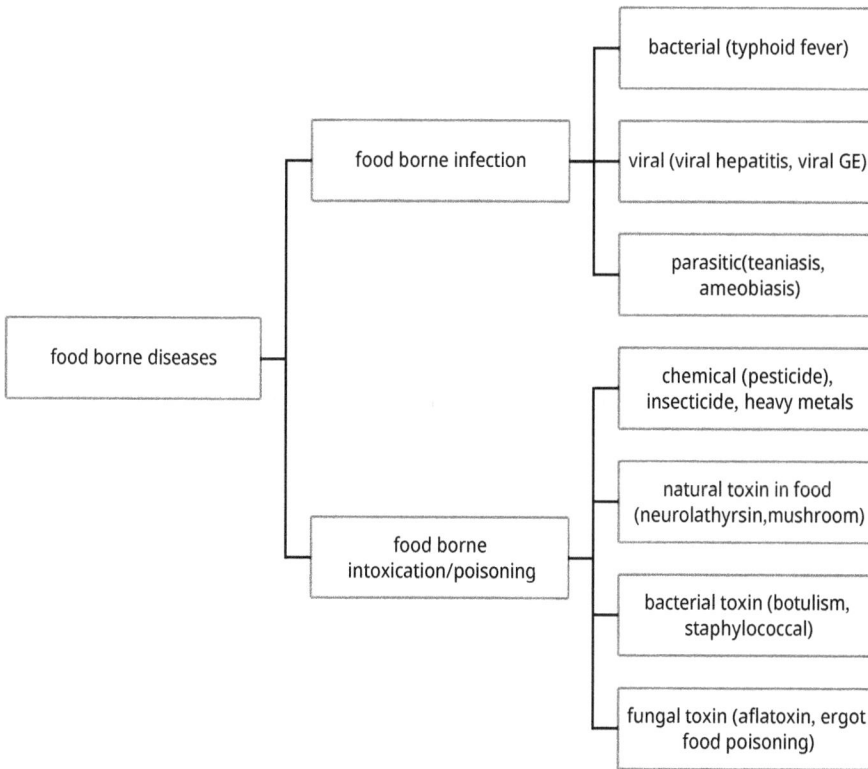

Figure 7.1: Classification of foodborne diseases [15].

7.4 Causes of foodborne diseases

Hazards of several origins have been identified to cause foodborne illnesses. A food safety hazard refers to an agent or a condition that could constitute a challenge to human health; the condition of the food itself can also constitute a source of hazard. Food safety hazards can broadly be categorized into three main divisions as indicated below [1, 17].

– Biological or natural hazards: these are of biological origin, such as viruses, bacteria, fungal, and parasitic organisms. Of these agents, bacteria and viruses constitute most of the foodborne diseases. Biological hazards pose a major challenge to food safety. They can be part of the product or come in contact with the product through improper handling such as wrong application of time and temperature. Toxins of these sources include peptides, proteins, or small molecular substances that cause diseases when they come in contact with body tissues by means of absorption or interaction with other biological structures such as cellular receptors and tissues. These toxins, which are produced by living organisms, are not injurious to the producers but are toxic to other organisms/creatures. Some plants also possess the ability to produce toxic compounds that when consumed result in adverse health implications [1]. Examples of these disease-causing agents and the toxins they produce are depicted in Tables 7.2 and 7.3.

– Chemical hazards: this group of contaminants refers to chemicals and residues that are present in the environment where several activities are executed, such as cultivation, harvesting, processing, packaging, storage, transportation, marketing, and consumption. In other words, chemicals hazards could be transmitted into food substances at varying points along the food chain. These chemicals can pose a threat when food processors do not adhere to stipulated regulations guiding permissible limits when these chemicals are added to food as additives with the aim of improving taste, shelf-life, and appearance of the food. Air, soil, and water are vectors of chemicals such as arsenic, nitrates, and mercury. The rampant utilization of chemicals in agricultural practices with the intention to promote crop yield at reduced cost has given rise to an increased rate in the occurrence of chemical residues in food. Examples of such chemicals include pesticides (herbicides, rodenticides, insecticides), veterinary drugs (chloramphenicol, malachite green, nitrofuran), plant growth hormones, and so on.

– Physical hazards such as metal chips from cans, plastic fragments, or broken glass speckles are also sources of food contamination.

Table 7.2: Foodborne infections, causative agents, and commonly affected food stuffs.

Disease type	Illness	Microbial pathogen	Food substrate
Bacteria type	Typhoid fever	*Salmonella typhi*	Fresh vegetables, salad, pastries
	Shigellosis	*Shigella* spp.	All foods handled by unhygienic workers
	Cholera	*Vibrio cholera*	Fruits and vegetables washed with contaminated water
	Salmonellosis	*Salmonella* spp.	Eggs, poultry, poorly cooked meals, non-pasteurized dairy products, marine foods, And sausages
	Brucellosis	*Brucella* spp., often *Brucella melitensis*	Milk and dairy foods from contaminated animals
	Bovine tuberculosis	*Mycobacterium bovis*	Untreated milk and meat from cows infected with tuberculosis
	Anthrax	*Bacillus anthracis*	Contaminated meat from unhealthy animals
	E. coli infection	*Escherichia coli*	Meat products, fresh produce, and sprouts.
	Listeriosis	*Listeria monocytogenes*	Dairy and meat products

Table 7.2 (continued)

Disease type	Illness	Microbial pathogen	Food substrate
Viral	Viral gastroenteritis (VGE)	Rotavirus, calicivirusesincluding norovirus, and astrovirus	Any viral-contaminated food
	Viral hepatitis	Hepatitis A and E viruses	Raw shellfish from infested water, salad products
	Poliomyelitis	Polio virus	Any food product infested with the virus
	Rift valley disease	Rift valley disease virus	Blood-contaminated foods. Sols from contaminated animal products or fetus
Parasite	Amoebiasis	*Entamoeba histolytica*	Any food contaminated with excreta Trichinosis *Trichinellaspiralis* Poorly cooked pork meat
	Ascariasis	*Ascaris lumbricoides*	Foods contaminated with soil microflora and fresh produce
	Giardiasis	*Giardia lamblia*	Any infested food commodity
Parasitic (multicellular)	Cryptosporidiosis Hydatid disease	*Cryptosporidium parvum* *Echinococcus granulosus*	Any infested food commodity Any food infested with dog excreta
	Diphyllobothriasis	*Diphyllobothrium latum*	Raw meat and fish product
	Taeniasis	Tapeworm infection *Taenia* species	
	Anisakiasis Roundworm	*Pseudoterranova decipiens*	Poorly cooked fish with larval stages
Unconventional agents	BSE or mad cow disease	Bovine spongiform encephalopathy (BSE)	Meat products from cows, sheep, etc.

Source: [7].

Table 7.3: Food poisoning or intoxication, toxin category, pathogen, and commonly affected foods.

Disease category	Disease	Toxin category/pathogen	Food substrate
Bacterial toxins	Staphylococcal food intoxication	Enterotoxins from *Staphylococcus aureus*	Dairy and meat products, legumes
	Perfringens food intoxication	Strain of *Clostridium welchii/ C. perfringens*	Poorly cooked meat, poultry, and legumes

Table 7.3 (continued)

Disease category	Disease	Toxin category/pathogen	Food substrate
	Botulism food intoxication	Toxin of *Clostridium botulinum*	Home-made can foods, high pH vegetables, and corn
	Escherichia coli food poisoning	Enterohemorrhagic *Escherichia coli* O157:H7	Comminuted beef, milk products, and uncooked beef
	Bacillus cereus food poisoning	Enterotoxins of *Bacillus cereus*	Cereals and milk products. Fresh produce, meats, and boiled rice
Fungal toxins	Ergotism	A toxin (ergot) synthesized by collection of fungi called *Claviceps purpurea*	Rye, wheat, sorghum, and barley
	Aflatoxin food poisoning	Aflatoxin generated by certain group of fungus (e.g., *Aspergillus flavus*, *Aspergillus parasiticus*)	Cereal grains, groundnuts, peanuts, cottonseed, and sorghum
Natural toxic substances in food	Neurolathyrism	Beta-oxalyl aminoalanine	*Lathyrus sativus* (Guaya)
	Mushroom intoxication	Phalloidine and alkaloids occurring in some toxic mushrooms	Poisonous mushrooms such as species of *Amanita phalloides* and *Amanita muscaria*
	Shellfish poisons, and tetrodotoxin (pufferfish)	Ingested poisonous dinoflagellates	Occasioned by eating of mussels, clams, and scallops
	Scombrotoxin ciguatera	Production of histamine amino acid because of bacterial contamination of fish toxin	Tropical fish
Chemical toxins	Chemical poisoning	Heavy metals (such as lead, mercury, and cadmium) Fish, canned food products Heavy metal contact with foods Pesticidal and insecticidal residue in foods	Residues on crops, vegetables, and fruits Accidental poisoning where some chemicals could be assumed for food recipe. When contaminated vessels are used to contain stored foods
		Additives	Several food commodities containing unapproved additives either as colorants, flavors, sweetening agents, and preservatives.
Additional toxins	Ecological contaminants	Dioxins Chlorinated biphenyls Furans	Environment contaminants in food chain through plants or animals sources

Table 7.3 (continued)

Disease category	Disease	Toxin category/pathogen	Food substrate
	Biotechnology	Food allergens Genetically modified foods	Gluten-based cereals such as wheat, rye, barley, oats, and hybrid cultivars of these cereals Genetically modified food commodity

Source: [7].

7.5 Sources of food contamination

Food value products have considerable nutrients that microbes require for prolifera-tion and can therefore become contaminated [31]. An in-depth knowledge of the sources of food contamination (Figure 7.2) is necessary in order to tackle the challenge of con-trolling foodborne illnesses. Contamination of food could occur from environment-based sources such as air, water and soil, and walls and floors of factories. Microorgan-isms and toxic substances can be found in soil. These agents could find their way to food through direct contact or dusts. Untreated sewage in the environment could con-tain disease agents that have been excreted from the human body as well as raw sew-age utilized in irrigation of vegetable crops. Unsafe or untreated water used in food processing and preparation is also a major source of food contamination, as well as un-safe drinking water. Microorganisms bound in the air are not left out in food contami-nation as these microbial contaminants are present in the environment where food is processed, packaged, or stored. Food processing equipment is often contaminated dur-ing production, and some materials are not properly cleaned. The walls, floors, and sur-faces of processing area are major sources of food contamination. Crevices and corners in buildings and equipment if not properly cleaned traps dirt and serve as a potential points for microorganisms to thrive and contaminate food which come in contact. Food processing operations such as washing, sorting, sterilization, and pasteurization that are not properly executed or done according to guidelines prescribed by food regula-tory bodies give rise to consumption of microorganisms that are not completely inacti-vated. Employees in the food processing area transfer microorganisms from their mouths, nostrils, hands, and hair into the food during handling by sneezing, coughing, inhaling or touching. Ingredients (particularly spices, flavoring and coloring agents, and preservatives) have the ability to transport hazardous or potentially harmful microbes and toxins. Insects and rodents are commonly found in houses, food restaurants, and food processing amenities, as well as lavatories, rubbish, and other dirt. Contaminants are transferred to food via these animals' waste products, mouths, fur, intestines, leg, and other parts of the body, as well as throwing up onto clean food after consumption.

Other animals' bodies: Microorganisms are transported straight to the soil and water from the digestive systems of other animals. They may then become embedded in plants, dust, utensils, and/or food. During the post-mortem processing of meat via exsanguination, cutting, evisceration, cold storage, and distribution, there could be possible contamination. Furthermore, contact of the carcass with the skin, feet, feces, mud, and internal fluids might cause further contamination. Drugs administered to prevent illness and enhance the growth process of animals could pose a danger to human health if they remain in meat or dairy products. Others include inadvertent use of a harmful substance as seasoning or sweeteners, or youngsters mistaking it for a drink, a person's intentional and malicious poisoning of food for some illogical purpose, water contaminated by farm chemicals and/or food tree spraying [32].

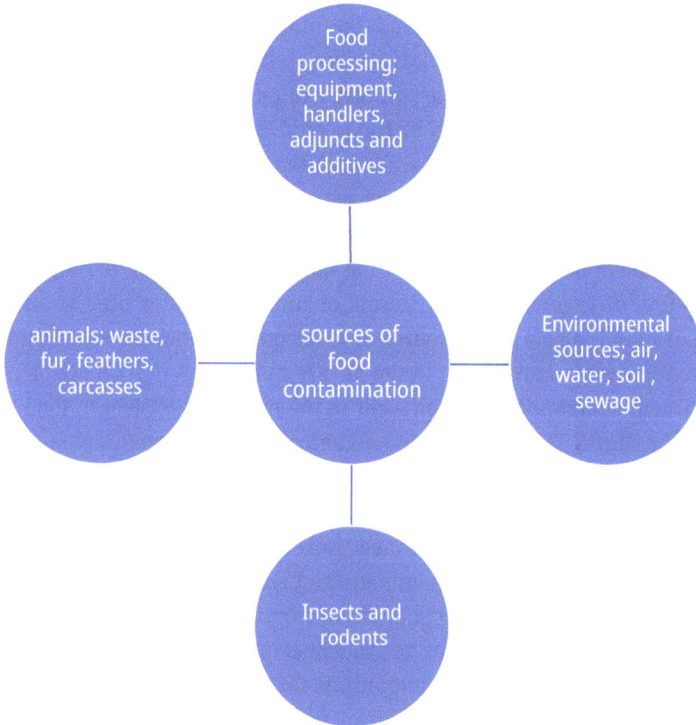

Figure 7.2: Sources of food contamination [33].

7.6 Seasonal variation of foodborne diseases

It is also important to remember that foodborne illnesses might have seasonal changes while studying and controlling them. Knowing this might aid in making precautions prior to an outbreak that has already occurred, lowering the risk rate.

Waterborne or foodborne enteric illnesses in temperate regions generally vary between moment of low endemic points and outbreaks, establishing a seasonal trend. For instance, *Salmonella* spp. or *Campylobacter jejuni* contamination and infection increase in summer but reduce during winter [34]. The protozoans *Giardia* and *Cryptosporidium* produce enteric infections, which have a seasonal fluctuation as well, but it shifts toward fall [35]. Hepatitis A and shigellosis, on the other hand, have no discernible periodicity [3].

Regular temporal variations for illnesses with comparable exposure sources or spread pathways point to the presence of environmental variables that synchronize with seasonal variation [36]. Differences from a well-established seasonal trend might reveal valuable information about the variables that influence the incidence of illness. Variation in the sources of contact and transmission, changes in the affected population, and variations in the microbial source itself are all possible variables. Ecological changes, such as those caused by climate change, may have an impact on the establishment and spread of parasite illnesses, including cryptosporidiosis and giardiasis. Short-term variations (weekly and monthly) in documented cases of food intoxication in the UK, often caused by *Salmonella* spp., have been linked to ambient temperature. Elevated temperatures and severe precipitation occurrences have equally been proven to affect health outcomes in the short run [18]. Diseases associated with change in climate have been estimated to involve 4.6% of all environmental risks. Change in climate during year 2000 accounts for nearly 2.4% of diarrhea epidemics globally, 6% of malaria cases in several developing nations, and 7% of the recorded cases of dengue disease of some industrialized nations. Total approximations reveal that the rate of mortality due to climatic changes was recorded to be 0.3% and the related disease burden was recorded to be 0.4% [37]. Foodborne pathogens that have been monitored have shown seasonal patterns (18). Emerging foodborne diseases have been screened and reported for *Salmonella* spp., verotoxigenic *Escherichia coli*, and *Campylobacter* spp.

Campylobacter infection caused by *Campylobacter* spp. is the causative agent of campylobacteriosis disease. The pathogen is a micro-aerophilic, heat-tolerant bacteria with a minimum infectious dosage of about 500 cfu/mL. Popular species implocated in human infection include *Camplobacter coli*, *Campylobacter jejuni*, and *Campylobacter lari* [19]. The outbreak of these infectious agents is linked with the consumption of contaminated water or food, while transmission of these agents from person to person is rarely reported [20, 21]. Seasonal patterns of campylobacteriosis have been studied in 18 different nations in 3 different continents. The results revealed that in Europe the peak periods were recorded around April and May, while record shows that the continent of North America, Canada, has a peak period between June and July. In Australia, there was no record of a peak period. It is normal to record more cases of *campylobacteriosis* in the warmer seasons of the year since the causative agent is heat resistant. Figure 7.3 shows the seasonal spread of confirmed campylobacteriosis reported cases by month, EU/EEA, 2008–2012 [19].

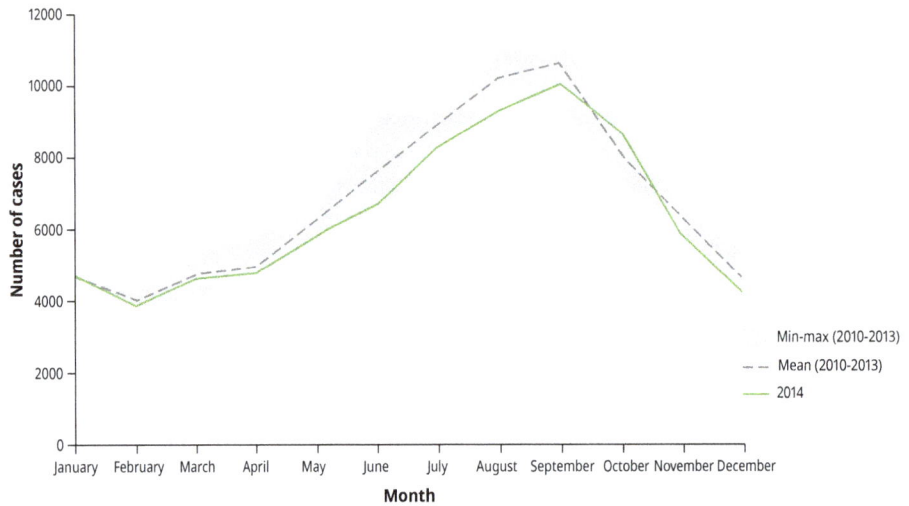

Figure 7.3: Distribution of confirmed campylobacteriosis reported cases by month, EU/EEA, 2008–2012 [19].

7.7 Epidemiology methods for studying foodborne diseases

Epidemiology describes the study of incidence and distribution of diseases in populations, as well as the variables that influence disease patterns. Foodborne illnesses are quite common and disproportionately reported among people; they exhibit seasonal and regional trends. They're linked to particular demographic features, or they're linked to consuming certain foods. Epidemiology has established ways for tracking disease in a community and determining its prevalence, as well as the causes of these disorders [8]. There are different measures applied in establishing epidemiology procedures. Descriptive epidemiology looks at the characteristics of a disease and answers questions like who got sick, what symptoms they had, when did they start, where did they become sick, and how many people got sick. When these data are shown graphically, an epidemic curve (Figure 7.4) emerges, with the commencement date of symptoms on the *x*-axis and the patient's size on the *y*-axis. Figure 7.5 shows the epidemiology curves for cases of foodborne diseases occasioned by specific pathogens by month, and Figure 7.6 is an epidemic curve showing the geographic locations of the spread of Covid-19 outbreak.

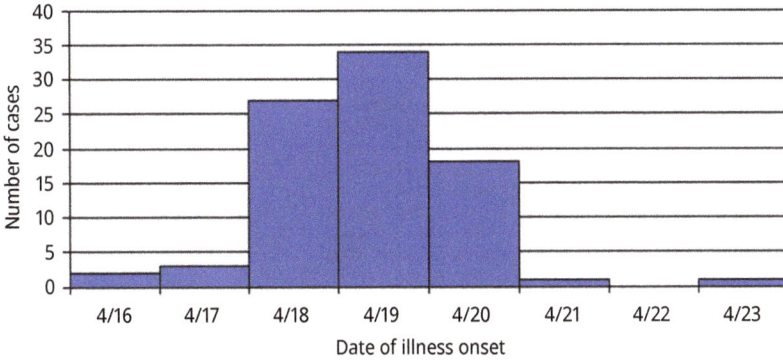

Figure 7.4: Epidemic curve of cases of foodborne disease X associated with eating at restaurant A, by date of illness onset, County Y, April 2011 (n=86). Source [8].

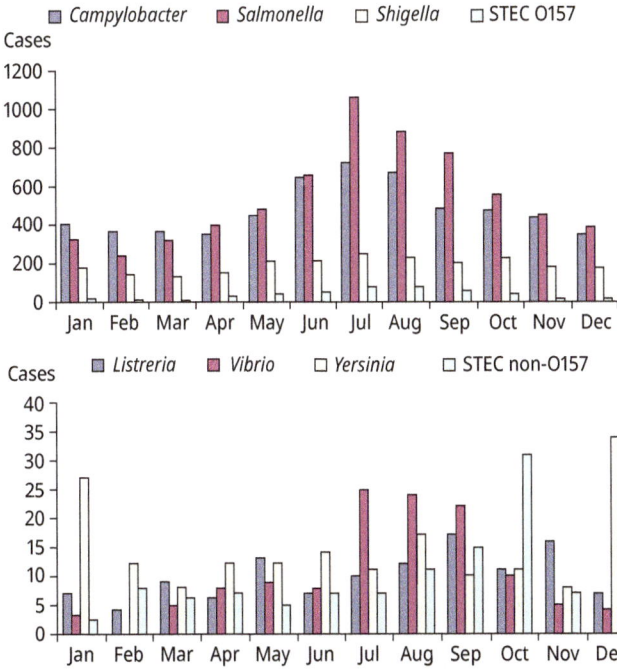

Figure 7.5: An epidemic curve showing cases of foodborne diseases caused by specific pathogens by month from foodnet surveillance report for 2004 [22].

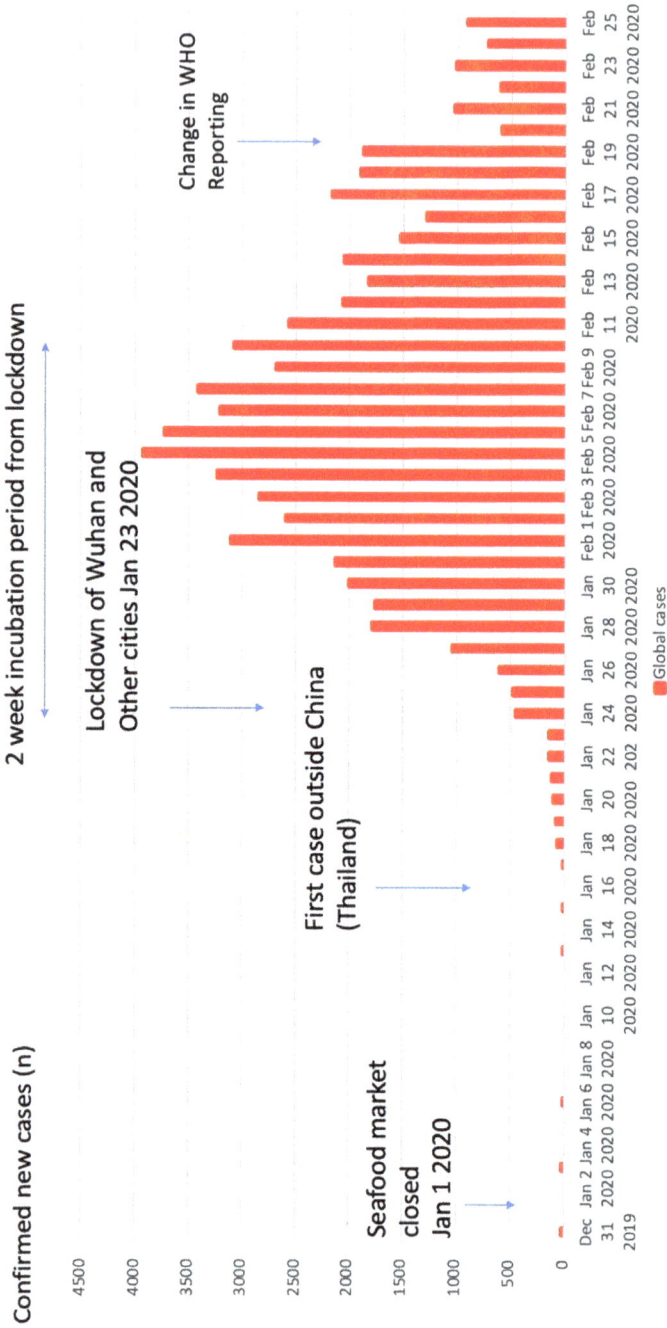

Figure 7.6: Geographic description of Covid-19 in Wuhan, China, and Thailand [23].

The interval between exposure and the commencement of symptoms is known as the incubation period. If a pathogenic agent was exposed at a particular location and period of time (a point source), the epidemic curve could have a unique peak. The geographic spread of illness can also be depicted on a map as represented in Figure 7.6. Figure 7.6 shows the spread of Covid-19 outbreak due to corona virus, which is the causative agent from Wuhan, China, to other outside China (Thailand).

Finally, descriptive epidemiology studies contain information about the afflicted persons, such as their age and gender status. Individuals with a significant level of exposure or who have grown ill are compared with those with no exposure. Often, an interview or a questionnaire is administered to collect data from sick and healthy people so that differences between those who are exposed and those who are not, ill and healthy people, can be evaluated and used to determine the potential factors that could be responsible for the disease, with particular reference to the foods consumed that may have been linked to illness. This procedure is referred to as analytical epidemiology cohort studies and case-control studies, which are two types of analytical studies widely utilized in foodborne illness epidemiology.

When a disease outbreak is attributed to a target group of individuals, cohort studies are commonly employed in outbreak studies to examine the incidence of illness in a precisely defined group of people (e.g., wedding occasion, church communion). In a cohort research, statistical approaches include calculating attack rates and relative risks. An attack ratio is the percentage of infected people after exposure to the illness during an outbreak. The outcome can be computed for all people who were exposed during an epidemic, or for a group of people with varied exposures, such as attack ratio among those who ate a particular meal. The comparative danger is determined by multiplying the exposed group's attack ratio by the non-exposed group's attack ratio. For instance, if 23 people became infected with campylobacteriosis disease after attending a church communion, with 20 of 50 chicken eaters and 3 of 30 non-chicken eaters, the comparative risk of illness among chicken consumers (exposed) relative to non-chicken eaters (non-exposed) is (20/50)/(3/30), or 4, indicating that chicken eaters were 4 times more likely to become ill. A relative risk of 1 implies that those who were exposed and those who were not also likely to acquire disease, whereas a comparative risk below 1 shows that exposure serves as a protective against disease [8, 24].

The study of particular characteristics of communicable pathogens at the molecular level is referred to as molecular epidemiology. Pulsed field gel electrophoresis, multiple locus variable number tandem repeat analysis, multilocus sequence typing, and full genome sequencing are some of the molecular subtyping techniques for foodborne infections. These methods help to determine if a group of people contracted the same or a varying microbial strain or subtype. A cluster or epidemic may be indicated by the discovery of a number of people with similar microbial strain or subtype over a period of time or in a specific environmental region. Molecular subtyping also aids in establishing a relationship between the suspected food vehicle and the illnesses under investigation [8, 25]. Surveillance and investigation are two major methods that have been applied in the control of foodborne illnesses.

7.8 Foodborne disease surveillance

Public health surveillance refers to a methodical gathering, analysis, interpretation, and distribution of data for public health action [8, 38]. Disease surveillance includes data from a number of sources, such as hospital records, physician visits, and death certificates, as well as continuing collecting and monitoring of morbidity and mortality. Surveillance data for foodborne diseases is based on laboratory diagnosis of foodborne infection cases. Foodborne infections and intoxication or poisoning exhibit similar symptoms and as a result, it is necessary to run laboratory diagnosis to determine the actual agent responsible for the disease [12].

Surveillance provides crucial data for the development and assessment of foodborne illness prevention strategies. This is achieved by taking records of illnesses that has occurred over time, demographic, temporal and geographic trends are identified via surveillance systems by making description of people with laboratory diagnosed ailments using personal information of characteristics as well as location and time of exposure to causative agent. This is aimed at providing summaries of important information regarding high risk population and potential exposures. The data achieved from surveillance can be used organizations to advance intervention schemes to diminish the incidence of disease as well as an evaluation of these interventions.

Foodborne disease surveillance can be established at regional, national and local strata by public health agencies [25]. Furthermore, surveillance systems are used to manage the occurrence of antimicrobial resistance in foodborne pathogens from different sources. This enables clinicians to make decisions concerning possible treatments and regulations for evaluation of the interlink between antibiotics used in animals, those available within the environment, and the resistance of the pathogens growing in humans [8].

7.9 Foodborne diseases outbreak and investigation

A foodborne illness resulting in outbreak is a public health hazard that requires the local health department's rapid attention. Some foodborne outbreaks are mild and may be treated locally, but others can spread across state lines and need the engagement of federal agencies. When two or more people (typically unrelated) get sick after eating the same meal, it is referred to as a foodborne outbreak. An outbreak, sometimes known as an epidemic, is a rapid increase in the population of people infected with a disease in which number of incidents of an ailment or disease in a specific location exceeds what is generally expected [12]. Foodborne diseases are a regular oc-currence around the world, despite the fact that many go unreported and/or un-treated. Each year, the World Health Organization reports that 600 million cases of people (almost 1 in 10) becoming ill and 420,000 death cases due to consumption of contaminated food [26]. Each year, hundreds of foodborne outbreaks are recorded in the United States. Moreover, over 900 foodborne reported disease outbreaks were ob-served by the Centers for Disease Control and Prevention (CDC) in 2015, involving bac-terial, chemical, toxin, parasite, or viral etiologies.

Why do so many food-related illnesses occur in the United States notwithstanding the high-quality food safety standards, and what are some of the primary variables that contribute to a foodborne illness? To begin with, there are a range of scenarios and reasons why foodborne illness occurs. However, food safety professionals and academ-ics generally agree that there are three main reasons why foodborne outbreaks occur;

- Contamination: Inefficient food preparation procedures that promote pathogens (e.g., natural poisons, toxic substances, uncovered hand contact) and other risks to enter into food – for instance, a sick food vendor who sells food with unwashed hands – are examples of contamination.
- Replication/proliferation of pathogens in food: Pathogens in food can proliferate swiftly if they are left at room temperature, cooled slowly, or left out in the sun for several hours.
- Survival of pathogens in food: This includes food that has not been cooked long enough or to the right temperature to kill pathogens [12].

Foodborne disease outbreaks can cause the death of large number of people within a short period of time, thus early discovery and management are critical. When the Health Team receives information of a disease outbreak, they are notified. If there is a risk of foodborne illness, action should be taken instantly. This action must be all-inclusive from the inception. The investigation and control of disease outbreaks calls for intensive effort of all health workers. Furthermore, being alert to such outbreaks in advance by gathering the essential details on foodborne illnesses and prior epidem-ics (specifically in the area) is critical [18]. The main goal of investigating a foodborne disease outbreak is expressed in Figure 7.7.

In investigating a foodborne disease outbreak, it is necessary to first identify the causative agent which resulted in the outbreak. After the causative agent has been identified, it is important that the source through which the causative agent was introduced, for example, the exact food, or food handlers responsible for the contamination of the food. The region or exact location from which the outbreak originated in order to monitor and curb further spread of the disease. It is also nec-essary to determine the circumstances and mechanisms that led to the contamina-tion of the identified food item after which the limitation and arrest of the impact of the progression of the outbreak should be attended to by applying the information

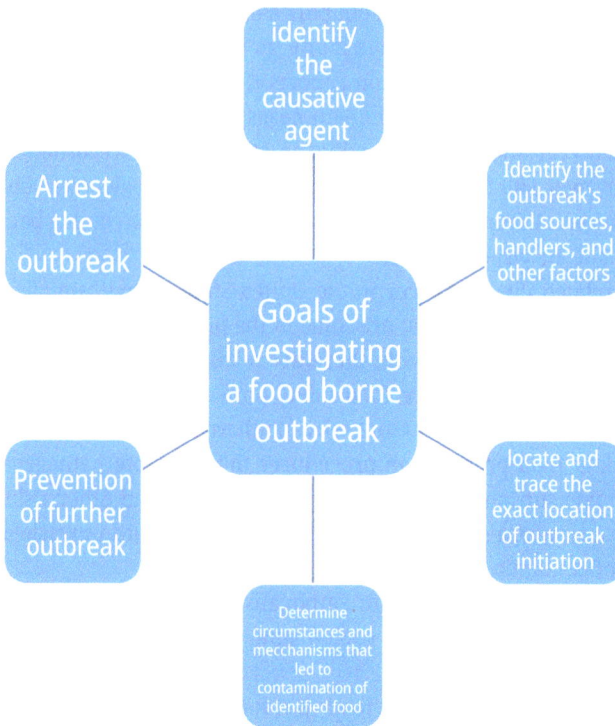

Figure 7.7: Goals of investigation of a foodborne disease outbreak.

and data obtained from investigation of the current outbreak to prevent further outbreaks as well as prepare for future outbreaks [24].

Having addressed the aim of investigating a foodborne disease outbreak, it is necessary for health workers to take certain steps in addressing possible foodborne disease outbreaks. These steps are listed as follows:

– Health workers should be able to obtain detailed information from all available informants. This entails the means of holding interviews with infected individuals, management and food handlers. It is necessary to further collate all the obtained information and register them systematically by means of prepared questionnaires.

– Attempts should be put in place to determine the average incubation period of the outbreak when information is being collected.

– The precise time and date when the suspected meal was consumed should be recorded as well as the number and proportion of those who ate against those who did not eat the food. In order to determine the attack rate, the proportion of those who were ill should be estimated. When assessing them, having and keeping a list of symptoms and indicators will be beneficial in determination of persons for the presence or absence of a foodborne illness. Nausea, vomiting, diarrhea, abdominal cramp, fever, headaches, and other signs are common in the outbreak of foodborne diseases.

– It is important to remember that the link between disease and exposure to the supposed food does not necessarily have to be "perfect"; in fact, it's seldom the case due to a variety of causes, one of which could be that the involved food not being contaminated at all times; additionally, host vulnerability fluctuates as well. There may be food reporting issues, as well as dose (the amount consumed). There may be inaccuracies in record keeping, as well as defective recall and doubt.

– If the epidemic is large enough that all participants cannot be interviewed, a random sample should be chosen and investigated about their symptoms and dietary exposure profile.

– Create a hypothesis that is centered on the preliminary clinical findings and other data on a possible foodborne illness in order to create a case. The sick people will be treated by management.

– Instruct informants and cases to keep or get all suspected food products, as well as their original containers and packaging.

– Collect suspicious food, feces, and vomit samples from sick people and forward them to a specified laboratory for investigation and analysis of the agent as soon as possible. Obtain and use proper sampling equipment, such as sterile containers and other instruments.

– Pay a visit to the place or location where the outbreak is might have originated. In the course of the tour, all the investigation team should avail themselves and conduct a systematic analysis of the environment and other scenarios, keeping track of every observation made.

– Analyze and interpret all of the data gathered using the various ways discussed above, and of necessity keep a track or record of the actual source of the contaminated food item.

– Finally, implement corrective measures and educate the public about prevention and control measures.

– Report your findings to the appropriate establishments, and maintain a record of them for onward reference.

It is critical to determine when to conduct a field examination of a suspected foodborne disease outbreak before launching a field investigation. There are several factors to deliberate in order to influence a decision to investigate or not. The factors include the following:

– A review of current resources for the investigation
– The complexity of the cases that have been reported
– Chances that the outbreak or epidemic is transmissible
– The pattern of the disease, if it is reported as being rare or an unusual
– Whether or if the epidemic was caused by a purposeful or unintentional conduct.

7.10 Prevention of foodborne diseases

The major way to prevent foodborne disease is to ascertain that food is safe for consumption. Ensuring food safety can be summarized in a number of steps which includes cleaning, separation, cooking and chilling.

7.10.1 Cleaning

Cleaning is an essential step in food safety, hands and food contact surfaces should be thoroughly washed as well as utensils used in food preparation or processing from time to time and in cases when they become polluted. Cleaning is said to be effective when soils and debris are properly removed or eradicated with hot foam-covered water and rinsed with clean water. The process of sanitizing aims at reducing or eliminating microorganisms to a level regarded as safe. Chemicals such as chlorine bleach are employed during sanitization. Washing of hands should not be replaced with alcohol-based sanitizers as they are not effective against some microorganisms such as *Norovirus*. Food items should be properly washed prior to cooking [14].

7.10.2 Separation

It is important that uncooked or raw food items are separated from cooked food. This is necessary to avert cross-contamination which refers to the movement of microorganisms from raw food materials, people, kitchen equipment, and counter tops to ready to eat food products. Separation should also be considered at the point of grocery shopping. Non-food items such as soaps, chemicals should not be kept with food items in the same bags. Never use recyclable bags for fresh meat, fish, or poultry for ready-to-eat items. Cloth-like bags should be cleaned and dried using air-drying machine. Before placing raw meat in a reusable bag, it should be separated into disposable plastic bags. Cross-contamination should also be avoided when foods are to be stored in the refrigerator. Fresh meats, poultry, fish should be kept beneath ready-to-eat foods when they are thawed in the refrigerator to prevent juices from them sipping into the ready to eat foods. Cross-contamination should also be avoided during handling, preparation and serving of foods [17].

7.10.3 Cooking

Food items should be thoroughly and properly cooked and at the right temperatures. Cooked food should be reheated prior to consumption. The USDA has advised that the following temperature stated below be used in cooking and that food products attain least possible internal temperature as specified in Table 7.4.

Table 7.4: USDA FSIS safe minimum internal temperature chart.

Product	Minimal internal temperature
Beef, pork, and mutton	145 °F (62.8 °C) and allow for resident time of at least 3 min
Comminuted meats	160 °F (71.1 °C)
Fresh and processed ham	145 °F (62.8 °C) and allowed for at least 3 min
Completely cooked ham	Reheat cooked hams contained in USDA-certified plants to 140 °F (60 °C);
All poultry (breasts, whole, legs, thighs, wings, and ground)	165 °F (73.9 °C)
Eggs	160 °F (71.1 °C)
Fish and shellfish	145 °F (62.8 °C)
Remnants	165 °F (73.9 °C)

Source: [26].

When cooking in a microwave oven, the food should be covered, stirred, and set in rotation for even cooking

7.10.4 Chilling

It is important that foods are properly stored. The two-hour rule for meat, poultry, seafood, eggs, and other perishables should be adhered to. The rule states that the above-named food product should be refrigerated within two hours of purchase or cooking. Food should never be thawed at room temperature. When refrigerating food products, they should be distributed in small containers for rapid and even freezing. Foods thawed in the microwave oven should be cooked immediately [14].

Other means of preventing foodborne diseases include the following:

Storage: When food products are stored above or below recommended holding temperatures, the proliferation and multiplication of bacterial cells is encouraged, growth of spores are catalyzed and toxins are produced to levels which are potentially dangerous. Certainly, consumer food safety studies have revealed about 93% of consumers are ignorant of proper low temperature (refrigeration) temperatures [39] and that 75% of customers are unaware of their personal refrigerator temperature. The dearth of information is matched by data documenting measurements of consumers' refrigerator temperatures. According to the findings, approximately 70% of customers' refrigerators surpassed approved temperatures [40], creating circumstances that might favor bacterial cell growth and raise the risk of foodborne illness.

Consumer awareness and education: Consumers remain the crucial last link in the food value chain when it comes to ensuring hygienic or safe food consumption and avoiding foodborne disease. Consumers have several food safety obligations since they not only buy items but also process and supply food for themselves and others. As processors of food, customers have obligations and must be aware of the characteristics and safety of food items. As a result, consumer food-handling habits in the home kitchen have an impact on pathogen existence and replication, and also cross-contamination of other goods.

7.11 Biocontrol of foodborne diseases

Food contamination by microorganisms results in the loss of around 25% of food output each year in the food industry [41]. Due to the minimal influence of hurdle technology/approach on the nutritional content and organoleptic qualities of food materials, the "hurdle approach" has become a new focal area of the food processing industry. Physical technologies that employ high-temperature processing techniques can degrade food quality [27].

7.11.1 Bacteriophages

Bacteriophages are obligatory bacteria parasites that multiply utilizing bacterial cells. These bacteriophages are usually host specific, limiting themselves to certain strains within a single bacterial species. Some bacteriophages, on the other hand, have a rather extensive host range, infecting numerous species within a genus and even members of closely related genera [28]. As a result, bacteriophages have been proposed as natural biocontrol agents against foodborne diseases that do not damage human cells, implying that they are harmless [42]. Bacteriophages are present in every living ecosystem and have been applied in human therapeutics, biopesticides, and food safety [28]. Bacteriophages as a tool for food ensuring food safety are applied in primary production as decontaminants for livestocks (*Campylobacter jejuni, Escherichia coli* and *Salmonella* spp.) and decontaminants for processed foods (*Campylobacter jejuni, Escherichia coli, Listeria monocytogenes, Shigella* spp., and *Salmonella* spp.). Bacteriophages are generally regarded as safe, ubiquitous, host specific, self-replicating, and do not alter the general composition of foods including color, odor, taste, and aroma. They can, however, undergo phage-mediated gene transfer, and resistance to phage might develop as a result of their use [29]. Microorganisms including phages mutate, and as a result of this mutation, they may lose genes that are responsible for biocontrol of pathogens, thereby making these pathogens resistant.

7.11.2 Bacteriocins

Bacteriocins are antibacterial chemicals synthesized by lactic acid bacteria (LAB) via ribosome synthesis. Bacteriocins generated by LAB often have antibacterial action against strains very similar to the producers, and also Gram-positive food spoilage and food-poisoning bacteria including *Bacillus* spp., *Staphylococcus* spp., *Listeria* spp., and *Clostridium* spp. [30]. LAB bacteriocins are usually resistant to elevated temperatures and low pH, and digesting enzymes can degrade them. In addition to these properties, LAB bacteriocins have received significant interest for future uses, such as food preservatives, because they are helpful and widely recognized as safe. In fact, nisin A, the best-studied bacteriocin, is applied as a food preservative in more than 50 nations by specific strains of *Lactococcus lactis* [30]. Furthermore, in the actual usage of LAB bacteriocins, adverse effects and formation of resistant bacteria have not been documented. One explanation is that they operate at extremely low concentrations by causing rapid development on the target cell membrane. Another advantage of LAB bacteriocins is their proteinaceous composition, which allows these antibacterial chemicals to be readily destroyed within human system and in the environment [29, 30].

7.12 Conclusion

Foodborne illnesses have been identified as a global challenge that requires immediate attention. In controlling a foodborne outbreak, the causes and sources of the foodborne illness have to be identified in order to develop measures to mitigate the effect of the illness on human populace. Effective control and prevention of foodborne illnesses are achievable by ensuring that the causative agents of these diseases are eliminated at every link across the food chain. Proper care should be taken during food processing and good manufacturing practices must be strictly adhered to taking into consideration regulations from regulatory bodies. Proper monitoring, surveillance, and investigation tools should be employed in order to curb the spread of an outbreak, thereby saving lives. The control and prevention of foodborne disease cuts across all individuals as everyone is a major participant of the food chain from processing to consumption; hence, all should be properly educated on their roles and expectation to reduce the burden of foodborne illnesses across the globe.

References

[1]   Jha, S. N. (2010). Food Safety and Quality. *Rapid Detection of Food Adulterants and Contaminants*, *2016*, 1–24. https://doi.org/10.1016/b978-0-12-420084-5.00001-9.

[2]   Morin, K. H. (2009). Foodborne infections. *MCN: The American Journal of Maternal/Child Nursing*, *34*(5), 325. https://doi.org/10.1097/01.NMC.0000360426.59851.0f.

[3]   Buckwell, A. (2003). Food Safety, Food Quality and the CAP. *Policy Vision for Sustainable Rural Economies in an Enlarged Euro*, 153–161. http://hdl.handle.net/10419/59979

[4]   Anwar, S. M., Uppal, S., Ganguly, G., Chitkara, A., & Gaur, M. (2009). *Human Ecology and family Sciences: Improve Food Quality and Food Safety* (First edition). http://www.worldbank.org/en/topic/agriculture/brief/improve-food-quality-and-food-safety

[5]   Braden, C. R., & Tauxe, R. V. (2013). Emerging trends in foodborne diseases. *Infectious Disease Clinics of North America*, *27*(3), 517–533. https://doi.org/10.1016/j.idc.2013.06.001.

[6]   Adley, C. C., & Ryan, M. P. (2016). The Nature and Extent of Foodborne Disease. In *Antimicrobial Food Packaging*. Elsevier Inc. https://doi.org/10.1016/B978-0-12-800723-5.00001-2.

[7]   Bari, M. L., & Yeasmin, S. (2018). Foodborne diseases and responsible agents. In *food safety and preservation*. Elsevier Inc. https://doi.org/10.1016/b978-0-12-814956-0.00008-1.

[8]   Hoffmann, S., & Scallan, E. (2017). Epidemiology, cost, and risk analysis of foodborne disease. In *Foodborne diseases: third edition* (Third Edition, Vol. 2010). Elsevier Inc. https://doi.org/10.1016/B978-0-12-385007-2.00002-4.

[9]   AL-Mamun, M., Chowdhury, T., Biswas, B., & Absar, N. (2018). Food poisoning and intoxication: A global leading concern for human health. In *Food Safety and Preservation*. Elsevier Inc. https://doi.org/10.1016/b978-0-12-814956-0.00011-1.

[10] Bintsis, T. (2017). Foodborne pathogens. *AIMS Microbiology*, *3*(3), 529–563. https://doi.org/10.3934/microbiol.2017.3.529.

[11] Getting, V. A. (2009). Food-borne diseases. *Nursing Outlook*, *13*(4), 364–367. https://doi.org/10.5005/jp/books/10560_29.

[12] Kearney, G. D. (2018). Introduction to Foodborne Illness Outbreak Investigations. *Environmental Public Health: The Practitioner's Guide*, *13*. https://doi.org/10.2105/9780875532943ch13.

[13] FAO. (2007). Food-borne Diseases: food Quality and Standards Service, FAO. *FAO/ OIE/ WHO Tripartite Meeting, Rome 31st Jan-2nd Feb*.

[14] Thobaben, M. (2010). Causes and prevention of foodborne illness. *Home Health Care Management & Practice*, *22*(7), 533–535. https://doi.org/10.1177/1084822310376611.

[15] Baraki, N., Wodajo, A., Abera, M., Oljira, L., Mitiku, H., Mengistu, S., & Ketema, F. (2005). *Food-borne Diseases*.

[16] Ombui, J. N. (2014). *Food borne diseases lectures*.

[17] Fraser, A. M., & Simmons, O. D. (2017). Food Safety Education. *Sustainability Challenges in the Agrofood Sector*, 643–659. https://doi.org/10.1002/9781119072737.ch27.

[18] Bhaskar, S. V. (2017). Foodborne diseases-disease burden. In *Food Safety in the twenty-first Century: Public Health Perspective* (Issue 1). Elsevier Inc. https://doi.org/10.1016/B978-0-12-801773-9.00001-7.

[19] Ahmet, K., Abdullah, D., Tolga, K., & Gökhan Kürşad, İ. (2017). Food Safety and Climate Change: Seasonality and Emerging Food Borne Pathogens. *Journal of Gastroenterology Research*, *1*(1), 24–29. https://doi.org/10.36959/621/584.

[20] ECDC (2014) European Centre for Disease Prevention and Control. Annual epidemiological report 2014 – Food- and waterborne diseases and zoonoses.

[21] Koluman, A. (2010). Detection of Campylobacter jejuni con- tamination in poultry houses and slaughterhouses. *Turk Hij Den Biyol Derg*, *67*, 57–64.

[22] http://scitechconnect.elsevier.com/waterillness/.

[23] https://jglobalbiosecurity.com/articles/55/print/#.

[24] Prieto, M., Colin, P., Fernández-escámez, P., & Alvarez-ordóñez, A. (2015). *Epidemiology, Detection, and Control of Foodborne Microbial Pathogens, 2015*, 2–4.

[25] WHO. (2008). Foodborne disease outbreaks. *Foodborne Disease Outbreaks: Guidelines for Investigation and Control.*

[26] World Health Organization. (2017). Food safety. Media Centre: Major foodborne illnesses and causes. Available at: http://www.who.int/mediacentre/factsheets. Accessed January 18, 2018.

[27] Ahmad, S. R., Kalam, A., & Ghosh, P. (2021). *Biocontrol Effect of Lytic Bacteriophages against Various Foodborne Diseases, 14*(June), 709–723.

[28] Kazi, M., & Annapure, U. S. (2016). Bacteriophage biocontrol of foodborne pathogens. *Journal of Food Science and Technology, 53*(March), 1355–1362. https://doi.org/10.1007/s13197-015-1996-8.

[29] Mahillon, J. (2017). *Biocontrol of foodborne pathogens : the pros and cons Foodborne illnesses.*

[30] Hayet, S., Sujan, K. M., Mustari, A., & Miah, M. A. (2021). Hemato-biochemical profile of turkey birds selected from Sherpur district of Bangladesh. *International Journal of Advanced Research in Biological Sciences, 8*(6), 1–5. https://doi.org/10.22192/ijarbs.

[31] Rawat, S. 2015. Food Spoilage: Microorganisms and their prevention. *Asian journal of plant science and Research*, 5(4): 47–56.

[32] Marriot N.G (1999): Principles of Food Sanitation. 4th ed. Aspen Publishers, Gaithersburg, MD.

[33] Ajayeoba, T. A., Kaka, O. M. and Ajibade, O. A. 2020. 7 Microbial food spoilage of selected food and food products. *Food Science and Technology: Trends and Future Prospects*: 173.

[34] Amin, O.M., 2002. Seasonal prevalence of intestinal parasites in the United States during 2000. Am. J. Trop. Med. Hyg. 66, 799–803.

[35] Birkhead, G. and Vogt, R. L. 1989. Epidemiologic surveillance for endemic Giardia lamblia infection in Vermont the roles of waterborne and person-to-person transmission. *American Journal of Epidemiology*, 129 (4): 762–768.

[36] Hald, T., Andersen, J.S., 2001. Trends and seasonal variations in the occurrence of Salmonella in pigs, pork and humans in Denmark, 1995–2000. Berl. Munch. Tierarztl. Wochenschr. 114, 346–349.

[37] WHO, (2002). The World Health Report 2002. WHO, Geneva.

[38] Sharma, S., Gittelsohn, J., Rosol, R. and Beck, L. 2009. Addressing the public health burden caused by the nutrition transition through the Healthy Foods North nutrition and lifestyle intervention programme. Journal of Human Nutrition and Dietetics, 23: 120–127.

[39] FOOD SAFETY AUTHORITY OF IRELAND. 1998. Public Knowledge and Attitudes to Food Safety in Ireland. October 1998, Research and Evaluation Services.

[40] Ndraha, N., Hsiao, H.-I., Vlajic, J., Yang, M.-F. and Lin, H.-T. V. 2018. Time-temperature abuse in the food cold chain: Review of issues, challenges, and recommendations. Food Control, 89: 12–21.

[41] Sarhan, W. A. and Azzazy, H. M. 2015. Phage approved in food, why not as a therapeutic? Expert review of anti-infective therapy, 13(1): 91–101.

[42] McCallin, S., Sarker, S. A., Barretto, C., Sultana, S., Berger, B., Huq, S., Krause, L., Bibiloni, R., Schmitt, B. and Reuteler, G. 2013. Safety analysis of a Russian phage cocktail: from metagenomic analysis to oral application in healthy human subjects. Virology, 443(2): 187–196.

Abimbola Motunrayo Enitan-Folami* and Feroz Mahomed Swalaha

Chapter 8
The applications of DNA-based techniques to food safety

Abstract: In the modern food trade, the safety of food products is a key subject because food and its products may encounter different health hazards along the production chain, which can result in contamination. Foodborne pathogens are among the leading cause of infection, outbreaks, and death, with significant economic impacts in both developing and developed countries. Interestingly, the applications of molecular and metagenomic analysis to determine food authenticity, investigate food microbiology, identify foodborne pathogens with other undesirable traits through identification and tracking of antibiotic resistance genes has gained considerable attention over the past few years. Hence, this chapter provides an overview of advanced molecular technologies that are currently in use for the identification and quantification of DNA fragments or whole genome of microorganisms so as to ensure good food quality and safety. The principles, advantages, disadvantages, and applications of polymerase chain reaction (PCR), quantitative polymerase chain reaction (qPCR), digital droplet PCR (ddPCR), and next-generation sequencing (NGS) in the identification and quantification of microbiome are discussed in this chapter. These techniques are important tools for quality assessment and mitigation of potential risks that are associated with the contamination of food resources. They have a wide range of sensitivity, varied specificity levels, robustness, time, and labor-saving to prevent food safety incidents.

8.1 Introduction

Foodborne illnesses often occur as a result of ingesting water or food that is contaminated with pathogenic bacteria, parasites, toxins, or viruses [1]. In particular, raw or ready-to-eat food products carry a higher risk of contamination, as they can harbor microorganisms like verocytotoxin-producing *Escherichia coli*, *Campylobacter* spp., *Listeria monocytogenes*, and *Salmonella* spp. If these microorganisms are ingested and enter the body through the gastrointestinal tract, they can give rise to various health-related

*Corresponding author: Abimbola Motunrayo Enitan-Folami, Department of Biotechnology and Food Technology, Durban University of Technology, P.O. Box 1334, Durban 4000, South Africa, e-mail: enitanabimbola@gmail.com
Feroz Mahomed Swalaha, Department of Biotechnology and Food Technology, Durban University of Technology, P.O. Box 1334, Durban 4000, South Africa

https://doi.org/10.1515/9783110748345-008

issues. These range from mild gastrointestinal discomfort to more severe conditions, including infections, food poisoning, and, in severe cases, life-threatening illnesses. To address these risks, the concept of food safety has emerged as a scientific discipline focusing on preventing infections or diseases that may arise at any stage of food preparation, handling, processing, storage, and packaging while ensuring adequate nutrient retention for a healthy diet. In the modern food trade, ensuring the safety of food products is of utmost importance, as food and its products can encounter different health hazards along the production chain, leading to contamination. The emergence and re-emergence of foodborne pathogens and diseases are also significant concerns for public health in the modern food trade.

The increasing occurrence of these pathogens highlights the need for heightened attention and stringent measures to ensure food safety and protect public health. According to the World Health Organization [2], food contamination can result in the development of more than 200 diseases, including diarrhea, gastroenteritis, meningitis, and certain types of cancer. Alarmingly, approximately 1 in 10 people worldwide fall ill each year due to consuming contaminated food, leading to an estimated 420,000 deaths annually. These illnesses impose a substantial burden on public health, causing the loss of 33 million healthy years of life measured by disability-adjusted life years (DALYs). The economic impact of foodborne illnesses is also significant, with approximately $10 billion lost annually in productivity and medical expenses in the United States alone [3]. The Centre for Disease Control and Prevention (CDC) has identified the top five most significant risk factors for foodborne illnesses (Figure 8.1), which contribute to one in six Americans falling sick each year, with approximately 3,000 deaths reported annually [4].

To address these challenges and mitigate the risks associated with foodborne pathogens, it is therefore crucial to prioritize food safety practices, regulations, and surveillance systems. This includes monitoring the effectiveness of hygienic processing, ensuring product quality and shelf-life stability, and verifying the absence of pathogens and their toxins to ensure food safety. Good manufacturing practices (GMPs) in the food industry play a crucial role in integrating and implementing various practices to ensure food safety. More so, GMP serves as a comprehensive monitoring process that helps detect the presence of pathogens and microbial toxins in food products, thereby preventing foodborne diseases and outbreaks.

Promptly detecting foodborne pathogens in food products is essential to prevent food poisoning or foodborne illnesses [5]. Therefore, surveillance, prevention, and control strategies are crucial in mitigating the risks associated with foodborne pathogens and minimizing their impact on public health and the economy. Utilizing cost-effective, rapid, and sensitive detection and quantification technologies is crucial for early identification and prevention of illnesses. Nucleic acid-based methods have emerged as valuable tools for the sensitive and rapid detection and identification of foodborne pathogens. These methods contribute to enhanced food safety and protection by enabling early identification and prevention of illnesses. By implementing these advanced techniques, we can effectively mitigate the risks associated with foodborne pathogens, safeguard public health, and reduce the financial burden caused by medical costs and product recalls. Hence, this chapter provides a review of early detection methods such as polymerase chain reaction (PCR), real-time PCR, multiplex PCR, and next-generation sequencing (NGS) for enhancing food safety and protection. By adopting and implementing these techniques, we can effectively identify and address risks associated with foodborne pathogens, ensuring the safety of food products and protecting public health. More so, the integration of these innovative methods guarantees the detection and mitigation of threats, securing our food supply and promoting overall well-being, thus preventing illnesses.

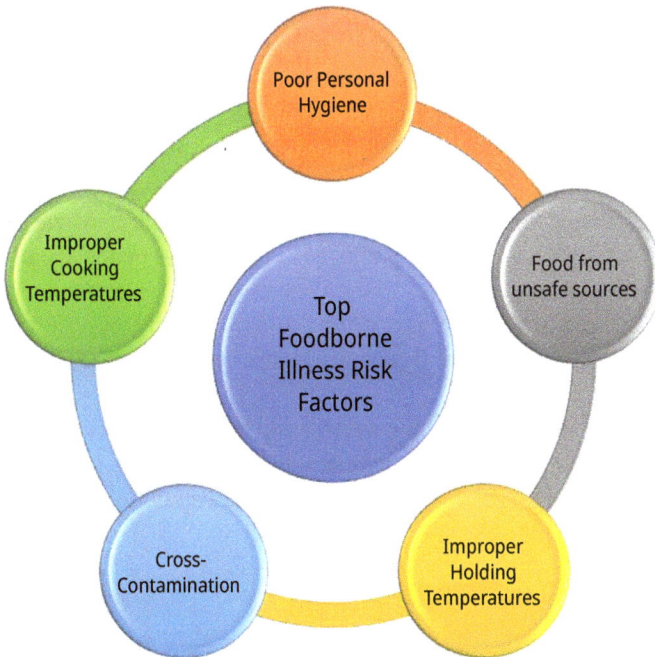

Figure 8.1: Top five foodborne illness risk factors.

8.2 Importance of rapid detection methods in food safety

There is a need for pathogens to be detected rapidly and accurately in a bid to achieve food safety and forestall foodborne disease outbreaks. Hence, the effective testing of pathogens such as bacteria and fungi (molds and yeasts) requires methods of analysis that can meet several challenging criteria. However, the traditional culture-dependent method as the oldest culturing and plating technique has been shown to be biased, inadequate, time-consuming, and gives an unambiguous result to be used in the food sector. It sometimes takes 4–9 days to obtain negative results and confirmatory results between 14 and 16 days based on the biochemical and morphological analysis of isolates [6]. This is apparently inconvenient for many industrial applications, especially in the food industry. The advancements in nucleic acid (DNA/RNA) based methods to detect microorganisms of interest offer better sensitivity over conventional microbiological techniques. The potential benefits of nucleic acid-based methods of high throughput give precise and reliable results with reduced time. Some of the unique characteristics of rapid and sensitive techniques that are newly adopted in the food industry include the following:

1. **Database development:** The technique should be able to detect and identify more than one strain of pathogens per assay or at a time. It should be able to predict microbial pathogens in a similar food product.
2. **Reduce time and cost:** The turnaround period for microbial analysis should be shorter than the traditional method. The techniques will not only reduce the time of analysis but also decrease human error and labor cost.
3. **Sensitivity and infectious dose determination:** It should provide immediate information on the presence or absence of pathogens in both raw materials and finished food products.
4. **Efficient monitoring process:** It should contribute to efficient process control, improve, and give a self-conscious manufacturing hygienic practice.

Therefore, in the subsequent sections, this chapter will delve into the discussion of commonly employed techniques for detecting the microbial community in food samples. It will provide valuable insights into the applications and capabilities of these techniques, highlighting their crucial role in ensuring food safety.

8.3 Trend in rapid detection techniques for food quality protection

The monitoring and detection of pathogenic bacteria play a crucial role in identifying and addressing food safety issues. According to Enitan-Folami and Swalaha [7], the essence of biotechnological tools is to improve and enhance food processing to provide efficient, quality, safe, and consistent food products within a short period of time. However, the traditional culture-dependent method has been shown to be biased inadequate, and time-consuming to be used in the food sector. The molecular-based techniques are becoming popular as the culture-independent approaches to detect and quantify foodborne pathogens [8]. Gene-based detection of pathogens presents several potential advantages compared to traditional methodologies, as it enables accurate, sensitive, and reliable differentiation of microbial communities in food samples. A diverse array of molecular techniques, like DNA microarrays utilizing gene chip technology, PCR, quantitative polymerase chain reaction (qPCR), loop-mediated isothermal amplification (LAMP), and NGS methods, which are DNA/RNA-based, have been successfully employed for the early detection, profiling, and characterization of foodborne pathogens in a wide range of food products, from fresh to ready-to-eat items (Figure 8.2).

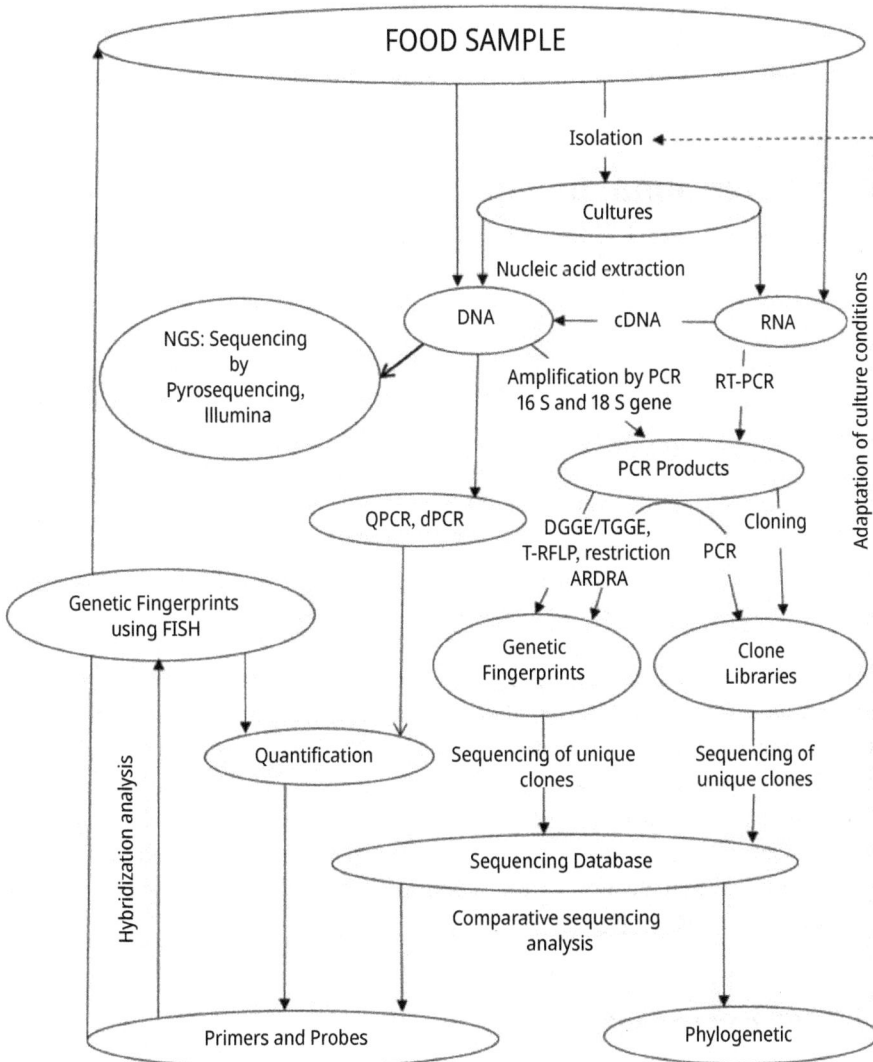

Figure 8.2: Biotechnological techniques employed in studying microbial communities in food samples to ensure safety [7].

In molecular biology, DNA or RNA extraction is often used as a starting point for all downstream molecular genomic techniques (aside from flow cytometry and fluorescence in situ hybridization (FISH), which require whole cells). In most cases, many laboratories develop their DNA/RNA extraction methods that are specific to their samples, while others use commercial kits. The topic of DNA/RNA extraction and optimization steps, as well as the synthesis of a complementary DNA (cDNA) strand from the RNA extracts, are too broad to be discussed in the context of this chapter. However, the focus will be on the application of PCR and its derivatives like qPCR, digital droplet PCR (ddPCR), and NGS in the food industry (Figure 8.2).

8.4 Conventional polymerase chain reaction (PCR) as powerful genetic fingerprint technique in food safety

In traditional (endpoint) PCR, detection of amplified sequences of a cDNA or gDNA template are carried out at the end of the PCR assay. The post-PCR analysis is often qualitative and the success of a reaction is often judged by comparing the amplified product with a known size standard. An approximate quantification of the initial target may be carried out by extrapolating from the final amplicon concentration.

8.4.1 Principle of polymerase chain reaction

In 1983, Mullis developed a new technology called polymerase chain reaction, which was later patented in 1985 [9]. The PCR process involves repetitive replication of a specific DNA region to produce numerous copies of a particular sequence through an enzymatic process. This technique relies on the utilization of DNA polymerase for in vitro replication of specific DNA sequences and mimics the natural DNA replication process of a cell in a test tube. It amplifies a single or a few copies of pieces of sequences in the nucleic acid, generating thousands to billions of copies of a desired DNA or RNA fragment of interest [9].

8.4.2 PCR components

Every PCR reaction must contain the following five main components (Figure 8.3); polymerase enzyme, oligonucleotide primers, deoxyribonucleotide triphosphate (dNTPs) buffer system, and the DNA template.

Polymerase enzyme: A thermostable polymerase enzyme such as DNA polymerase or Taq polymerase that does not denature at very high temperature is incorporated into the PCR master mix for replication. Taq polymerase was obtained through the isolation, purification, or cloning of the enzyme from an extremophilic bacterium called *Thermus aquaticus,* by Thomas Brock in 1965 [10, 11]. This bacterium resists temperatures above 100 °C in hydrothermal vents and survives hot springs due to its ability to produce polymerase enzymes as compared with DNA polymerase produced from *Escherichia coli* that has an optimum temperature between 75 and 80 °C. Taq polymerase has a remarkable characteristic to withstand high temperatures above 100 °C for the denaturation of protein during PCR [9, 12]. In addition to Taq and DNA polymerases, other thermostable polymerases have been discovered in thermophilic archaea and bacteria. One example is Pfu DNA polymerase, which possesses a "proof-reading" activity and is used for high-fidelity DNA amplification instead of (or in conjunction with) Taq polymerase [13]. The use of this thermostable DNA polymerase eliminates the need for additional polymerase enzyme in the PCR master mix, resulting in successful amplification.

Buffer system: Divalent cations such as magnesium and potassium are utilized as catalysts for polymerase enzymes to facilitate optimum denaturation and annealing of DNA. These cations play a crucial role in maintaining the fidelity and stability of polymerase activity by serving as a cofactor to stabilize two single strands of DNA (ssDNA). However, it is essential to maintain the correct concentration of cations because excessive amounts can lead to greater promiscuity of the Taq polymerase. For instance, approximately 2.5 µM of Magnesium (Mg^{2+}) concentration is usually used for PCR reaction [14].

Oligonucleotide primers: Primers are short oligonucleotides that are synthesized and added as single-stranded DNA fragments. They contain sequences that are complementary to the 3'- and 5'-terminal regions of a target DNA sequence of interest. Typically, primers are 20–25 bases long, but not more than 30 nucleotides in length [15]. The primers are crucial components in the amplification of target DNA fragments from a pool of DNA. They play an essential role in forming hybrids or duplexes of a higher order.

Deoxyribonucleotide triphosphate: Deoxyribonucleotide triphosphate (dNTPs) are single units of bases (A, C, G, and T) that provide energy for polymerization and building blocks for the synthesis of a new strand of DNA. The addition of new DNA bases requires the use of the right concentration of buffer solution to ensure the optimum conditions for PCR. Some of the important functions of oligonucleotide primers in DNA amplification, including:
1. The initiation of DNA synthesis.
2. Facilitate the elongation of strands by DNA polymerase and contribute to the synthesis of complementary strand.

3. During the amplification process, DNA polymerase plays a crucial role in catalyzing the assembly of dsDNA from dNTPs, which act as the building blocks. DNA polymerase begins at the 3'-end of the primer and proceeds to complement the opposite strand during DNA replication. The DNA polymerase utilizes a ssDNA template as a guide to synthesize a complementary DNA strand. The dNTPs are then added sequentially and joined together by the polymerase, resulting in the formation of a new dsDNA strand. This newly synthesized dsDNA strand consists of the molecular units necessary for the structure of RNA and DNA [16].
4. Ensure that the primer's annealing temperature falls within an appropriate range to prevent issues such as miss-hybridization or failure of annealing, which could lead to incorrect DNA extension during the amplification process [15].
5. Also, during the process of designing primers, it is of utmost importance to carefully select sequences that specifically target a particular DNA region. This precaution is necessary to prevent the possibility of miss-hybridization to similar nearby sequences. By ensuring the unique binding of primers to the intended target to minimize the chance of nonspecific interactions.
6. Another critical consideration in primer design is to avoid self-annealing or annealing between different primers within the PCR mixture. This is to prevent the formation of undesired "primer dimer" products that can contaminate the PCR. Therefore, it is essential to design primers in a way that minimizes potential annealing interactions.
7. The use of degenerate primers to reduce PCR specificity in order to enable gene amplification from uncultivated microorganisms or those lacking genomic information.

8.4.3 Steps in PCR process

The majority of PCR machines now employ thermal cycling to amplify extracted nucleic acids. This technique involves a series of temperature steps that are precisely controlled to facilitate the amplification process. The major PCR temperature steps are carried out in repeated cycles (Figure 8.3). This involves denaturation, annealing, and elongation steps. The PCR tube consisting of reaction mixtures is inserted into a metal thermal block well in a thermal cycler system. The thermal cycler employed in PCR assays is programmed to raise and lower the temperature of the heating block in a series of predefined steps throughout the reaction process. This meticulous temperature control is essential for the specific stages of DNA denaturation, annealing, and extension that occur during PCR [15].

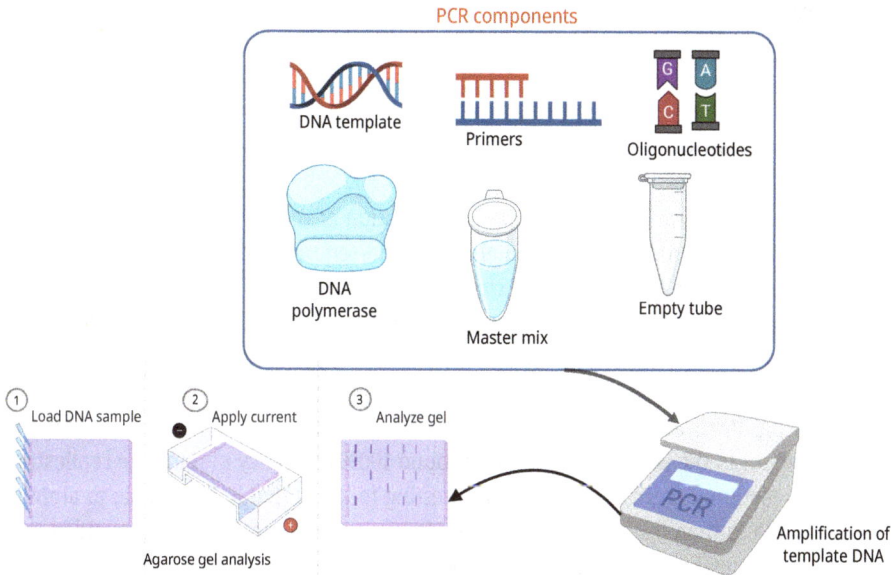

Figure 8.3: Overview process for PCR assay and analysis of generated products. Created with BioRender.com.

8.4.3.1 The denaturation step

The initial step in thermal cycling is denaturation or separation of dsDNA. At this stage, the heating and melting of dsDNA are performed at temperatures between 92 and 96 °C. The denaturation temperature unwinds and separates dsDNA into ssDNA by breaking down the hydrogen bonding between the base pairs within a short period (between 10 and 30 s) [15]. Then, every single strand is used as a template to build a new strand of dsDNA.

8.4.3.2 Annealing stage

The annealing stage is where hybridization of primers takes place between 40 and 70 °C. During the annealing step of PCR, the temperature is lowered to allow the primers to anneal or bind to their complementary sequences on the ssDNA template. Each primer anneals to its respective forward or reverse strand of the DNA molecule. This allows the formation of hydrogen bonds to reform, thus resulting in two new dsDNA. During amplification short single-strand primer sequences complementary to the region that could align to the ends of the new DNA sequence of interest are used. The higher the annealing temperature, the more selective and specific the hybridization; therefore, short-chain nucleotides in the primers make it more easier to hybridize than the long strand of DNA [9].

8.4.3.3 Elongation step

The last step in the cycling is the enzymatic replication of the DNA at an elongation temperature of 72 °C [9]. At elongation temperature, the required volume of the dNTPs and polymerase enzyme in the reaction mixture binds with the ssDNAs to start DNA replication at the primer location. In the next cycle, the previously synthesized dsDNA fragment is heated to separate the strands (Figure 8.4), and this happens for about 20–40 cycles of PCR process per assay in order to synthesize an analyzable amount of DNA of approximately 0.1 µg. With each cycle of PCR, the amount of DNA theoretically doubles the amount present in the previous cycle with around 100% efficiency of template replication in a chain reaction (Figure 8.4) [15]. It is interesting to know that, it is possible to amplify sequences of less than 6 kilobases with PCR. However, it is therefore recommended that an extended elongation step should be included in a cycle at 72 °C at a rate of 2 min per kilobase, if a sequence of interest is greater than 1 kilobase [9, 17]. It is important to note that about 35–40 cycles per PCR assay are generally recommended to provide sufficient synthesis of dsDNA in a sample for efficient further analysis and identification.

Overall, PCR provides a versatile and efficient tool for DNA amplification, offering several advantages:

(1) **Simplicity and power**: PCR is a simple yet powerful technique that enables the amplification of specific DNA regions, resulting in the generation of numerous copies of a desired sequence.
(2) **Signal amplification**: PCR can amplify a signal of interest from background noise, enhancing the detection sensitivity of target DNA or RNA molecules.
(3) **Enhanced sensitivity and rapid diagnosis**: PCR allows for faster diagnosis and identification of genetic material, providing enhanced sensitivity for detecting even small amounts of target nucleic acids while maintaining specificity.
(4) **Molecular cloning**: PCR serves as a molecular cloning method, facilitating the purification and amplification of cloned DNA fragments, which can be further analyzed and identified using biochemical techniques.
(5) **Speed**: PCR is a rapid process that typically takes a few hours, with 30 cycles completed in approximately 2–3 h. High-speed machines and specialized master mixes can further shorten the reaction time to less than an hour [15].
(6) **Amplification of single copy**: PCR can amplify a single copy of a nucleic acid target, enabling the detection of targets that may be undetectable using standard hybridization methods. This amplification can generate millions or more copies of the target sequence within a relatively short time frame. PCR wide range of applications makes it invaluable in various fields, including research, diagnostics, and molecular biology.

Nucleic acid amplification technology using PCR techniques has certain limitations. It is important to consider these limitations and address them appropriately to ensure accurate and reliable results when using this technology. Some of the limitations are:

(1) **Specialized skills and expertise:** PCR requires specialized skills and expertise in molecular biology techniques, including primer design, optimization, and protocol setup.

(2) **Primer specificity**: PCR primers must be designed to be unique to the target organism or sequence of interest, hence, requiring careful primer design and optimization.

(3) **Limitations in quantitative analysis**: While PCR has revolutionized nucleic acid detection, its application to quantitative analysis is limited. PCR has low sensitivity, and the analysis is often performed at the endpoint, which may not accurately reflect the initial concentration of the template DNA. PCR is also associated with inherent limitations and biases that can affect the quantification accuracy [18].

(4) **Further analysis of PCR amplicons**: The PCR amplicons generated need further analyzes, typically through techniques like gel electrophoresis, to visualize the amplicon and determine its quantity and size.

(5) **Additional analysis for target identification**: For complete identification of the target, additional technical analysis such as DNA sequencing may be required to determine the specific sequence of the amplified DNA.

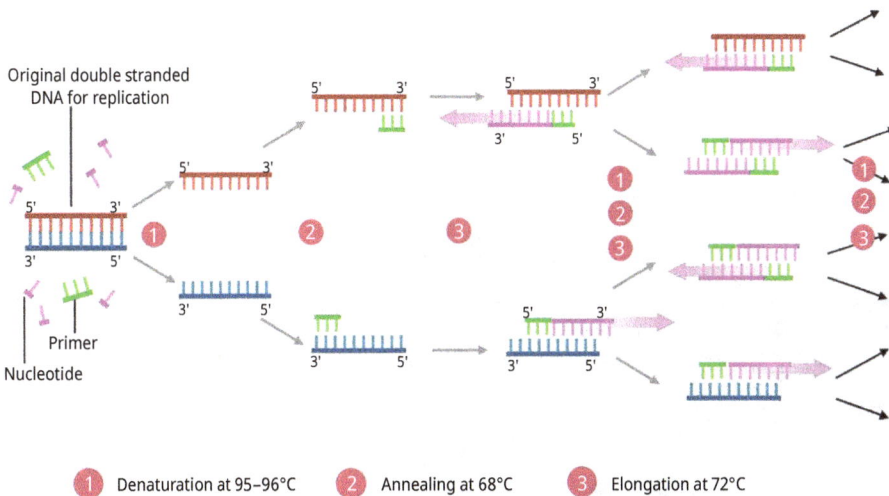

Figure 8.4: Exponential amplification of DNA fragments containing sequences that are complementary to the terminal part (3'- and 5'-ends) of a sequence of interest or target region in the extracted DNA during polymerase chain reaction. Created with BioRender.com.

8.4.4 Gel electrophoresis

Gel electrophoresis is indeed a commonly used analytical technique following PCR. It is used to resolve PCR amplicon on gel matrix, such as agarose or polyacrylamide in order to check the quality and size of the DNA fragments produce during the amplification [19]. It allows for the separation and visualization of DNA fragments based on their size, which provides valuable information about the amplified products. By applying an electric field to a gel matrix, such as agarose or polyacrylamide, the negatively charged nucleic acid molecules migrate through the gel. Smaller DNA fragments move more quickly through the gel matrix, while larger fragments migrate at a slower rate.

Gel electrophoresis is particularly useful for determining the presence or absence of a specific target population within a complex mixture of DNA or PCR products. This separation by size enables the sorting and visualization of DNA fragments according to their number of base pairs (Figure 8.5). By comparing the migration distance of the amplified DNA fragments with known DNA size markers, the size of the target fragments can be estimated when viewed under ultraviolet (UV) light as illustrated in Figure 8.5. At this stage, gel electrophoresis aids in the identification of unwanted PCR products, such as off-target amplifications and primer dimers (Figure 8.6). These artifacts can be distinguished from the desired target fragments based on their distinct size profiles. Therefore, gel electrophoresis is an essential tool in molecular biology as it allows for the separation, visualization, and identification of DNA fragments. It helps in accurate detection of target populations, exclusion of unwanted products, and detection of artifacts, enhancing the reliability and specificity of PCR-based experiments.

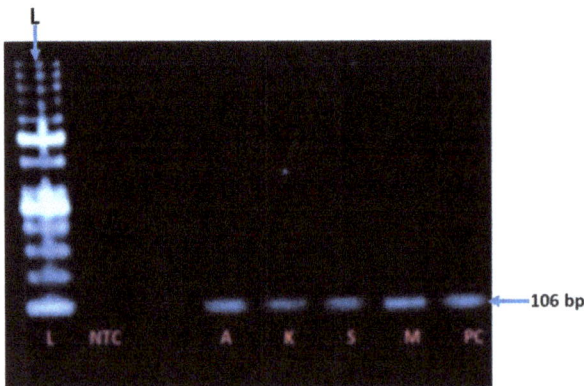

Figure 8.5: Clear bands of amplified PCR products of *Salmonella* spp. specific gene that was resolved on 1% (w/v) agarose gel electrophoresis at 80 V for 60 min and visualized under ultraviolet (UV) light. L is the molecular ladder, PC is a positive control, A, K, S, and M are the target amplified in the sample, while NTC is the no template control used in the PCR assay.

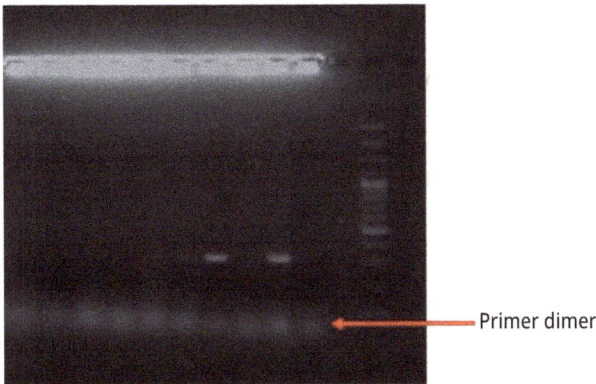

Primer dimer

Figure 8.6: Gel electrophoresis showing primer dimers formed from resolved PCR products on agarose gel.

8.5 Enumeration of microbial concentration in food samples using quantitative PCRs

From the onset of PCR technology, numerous derivatives of PCR, such as nested, inverse and in situ PCR, digital PCR (dPCR), fluorescent quantitative PCR, reverse transcription PCR (RT-PCR), anchored PCR, asymmetric PCR, immune PCR (IPCR), long fragment PCR, multiplex PCR, and multiplex allele PCR have emerged (Figure 8.7). Like conventional PCR technique, qPCR is a powerful technique for the amplification and quantification of target gene sequence to multiple detectable copies. The principle of qPCR is similar to that of conventional PCR, as both techniques involve the amplification of a target gene over a defined number of PCR cycles [20]. However, qPCR differs from conventional PCR in several key aspects. Unlike conventional PCR, qPCR does not require agarose gel electrophoresis for the detection of PCR products. Instead, it utilizes fluorescence-based detection to monitor the amplification in real time. This is achieved by using specific dual-labeled probes or intercalating dyes that produce fluorescent signals during the amplification process. The intensity of the fluorescence generated during the qPCR amplification process is directly proportional to the amount or level of the target DNA present in the sample [21, 22].

Another advantage of qPCR over conventional PCR is its increased sensitivity and accuracy. In recent years, qPCR has emerged as a rapid and reliable method for the simultaneous detection of foodborne pathogens in dairy products. This real-time detection is crucial for the absolute and accurate quantification of targeted DNA sequences [23]. The lower limit of detection is enhanced, and quantification of the target DNA can be achieved by comparing the fluorescence signals to standard curves generated from known concentrations of target DNA. In addition, qPCR offers a lower risk of cross-contamination compared to conventional PCR. This is because the presence of the target in the sample is indicated by an increased fluorescent signal, eliminating the need for post-amplification analysis, hence, minimizes the possibility of false-positive results.

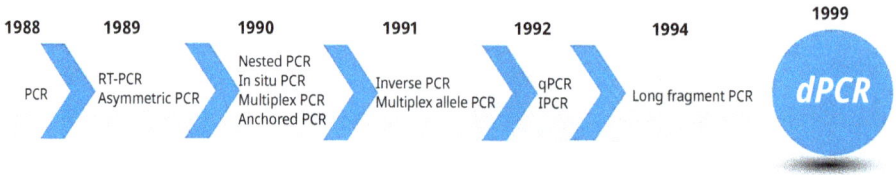

Figure 8.7: The progressive timeline of PCR technology in molecular genetics [24].

There are several detection chemistries available, like intercalating dyes, hybridization probes, and hydrolysis probes that are used in the real-time assay [25]. Unlike the conventional PCR technique that uses only endpoint detection, the amount of amplified DNA in the food sample is measured after each cycle in the real-time assay based on the fluorescence resonance emission from the fluorescent dye or probe [20, 26]. The instrument measures the accumulating fluorescent signal, which is directly proportional to the number of PCR amplicons produced. The observable variations in the intensity of fluorescence provide real-time information about the concentration of the amplified gene [20, 27, 28].

Other benefits of using qPCR over endpoint PCR analysis include: (1) speed and specificity of detection of PCR amplification products; (2) a high sensitivity, capable of detecting 1–10 colony-forming units (CFUs) per assay, that can be achieved by implementing a bacterial enrichment step for a minimum duration of 24 h in specific food matrices [14, 29]; (3) the collection of data during exponential amplification phase in real-time; (4) automation potential [14]; (5) it allows users to amplify, estimate, and simultaneously quantify the initial starting concentration of target gene with high level of precision. It also enables simultaneous detection of multiple targets and/or different organisms [25]. In other words, the quantity of amplified gene copies derived from bulk DNA is indicative of the relative abundance of microorganisms present in the food sample. This quantitative analysis provides valuable information about the quantity and relative abundance of specific genes or target sequences, contributing to a better understanding of the DNA composition in food samples. Hence, the quality of extracted genomic DNA strongly affects the efficiency and reliability of qPCR and other molecular techniques results [30].

Using qPCR, the quantification of either a DNA or RNA target is done in two ways: absolute quantification (AQ) and relative quantification (RQ). For total microbial quantification, the commonly used method is AQ as compared to RQ. The application of RQ is routinely used for monitoring changes in gene expression because it compares changes in the expression of a specific functional gene in relation to a constitutively expressed housekeeping gene [28, 31]. RQ determines the actual transcript number and the level of gene expression. It can be used to compare gene expression between two treatments in a given sample relative to another reference sample. The template is recognized to contain the targeted message in a high concentration; however, its exact quantity may not be precisely determined.

Moreover, AQ involves the construction of a linear standard curve for the direct measurement of samples along with the curve. It is used to measure a known concentration of template used for the construction of a standard curve. The qPCR machine will generates an amplification plot by delimiting fluorescence signals from each sample against the cycle number to create accumulation plots of products, and the signal should double in every cycle. The concentration of an unknown target gene in a sample is determined by comparing the observed fluorescence intensity during real-time PCR with known copy numbers of standard references. By analyzing the amplification curves, the fluorescence intensity of the unknown samples can be compared to the standard curve, allowing for an estimation of their target gene concentrations. This method accurately quantifies target gene concentrations in samples by utilizing the relationship between known standards and their fluorescence intensities throughout the real-time PCR assay [20, 26, 32]. Therefore, early detection of a peak in the amplification cycle suggests the abundance of target DNA or RNA in the sample, while late observation shows that the target sequence is scarce [26].

The quantification cycle (C_q), also known as the threshold cycle (C_t), holds significant importance in qPCR assays. It represents the number of amplification cycles required for the template DNA to reach a detectable level or the cycle at which the amplification curve intersects the amplification threshold [33]. Both quantification methods utilize the C_t value to measure the target concentration in the PCR reaction. A lower C_t value indicates a higher concentration of genetic material in the sample. Therefore, determining the absolute concentration of unknown sample, a standard curve needs to be generated using a known amount of target template by plotting C_q against concentration (log concentration). Then, the derived linear regression analysis equation of the standard curve is used for the calculation. The qPCR has emerged as a preferred method for studying microbial ecology in various ecosystems, including food samples. This is primarily due to its key advantages, including reproducibility, specificity, sensitivity, and the ability to quantify over a wide range of target. These characteristics make qPCR a valuable tool for generating valuable insights and understanding the microbial dynamics and populations within different environments, including the analysis of microorganisms in food samples.

Despite its widespread use in detecting food pathogens, qPCR is limited by its reliance on cycle threshold (C_t) measurements, which offer RQ [29]. Other issues and critical points that can significantly affect qPCR results, particularly when dealing with complex sample matrices and a low range of microbial concentration include the need for consideration to use group-specific primers and florescent probes followed by nucleic acid extraction efficiency and the quality of nucleic acid templates. Further discussion on the efficiency, detection of qPCR chemistry, and sample preparation for qPCR assay, see Gokal et al. [20].

Challenges in viability detection of foodborne pathogens have led to the use of multiplex PCR (mPCR) method for simultaneous and rapid quantification of multiple pathogenic organisms in food samples [34]. Several mPCR assay protocols have been applied to the detection of *L. monocytogenes*, *S. aureus*, and *Salmonella* spp. in food products [35, 36]. However, to address the challenge of differentiating viable cells from dead cells and reducing false-positive results in food analysis, researchers have developed techniques such as multiplex mPCR and real-time PCR combined with propidium monoazide (PMA) for internal amplification [34, 37–39]. By applying PMA as a pretreatment to the sample before PCR amplification, PMA selectively enters dead cells. Upon exposure to strong visible light, it forms a cross-link with DNA through its azide group [34]. The azide group in PMA is converted to a reactive nitrene under halogen light, which reacts with DNA bases and forms a stable covalent nitrogen–carbon bond. This modification alters the DNA structure, preventing subsequent PCR amplification [40, 41]. Incorporating PMA into the qPCR process, known as qPCR-PMA, helps overcome viability detection challenges and avoids amplification of DNA from nonviable bacteria.

8.6 Digital polymerase chain reaction (dPCR): an innovative technique for precise and absolute quantification

The amplification steps during qPCR assay could significantly affect the accuracy of template and quantification of a very low pathogen in complex biologic and food matrices [42]. To circumvent this problem, dPCR as an emerging technology was recently developed as a more advanced version of qPCR and reverse transcriptase qPCR for the detection and quantification of single or multiple targets simultaneously without the use of a standard curve [24, 43, 44]. The dPCR is categorized into chip digital PCR (cdPCR) and ddPCR; however, it is challenging to use the latter for high-throughput quantification [24].

Hence, high-throughput ddPCR, the third generation of PCR, was developed to overcome the shortcoming of sample apportioning, sensitivity, and reproducibility during AQ of low levels of nucleic acid in a sample [45–47]. In contrast to analog qPCR, which utilizes rate-based measurements (C_t values) and calibration curves, ddPCR is an endpoint and AQ method that relies on limiting dilution and Poisson distribution principles [45, 48]. It offers improved sensitivity and accuracy compared to qPCR, particularly for quantifying rare target molecules with low copy numbers [49]. Additionally, ddPCR exhibits exceptional efficiency and high-throughput capabilities, enabling the simultaneous processing of up to 96 samples [50]. Figure 8.8 illustrates

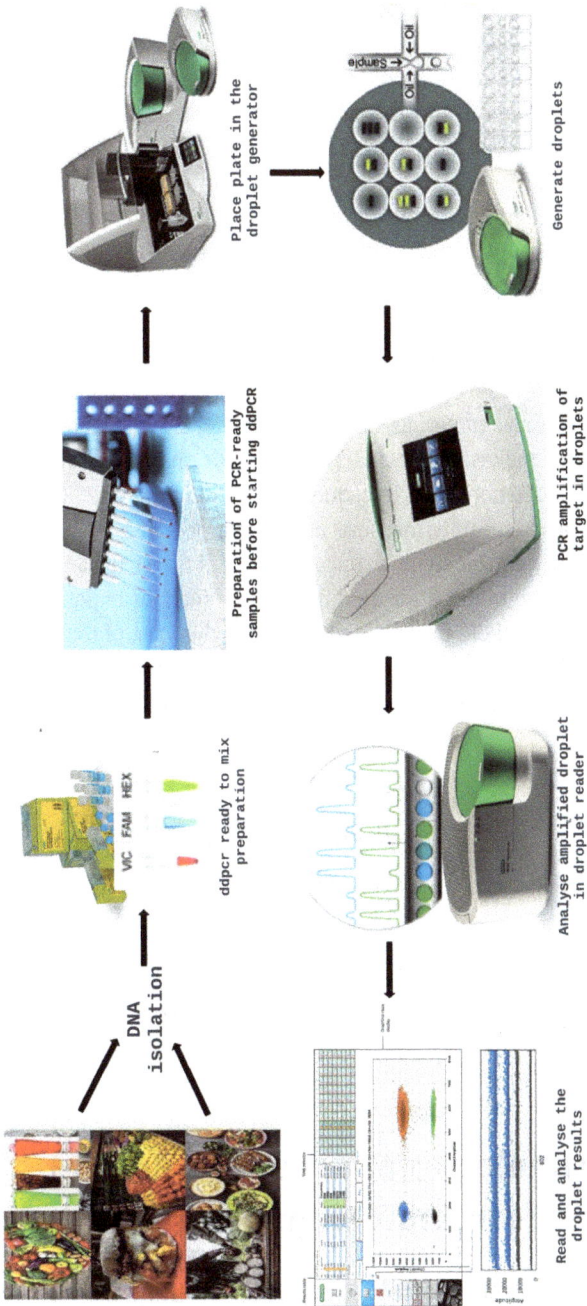

Figure 8.8: Overview of the droplet digital PCR assay workflow analyzing foodborne pathogens. Created with Lucidchart.com.

the schematic workflow of ddPCR technology and has been widely utilized for analyzing microbial composition in food samples [51, 52].

dPCR is a method similar to qPCR in terms of reaction components, assembly, and amplification protocols, but it differs in the way the target is measured. In dPCR, the sample is divided into numerous water-in-oil droplets before thermal cycling [53, 54]. Each droplet chamber may contain either one or zero copies of the target gene (Figure 8.9). Following endpoint PCR amplification, the droplets are examined for positive amplification using fluorescent target-specific hydrolysis probes [55]. The presence or absence of amplified targets in the droplets is determined by applying thresholds based on their fluorescence signals. Without the need for a standard curve, the initial concentration of the target gene is calculated using Poisson statistical analysis.

Compared to qPCR, ddPCR offers several advantages. It exhibits higher sensitivity and does not rely on standard reference materials, calibration curves, or endogenous controls for target quantification. This makes it particularly suitable for quantifying rare genetic mutations, duplications, and deletions in DNA sequences. Additionally, ddPCR is less affected by inhibitors [56, 57], can accurately measure multiple targets in a single reaction, and has the ability to quantify extremely low concentrations of target nucleic acids (DNA/RNA) [58, 59]. However, the use of dPCR for quantification of foodborne pathogens is still relatively limited compared to qPCR, primarily due to the higher cost of the equipment. For more detailed discussions on AQ using ddPCR compared to analog real-time PCR and its broad applications in microbiological samples, refer [50, 57, 60].

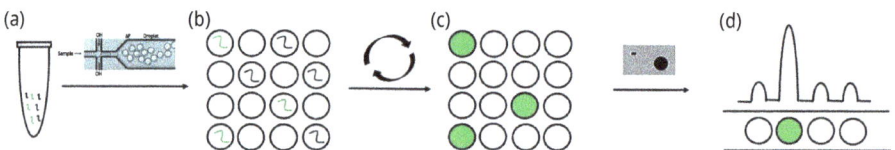

Figure 8.9: Digital droplet PCR workflow of absolute quantification: (a) preparation and mixing of nontargets, targets, probes primers, and other ddPCR reagents (b) dispersed into massive water-in-oil droplets. (c) After ddPCR, amplified relative templates (d) receiving fluorescence signals for the measurement report [24].

8.7 Sequencing technologies and next-generation sequencing (NGS) for improving food safety

Sequencing of PCR product through Sanger sequencing is acknowledged for doing quicker and more precise work. However, other improved techniques have evolved over time and are collectively named as next-generation sequencing (NGS) (Figure 8.10). The NGS technology in food microbiology is majorly used to determine the whole-genome sequencing (WGS) of a single cultured isolate and metagenomics in generating

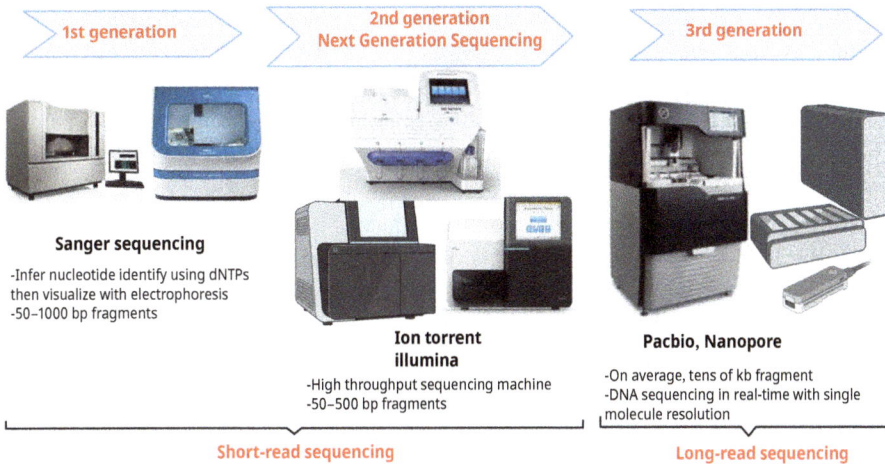

Figure 8.10: The evolution of sequencing technology. Created with BioRender.com.

sequences of multiple (if not all) microorganisms in that sample [61]. The technology is developing at a fast rate, with continuous cost reduction and improvement of output quality [61, 62]. The WGS has been accepted as a common molecular typing tool for food-borne illness surveillance, gradually replacing the conventional microbial typing and characterization techniques because it determines the WGS of a single cultured isolate either as a virus, bacterial colony, or any other organism with high discriminatory power [63, 64]. On the other hand, metagenomics as a promising form of NGS when applied to a biological sample is capable of generating sequences of multiple (if not all) microorganisms at the same time.

NGS technology represents a significant advancement over automated Sanger sequencing, enabling the simultaneous sequencing of up to 1 kilobase (kb) for 96 individual samples [65]. Its application in food analysis allows for the determination of the microbial population's composition and gene content. NGS provides a comprehensive understanding of the genetic makeup of microorganisms present in food samples, facilitating detailed analysis of their diversity and functional potential. By generating large volumes of sequence data in a single run, NGS offers valuable insights into the genetic characteristics of foodborne microorganisms and their potential implications for food safety and quality [66]. There are different commercially available high-throughput sequencing devices with a variety of chemistries and detection methods. NGS technology utilizes a distinct detection system and follows a consistent two-step process across various applications. The first step involves the fragmentation or preparation of amplicon libraries, while the second step involves the detection of incorporated nucleotides. NGS methods can be broadly classified into two main groups: PCR-based technology and non-PCR-based technology, also known as single-molecule sequencing (SMS) technology.

In PCR-based NGS, amplification is performed prior to sequencing, whereas SMS technology does not require amplification before sequencing. These two approaches offer different advantages and are suitable for various research or diagnostic applications. The general process of NGS involves isolating DNA from the target organism and fragmenting it into small pieces [68] (Figure 8.11). These fragments are then sequenced using NGS systems. The sequenced fragments are subsequently assembled into larger pieces to identify overlaps in the sequence [67]. NGS methods have revolutionized the detection of over 1,000 microbes and multicellular species, significantly enhancing our understanding of biology and evolutionary processes.

NGS technologies have the capacity to generate a large number of sequencing reads, ranging from several hundred thousands to tens of millions, in a single run. This capability has made mass sequencing of samples from complex environmental matrices a topic of great interest, particularly in the field of ecological biodiversity research. NGS methods have significantly contributed to the analysis of diverse environmental samples derived from various ecosystems. However, the application of NGS technologies also comes with certain challenges. First, the output of sequencing in terms of read length and accuracy needs to be carefully considered to ensure reliable results. Second, the total output of a sequencing experiment should be balanced with the associated costs and labor required. Finally, there are challenges related to PCR bias, the formation of chimeric sequences, and issues stemming from the amplifi-

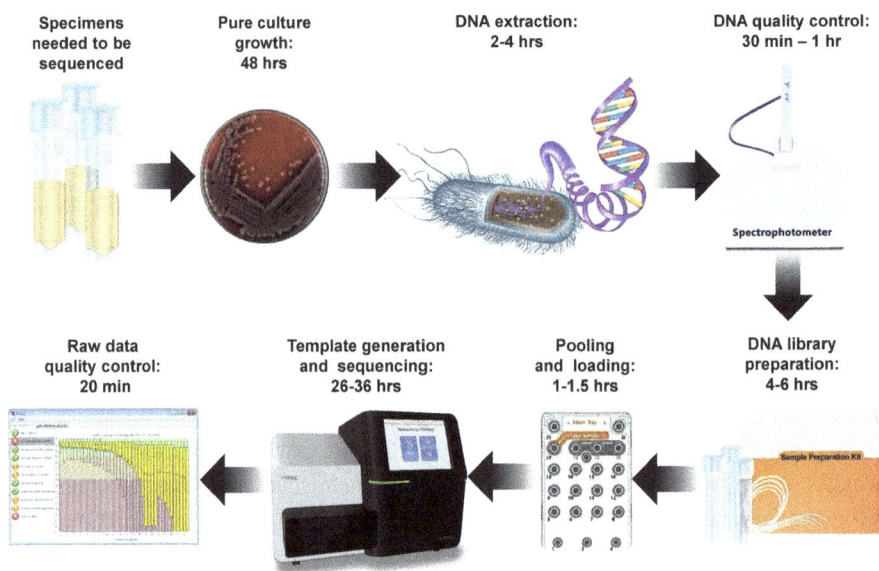

Figure 8.11: Typical whole-genome sequencing workflow in a clinical or public health laboratory. Permission to reuse under Creative Common CC BY-NC-ND license (http://creativecommons.org/licenses/by-nc-nd/4.0/) [67].

cation step prior to sequencing, such as secondary structure-related problems. Overcoming these challenges is essential to obtain accurate and high-quality data from NGS experiments.

8.8 Application of molecular tools in food safety

Due to the continuous technological advancement and development of contemporary society, the importance of having high-quality and safe food has significantly increased. As described in the subsections above, different advanced molecular techniques can be used to ensure safe food consumption. Hence, some of the importance of microbiology in food safety and different case studies were reviewed to determine essential tools that have been used in monitoring, quantifying, and identifying microbiomes in food materials.

8.8.1 The role of PCR-based techniques in detection, characterization and quantification of foodborne pathogens

The prevalence of foodborne diseases is on the rise globally, and PCR-based techniques for in vitro amplification of nucleic acids continue to be the most commonly used microbiology technique for the detection, identification, and quantification of foodborne pathogens [68]. Both conventional and qPCR techniques have been employed to detect food pathogens [52, 69–71]. Food products at different stages of the food chain have been reported in various countries to be associated with foodborne diseases [29, 35, 72–75]. Pathogens include *Cyclospora,* Shiga toxin-producing *E. coli* (STEC) O157, *C. jejuni, L. monocytogenes, Salmonella* spp., *Shigella* spp., and *S. aureus* [72, 76–80], and viruses like noroviruses have been reported to be responsible for food poisoning [52, 81]. Based on the CDC on several reports, these foodborne pathogens often linked to leafy greens illness [82–84]. The connection of leafy greens with foodborne illnesses has been mentioned in different countries [71, 72, 85, 86].

To understand the most probable sources of contamination in farm produce, several investigations have probed the prevalence of foodborne pathogens in different irrigation water sources, leafy greens, agricultural soil, and ready-to-eat foods using other PCR-based methods [73, 87, 88]. Using these methods, studies have revealed that agricultural water and farm vegetables have a high rate of foodborne DNA pathogens [70, 71, 85]. Moyne et al. [89] evaluated the risk and mechanisms of preventing an outbreak of foodborne illness, total and viable *E. coli* O157:H7, preharvest contamination of farm foods, and laboratory-grown lettuce. The authors found that this bacteria can colonize or viable cells survive on field-grown plants in low quantities. In another study, Kim and Oh [85], reported the sensitivity of qPCR without enrichment in detecting *S. typhimurium* and *E. coli* O157: H7 in iceberg cabbage and lettuce. In determining the microbiological safety of fresh vegetables in commercial supply chains, multidrug resistant *E. coli* was also monitored using PCR, and high concentrations were found in ready to eat spinach leaves [90, 91], lettuce [91], and food products of animal origin [92]. Richer et al. [72] reported a high occurrence of multidrug-resistant *E. coli* in South Africa, specifically in fresh vegetables sold by some informal and formal traders in densely populated areas. This finding highlights the potential public health risks associated with the contamination of fresh produce with multidrug-resistant (MDR) bacteria, emphasizing the need for effective surveillance and control measures to ensure food safety and prevent the spread of antibiotic resistance.

Poultry meat is a popular food choice globally, but it presents a significant public health concern due to the potential presence of MDR bacteria. Contaminated meat can lead to the transmission of pathogens to humans, resulting in zoonotic infections and diseases when consumed [93, 94]. Several studies have highlighted the rapid emergence of multidrug-resistant *E. coli* strains in retail poultry products, which have caused considerable illness and even death in individuals who consumed contaminated animal products. In a study conducted by Kar et al. [93], a total of eighteen MDR *E. coli* strains producing extended-spectrum beta-lactamase (ESBL) were isolated from both cattle and poultry. These strains possessed genes associated with β-lactamase, sul resistance, plasmid-mediated quinolone resistance (PMQR), and extraintestinal pathogenic *E. coli* (ExPEC) virulence. Similar studies conducted in different countries have also documented the genomic diversity among isolated *E. coli* strains and the prevalence of AmpC- and ESBL- positive *E. coli*, along with their corresponding resistance genes, in various types of meat, including organic meat [95–98].

Furthermore, in Bangladesh, the molecular profiles of *E. coli* isolated from layer and broiler chicken, meat obtained from retail poultry shops in the Sylhet division were examined and confirmed using multiplex and uniplex PCR assays [99]. These findings emphasize the significance of monitoring antibiotic resistance in poultry products and the need for stringent control measures to ensure food safety and prevent the dissemination of resistant bacteria to humans. The isolates were reported to be highly resistant to most commonly used antibiotics, such as tetracycline, erythromycin, and ampicillin, with about 13.91% (53/381) of all isolates containing ESBL and AmpC β-lactamase (AmpC) producing genes [99]. Furthermore, the transmission of AmpC and ESBL-positive *E. coli* from meat to humans were reported in Germany [100], Bangladesh [99], and South Korea [101], among other countries. In addition, the prevalence of MDR meat-borne *Salmonella* serotypes, such as *Salmonella* spp. was isolated from retail markets broiler meat [77]; chicken and pork meat [102] as well as raw meat of buffalo and chicken [103]. In addition, the fingerprint of MDR *Salmonella enterica* serovar Heidelberg (S. Heidelberg) was found in imported poultry meat [104] and *S. enterica* serovar infantis clone found in retail chicken meat [105].

In several studies, qPCR is effectively used to identify and spontaneously quantify different food microbes and genes at the same time [35, 37, 52, 85]. Real-time qPCR was used to identify and quantify *L. monocytogenes* in clinical and food samples and have become a more popular detection method for Listeria. Day and Basavanna [29] developed a quick and highly sensitive detection method for *L. monocytogenes* in lettuce and infant formula. This method combines real-time PCR with a macrophage-based enrichment protocol, providing both presumptive and confirmatory identification of the pathogen. In recent years, the combination of real-time PCR and PMA has gained significant popularity for the detection of diverse pathogenic bacteria. This approach has been successfully utilized for the detection of *Salmonella* spp., *Pseudomonas aeruginosa, Bacillus cereus,* and *E. coli* O157:H7 [93–95]. By incorporating PMA into the real-time PCR workflow, it becomes possible to selectively target and detect viable cells of these pathogenic bacteria, enhancing the accuracy and reliability of the detection process. This combination offers a valuable tool for ensuring food safety and effectively monitoring the presence of these harmful pathogens. This technique has been successfully applied to detect viable but nonculturable *S. typhimurium* and *E. coli* O157:H7 in cabbage, iceberg lettuce, milk, and dairy products [75, 96].

Additionally, Bansal et al. [74] developed a multiplex PCR method for the direct and rapid identification of *Salmonella* species, as well as specific serotypes such as enteritidis, typhimurium, and typhi, in poultry carcasses and viscera. The study further showed that the mPCR detected about 2.74% genus *Salmonella* in refrigerated viscera, and 1.37% of serotype enteritidis in one of the samples that tested positive for *Salmonella* sp. Their results show that mPCR could be used to detect pathogens within a short period and identify bacterial serotypes in a sample, hence, enabling efficient monitoring and control of contamination in poultry products.

8.8.2 Use of ddPCR in the quantification of foodborne pathogens as rapid food safety

ddPCR is more accurate in identifying rare and low concentrations of target organisms at low copy numbers [24, 50, 57]. The application of ddPCR as a surveillance tool for quantifying foodborne pathogens has been confirmed by numerous researchers. [97–99]. These authors have successfully demonstrated the efficacy and reliability of ddPCR in accurately quantifying various foodborne pathogens in different food samples. Cai et al. [69] reported the initial application of dPCR for quantifying specific nucleic acids in meat products, specifically chicken and pork. They demonstrated the effectiveness of dPCR in detecting and quantifying targeted nucleic acids in these meat samples. Furthermore, the prevalence and characterization of *L. ivanovii and L. monocytogenes* in food and environmental samples were conducted by employing both conventional real-time qPCR and ddPCR techniques [14]. The versatility and potential of dPCR as a valuable tool in the field of food safety, enabling accurate detection and quantification of specific pathogens in various samples was highlighted.

The robustness and sensitivity of ddPCR make it a promising technique for surveillance and monitoring purposes in the field of food safety. Its ability to provide AQ and overcome limitations of traditional methods has established ddPCR as a valuable tool for pathogen surveillance in the food industry. Cremonesi et al. [97], developed highly sensitive qPCR and ddPCR assays for the detection and quantification of total eumycetes, specifically *Wickerhamomyces anomalus* and *Saccharomycopsis fibuligera*, in bread samples. The authors demonstrated the efficacy of these assays in early detection and monitoring of contamination by these fungi, providing valuable tools for high-throughput approaches in assessing bread quality. Likewise, the same methods were utilized to assess the behavior of *Yersinia enterocolitica*, a pathogenic bacterium, in experimentally contaminated *Lactuca sativa L.* [100]. The authors compared the detection rate after an eleven-day incubation period at different temperatures and observed notable differences between the two technical approaches. Specifically, only ddPCR demonstrated the ability to rapidly detect the pathogen at low concentrations in the leafy greens. Similarly, in another study, four different techniques, including real-time PCR, ddPCR and culture-based approach with or without membrane filtration were compared for the detection of *Campylobacter* spp. in lettuce salad, alfalfa and clover sprouts [101]. The efficiency and performance of these methods were evaluated, highlighting the potential of ddPCR as a reliable and sensitive technique for Campylobacter detection in various food samples.

In addition, multiplex ddPCR was employed as a reliable quantification method to quantify *tlh, tdh* and *ureR* genes of *Vibrio parahaemolyticus* in seafoods [102]. Chon, Jung [101] and Lei, Gu [102] compared different technical methods as an accurate, convenient, with specific, improved sensitivity and reproducibility for quantification of food pathogens at lower detection limit. Several other studies have reported the use of ddPCR or in combination with qPCR and other methods as quick detection techniques for screening or monitoring food pathogens [103–105]. Overall, these studies showcase the utility and effectiveness of qPCR and ddPCR assays in detecting and quantifying specific pathogens or contaminants in different food matrices, enabling improved monitoring and quality control in the food industry.

8.8.3 The role of next-generation sequencing in the study and control of foodborne pathogens

Microbiomes are defined as the collection of organisms that inhabit a specific environment. The advent of high-throughput NGS techniques (HT-NGS) has revolutionized the identification of genetic material within various matrices [66]. Numerous studies have utilized NGS to investigate the microbial diversity associated with food samples [66]. NGS has proven to be a valuable tool for screening fresh produce to detect the presence of human pathogens while simultaneously providing information about the associated microbiome [106]. The prevalence of MDR pathogens in food samples has also been investigated using NGS. For example, multidrug-resistant *E. coli* strains carrying CTX-M, MCR-1, and FosA genes have been isolated from both retail food and humans using WGS [107].

Indeed, the use of multilocus sequence typing (MLST) combined with WGS, known as MLST-WGS approach, has further revolutionized salmonella monitoring and identification. This approach has replaced traditional serotyping methods and provides valuable insights into the genetic diversity and relatedness of different *Salmonella* species. MLST analyzes specific genetic markers to determine the variation and relatedness among strains, while WGS offers a comprehensive view of the entire salmonella genome, detecting additional genetic variations [108, 109]. The combination of MLST and WGS enables detailed analysis of genetic diversity, aiding in outbreak investigations, source tracking, and understanding transmission patterns. By comparing genetic profiles, common sources of contamination can be identified and the spread of specific strains can be tracked. The adoption of MLST-WGS has significantly improved the precision and speed of salmonella identification, becoming a vital tool in public health surveillance for identifying emerging strains, assessing virulence potential, antimicrobial resistance profiles, and informing targeted interventions to prevent and control *Salmonella* infections [107]. For further exploration of the future potential of NGS in food microbiology and safety, additional references such as [62, 67, 110, 111] can provide comprehensive reviews on the subject. These sources delve into the ongoing advancements and possibilities that NGS technology offers in the field of food microbiology.

8.9 Conclusion

In conclusion, molecular techniques have significantly contributed to the identification and detection of microorganisms in various microbiomes, including those associated with food. These techniques, such as PCR, qPCR, ddPCR, and NGS, offer advantages in terms of sensitivity, speed, and accuracy compared to traditional methods in food safety. The early detection of foodborne pathogens is crucial for preventing outbreaks and protecting public health. Molecular tools have played a key role in enabling rapid and reliable detection of these pathogens, allowing for timely interventions and effective control measures. The use of PCR, qPCR, and ddPCR has provided sensitive and specific detection of nucleic acid sequences, facilitating the identification of pathogens in food samples. NGS, on the other hand, has revolutionized our understanding of the diversity and composition of microbiomes associated with foodstuffs. It has enabled comprehensive profiling of microbial communities and the identification of potential pathogens. NGS has also been used for the screening of fresh produce, providing insights into the presence of both human pathogens and the associated microbiome. Implementing a molecular-based surveillance system is essential for monitoring the international spread of foodborne pathogens and ensuring food safety. These techniques offer improved sensitivity, specificity, and rapidity compared to traditional methods. By adopting molecular tools in food safety practices, it becomes possible to enhance the early detection, surveillance, and control of foodborne microorganisms, ultimately safeguarding the health and well-being of consumers.

References

[1]   European Centre for Disease Prevention and Control. (2015). The European Union summary report on trends and sources of zoonoses, zoonotic agents and food-borne outbreaks in 2013. *European Food Safety Authority Journal, 13*(1), 3991.

[2]   World Health Organization. (07 April 2020). COVID-19 and food safety: guidance for food businesses: interim guidance.

[3]   Mandal, P. K., & Biswas, A. K. (2020). Chapter 16 – Modern techniques for rapid detection of meatborne pathogens. In A. K. Biswas & P. K. Mandal (Eds.), *Meat Quality Analysis: Advanced Evaluation Methods, Techniques, and Technologies*. Academic Press: Cambridge, MA, USA, 287–303.

[4]   Silva, F., & Domingues, F. C. (2017). Antimicrobial activity of coriander oil and its effectiveness as food preservative. *Critical Reviews in Food Science and Nutrition, 57*(1), 35–47.

[5]   Bolek, S. (2020). Consumer knowledge, attitudes, and judgments about food safety: A consumer analysis. *Trends in Food Science and Technology, 102*, 242–248.

[6]   Brooks, B., Devenish, J., Lutze-Wallace, C., Milnes, D., Robertson, R., & Berlie-Surujballi, G. (2004). Evaluation of a monoclonal antibody-based enzyme-linked immunosorbent assay for detection of Campylobacter fetus in bovine preputial washing and vaginal mucus samples. *Veterinary Microbiology, 103*(1–2), 77–84.

[7]   Enitan-Folami, A. M., & Swalaha, F. M. (2020). Application of biotechnology in the food industry. *In food science and technology: Trends and future prospects*. Walter de Gruyter, GmbH, Berlin/Boston, 235–277.

[8]   Zeng, D., Chen, Z., Jiang, Y., Xue, F., & Li, B. (2016). Advances and challenges in viability detection of foodborne pathogens. *Frontiers in Microbiology, 7*, 1833.

[9]   Kadri, K. (2019). Polymerase chain reaction (PCR): Chapter 9 - Principle and applications. *Synthetic biology-new interdisciplinary science*. IntechOpen, London, United Kingdom, 147–163, DOI: 10.5772/intechopen.86491

[10]   Brock, T. D. (1997). The value of basic research: Discovery of Thermus aquaticus and other extreme thermophiles. *Genetics, 146*(4), 1207.

[11]   de Miguel Bouzas, T., Barros-Velázquez, J., & Gonzalez Villa, T. (2006). Industrial applications of hyperthermophilic enzymes: A review. *Protein and Peptide Letters, 13*(7), 645–651.

[12]   Saiki, R. K., Gelfand, D. H., Stoffel, S., Scharf, S. J., Higuchi, R., Horn, G. T., et al. (1988). Primer-directed enzymatic amplification of DNA with a thermostable DNA polymerase. *Science, 239*(4839), 487–491.

[13]   Śpibida, M., Krawczyk, B., Olszewski, M., & Kur, J. (2017). Modified DNA polymerases for PCR troubleshooting. *Journal of Applied Genetics, 58*(1), 133–142.

[14]   Chen, J-Q., Healey, S., Regan, P., Laksanalamai, P., & Hu, Z. (2017). PCR-based methodologies for detection and characterization of Listeria monocytogenes and Listeria ivanovii in foods and environmental sources. *Food Science and Human Wellness, 6*(2), 39–59.

[15] Klancnik, A., Kovač, M., Toplak, N., Piskernik, S., & Jeršek, B. (2012). PCR in food analysis. In Polymerase chain reaction, Dr Patricia Hernandez-Rodriguez (Ed.), Rijeka: Intech, cop, 195–220.

[16] Bartlett, J. M., & Stirling, D. (2003). A short history of the polymerase chain reaction. In PCR protocol: Methods in molecular biology. 226, 3–6. Humana Press, NJ, 226, 3–6.

[17] van Pelt-Verkuil, E., van Belkum, A., & Hays, J. P. (2008). A brief comparison between in vivo DNA replication and in vitro PCR amplification. In: Principles and technical aspects of PCR amplification. Springer, Dordrecht. 9–15, https://doi.org/10.1007/978-1-4020-6241-4_2.

[18] Zhang, T., & Fang, H. H. (2006). Applications of real-time polymerase chain reaction for quantification of microorganisms in environmental samples. *Applied Microbiology and Biotechnology, 70*(3), 281–289.

[19] Sambrook, J., & Russell, D. W. (2001). Molecular cloning: Ch. 8. In *Vitro amplification of DNA by the polymerase chain reaction*. Cold Spring Harbor Laboratory Press, Cold Spring Harbor, NY8. 2, 18–8.50

[20] Gokal, J., Awolusi, O., Enitan, A., Kumari, S., & Bux, F. (2016). *Molecular characterization and quantification of microbial communities in wastewater treatment systems. Microbial biotechnology: An interdisciplinary approach* (pp. 59–114). Boca Raton: Taylor & Francis.

[21] Omiccioli, E., Amagliani, G., Brandi, G., & Magnani, M. (2009). A new platform for real-time PCR detection of *Salmonella* spp., *Listeria monocytogenes* and *Escherichia coli* O157 in milk. *Food Microbiology, 26*(6), 615–622.

[22] Zhao, X., Lin, C-W., Wang, J., & Oh, D. H. (2014). Advances in rapid detection methods for foodborne pathogens. *Journal of Microbiology and Biotechnology, 24*(3), 297–312.

[23] Williams, M. (2009). Real-time polymerase chain reaction. In S. A. Bustin (Ed.), *The PCR revolution: Basic technologies and applications* (pp. 3–11). Cambridge: Cambridge University Press.

[24] Lei, S., Chen, S., & Zhong, Q. (2021). Digital PCR for accurate quantification of pathogens: Principles, applications, challenges and future prospects. *International Journal of Biological Macromolecules, 184,* 750–759.

[25] Kim, J., Lim, J., & Lee, C. (2013). Quantitative real-time PCR approaches for microbial community studies in wastewater treatment systems: Applications and considerations. *Biotechnology Advances, 31*(8), 1358–1373.

[26] Enitan, A. M., Kumari, S., Swalaha, F. M., & Bux, F. (2014). Real-time quantitative PCR for quantification of methanogenic Archaea in an UASB reactor treating brewery wastewater. *Conference of the International Journal of Arts and Sciences, 07*(3), 103–106.

[27] Musa, J. J. (2014). Effect of domestic waste leachates on quality parameters of groundwater. *Leonardo Journal of Sciences, 24,* 28–38.

[28] Alvarado, A., Montañez-Hernández, L. E., Palacio-Molina, S. L., Oropeza-Navarro, R., Luévanos-Escareño, M. P., & Balagurusamy, N. (2014). Microbial trophic interactions and mcrA gene expression in monitoring of anaerobic digesters. *Frontiers in Microbiology, 5,* 597.

[29] Day, J., & Basavanna, U. (2014). Real-time PCR detection of Listeria monocytogenes in infant formula and lettuce following macrophage-based isolation and enrichment. *Journal of Applied Microbiology, 118*(1), 233–244.

[30] Bergmann, I. (2012). Characterization of methanogenic Archaea communities in biogas reactors by quantitative PCR: der Technischen Universität Berlin.

[31] Yu, L., Wensel, P. C., Ma, J., & Chen, S. (2013). Mathematical modeling in anaerobic digestion (AD). *Journal of Bioremediation and Biodegradation, 4,* 2.

[32] Dhanasekaran, S., Doherty, T. M., Kenneth, J., & Group, T. T. S. (2010). Comparison of different standards for real-time PCR-based absolute quantification. *Journal of Immunological Methods, 354*(1–2), 34–39.

[33] Rodríguez, A., Andrade, M. J., Rodríguez, M., & Córdoba, J. J. (2017). Real-time PCR for the detection of pathogens in meat and meat products. In *Advanced technologies for meat processing* (pp. 113–152). CRC Press, Taylor & Francis Group, Boca Raton.

[34] Xie, G., Yu, S., Li, W., Mu, D., Aguilar, Z. P., & Xu, H. (2020). Simultaneous detection of Salmonella spp., Pseudomonas aeruginosa, Bacillus cereus, and *Escherichia coli* O157:H7 in environmental water using PMA combined with mPCR. *Journal of Microbiology, 58*(8), 668–674.

[35] Ding, T., Suo, Y., Zhang, Z., Liu, D., Ye, X., Chen, S., et al. (2017). A multiplex RT-PCR assay for S. aureus, L. monocytogenes, and Salmonella spp. detection in raw milk with pre-enrichment. *Frontiers in Microbiology, 8*, 989.

[36] Parichehr, M., Mohammad, K., Abbas, D., & Mehdi, K. (2019). Developing a multiplex real-time PCR with a new pre-enrichment to simultaneously detect four foodborne bacteria in milk. *Future Microbiology, 14*(10), 885–898.

[37] de Freitas, C. G., Santana, Â. P., da Silva, P. H. C., Gonçalves, V. S. P., Barros, M. D. A. F., Torres, F. A. G., et al. (2010). PCR multiplex for detection of *Salmonella Enteritidis, Typhi* and *Typhimurium* and occurrence in poultry meat. *International Journal of Food Microbiology, 139*(1–2), 15–22.

[38] Forghani, F., Langaee, T., Eskandari, M., Seo, K-H., Chung, M-J., & Oh, D-H. (2015). Rapid detection of viable Bacillus cereus emetic and enterotoxic strains in food by coupling propidium monoazide and multiplex PCR (PMA-mPCR). *Food Control, 55*, 151–157.

[39] Li, F., Xie, G., Zhou, B., Yu, P., Yu, S., Aguilar, Z. P., et al. (2016). Rapid and simultaneous detection of viable *Cronobacter sakazakii, Staphylococcus aureus*, and *Bacillus cereus* in infant food products by PMA-mPCR assay with internal amplification control. *LWT, 74*, 176–182.

[40] Nocker, A., Cheung, C-Y., & Camper, A. K. (2006). Comparison of propidium monoazide with ethidium monoazide for differentiation of live vs. dead bacteria by selective removal of DNA from dead cells. *Journal of Microbiological Methods, 67*(2), 310–320.

[41] Hixon, S. C., White Jr, W. E., & Yielding, K. L. (1975). Selective covalent binding of an ethidium analog to mitochondrial DNA with production of petite mutants in yeast by photoaffinity labeling. *Journal of Molecular Biology, 92*(2), 319–329.

[42] Ramakers, C., Ruijter, J. M., Deprez, R. H. L., & Moorman, A. F. (2003). Assumption-free analysis of quantitative real-time polymerase chain reaction (PCR) data. *Neuroscience Letters, 339*(1), 62–66.

[43] Huggett, J. F., Cowen, S., & Foy, C. A. (2015). Considerations for digital PCR as an accurate molecular diagnostic tool. *Clinical Chemistry, 61*(1), 79–88.

[44] Rocchigiani, A. M., Tilocca, M. G., Portanti, O., Vodret, B., Bechere, R., Di Domenico, M., et al. (2020). Development of a digital RT-PCR method for absolute quantification of Bluetongue virus in field samples. *Frontiers in Veterinary Science, 7*, 170.

[45] Hindson, B. J., Ness, K. D., Masquelier, D. A., Belgrader, P., Heredia, N. J., Makarewicz, A. J., et al. (2011). High-throughput droplet digital PCR system for absolute quantitation of DNA copy number. *Analytical Chemistry, 83*(22), 8604–8610.

[46] He, L., Simpson, D. J., & Gänzle, M. G. (2020). Detection of enterohaemorrhagic Escherichia coli in food by droplet digital PCR to detect simultaneous virulence factors in a single genome. *Food Microbiology, 90*, 103466.

[47] Porcellato, D., Narvhus, J., & Skeie, S. B. (2016). Detection and quantification of Bacillus cereus group in milk by droplet digital PCR. *Journal of Microbiological Methods, 127*, 1–6.

[48] Lei, S., Gu, X., Xue, W., Rong, Z., Wang, Z., Chen, S., et al. (2020). A 4-plex droplet digital PCR method for simultaneous quantification and differentiation of pathogenic and non-pathogenic Vibrio parahaemolyticus based on single intact cells. *Frontiers in Microbiology, 11*:1727, 1–11.

[49] Bahar, M. H., Wist, T. J., Bekkaoui, D. R., Hegedus, D. D., & Olivier, C. Y. (2018). Aster leafhopper survival and reproduction, and Aster yellows transmission under static and fluctuating temperatures, using ddPCR for phytoplasma quantification. *Scientific Reports, 8*(1), 1–9.

[50] Hindson, C. M., Chevillet, J. R., Briggs, H. A., Gallichotte, E. N., Ruf, I. K., Hindson, B. J., et al. (2013). Absolute quantification by droplet digital PCR versus analog real-time PCR. *Nature Methods, 10*(10), 1003–1005.

[51] Elmahalawy, S. T., Halvarsson, P., Skarin, M., & Höglund, J. (2018). Droplet digital polymerase chain reaction (ddPCR) as a novel method for absolute quantification of major gastrointestinal nematodes in sheep. *Veterinary Parasitology*, *261*, 1–8.

[52] Shirima, R. R., Maeda, D. G., Kanju, E., Ceasar, G., Tibazarwa, F. I., & Legg, J. P. (2017). Absolute quantification of cassava brown streak virus mRNA by real-time qPCR. *Journal of Virological Methods*, *245*, 5–13.

[53] Nyaruaba, R., Xiong, J., Mwaliko, C., Wang, N., Kibii, B. J., Yu, J., et al. (2020). Development and evaluation of a single dye duplex droplet digital PCR assay for the rapid detection and quantification of *Mycobacterium tuberculosis*. *Microorganisms*, *8*(5), 701.

[54] McDermott, G. P., Do, D., Litterst, C. M., Maar, D., Hindson, C. M., Steenblock, E. R., et al. (2013). Multiplexed target detection using DNA-binding dye chemistry in droplet digital PCR. *Analytical Chemistry*, *85*(23), 11619–11627.

[55] Floren, C., Wiedemann, I., Brenig, B., Schütz, E., & Beck, J. (2015). Species identification and quantification in meat and meat products using droplet digital PCR (ddPCR). *Food Chemistry*, *173*, 1054–1058.

[56] Gobert, G., Cotillard, A., Fourmestraux, C., Pruvost, L., Miguet, J., & Boyer, M. (2018). Droplet digital PCR improves absolute quantification of viable lactic acid bacteria in faecal samples. *Journal of Microbiological Methods*, *148*, 64–73.

[57] Huggett, J. F., Foy, C. A., Benes, V., Emslie, K., Garson, J. A., Haynes, R., et al. (2013). The digital MIQE guidelines: Minimum information for publication of quantitative digital PCR experiments. *Clinical Chemistry*, *59*(6), 892–902.

[58] Brink, B. G., Meskas, J., & Brinkman, R. R. (2018). ddPCRclust: An R package and Shiny app for automated analysis of multiplexed ddPCR data. *Bioinformatics*, *34*(15), 2687–2689.

[59] Zhong, Q., Bhattacharya, S., Kotsopoulos, S., Olson, J., Taly, V., Griffiths, A. D., et al. (2011). Multiplex digital PCR: Breaking the one target per color barrier of quantitative PCR. *Lab on a Chip*, *11*(13), 2167–2174.

[60] Salipante, S. J., & Jerome, K. R. (2020). Digital PCR – An emerging technology with broad applications in microbiology. *Clinical Chemistry*, *66*(1), 117–123.

[61] Jagadeesan, B., Gerner-Smidt, P., Allard, M. W., Leuillet, S., Winkler, A., Xiao, Y., et al. (2019). The use of next generation sequencing for improving food safety: Translation into practice. *Food Microbiology*, *79*, 96–115.

[62] Haynes, E., Jimenez, E., Pardo, M. A., & Helyar, S. J. (2019). The future of NGS (Next Generation Sequencing) analysis in testing food authenticity. *Food Control*, *101*, 134–143.

[63] Allard, M. W., Luo, Y., Strain, E., Pettengill, J., Timme, R., Wang, C., et al. (2013). On the evolutionary history, population genetics and diversity among isolates of *Salmonella Enteritidis* PFGE pattern JEGX01. 0004. *PLoS One*, *8*(1), e55254.

[64] Ashton, P. M., Nair, S., Peters, T. M., Bale, J. A., Powell, D. G., Painset, A., et al. (2016). Identification of Salmonella for public health surveillance using whole genome sequencing. *Peer Journal*, *4*, e1752.

[65] Shokralla, S., Spall, J. L., Gibson, J. F., & Hajibabaei, M. (2012). Next-generation sequencing technologies for environmental DNA research. *Molecular Ecology*, *21*(8), 1794–1805.

[66] Cao, Y., Fanning, S., Proos, S., Jordan, K., & Srikumar, S. (2017). A review on the applications of next generation sequencing technologies as applied to food-related microbiome studies. *Frontiers in Microbiology*, *8*. article 1829, 1–16.

[67] Besser, J., Carleton, H. A., Gerner-Smidt, P., Lindsey, R. L., & Trees, E. (2018). Next-generation sequencing technologies and their application to the study and control of bacterial infections. *Clinical Microbiology and Infection*, *24*(4), 335–341.

[68] Postollec, F., Falentin, H., Pavan, S., Combrisson, J., & Sohier, D. (2011). Recent advances in quantitative PCR (qPCR) applications in food microbiology. *Food Microbiology*, *28*(5), 848–861.

[69] Cai, Y., Li, X., Lv, R., Yang, J., Li, J., He, Y., et al. (2014). Quantitative analysis of pork and chicken products by droplet digital PCR. *BioMed Research International*, *2014*, 810209.

[70] Decol, L. T., Casarin, L. S., Hessel, C. T., Batista, A. C. F., Allende, A., & Tondo, E. C. (2017). Microbial quality of irrigation water used in leafy green production in Southern Brazil and its relationship with produce safety. *Food Microbiology*, *65*, 105–113.

[71] Ssemanda, J. N., Reij, M. W., van Middendorp, G., Bouw, E., van der Plaats, R., Franz, E., et al. (2018). Foodborne pathogens and their risk exposure factors associated with farm vegetables in Rwanda. *Food Control*, *89*, 86–96.

[72] Richter, L., Plessis, E. D., Duvenage, S., & Korsten, L. (2021). High prevalence of multidrug resistant Escherichia coli isolated from fresh vegetables sold by selected formal and informal traders in the most densely populated Province of South Africa. *Journal of Food Science*, *86*(1), 161–168.

[73] Acheamfour, C. L., Parveen, S., Hashem, F., Sharma, M., Gerdes, M. E., May, E. B., et al. (2021). Levels of Salmonella enterica and Listeria monocytogenes in alternative irrigation water vary based on water source on the Eastern Shore of Maryland. *Microbiology Spectrum*, *9*(2), e00669–21.

[74] Ali, M., Faruque, M., Molla, M., Khanam, R., Mahmud, S., & Mohiuddin, A. K. M. (2015) Antibacterial activity of eight medicinal plants against multidrug resistant *Escherichia coli* and Salmonella spp. isolated from broiler meat. *Grassroots Journal of Natural Resources*, *3*(4), 28–48.

[75] Diem Do, T., Lao, T., & Huyen Le, T. (2021). Establishment of PMA real-time PCR method to detect viable cells of Listeria monocytogenes and Salmonella spp. in milk and dairy products. *Asian Journal of Pharmaceutical Research and Health Care*, *13*(2), 146–156.

[76] Cheah, Y-K., Salleh, N. A., Lee, L-H., Radu, S., Sukardi, S., & Sim, J-H. (2008). Comparison of PCR fingerprinting techniques for the discrimination of Salmonella enterica subsp. enterica serovar Weltevreden isolated from indigenous vegetables in Malaysia. *World Journal of Microbiology and Biotechnology*, *24*(3), 327–335.

[77] Alves, J., Marques, V. V., Pereira, L. F. P., Hirooka, E. Y., & De Oliveira, T. C. R. M. (2012). Multiplex PCR for the detection of *Campylobacter spp.* and *Salmonella spp.* in chicken meat. *Journal of Food Safety*, *32*(3), 345–350.

[78] Chiang, Y-C., Tsen, H-Y., Chen, H-Y., Chang, Y-H., Lin, C-K., Chen, C-Y., et al. (2012). Multiplex PCR and a chromogenic DNA macroarray for the detection of *Listeria monocytogens, Staphylococcus aureus, Streptococcus agalactiae, Enterobacter sakazakii, Escherichia coli* O157: H7, *Vibrio parahaemolyticus, Salmonella spp.* and *Pseudomonas fluorescens* in milk and meat samples. *Journal of Microbiological Methods*, *88*(1), 110–116.

[79] Zhou, Y., Pan, F-G., Li, Y-S., Zhang, Y-Y., Zhang, J-H., Lu, S-Y., et al. (2009). Colloidal gold probe-based immunochromatographic assay for the rapid detection of brevetoxins in fishery product samples. *Biosensors and Bioelectronics*, *24*(8), 2744–2747.

[80] Lee, N., Kwon, K. Y., Oh, S. K., Chang, H-J., Chun, H. S., & Choi, S-W. (2014). A multiplex PCR assay for simultaneous detection of *Escherichia coli* O157: H7, *Bacillus cereus, Vibrio parahaemolyticus, Salmonella spp., Listeria monocytogenes*, and *Staphylococcus aureus* in Korean ready-to-eat food. *Foodborne Pathogens and Disease*, *11*(7), 574–580.

[81] Shaheen, M. N., Elmahdy, E. M., & Chawla-Sarkar, M. (2019). Quantitative PCR-based identification of enteric viruses contaminating fresh produce and surface water used for irrigation in Egypt. *Environmental Science and Pollution Research*, *26*(21), 21619–21628.

[82] Bennett, S., Sodha, S., Ayers, T., Lynch, M., Gould, L., & Tauxe, R. (2018). Produce-associated foodborne disease outbreaks, USA, 1998–2013. *Epidemiology and Infection*, *146*(11), 1397–1406.

[83] Aworh, O. C. (2021). Food safety issues in fresh produce supply chain with particular reference to sub-Saharan Africa. *Food Control*, *123*, 107737.

[84] Turner, K., Moua, C. N., Hajmeer, M., Barnes, A., & Needham, M. (2019). Overview of leafy greens-related food safety incidents with a California link: 1996 to 2016. *Journal of Food Protection*, *82*(3), 405–414.

[85] Kim, J-H., & Oh, S-W. (2020). Rapid and sensitive detection of *E. coli O157: H7 and S. Typhimurium* in iceberg lettuce and cabbage using filtration, DNA concentration, and qPCR without enrichment. *Food Chemistry, 327*, 127036.

[86] Herman, K., Hall, A., & Gould, L. (2015). Outbreaks attributed to fresh leafy vegetables, United States, 1973–2012. *Epidemiology and Infection, 143*(14), 3011–3021

[87] Iwu, C. D., & Okoh, A. I. (2020). Characterization of antibiogram fingerprints in Listeria monocytogenes recovered from irrigation water and agricultural soil samples. *PLoS One, 15*(2), 1–22.

[88] Truchado, P., Hernandez, N., Gil, M. I., Ivanek, R., & Allende, A. (2018). Correlation between E. coli levels and the presence of foodborne pathogens in surface irrigation water: Establishment of a sampling program. *Water Research, 128*, 226–233.

[89] Moyne, A-L., Harris, L. J., & Marco, M. L. (2013). Assessments of total and viable Escherichia coli O157: H7 on field and laboratory grown lettuce. *PLoS One, 8*(7), e70643.

[90] Vengarai Jagannathan, B., Kitchens, S., Priyesh Vijayakumar, P., Price, S., & Morgan, M. (2021). Efficacy of bacteriophage cocktail to control *E. coli* O157: H7 contamination on baby spinach leaves in the presence or absence of organic load. *Microorganisms, 9*(3), 544.

[91] Ratshilingano, M. T., du Plessis, E. M., Duvenage, S., & Korsten, L. (2022). Characterization of multidrug-resistant *Escherichia coli* isolated from two commercial lettuce and spinach supply chains. *Journal of Food Protection, 85*(1), 122–132.

[92] Nagy, B., Szmolka, A., Možina, S. S., Kovač, J., Strauss, A., Schlager, S., et al. (2015). Virulence and antimicrobial resistance determinants of verotoxigenic *Escherichia coli* (VTEC) and of multidrug-resistant *E. coli* from foods of animal origin illegally imported to the EU by flight passengers. *International Journal of Food Microbiology, 209*, 52–59.

[93] Yu, S., Yan, L., Wu, X., Li, F., Wang, D., & Xu, H. (2017). Multiplex PCR coupled with propidium monoazide for the detection of viable *Cronobacter sakazakii, Bacillus cereus, and Salmonella* spp. in milk and milk products. *Journal of Dairy Science, 100*(10), 7874–7882.

[94] Liang, T., Zhou, P., Zhou, B., Xu, Q., Zhou, Z., Wu, X., et al. (2019). Simultaneous quantitative detection of viable Escherichia coli O157: H7, *Cronobacter spp., and Salmonella sp*p. using sodium deoxycholate-propidium monoazide with multiplex real-time PCR. *Journal of Dairy Science, 102*(4), 2954–2965.

[95] Golpayegani, A., Douraghi, M., Rezaei, F., Alimohammadi, M., & Nodehi, R. N. (2019). Propidium monoazide-quantitative polymerase chain reaction (PMA-qPCR) assay for rapid detection of viable and viable but non-culturable (VBNC) *Pseudomonas aeruginosa* in swimming pools. *Journal of Environmental Health Science and Engineering, 17*(1), 407–416.

[96] Kesmen, Z., & Aslan, H. (2017). Determination of viable Salmonella Typhimurium cells in heat treated milk by PMA/real-time PCR method. *Turkish Journal of Agriculture-Food Science and Technology, 5*(5), 518–524.

[97] Cremonesi, P., Garofalo, C., Picozzi, C., Castiglioni, B., Mangieri, N., Milanović, V., et al. (2022). Development of quantitative real-time PCR and digital droplet-PCR assays for rapid and early detection of the spoilage yeasts Saccharomycopsis fibuligera and Wickerhamomyces anomalus in bread. *Food Microbiology, 101*, 103894.

[98] Cremonesi, P., Cortimiglia, C., Picozzi, C., Minozzi, G., Malvisi, M., Luini, M., et al. (2016). Development of a droplet digital polymerase chain reaction for rapid and simultaneous identification of common foodborne pathogens in soft cheese. *Frontiers in Microbiology, 7*, 1725.

[99] Witte, A. K., Fister, S., Mester, P., Schoder, D., & Rossmanith, P. (2016). Evaluation of the performance of quantitative detection of the Listeria monocytogenes prfA locus with droplet digital PCR. *Analytical and Bioanalytical Chemistry, 408*(27), 7583–7593.

[100] Cristiano, D., Peruzy, M. F., Aponte, M., Mancusi, A., Proroga, Y. T. R., Capuano, F., et al. (2021). Comparison of droplet digital PCR vs real-time PCR for Yersinia enterocolitica detection in vegetables. *International Journal of Food Microbiology, 345*:109321, 1–6.

[101] Chon, J. W., Jung, J. Y., Ahn, Y., Bae, D., Khan, S., Seo, K. H., et al. (2021). Detection of campylobacter jejuni from fresh produce: Comparison of culture-and pcr-based techniques, and metagenomic approach for analyses of the microbiome before and after enrichment. *Journal of Food Protection*, *84*(10), 1704–1712.

[102] Lei, S., Gu, X., Zhong, Q., Duan, L., & Zhou, A. (2020). Absolute quantification of Vibrio parahaemolyticus by multiplex droplet digital PCR for simultaneous detection of tlh, tdh and ureR based on single intact cell. *Food Control*, *114*, 107207.

[103] Peruzy, M. F., Proroga, Y., Capuano, F., Corrado, F., Santonicola, S., De Medici, D., et al. (2020). Detection and quantification of Campylobacter in foods: New analytic approaches to detect and quantify Campylobacter spp. in food samples. *Italian Journal of Food Safety*, *9*:8591, 88–92.

[104] Lv, X., Gu, X., Wang, L., He, X., He, C., Zhang, J., et al. (2021). Rapid and absolute quantification of VBNC Cronobacter sakazakii by PMAxx combined with single intact cell droplet digital PCR in infant foods. *LWT*, *145*:111388, 1–9.

[105] Öz, Y. Y., Sönmez, Ö. İ., Karaman, S., Öz, E., Unal, C. B., & Karataş, A. Y. (2020). Rapid and sensitive detection of Salmonella spp. in raw minced meat samples using droplet digital PCR. *European Food Research and Technology*, 246, 1895–1907.

[106] Leonard, S. R., Mammel, M. K., Lacher, D. W., & Elkins, C. A. (2015). Application of metagenomic sequencing to food safety: Detection of Shiga toxin-producing Escherichia coli on fresh bagged spinach. *Applied and Environmental Microbiology*, *81*(23), 8183–8191.

[107] Ramadan, H., Soliman, A., Elbediwi, M., Hiott, L., Woodley, T., Chattaway, M., et al. (2021). Emergence of multidrug-resistant *Escherichia coli* producing CTX-M, MCR-1, and FosA in retail food from Egypt. *Frontiers in Cellular and Infection Microbiology*, *11*:681588, 1–9.

[108] Ford, L., Carter, G. P., Wang, Q., Seemann, T., Sintchenko, V., Glass, K., et al. (2018). Incorporating whole-genome sequencing into public health surveillance: Lessons from prospective sequencing of *Salmonella Typhimurium* in Australia. *Foodborne Pathogens and Disease*, *15*(3), 161–167.

[109] Li, W., Cui, Q., Bai, L., Fu, P., Han, H., Liu, J., et al. (2021). Application of whole-genome sequencing in the national molecular tracing network for foodborne disease surveillance in China. *Foodborne Pathogens and Disease*, *18*(8), 538–546.

[110] Jagadeesan, B., Gerner-Smidt, P., Allard, M. W., Leuillet, S., Winkler, A., Xiao, Y., et al. (2019). The use of next generation sequencing for improving food safety: Translation into practice. *Food Microbiology*, *79*, 96–115.

[111] Sabater, C., Cobo-Díaz, J. F., Álvarez-Ordóñez, A., Ruas-Madiedo, P., Ruiz, L., & Margolles, A. (2021). Novel methods of microbiome analysis in the food industry. *International Microbiology*, *24*(4), 593–605.

Titilayo A. Ajayeoba*, Ebunoluwa J. Atolagbe, Oluwatosin D. Ayannuga,
Opeyemi Lala and Oluwatosin A. Ijabadeniyi

Chapter 9
Consumer's handling of food and food safety knowledge

Abstract: Knowledge on the implementation of food safety is an important aspect along the food value chain, and it represents the overall food wholesomeness, irrespective of the precautionary measures taken during harvesting, processing, storage, and transportation. Consumers' understanding of food handling and practices is influenced by several interwoven factors, including the level of knowledge about the type and nutritional component of food during processing, food hazards that compromise food quality, cultural practices, educational status, prevailing food policies and public health safety, and other physiological issues. Several factors including climate change, antimicrobial residues, socioeconomic influences, prevailing laws and regulations on agricultural practices processing and storage affect the overall knowledge, attitude, and food safety habits inculcated by consumers. Teaching proper consumers' attitude such as hazard identification, exposure evaluation, risk characterization, and management in all consumers' age groups is an effective strategy, and these are important components that must be considered by food safety management systems to reduce the public threats since consumers have responsibilities for acquiring, processing and storing food and food products.

9.1 Introduction

The consumption of contaminated food affects millions of individuals each year and has resulted in the spread of foodborne illnesses. This is a global emerging public health challenge as some of the variables linked to the rising trends of foodborne diseases include unsanitary settings and lack of suitable food safety standards and regulations

*Corresponding author: Titilayo A. Ajayeoba, Department of Microbiology, Faculty of Science, Adeleke University, Ede, Osun State, Nigeria: Department of Public Health Sciences, Faculty of Basic Medical Sciences, Adeleke University, Ede, Osun State, Nigeria,
e-mail: ajayeoba.titilayo@adelekeuniversity.edu.ng
Ebunoluwa J. Atolagbe, Department of Public Health Sciences, Faculty of Basic Medical Sciences, Adeleke University, Ede, Osun State, Nigeria
Oluwatosin D. Ayannuga, Opeyemi Lala, Department of Microbiology, Faculty of Science, Adeleke University, Ede, Osun State, Nigeria
Oluwatosin A. Ijabadeniyi, Department of Biotechnology and Food Technology, Durban University of Technology, South Africa

https://doi.org/10.1515/9783110748345-009

along the food supply chain. Furthermore, improper awareness, use of contaminated water on food products, unstable food storage temperature, and pesticide residues may enhance foodborne illness [1]. Consumer handling, on the other hand, is a significant part of food wholesomeness because the way it is handled impacts the quality and nourishment consumers receive. Consumers contribute to food safety by handling food after purchase, and suitable practices must be communicated by appropriate authorities.

Several factors are associated with consumers handling (Figure 9.1). These factors are interwoven, ranging from the level of knowledge and awareness of consumers to attitude, to personal hygiene and awareness of standards associated with food. Consumer education and an explicit explanation of the nature of food and critical stages that make food vulnerable to different forms of contamination is necessary to promote safer handling since food handling behaviors vary and can be determined by socioeconomic and cultural differences [2]. When contemplating a 'From Farm to Table' strategy, consumers are a key component in the food safety chain. Appropriate knowledge of thawing and refrigeration temperatures of food products, possible cross-contamination between food products during handling at different temperatures, and the prevention/control are all examples of home food safety practices regarding consumer awareness [3].

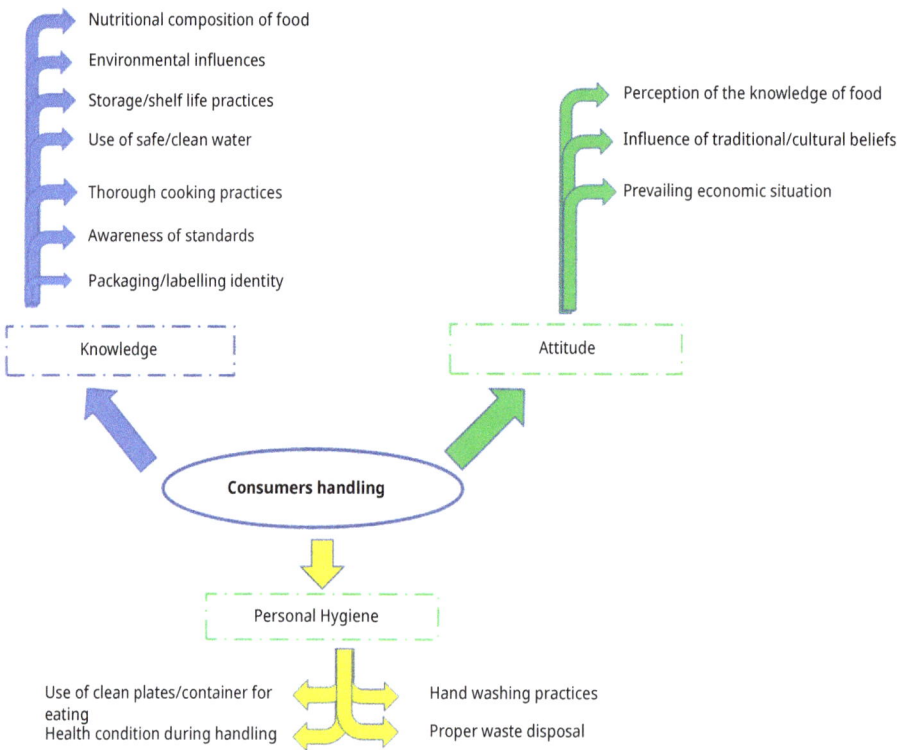

Nutritional composition of food

Environmental influences

Storage/shelf life practices

Use of safe/clean water

Thorough cooking practices

Awareness of standards

Packaging/labelling identity

Perception of the knowledge of food

Influence of traditional/cultural beliefs

Prevailing economic situation

Knowledge

Attitude

Consumers handling

Personal Hygiene

Use of clean plates/container for eating
Health condition during handling

Hand washing practices

Proper waste disposal

Figure 9.1: Some factors associated with consumers' handling practices.

Foodborne infections are generally on the rise and have been reported in private home setting due to improper food preparation practices in consumers' homes and these have raised international concerns to evaluate domestic food-handling practices but a significant proportion have been attributed to inadequate personal and/or good environmental hygienic practices [4, 5]. Observation studies indicate that a significant proportion of customers regularly implement unhygienic food-handling practices and improved performance in consumer food-handling attitude is necessary to decrease the threats and occurrence of foodborne disease [4].

The most successful way to improving food handling practices may be to combine measures to promote understanding of real and perceived food safety issues in the home with strategies that target certain demographic groups. Food safety education tools and messages could be more effective if we had a deeper grasp of customers' risk perceptions and behaviors [6]. Furthermore, educational efforts aimed at improving an inclusive foodborne illness responsiveness as well as cross-contamination prevention procedures could provide useful information for determining where handlers do not pay critical attention to the food safety continuum and result in better compliance with actual safe food handling [7]. It is critical to provide consumer education about safe food handling procedures in the home and the prerequisite to prevent foodborne illnesses at any stage before consumption. Television and radio programs are significant mediums for disseminating food safety information to consumers [8], and other internet-based tools could be useful as well.

9.2 Food safety management system

Consumers' concern about food-related indulgences is growing. Food safety practices are the outcome of human culture/tradition, socioeconomic disposition, and lifestyle. Consumers may not be directly connected to the food supplied from the farm because it may require certain processing or branding; thus, it is essential for these "intermediaries" to understand good food practice principles. Good food practices and its safety are broadly classified in connection with food technology (standard manufacturing practice), food issues (studying the nature of food, providing timely correct awareness, providing appropriate training and personnel), and all activities regarding consumers' food handling (Good Housekeeping Practice) [9]. Food commodity can only be declared safe when it is free from both microbiological, biological, chemical, and physical hazards that may ultimately result in health-related hazards and/or deter the nutritional quality [10].

Consumer priorities and preferences must be taken into account when developing and implementing effective methods to identify critical points in the food chain that can possibly expose food products different pollutants. Consumer perceptions of pollutants in certain meals were strongly connected to chemical contaminants in a study [11] with the possibility for severe repercussions, long-term health effects, and even death

[11]; thus, traceability is seen as a beneficial approach that can improve consumer self-confidence in food safety. Quantitative risk analysis is a useful technique with growing application for the production of safe products, which is a stepwise approach to investigate associated health risks in a specific food product, resulting in an assessment of the possibility of negative health implication(s) following the consumption of that food and the nature of the associated risks. To identify and tackle such associated risks, it is essential to identify potential associated hazard, sources and level of exposure to such hazard, risk description and classification, and risk management [12].

Globally, there are some high-valued food and food products whose demand is gradually doubling on the national and international trade because it both retained the nutritional quality and convenience characteristics. The consumer's quest for food safety requires continual surveillance on possible hazards and the ability to develop effective control measures if challenges arise [13].

Food products have distinct properties, resulting in diverse demands that food supply chain operators must fully comprehend. To demonstrate that effective consumer response adoption enhances both operational and financial performance, corporate social responsibility is essential in food supply chains between industrial producers, distributors, retailers, and consumers [14]. While there are continual measures to care for consumers' safe food consumption along the food supply chain to prevent foodborne illnesses, the final consumption stage, which is the final phase is an unmonitored segment in the "farm to fork" chain since preparation and eating pattern are subjective to the consumer and beyond authorized tests carried out by the skilled and knowledgeable personnel involved with food quality. The unmonitored phases that consumers have the responsibilities may include purchasing, storing, and processing food; thus, there must be awareness of the risk and nature of food nutritional composition, and teaching proper food handling behavior in all age groups is an effective strategy [10].

Consumer behavior and practice toward food safety have shown the need to increase their levels of understanding, motivation, and trust in the food product. A sound nutritional practice is necessary to rebuild the existing food safety system, and it primarily entails consumer education, continual education, and information exchange [9]. The establishment of hazard analytical critical control points (HACCP) as the initial food safety management system in the food business, as well as for consumers, is a significant step in the production cycle as is maintaining food processing and storage under circumstances that avoid recontamination. In addition, the HACCP approach focuses on risk management and prevention, tying operational management and food chain quality assurance together [15].

While industries must strategize, implement, function, sustain, and regularly update food safety management systems whose responsibility is aimed at providing products that are safe for the consumer based on their intended use, consumers' must demonstrate compliance with food safety requirements, and customer requirements must be continually evaluated and accessed to demonstrate compliance, according to the international standard on food safety [16]. The role of consumers in food system elements is enormous (Figure 9.2). There are various standards associated with food safety and handling. An understanding of the standards of different food products is very essential; else, the implementation of different food standards by consumers will be impossible. Consumers must understand the quality and good manufacturing practices of food products to ensure that the food purchased is in the right condition and they do not bargain for less quality. Once the consumer has the food product, it is essential to adopt the principles of HACCP (Figure 9.2) by implementing the right food processing and preparation methods before consumption, proper food storage techniques to avoid continuous thawing of freezing of certain foods, and the correct storage method for perishable and nonperishable foods as well as proper food disposal methods to prevent contamination and re-contamination for food product and processing environment.

In terms of healthy lifestyles and illness prevention, consumer understanding of food safety and nutrition is crucial. Some reported incidences of the foodborne disease are linked to poor consumer food management. Consumers must adjust behaviors that are inconsistent with safe food storage and preparation methods to lower the risk of foodborne illness. Consumer knowledge of good food handling techniques is closely linked to changes in such behaviors [10].

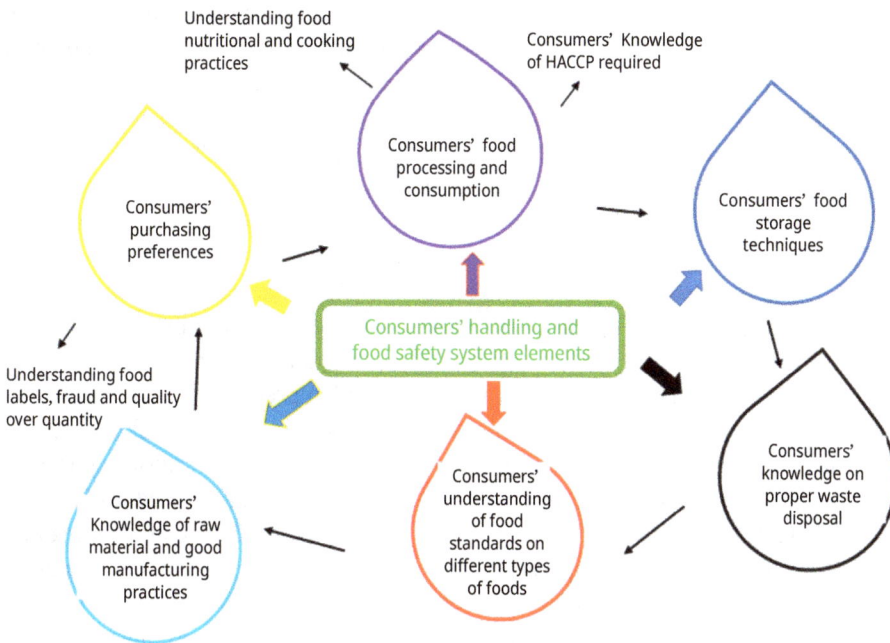

Figure 9.2: Interrelationship between consumers' handling and food safety systems.

9.3 Food safety chain

The importance of the customer in engaging in self-protective behavior is now being emphasized in food safety. As a result, food safety goals are defined at the point of consumption rather than purchase, with consumers being addressed as the critical final link in the food chain for ensuring safe food consumption and avoiding foodborne disease. The current state of consumers' hygiene in the home has been emphasized through domestic food preparation methods. In addition, inadequate hand and surface hygiene is a major contributor to major household food poisoning outbreaks. Overall, research has indicated that good hand washing and personal hygiene, safe and enough cooking of food, storing foods at safe temperatures, and efficiently washing surfaces and equipment to minimize cross-contamination are the most common behaviors affecting the control of various infections. The lack of adequate consumer hygiene at home underscores the need for effective psychosocial treatments aimed at changing food safety behaviors, attitudes, or knowledge through educational materials or training. Several studies have indicated that if food hygiene standards are to improve, education about the prevention of foodborne disease is essential [17].

Since customers are free to handle food as they deem fit, the house is the final checkpoint for food safety. Although the home is the most common setting of exposure for verified outbreaks, many consumers do not accept the role of the home in foodborne infections, and failure to accept personal responsibility for food safety is a prerequisite for implementing appropriate food safety behaviors to reduce microbiological risks [18, 19]. Intestinal disease prevention is based on "targeted hygiene," which entails identifying infection transmission routes in the household and community and implementing hygiene measures at "critical spots" to break the transmission chain. Critical considerations include (i) proper cooking temperatures for various meals, (ii) minimizing cross-contamination between foods or food and cooking surfaces to limit the danger of opportunistic pathogenic bacteria, and (iii) proper storage temperatures [19]. A lack of thoroughly washing contaminated hands, knives, and chopping boards both during and after meal preparation predicts the spread of pathogenic microorganisms such as bacteria, fungi, and even viruses around the kitchen environment and onto prepared meals; thus, a higher level of perceived importance of proper food handling behavior is associated with higher levels of educational attainment [20]. It is crucial to minimize the threat of inappropriate food handling and storage because the house represents an important site for the spread of pathogenic microorganisms responsible for foodborne diseases, overall safety procedures of food handling, storage and preparation have to be indicated to consumers [5].

The majority of microbial foodborne illnesses are thought to be preventable if food safety principles are understood and practiced throughout the entire food chain and especially by food handlers.

9.4 Factors influencing food safety

Most microbiological foodborne infections are regarded to be preventable if food safety principles are recognized and followed throughout the food chain, particularly by consumer handlers. Food safety is influenced by several factors, many of which are related to consumer handling. One factor is temperature control of several food product categories to reduce contamination, cross-contamination, and food waste, particularly in food cold chains [21]. Another factor is food labels. Food labels are critical for disseminating information about food products to consumers and for conveying food safety messages, especially since improper domestic food handling practices generate a large proportion of foodborne illnesses. However, since labels contain so much information, it can be difficult for consumers to figure out which statements are most relevant in terms of food safety. There is some misunderstanding about terms related to spoilage dates, and information such as cooking requirements is sometimes overlooked. Labels are influenced by a variety of factors, including attitudes, knowledge, experience, sociodemographic traits, trust in message providers, and label design, as is the case with many other behaviors [22].

There are several pointers, including animals such as household pets and materials such as old newspapers, filthy clothing and even soil that some consumers use traditionally within the kitchen space which have direct/indirect contact with food products either during food preparation or storage. Furthermore, filthy dishes that are packed alongside with clean plates on kitchen counters, improper placement of raw and uncleaned vegetables, drippings from raw meat/fish, cooked ready-to-eat meals in household refrigerators are factors contributing to cross-contamination and development of pathogens like coliforms, *S. aureus*, *Salmonella*, *Candida*, and *Aspergillus* spp. and other fungal species in food and food products [23]. In another study, kitchen door knobs, abuse of temperature of food leading to unclean refrigerators, dishwashers of inferior quality, cooking stove knobs, kitchen surfaces for food preparation or where food are kept, taps and unclean kitchen towels, as well as water closet knobs and taps have been identified as reservoirs of pathogenic microorganism with indirect link to consumer's infection, thus, showing necessitating educational campaigns targeted towards good hygiene and sanitary practices at home [24].

Aside diverse microorganisms that can contaminate and reduce the quality of food, the nature and structure of the food product, physiological adaptation of microorganism to the food-processing environment, their level of pathogenicity and virulence, level of exposure to humans, synergetic ability of different classes of microorganism to cause infection, ability to persist, antibiotic susceptibility pattern and biofilm formation are factors that raises challenges with food safety [25].

After consumer purchases a specific food item, there are numerous key phases in the food preparation. Cooking procedures, contamination from cooking equipment, and exposure of cooked foods to raw foods are all important control parameters to maintain the safety of food items. The virulence of the microorganism present, the microbial load present in the food, and the immune system conditions of the consumers all have a role in the development of foodborne disease. In addition to microbiological pollutants, the presence of other forms of contaminants can harm consumer health. As a result, strict adherence to hygienic guidelines by people who come into touch with food during the manufacturing, processing, transportation, and storage phases is critical to ensuring food safety and preventing foodborne disease [26].

9.5 Knowledge, attitude, and perception of food safety

Food contamination and possible side effect continue to be a challenge to worldwide public health, especially the outbreaks majority reported among consumers. The way consumers respond to food safety issues both developed, developing and underdeveloped countries is very crucial to the quality of the final food product consumed. Although underdeveloped and developing countries are likely to be affected by food safety challenges and foodborne diseases due to certain socioeconomic factors, poor personal hygiene, inadequate or unavailable medical treatment, unskilled food handlers, poor storage practices, lack of food safety monitoring and surveillance personnel and equipment, poor knowledge on standard methods for prevention and control of food-related microorganisms, food fraud and lack of accurate and updated data to monitor foodborne diseases [27].

Consumers' food safety attitudes are influenced unidirectionally by food safety knowledge, leading to the practices adopted (Figure 9.3). Consumers can only make logical decisions if they are aware of the implications connected with their behaviors, thus it is critical to first understand their attitudes and behaviors before delivering information and instilling familiarity with food safety issues and safe food handling techniques. This strategy would result in instructional programs that encourage consumers to accept experts' opinions [23]. Different countries have absorbed various food cultures, which have a direct impact on consumer attitudes. Consumers are now more concerned about food hazards that they believed were caused by technology. Females have also been recognized as having issues regarding the efficacy of household hygiene measures than males, as well as an understanding of food safety problems linked with the use of ineffective domestic hygiene procedures [5, 23], Younger adults, particularly those in higher education, are more likely to engage in unsafe eating habits, making them more susceptible to foodborne infections and engaging in risky food handling practices [28]. Furthermore, food safety attitudes of older adults who cling to certain beliefs and traditional practices about method of food preparation, storage and improper handling and eating habit can increase the risk of foodborne illness and diseases [29].

Apart from the factors that are likely to contribute to foodborne illness outbreaks in home settings, affluence, demand for food convenience, quest for product development and even traditional beliefs and rites have instigated deliberate consumption of certain "high-risk" foods, raw and undercooked foods in specific locations. These practices are likely to expose certain classes of people (pregnant women, toddlers, the elderly, and immunocompromised) to food-related infections.

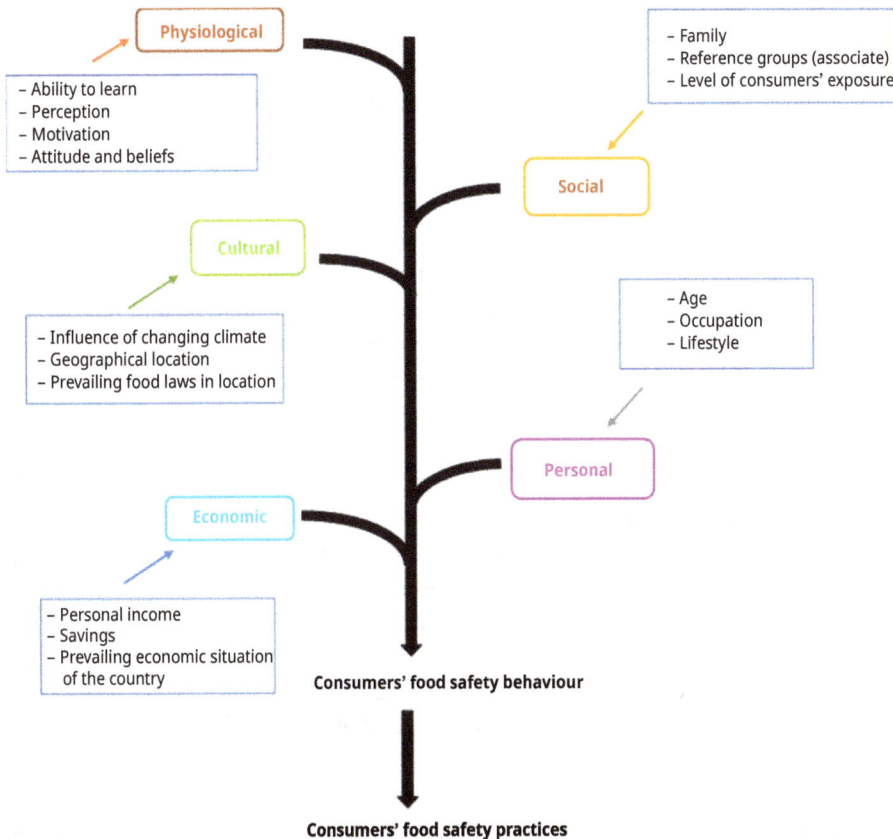

Figure 9.3: Factors influencing consumers' behavior and food handling practices.

Food safety messages should consequently reinforce and underline consumers' understanding of their role and relevance in the safety of foods made at home [30].

Processors, marketers, and handlers all share a set of ideals when it comes to food safety. Maintaining a food safety culture requires these participants to be aware of the hazards connected with the products or meals they make, understand why risk management is critical, and effectively provably manage those risks. A shift in organizational culture can have a direct impact on front-line employee behavior. Many food-handling behavior therapies focus on education and training, however, the effects of food handler training programs are variable, and program evaluation is rarely undertaken on real food handler practices [31]. Therefore, acknowledging personal responsibility for food safety is a prerequisite for the implementation of proper food safety behaviors.

9.6 Emerging diseases and knowledge of food safety

The ever-increasing human population has necessitated a continual increase for food production. To support this quest, pesticides and fertilizers are used heavily in certain locations. During pesticide applications, occurrences of intoxication and poisoning of farmers and rural workers either through direct inhalation, skin contact have been reported. Excessive exposure and inappropriate application of toxic chemicals can lead to health-related changes such as cancer, obesity, endocrine disruption, and other disorders in humans. Furthermore, pesticide residues in soil can be washed off into coastal marine systems, and their harmful effects on humans and nonhuman biota have a huge influence upon sea food which human consume. Unfortunately, consumers and the general public may not be fully aware of the serious environmental and health consequences of toxic chemicals in seafoods and other related food product [32].

Antimicrobial residues in food are an increasing source of food safety and public health problems. Their prevalence in animal-based foods and products poses socioeconomic concerns. Antimicrobial residues cause antimicrobial drug resistance, hypersensitivity, carcinogenicity, mutagenicity, teratogenicity, bone marrow suppression, and disturbance of normal intestinal flora, among other things. Since antimicrobials can accumulate as residues in tissues before they are completely metabolized or excreted from the body, prudent use of antibiotics, as well as the use of alternatives to antibiotics such as vaccination, probiotics, phage therapy, and essential oils, is recommended in ensuring consumer food safety [33].

The occurrence of antibiotic resistance microorganisms along the food chain is another public health concern. Food and food products, whether animal or plant origin, can readily be colonized and contaminated by antibiotic-resistant microorganisms, particularly of bacteria origin [34]. For those of plant origin, the antibiotic-resistant bacteria (AnRB) may result from contact with soil debris as soil is a reservoir for AnRB. Antibiotic-resistant bacteria can colonize and proliferate in food of animal origin and enter the human body primary or secondary contact with diseased animals or their bodily elements such as blood, urine or feces. To efficiently handle issues regarding food safety, One Health approach is required. This involves the coordinated process of ensuring that the safety of all animated and non-animated objects in contact with food and food products. This process will enhance reduced transfer of microorganisms from infectious sources, the spread of AnRB and development of new AnRB [35].

Agriculture, food processing, food packaging, and the surveillance of the movement and sale of food and food ingredients are all becoming more reliant on nanotechnology. Many of them are optimistic about nanotechnology's potential benefits, but they are concerned about its application in the food industry. Nanoparticles have far more access to the cells, tissues, and organs of the body than bigger particles [36]. Several food-related applications of engineered nanomaterials (ENMs) have the potential to be swallowed, absorbed, metabolized, and excreted/ removed after the internal exposure and toxicity of ENMs. Basic cellular functions such as proliferation, metabolism, and death can be impacted by ENMs due to their unique physicochemical features and high reactivity. The stimulation of reactive oxygen species production is the most important pathogenetic mechanism linking ENM exposure to tissue damage [37, 38]. The latter can influence intracellular calcium levels, activate transcription factors, and trigger cytokine production. Cytotoxicity, genotoxicity, and induction of an inflammatory and/or immunological response are all frequent toxicity outcomes [38]. Consumer understanding of nanotechnology applications is restricted since the anticipated hurdles to nanotechnology adoption were potentially unknown in terms of human health, environmental implications, and knowledge of how these food products should be handled and consumed [39]. To avoid future catastrophes caused by the use of nanotechnologies in food processing, regulatory systems capable of regulating any dangers related to nano-foods and the use of nanotechnologies in food processing and product development are required.

Climate change has aided the spread of zoonotic illnesses such as Covid-19, H1N1, swine flu, Ebola, and the Nipah virus, all of which have endangered human health and livelihoods. Because of a lack of properly enforced food-safety requirements in managed agricultural production systems, illnesses can mutate into extremely contagious strains, and consumers have become victims of improper handling [40]. Due to the interconnections between environmental conditions and food contamination, climate change may have an impact on the occurrence of food safety issues, which consumers are unaware of. Extreme weather circumstances, such as floods and droughts, that have never occurred before in some countries may be contributing factors to crop contamination by diverse species of toxigenic fungi and related mycotoxins, exposing consumers to natural hazards at a high risk. Mycotoxins are a class of naturally occurring harmful chemical compounds produced mostly by tiny filamentous fungus species found on a variety of crops causing adverse health effects when consumed by humans and animals [41].

Climate change impacts food security, and consequently human diets and nutrition, in three ways: household food security (access to safe, inexpensive, and sufficient food), child feeding and care practices, and environmental health and access to health services. Climate change impacts what food is available and at what price, as well as consumers' attitudes toward food, calorie consumption, and healthful food consumption. Other factors, such as personal preference and convenience, influence what consumers buy and consume. The variety of foods consumed is important for people's nutritional condition and health. Dietary diversity is linked to appropriate nutrition, particularly micronutrient sufficiency, as evaluated by the number of food types and the amount of nutritious vegetables, fruits, and animal source food intake [42].

Foodborne pathogens are causing a great number of diseases with significant effects on human health and the economy. The characteristics of the most common pathogenic bacteria (*Clostridium perfringens, Escherichia coli, Bacillus cereus, Clostridium botulinum, Cronobacter sakazakii, Staphylococccus aureus, Shigella* spp., *Listeria monocytogenes*, viruses (hepatitis A and noroviruses), *Salmonella* spp., *Campylobacter jejuni, Vibrio* spp., and *Yersinia enterocolitica*), and parasites (*Cyclospora cayetanensis, Toxoplasma gondii*, and *Trichinella spiralis*), and some outbreaks have been affecting consumers because they have little knowledge on the opportunistic nature of these pathogens. Food safety management systems based on the classical hazard-based approach are ineffective, and a risk-based food safety strategy is more effective in identifying, selecting, and implementing mitigation measures to control and decrease these risks [43].

The consumer's socioeconomic status (SES) is a significant indicator of how they handle food. Although the low socioeconomic level is generally associated with negative health outcomes, it can be defined in a variety of ways. It is frequently measured based on individual and community-level education, income, wealth, employment, and family background compared to other individuals or groups [44] because consumers will likely go for quantity over quality. In different countries, there exist correlations between foodborne illness and causal bacteria, SES, and level of development/exposure. SES had a different influence on different pathogens: for *Campylobacter*, salmonellosis, and *E. coli* infection, the majority of research found a link between high SES and illnesses, and SES should be taken into account when targeting consumer-level public health interventions for foodborne pathogens [44].

9.7 Future perspectives on food safety and consumer handling

As a result of multiple food scandals and incidents, consumers have become increasingly concerned about food safety. Therefore, consumers should be educated about the procedures that take place across the food supply chain. Food companies and governments have embraced using a traceability system to track food from farm to fork as a vital tool for restoring and increasing public confidence in food safety [45]. Consumers will desire food traceability systems, particularly for imported food products. Appropriate measures in terms of equipment, skilled personnel, timely reporting of food incidences and monitoring pattern as well as platforms providing feedbacks are important to cope with the increasing complexity and globalization of the food supply chains and prevent food hazard and food fraud and authenticity. The use of artificial intelligence, networking with other food-related platforms is crucial to predict issues of food safety and food authenticity. When this information has been gathered, handling structured and unstructured data recovered from food-related sources and origins to ensure the health of consumers and to minimize economical losses will not be difficult [46].

In low- and middle-income nations, foodborne infections take a heavy toll on people's health. Several factors contribute to the food safety challenge in these locations. Apart from being ignorant of food safety issues, consumers lack the ability to pay for food safety as they prefer quantity over quality. Furthermore, the agents along the food supply chain (farmers, retailers, processors, and other service providers) sometimes require incentives from consumers in exchange of quality and the responsiveness of agencies saddled with the responsibility of enforcing standard food regulations is not optimally functional. Programs that involve midsize and larger enterprises should have a common regulation and compensate farmers and businesses for investing in food safety are examples of possible solutions [47].

Two key characteristics that influence food handling behavior are self-protection awareness and food safety knowledge. Personality, frugality, the physical environment, the social environment, and the pursuit of quality of life all play a role. The self-protection habit contains the consciousness superiority trap and the knowledge superiority trap. As a result, comprehending consumers' motivations for adopting safe food handling habits and providing a theoretical foundation for the government to create effective intervention programs are essential [48].

References

[1]   Sanlier, N. (2009), The knowledge and practice of food safety by young and adult consumers. *Food Control, 20*(6), 538–542.
[2]   Patil, S. R., Cates, S., & Morales, R. (2005), Consumer Food Safety Knowledge, Practices, and Demographic Differences: Findings from a Meta-Analysis. *Journal of Food Protection, 68*(9), 1884–1894.
[3]   Jevšnik, M., Hlebec, V., & Raspor, P. (2008), Consumers' awareness of food safety from shopping to eating. *Food Control, 19*(8), 737–745.
[4]   Redmond, E. C., & Griffith, C. J. (2003), Consumer food handling in the home: a review of food safety studies. *Journal of Food Protection, 66*(1), 130–161.
[5]   Langiano, E., Ferrara, M., Lanni, L., Viscardi, V., Abbatecola, A. M., & De Vito, E. (2012), Food safety at home: knowledge and practices of consumers. *Zeitschrift Fur Gesundheitswissenschaften, 20*(1), 47–57.
[6]   Roseman, M., & Kurzynske, J. (2006), Food Safety Perceptions and Behaviors of Kentucky Consumers. *Journal of Food Protection, 69*(6), 1412–1421.
[7]   Abbot, J. M., Byrd-Bredbenner, C., Schaffner, D., Bruhn, C. M., & Blalock, L. (2009), Comparison of food safety cognitions and self-reported food-handling behaviors with observed food safety behaviors of young adults. *European Journal of Clinical Nutrition, 63*(4), 572–579.
[8]   Ergönül, B. (2013), Consumer awareness and perception to food safety: A consumer analysis. *Food Control, 32*(2), 461–471.

[9]   Raspor, P. (2008), Total food chain safety: how good practices can contribute? *Trends in Food Science and Technology, 19*(8), 405–412.

[10] Losasso, C., Cibin, V., Cappa, V., Roccato, A., Vanzo, A., Andrighetto, I., et al. (2012). Food safety and nutrition: Improving consumer behaviour. *Food Control, 26*(2), 252–258.

[11] Kher, S. V., De-Jonge, J., Wentholt, M., Deliza, R., De andrade, J. C., Cnossen, H. J., et al. (2013). Consumer perceptions of risks of chemical and microbiological contaminants associated with food chains: a cross-national study. *International Journal of Consumer Studies, 37*(1), 73–83.

[12] Notermans, S., & Teunis, P. (1996), Quantitative risk analysis and the production of microbiologically safe food: an introduction. *International Journal of Food Microbiology, 30*(1), 3–7.

[13] Quested, T. E., Cook, P. E., Gorris, L. G. M., & Cole, M. B. (2010). Trends in technology, trade and consumption likely to impact on microbial food safety. *International Journal of Food Microbiology, 139*, S29–S42.

[14] Fredriksson, A., & Liljestrand, K. (2015), Capturing food logistics: a literature review and research agenda. *International Journal of Logistics: Research and Applications, 18*(1), 16–34.

[15] Papademas, P., & Bintsis, T. (2010), Food safety management systems (FSMS) in the dairy industry: A review. *International Journal of Dairy Technology, 63*(4), 489–503.

[16] Frestedt, J. L. (2017). Chapter 2 – introduction to quality systems. In J. L. Frestedt (Ed.), *FDA warning letters about food products* (pp. 23–50). Academic Press.

[17] Milton, A., & Mullan, B. (2010), Consumer food safety education for the domestic environment: a systematic review. *British Food Journal, 112*(9), 1003–1022.

[18] Authority, E. F. S. (2009), The community summary report on food-borne outbreaks in the European Union in 2007. *EFSA Journal, 7*(5), 271r.

[19] Kennedy, J., Gibney, S., Nolan, A., O'Brien, S., McMahon, M. A. S., McDowell, D., et al. (2011). Identification of critical points during domestic food preparation: an observational study. *British Food Journal*.

[20] Kennedy, J., Nolan, A., Gibney, S., O'Brien, S., McMahon, M. A. S., McKenzie, K., et al. (2011). Deterninants of cross-contamination during home food preparation. *British Food Journal, 113*(2), 280–297.

[21] Ndraha, N., Hsiao, H. I., Vlajic, J., Yang, M., & Lin, H. V. (2018). Time-temperature abuse in the food cold chain: Review of issues, challenges, and recommendations. *Food Control, 89*, 12–21.

[22] Hall, C., & Osses, F. (2013), A review to inform understanding of the use of food safety messages on food labels. *International Journal of Consumer Studies, 37*(4), 422–432.

[23] Byrd-Bredbenner, C., Berning, J., Martin-Biggers, J., & Quick, V. (2013), Food Safety in Home Kitchens: A Synthesis of the Literature. *International Journal of Environmental Research and Public Health, 10*(9), 4060–4085.

[24] Azevedo, I., Albano, H., Silva, J., & Teixeira, P. (2014). Food safety in the domestic environment. *Food Control, 37*, 272–276.

[25] Schirone, M., Visciano, P., Tofalo, R., & Suzzi, G. (1974). Editorial: Foodborne pathogens: Hygiene and safety. *Frontiers in Microbiology, 2019*, 10.

[26] Gallo, M., Ferrara, L., Calogero, A., Montesano, D., & Naviglio, D. (2020). Relationships between food and diseases: What to know to ensure food safety. *Food Research International, 137*, 109414.

[27] Odeyemi, O. A., Sani, N. A., Obadina, A. O., Saba, C. K. S., Bamidele, F. A., Abughoush, M., et al. (2019). Food safety knowledge, attitudes and practices among consumers in developing countries: An international survey. *Food Research International, 116*, 1386–1390.

[28] Sanlier, N., & Konaklioglu, E. (2012), Food safety knowledge, attitude and food handling practices of students. *British Food Journal, 114*(4), 469–480.

[29] Evans, E. W., & Redmond, E. C. (2014), Behavioral Risk Factors Associated with Listeriosis in the Home: A Review of Consumer Food Safety Studies. *Journal of Food Protection, 77*(3), 510–521.

[30] Nesbitt, A., Thomas, M. K., Marshall, B., Snedeker, K., Meleta, K., Watson, B., et al. (2014). Baseline for consumer food safety knowledge and behaviour in Canada. *Food Control*, *38*, 157–173.

[31] Powell, D. A., Jacob, C. J., & Chapman, B. J. (2011), Enhancing food safety culture to reduce rates of foodborne illness. *Food Control*, *22*(6), 817–822.

[32] Carvalho, F. P. (2017), Pesticides, environment, and food safety. *Food and Energy Security*, *6*(2), 48–60.

[33] Okocha, R. C., Olatoye, I. O., & Adedeji, O. B. (2018), Food safety impacts of antimicrobial use and their residues in aquaculture. *Public Health Reviews*, *39*(1), 21.

[34] Al Bayssari, C., Dabboussi, F., Hamze, M., & Rolain, J. (2015), Emergence of carbapenemase-producing Pseudomonas aeruginosa and Acinetobacter baumannii in livestock animals in Lebanon. *Journal of Antimicrobial Chemotherapy*, *70*(3), 950–951.

[35] Founou, L. L., Founou, R. C., & Essack, S. Y. (1881). Antibiotic Resistance in the Food Chain: A Developing Country-Perspective. *Frontiers in Microbiology*, *2016*, 7.

[36] Bánáti, D. (2011), Consumer response to food scandals and scares. *Trends in Food Science and Technology*, *22*(2), 56–60.

[37] Donaldson, K., Borm, P. J., Castranova, V., & Gulumian, M. (2009), The limits of testing particle-mediated oxidative stress in vitro in predicting diverse pathologies; relevance for testing of nanoparticles. *Particle and Fibre Toxicology*, *6*(1), 1–8.

[38] Martirosyan, A., & Schneider, Y. J. (2014), Engineered nanomaterials in food: implications for food safety and consumer health. *International Journal of Environmental Research and Public Health*, *11*(6), 5720–5750.

[39] Handford, C. E., Dean, M., Spence, M., Henchion, M., Elliott, C. T., & Campbell, K. (2015). Awareness and attitudes towards the emerging use of nanotechnology in the agri-food sector. *Food Control*, *57*, 24–34.

[40] Aiyar, A., & Pingali, P. (2020), Pandemics and food systems – towards a proactive food safety approach to disease prevention & management. *Food Security*, *12*(4), 749–756.

[41] Milicevic, D., Lakicevic, B., Petronijevic, R., Petrovic, Z., Jovanovic, J., Stefanovic, S., et al. (2019). Climate change: impact on mycotoxins incidence and food safety. *Theory and Practice of Meat Processing*, *4*, 1, 9–16.

[42] Fanzo, J., Davis, C., McLaren, R., & Choufani, J. (2018). The effect of climate change across food systems: Implications for nutrition outcomes. *Global Food Security, Amsterdam, The Netherlands*, *18*, 12–19.

[43] Bintsis, T. (2017), Foodborne pathogens. AIMS Microbiology, 3(3), 529–563.

[44] Newman, K. L., Leon, J. S., Rebolledo, P. A., & Scallan, E. (2015), The impact of socioeconomic status on foodborne illness in high-income countries: a systematic review. *Epidemiology and Infection*, *143*(12), 2473–2485.

[45] Zhang, A., Mankad, A., & Ariyawardana, A. (2020), Establishing confidence in food safety: is traceability a solution in consumers' eyes?. *Journal of Consumer Protection and Food Safety*, *15*(2), 99–107.

[46] Fritsche, J. (2018), Recent Developments and Digital Perspectives in Food Safety and Authenticity. *Journal of Agricultural and Food Chemistry*, *66*(29), 7562–7567.

[47] Hoffmann, V., Moser, C., & Saak, A. (2019). Food safety in low and middle-income countries: The evidence through an economic lens. *World Development*, *123*, 104611.

[48] Wang, M., Bai, L., Gong, S., & Huang, L. (2020). Determinants of consumer food safety self-protection behavior– an analysis using grounded theory. *Food Control*, *113*, 107198.

James A. Elegbeleye*, Kehinde O. Oyeneye, Olusola A. Akinboboye
and Emmanuel G. Abere

Chapter 10
The economic cost of food recall

Abstract: This chapter focuses on the economic costs of food safety and food recall. In this chapter the reader will be exposed to various definitions of key concepts, forms and types of cost classification as it relates to food safety and food recall. Furthermore, the estimation of the impact of an outbreak and costing prevention strategies will also be discussed. How food recalls affect individual, firms, and a country's economy as a whole will be discussed. Finally, the chapter will give a brief summary and conclusion on the cost of food safety and food recall. This chapter will attempt to review the incidence of foodborne outbreaks, their economic and social impacts, and inherent challenges in the determination of economic cost of food recalls on a global scale as well measures that can be used to mitigate these impacts.

10.1 Introduction

As the global food supply systems increasingly expand, there is a corresponding increase in pressure from anthropological sources as well as increase in consumer demands such as the demand for fresh-like, minimally processed food products [1]. Other factors such as climate change, degradation of land, pesticides, environmental hormones, and, more importantly, the evolution of antibiotic-resistant microorganisms in food also present an overwhelming challenge to global food supply chain [2–4]. All these factors, with the frequent occurrence of outbreaks of foodborne diseases (FBDs) conspire to negatively impacts and compromise global food security and health [5]. FBDs encompass a range of diseases that arise from the ingestion of contaminated foods or water. The issue of food safety is becoming an increasingly important one on a global scale, therefore increasing FBDs and their attending consequences. It is impor-

*Corresponding author: James A. Elegbeleye,** Department of Consumer and Foodrecallfood Sciences, Faculty of Natural and Agricultural Sciences, University of Pretoria, Pretoria, South Africa; Department of Economics, Crawford University, Faith City, P.M.B. 2001, Igbesa, Ogun, Nigeria,
e-mail: ayoelegs@yahoo.co.uk
Kehinde O. Oyeneye, Department of Economics, Crawford University, Faith City, P.M.B. 2001, Igbesa, Ogun, Nigeria
Olusola A. Akinboboye, Department Microbiology, Faculty of Pure and Applied Sciences, Southwestern University Nigeria, Kilometre 20 Sagamu-Benin Expressway, Okun-Owa, Ogun, Nigeria
Emmanuel G. Abere, Department of Economics, Faculty of Social and Management Sciences, Southwestern University Nigeria, Kilometre 20 Sagamu-Benin Expressway, Okun-Owa, Ogun, Nigeria

https://doi.org/10.1515/9783110748345-010

tant because of the formidable threat it presents to public health and socioeconomic development of the nations where it occurs [6].

The dearth of accurate and reliable data has made it difficult to adequately address the concerns on FBDs on a global scale, the attendant burden, and socio-economic cost. In most low- and middle-income countries (LMIC), epidemiological data that can be used to track, control, and carry out disease surveillance during an outbreak of food pathogens are very hard to come by. Therefore, most outbreaks may go unnoticed, under-reported, or entirely unreported, especially in a situation where the outbreak is not linked with a major public health or economic impact. The gap in available quantitative data between LMIC and high-income countries (HIC) gave rise to an initiative by WHO Foodborne Disease Burden Epidemiology Reference Group (FERG) to estimate the global disease burden using meta-analysis and reviews [7].

The effect of FBDs ranges from mild health effects such as vomiting and diarrhea to more severe ones such as kidney failure and even death. FBD is one of the most serious public health problems in our world of today, especially in developing nations mostly due to poor hygiene and lack of basic sanitation. The Foodborne Disease Burden Epidemiology Reference Group (FERG) of the WHO estimated the global burden of FBDs to be approximately 600 million cases and 420,000 deaths annually emanating from 31 foodborne hazards [8]. This estimate translates to the fact that approximately one out of every 10 people will be affected by FBD on an annual basis.

These foodborne hazards are usually biological and chemical hazards that gained entry into the food chain. There are over 250 different FBDs in existence, and most of these diseases are caused by bacteria, viruses, parasites, and chemicals such as mycotoxins commonly associated with food. Diarrhea is the most common symptom among FBDs, while *Salmonella, Campylobacter*, and enterohemorrhagic *Escherichia coli* are the most common foodborne pathogens causing enormous impacts on public health on an economic scale [9]. The foodborne outbreaks and public health and socio-economic impacts of FBDs have taken a quantum leap with increasing international food trade, extensive production, and multifaceted supply food chain. The increase in the trade volume may be attributed to the mounting pressure on the food industry to meet up global demand, thereby causing unintended consequences of FBDs and economic loss arising from food recalls [10].

Table 10.1: 31 Hazards included in the final FERG analyses [8].

Parasites	Chemicals	Invasive bacteria/virus	Diarrheal/disease
Ascaris spp.	Aflatoxin	*Brucella* spp.	*Campylobacter* spp.
Taenia solium	Cassava cyanide	Hepatitis A virus	Cryptosporidium spp
Paragonimus spp.	Dioxin	*Listeria* spp.	*Entamoeba histolytica*

Table 10.1 (continued)

Parasites	Chemicals	Invasive bacteria/virus	Diarrheal/disease
Opisthorchis spp.		Mycobacterium bovis	Enteropathogenic Escherichia coli
Clonorchis sinensis		Invasive nontyphoid Salmonella	Enterotoxigenic Escherichia coli
Fasciola spp.		Salmonella paratyphi A	Giardia spp.
E. granulosus		Salmonella typhi	Norovirus
E. multilocularis			Nontyphoid Salmonella
Toxoplasma gondii			Shigella spp.
Trichinella spp.			Shiga toxin-producing Escherichia coli (STEC)
Intestinal flukes			Vibrio cholerae

10.2 Food safety and food recalls

Food is safe when it is processed, stored, distributed, and prepared in a way without causing either acute or chronic adverse effects on consumers [11]. Conversely, food is considered to be unsafe food when the food is contaminated with physical, microbiological or chemical hazard that can negatively impact the health of the consumers [12]. Biological hazards in food include pathogenic bacteria, fungi, viruses and parasites causing either food infections or intoxication upon consumption [13]. The persistence of biological and chemical hazards throughout the food chain is attributed to multiple factors at varying degrees of occurrence [55]. Notable among them are the poor handling and storage of raw material, poor product processing, unsanitary processing environment, and poor application of good manufacturing practices and food safety standards among others [56]. The failure of farmers or processors to adhere to one or more of these factors provides a favorable environment for pathogens to thrive in a specific food type. The pathogen exerts its dominance in the food environment by adapting to the environmental stresses during processing using genetic and physiological modifications which make it develop some measures of resistance [53, 54].

Outbreak of FBDs is known to cause some morbidity and mortality depending on the causative agent among the population. For the food industry, it has resulted in lawsuits, economic loss because of product recalls and declining confidence of consumers in the product associated with the FBD. The United States Centre for Disease

Control and Prevention identified *Salmonella typhi, Clostridium perfringens, Campylo-bacter,* and *Staphylococcus aureus* as the main etiologic bacteria responsible for most outbreaks [50, 51]. The common food categories implicated in the outbreaks are mostly fish, unpasteurized dairy products, seeded vegetables, leafy vegetables, pork, and chicken, causing about 10,008 illnesses and 896 deaths in 2015 alone [14]. In the European Union, FBDs affect more than 23 million people annually with approximately 5,000 deaths. Diarrheal diseases from norovirus and *Campylobacter* account for about 20 million of the illnesses [8].

Large-scale global trades between countries and regions of the world coupled with an extensive production and supply chain network is a major contribution to the increased amount of food safety outbreaks [10]. Such outbreaks in FBDs have resulted in economic losses especially when it is above a certain predetermined threshold. The impacts of such biological and chemical hazards in foods affect not just the food processing industries but also farmers, retailers, consumers and the society at large [15]. Therefore, the importance of ensuring safe foods and compliance with known local and international food safety standards cannot be overemphasized.

This economic loss can be direct or indirect costs and may come in the form of loss to the industry and victims of the FBDs besides that loss at the national level when it involves international trades. According to a scoping review by McLinden et al. [16], the direct cost of FBD at the individual level may include the following medical, laboratory cost, personnel cost (consultant, specialist, etc.), cost of hospitalization depending on the severity, out-patient cost besides numerous indirect costs. The cost to the society often takes the form of government and regulatory costs, cost of litigation, cost to the public health system and cost to the industry that is implicated. At the industry level, the loss can be loss to the business because of decline in the demand, product recall, plant closure, adjustment in the manufacturing process and other related cost such as herd slaughter as in the case of swine flu or avian flu. In the United States, United States Department of Agriculture (USDA) Economic Research Service (ERS) has estimated the total cost of 15 leading foodborne illnesses at $15.5 billion and $17.6 billion in 2013 and 2018, respectively [17].

Food recalls are a tremendous challenge to the food industry because they can cause indelible and irrecoverable damages to a firm's reputation [18]. An example is the Hudson Foods Co's recall of 25 million pounds of ground beef in 1997 which happened to be the largest food recalls in the United States till date. This occurrence resulted in the company's acquisition by Tyson Foods, after losing its largest customer, Burger King [19]. The figure of reported cases of contaminated food products that have led to product recalls have increased over the years with a great loss to the food industry [20]. In the United States, there were twenty outbreaks that resulted in product recall in 2015. The food products indicted in the outbreaks were chicken, ground beef, pork, raw tuna, bread, and unpasteurized milk, lettuce, alfalfa seeds among others [14].

Quantifying the specific cost of a recall to a food industry is a difficult challenge because the process requires data at the firm-level that may be private or difficult to assess by a third party due to nondisclosure agreement [18]. 18,was able to overcome this limitation in calculating the economic impacts of food recalls in the meat and poultry industry by analyzing reactions of the financial markets. They discovered that the average wealth of shareholders is contracted by 1.15%. This value depreciation is approximately $109 million within 5 days of the recall, depending on the severity of the food safety hazard, experience of the organization and the size of the organization. Beyond the consumers and stakeholders' reactions to a product recall, there is also the impact of food recall at the national and global scale.

The costs usually associated with a foodborne outbreak, which can also be applied in the case of food recall, have been identified to be as follows:

- The cost of assembling a team to collect information for preliminary report. This initial report may contain the report of interviews of suspected victims, product samples, microbial analyses, laboratory diagnoses, etc.
- Organization of the results from the inspection of sites of the outbreaks, microbiological and laboratory analyses to determine the cause of illnesses etc.
- Efforts to monitor the specific outbreak for surveillance and containment to prevent future occurrence, improvement in the food safety management program of the indicted firm, as well as extensive trainings specific groups and the general population on how to remove the threat of foodborne illnesses.
- The cost documenting the findings of the detailed investigation into the causes, impacts, and mitigating the outbreak [21, 22].

The method commonly adopted to empirically determine the economic and business costs/benefit range from cost-benefit analysis (CBA) to cost-effectiveness analysis (CEA), and to quality-adjusted life years (QALYs) in healthcare which are mostly quantified in monetary value [23]. Regardless of the model used in estimating the cost of food recalls, it is often challenging to obtain publicly available data that can be used quantifying the cost.

10.3 Global food recalls and foodborne outbreaks

A food recall or withdrawal occurs when a producer or an agency of the government recalls a product because they believe it could cause illnesses when such indicted food product is consumed. Food recalls is usually done to safeguard the health of the consumers and guarantee food safety by removing the products from the supply chain. The reasons for such recalls may be the presence of a pathogenic organisms, foreign particles such as a metal or glass and mislabeling especially in the case of an allergen [24]. Conversely, a foodborne outbreak is associated with the consumption of contaminated food or drink by at least two or more people. Other reasons for food recalls may include misbranding, unreported allergens, product without benefit of inspection, processing defect, import violation, production under an unhygienic condition among others.

The process of recalling food products can be grouped into three classes:

1. Recalls for product which could cause serious injury or death on consumption
2. Recalls for product which could cause serious injury or temporary illness
3. Recalls for product which are unlikely to cause injury or illness but that violate the established regulations such as mislabeling of a product in the case of an allergen [25, 18]

Some important factors to be considered during a food recall or foodborne outbreak is the attendant cost to the victims, the public, the firm/food industry linked with the contaminated food product and the government.

10.4 Food recalls in the United States

In the United States, the USDA and the FSIS (Food Safety Inspection Service) are the federal agencies that oversee the safety and protection of the US food supply. They mainly deal with meat and poultry recalls. Recalls are initiated by the meat and poultry industry when they encounter food safety issues. These actions are usually carried out in the interest of public health [26]. According to the USDA website, there is a noticeable decline in recorded food recalls since 2017 to 2020 (Figure 10.4) from 145 recalls in 2017, 133 recalls in 2018 and 50 recalls in 2020. According to the USDA, the cost of 15 foodborne illnesses increased by about US $2 billion from US $15.5 billion 2013 to US $17.6 billion in 2018 (Figure 10.3). There is a significant drop in the 2020 incidence of food recalls, we opined, is due to the Corona virus pandemic and the eventual lockdowns which prevented regulators from carrying out their oversight functions [27].

As at October 2021, there are 53 product recalls with the latest being the recalls of 11,095 kg frozen raw lamb imported by Affco USA though there is no report of any adverse reactions from consumers thus far [27]. In the history of food recalls in the United States, two notable devastating food safety occurrences were recorded in 2008 and 2009. There was an outbreak of *Salmonella* food contamination in the processing plant of Peanut Corporation of America which affected about 700 people with serious illness and at least 9 deaths. This created a domino effect, as in other outbreaks and recalls, affecting food giants such as Kellogg with a declared losses of $70 million and the bankruptcy of Forward Foods of Minden [10]. Another notable recall is the largest meat recall involving Westland/Hallmark Meat Packing Company in 2008. About 65,000 tonnes of beef was recalled which cost the company over $117 million and the company was declared bankrupt in 2012 after settling with the government for $500 million [28]. There were also recalls of over a half-billion fresh eggs associated with an outbreak of *Salmonella* which caused about 1,900 reported cases of illness but without any death

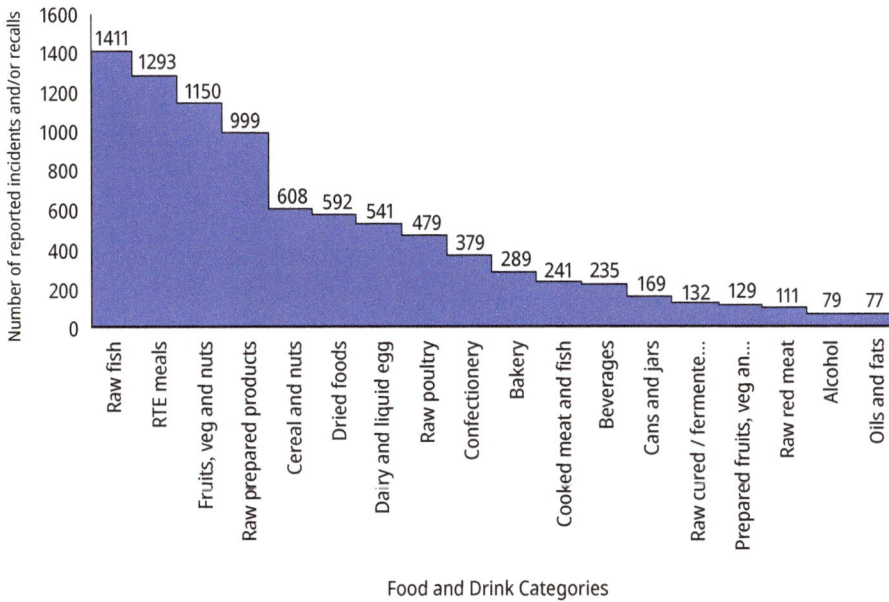

Figure 10.1: The incidents of global food safety and/or recalls from the year 2008 to 2018 (*n = 8,914) [24].

recorded. The firms implicated in the outbreak (Wright County/Hillandale Farms Eggs) were fined were also mandated to improve the sanitation and animal health practices before they could continue operations in their facilities [29].

Figure 10.4 gives the yearly number of food recalls as listed by USDA/FSIS. Generally, there is an increased level of detection from 2015 possibly as high-throughput tools such as whole-genome sequencing becomes cheaper and affordable to major food producing companies among other measures [30].

10.5 Recent food recalls in Europe

The monthly average of food recalls in Europe was 97.3 in the year 2020 according to the Rapid Alert System for Food and Feed Annual Report 2020 [31]. Food safety risks are monitored in Europe using the Rapid Alert System for Food and Feed (RASFF) which is a communication tool [32]. There has been a steady increase in the yearly average of food recalls and outbreaks in Europe generally since the adoption and integration of genomics in surveillance of foodborne pathogens such as *Salmonella* [33]. There is also a Europe-wide cross-sectoral surveillance project for the use of whole-

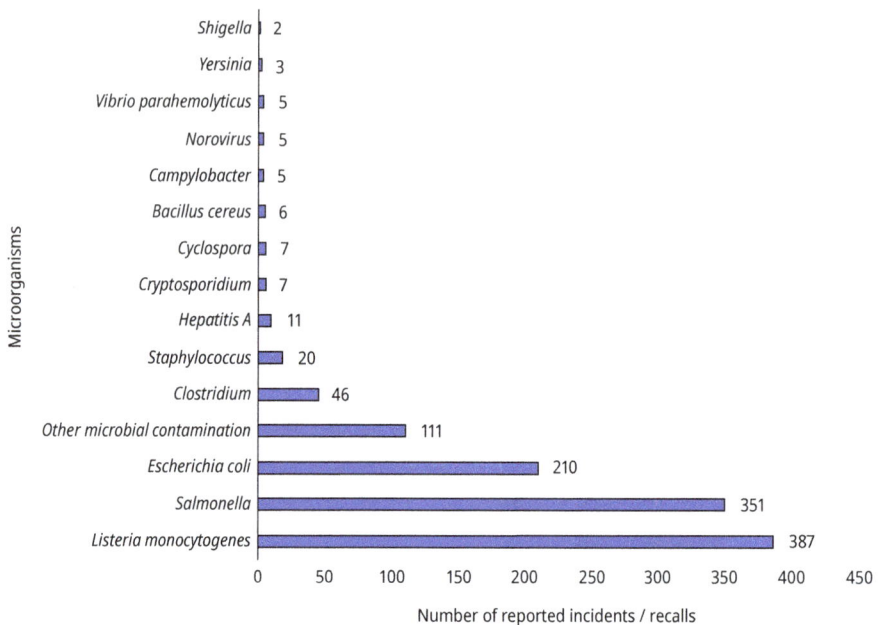

Figure 10.2: The number of recalls linked to biological hazards from 2008 to 2018 ($n = 1,176$) [24].

genome sequencing (WGS) in the surveillance, outbreak detection and investigation of foodborne pathogens targeted at countries with limited resources [34, 35].

There was a noticeable drop in the average incidence of food recalls in 2020 within Europe possibly due to the effect of the pandemic. As at January 2021, there were about 73 product recalls according to the RASFF notification system with the top five products implicated being nuts and nut products, cereals and bakery products, fruits and vegetables, dietetic supplements, and soups and sauces. The reasons for the recalls are the presence of unauthorized substances, toxic chemicals such as ethylene oxide, *Salmonella*, *Listeria*, and foreign substances such as metals (Foodwatch, https://www.foodwatch.org/en/foodwatch-international).

The countries with the highest incidences of reported recalls in Europe are usually Germany, Belgium, France, Denmark, and Spain in that particular order. In terms of reasons for recall, the largest part of recalls was due to unauthorized substances, which were dominated by the presence of ethylene oxide. About 36% of the recalls in January 2021 were linked to the high concentrations of the pesticide and others in fruits imported from Turkey [27].

Costs of 15 major foodborne illnesses in the United States totaled $17.6 billion in 2018

Billions $

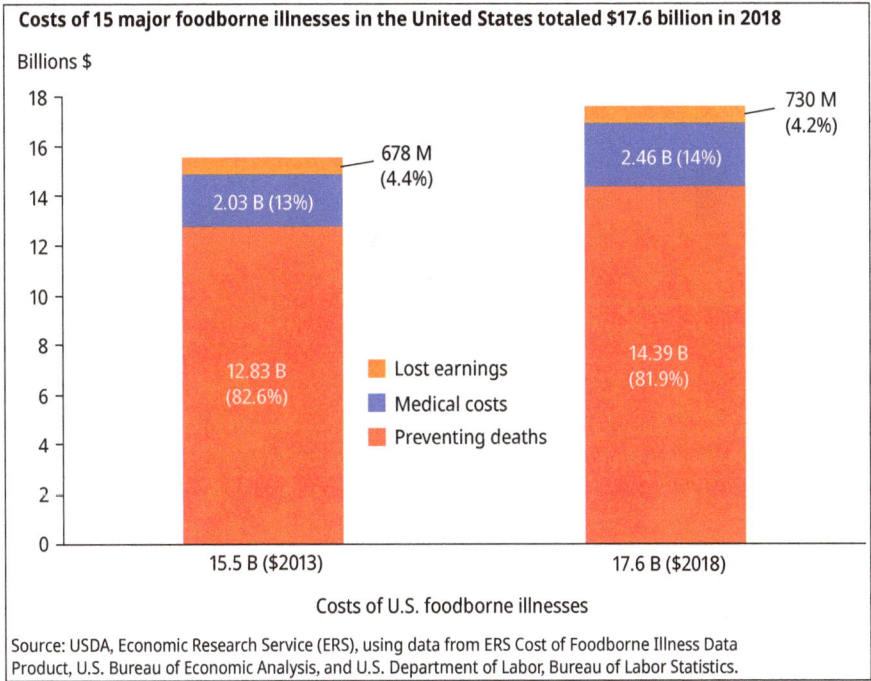

- Lost earnings
- Medical costs
- Preventing deaths

Costs of U.S. foodborne illnesses

Source: USDA, Economic Research Service (ERS), using data from ERS Cost of Foodborne Illness Data Product, U.S. Bureau of Economic Analysis, and U.S. Department of Labor, Bureau of Labor Statistics.

Figure 10.3: Cost of foodborne illness in the United States from 2013 to 2018 (source: USDA, Economic Research Service).

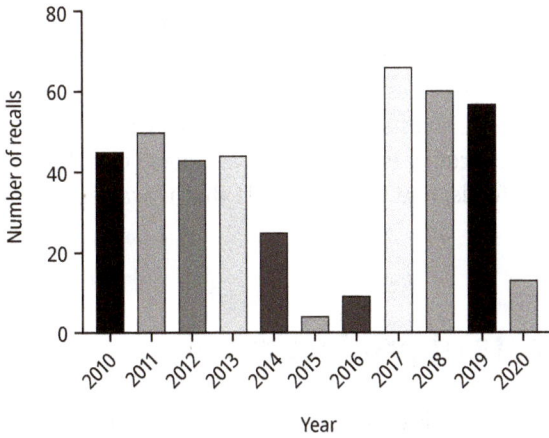

Figure 10.4: The number of recalls in the United States between 2010 and 2020 [27].

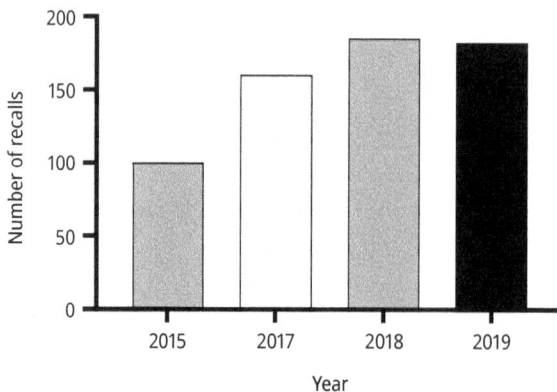

Figure 10.5: The average number of food recalls per year in Germany between 2015 and 2019. Source: Lebensmittelwarnung https://www.lebensmittelwarnung.de/bvl-lmw-de/liste/lebensmittel/deuts chlandweit/10/0

10.6 Cost of foodborne illness in low- and middle-income countries

It is generally challenging to estimate FBDs and related costs in LMICs especially in sub-Saharan African due to dearth of data and unequal distribution in the occurrences. These LMICs (in South Asia, Southeast Asia, and sub-Saharan Africa) account for about 41% of the global population but 53% of all foodborne illness and 75% of associated deaths according to the World Bank report [8]. This translated to an estimated total cost of US $95.2 billion per year, besides the cost of treatment (estimated at US $15 billion) and other nonmonetizable costs such as losses of farm, sales of company and bankruptcy, foregone trade income, impact on shareholders" wealth, consumer avoidance of the implicated foods, and the burden of that is wasted to the environment [50].

10.7 Evaluating the costs of product recalls

The methods of valuing the cost of food recalls differ from country to country depending on the goal of the decision makers. This may be narrow or may factor in the impacts of a specific recall on the wider society just as in the case of the evaluation of cost for food safety improvement [36]. Nonetheless, it is important to examine how costs are classified in economics. Firstly, the definition of cost of any item, activity, occurrence of events, and nonoccurrence of events depends on the perspectives of the economic agent involved. The key variables that must be considered when evaluating the costs of food recalls may include the following: size of the recall, severity of the threats, the pathogen or hazard implicated in the recall, the level of media coverage that the recall receives, the size and experience of the firm implicated in the recall, among others [18].

On an economic perspective, cost associated with food safety and by extension food recalls can be classified into category A and category B. For category A, cost can be explicit or implicit and for category B, cost is divided into private costs and social costs [37, 36].

10.8 Category A: explicit and implicit costs

These are identifiable and actual payments made to procure goods and services, settle disputes among others during a food recall [38, 36]. This may also include the total cost of ensuring food safety in the supply chain such as the cost of tracing and tracking contaminated foods, cost to government agencies such as the USDA involved in monitoring food safety issues and managing the recall in the case of an outbreak or contamination, the compensation paid to the victims of foodborne illnesses, cost of litigation and cost of treatment and hospitalization of victims among others.

In contrast, implicit costs are opportunity costs. Implicit cost is defined as the cost of alternative forgone [N. I. 36]. These costs are sources of potential benefit that an individual, firm or government lost when choosing a compelling alternative. As an example, if a processing plant A shut down for two days in order to carry out routine maintenance, if the estimated cost of US $10,000 was used to carry out the maintenance and the value of production/sales lost within that period is US $100,000, the explicit costs to the processing plant A will be US $10,000 while the implicit cost will be an amount (profit/interest) that would have accrued to the firm if 110,000 were invested in another venture plus US $100,000.

This concept is applicable to cost of food safety and food recall. For instance, when there is an outbreak, firms involved are likely to temporarily decrease or stop production altogether. This will definitely come with a financial cost to the firm such as the loss of production and sales. Again, money, time and other resources that should be applied in production and sales will be channeled to ensure recall of the contaminated foods and interaction with regulatory authorities. Finally, the victims of the outbreak and the public will spend resources in treatment and hospitalization of victims beside the man-hour that is lost by the victims and caregivers which all involved the use of resources.

Computation of the implicit cost from the scenario above will involve the addition of all other associated costs. Assuming the total cost of the food recall and other losses is US $4 million, the implicit cost is calculated as if the loss incurred is rather invested in any government bond. The implicit costs will be US $4 million in addition to the yield from the investment (i.e., bond). Consequently, for a realistic computation of economic costs of food safety and food recall, it is important that both explicit and implicit costs are considered. Hence, there is a need to delineate between *appearance and reality*. Appearance will be the real monetary expenditure which is the explicit costs while the reality is the summation of both implicit and explicit costs.

10.9 Category B: private cost and social cost

Economic theory defines private costs as the cost incurred by a consumer of goods and services and/or producer of goods and services [39]. It is entirely private to the concerned economic agent. This could be an individual, household, business organization or government. It is measured by the market prices of goods and services and is the summation of explicit and implicit costs. On the other hand, a social cost is the total cost to the society at large [40]. It can occur during production activities or consumption activities. These costs are external costs to the producer or the consumer. In a broader sense, social cost includes both the private cost and the external cost. This classification is also applicable in the area of food safety since ensuring safe foods and mitigating the incidence of an outbreak comes with a cost and public health concern to the society at large. Thus, in calculating the economic cost of food recall, greater care must be taken in ensuring that all related costs; private and social are taken into consideration.

According to [41], cost associated with food safety and food recall was broadly divided into monetized costs and nonmonetized costs. The cost associated with foodborne illness should include monetizable direct costs obligatory to both the consumer (victim), and the firm (producer/seller). Thus, the monetizable direct costs includes loss of wages and medical costs of the sick person, loss of sales, product recall, legal cost and many more. The nonmonetizable direct costs can be the cost of pain and inconvenience to the victim affected by the foodborne illness. Succinctly put, costs affecting parties outside of the market transaction was referred to as indirect cost. It is indirect because it affects those parties who did not consume the product nor become sick as a result of it. This cost is also divided into monetizable indirect costs and nonmonetizable indirect costs. The former includes cost incurred by relatives and acquaintances transiting to the hospitals and the attendant costs of stress and inconveniences. This classification is summarized in Figure 10.6.

It is important to note that both monetizable and nonmonetizable costs are underestimated in the economic sense because it excludes other aspect of implicit cost from the model. It is important to note that both monetizable cost and nonmonetizable cost is underestimated in economic sense because it excludes + other aspect of implicit cost from the model. 42,discussed the following classification of economic

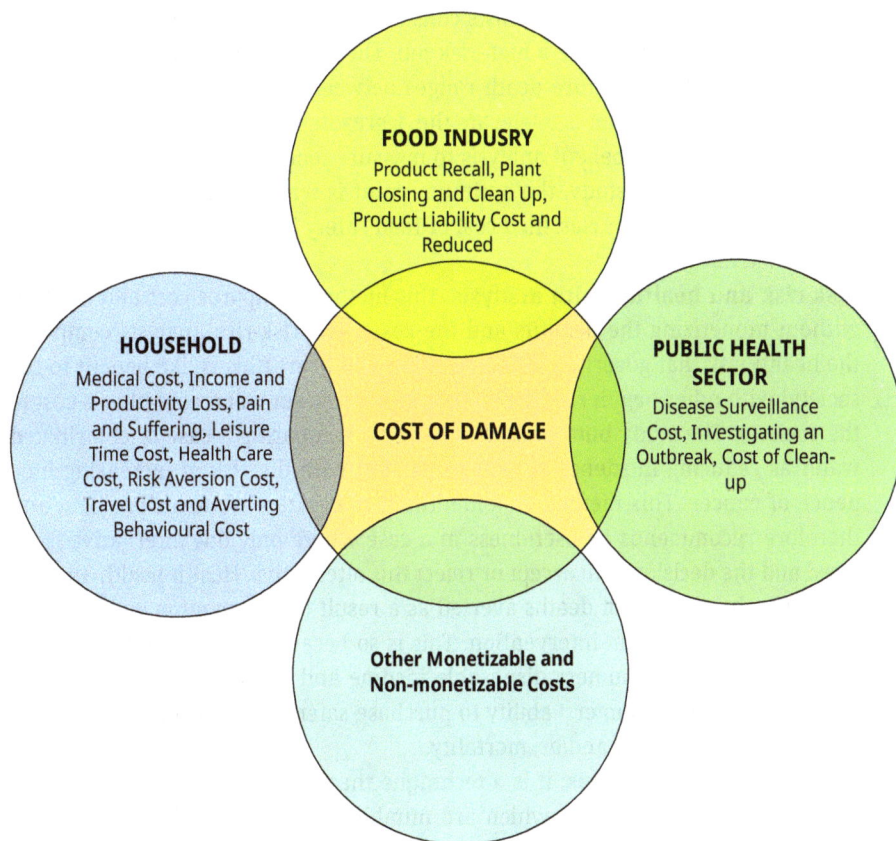

Figure 10.6: The different costs associated with foodborne outbreaks involving monetizable and nonmonetizable costs.

costs of food safety. In their submission, they proposed the optimal level of food safety to be where the marginal cost of one more unit of food safety equals its marginal benefit. The cost will be the total explicit costs of ensuring food safety while the benefits are indices such as reduced illness and death associated with unsafe food. However, they noted that it is very difficult to account for the benefit of safer food since this is not traded in the market. They further identified the **"*cost of illness*"** as a way of measuring the cost of food safety. They concluded that the approach is crudely economics because it only values lost income and associated consumption expenditure and fail to recognize other implicit costs like what consumers will be willing to pay for feeling healthy, avoiding pain and use of their free time.

Landefeld and Seskin [43] proposed the **human capital** approach as a method that measures the willingness to pay for the value for a quality-adjusted life year (QALY). The approach evaluates premature death using existing dollar value of all future income that would have accrued to the victim if he or she were still alive.

The second approach used by economists considers the amount an individual will be willing to accept in order to take a high-risk job. The amount to be paid for high-risk job which can lead to premature death ranges between US $3 million and 7 million [44]. This range in the amount is to elevate the aggregate risk of death by one. Ivanek, R. et al. [49] uses cost and benefit analysis to measure food safety in *Listeria monocytogenes*. According to the study, the optimum point is where benefits of food safety equal costs of food safety. 45,classified cost of food safety into three groups which are discussed briefly below:

(1) **Risk-risk and health-health analysis**: this method compares cost and benefits without monetizing the benefits and the costs. The risk-risk analysis compares the health risk that government intervention will avert (this is like benefit to the society] with other health risk the intervention program will create [this is cost to the society). The study buttresses this point by comparing a case of chlorinated water as reducing incidence of water infection with the risk of increasing incidence of cancer. This method was identified to create a trade-off problem and therefore recommends its usefulness in a case where only one alternative is offered and the decision is to accept or reject this alternative. Health-health analysis compares the number of deaths averted as a result of intervention with number of deaths resulting from intervention. This is so because costs in terms of higher taxes will reduce consumers disposable income and higher prices on the other hand will reduce consumers' ability to purchase safer food which in the long run may increase morbidity and/or, mortality.

(2) **Cost-effectiveness analysis**: it is a technique that relates monetized costs to the number of physical benefits which are number of prevented illness from food borne diseases. This method will help policy maker in ranking different interventions by the government, such that the one with least *"cost-benefit"* is the most *cost effective*. Since this approach uses different scale of measurement cost and benefit, it makes it difficult to calculate the optimum point where net benefit is highest.

(3) **Cost–benefit analysis**: this approach compares cost and benefit in monetary terms. It makes it easier to calculate the trade-off of different interventions across sectors. An intervention is counted as worthwhile if the net benefit is positive.

Antle [46] identified three popular approaches for estimating food safety regulation cost at plant-level which may likely be imposed after an outbreak or a recall. These are the ***accounting approach, economic-engineering*** approach and the ***econometric approach***. The accounting approach estimate cost based on extra labor requirement and capital stock in order to ensure food safety. The economic-engineering approach combine detailed engineering data with cost of input to construct quantitative model for the production process. A parametric cost function is derived using the plant production function. To implement this system of costing at the firm-level is however expensive. This method generates the potential costs of food safety regulation using econometrically estimated cost function.

Olanya et al [22] applied an incidence-based method to compute cost of listeriosis outbreak in South Africa which occurred between 2017 and 2018 (Figure 10.7]. In their analysis, cost was broadly divided two; these are direct and indirect costs. The direct costs are quantifiable while the indirect costs are nonquantifiable. Included in the direct costs are value of human life, cost of hospitalization, cost of contamination colony, cost of lost work, current and future value of outpatient care and medicines, future value of human suffering and lost productivity, domestic cost like loss of market share, product substitution, value of lost polony consumption, and current and future values of human life. For indirect costs we have, administrative costs of press-release and reports, surveillance, sample processing cost, plant sanitization costs and long- term cost to the people, the firm and the location. This costing classification appears comprehensive because it adds up both explicit and implicit costs of food safety.

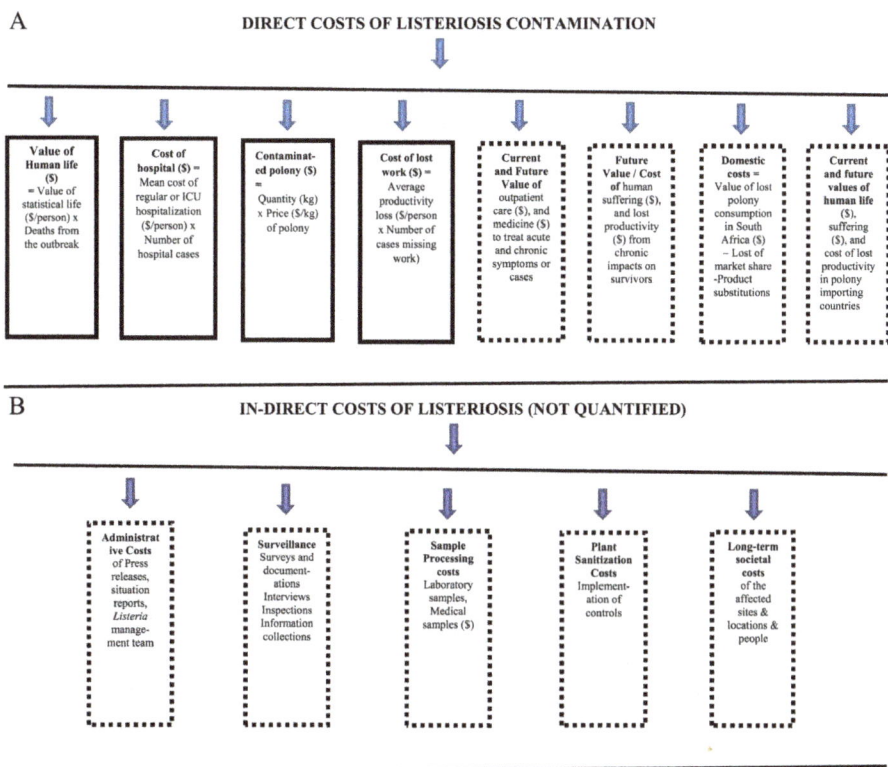

Figure 10.7: The indirect and direct costs of listeriosis outbreak in South Africa which occurred between 2017 and 2018 [30].

Considering the importance of food safety to the society, it is important that there should be a uniform food safety cost classification so that all hidden and obvious cost content are factored in modelling economic cost of food safety and food recall.

10.10 Conclusion

There is a pressing demand to harmonies the current systems in individual countries and regions and established model to calculate the specific costs of food recalls on a global scale. This cannot be overemphasized with the mounting volume of trades among countries, the attendant preponderance perpetuation of food frauds and the inclination to circumvent established food safety regulations and standards by some producers. This will require the adoption of appropriate technologies and systems that can monitor and integrate all the product information across board on a global scale. Examples of such technologies are sensor and blockchain technologies which will greatly improve food traceability across the food chain since product recall is inherently challenging. This may mean the interoperability of existing national and international agencies and organizations saddled with such functions or the creation of another designated with this specific task. Blockchain networks have the advantages for this type of adoption in the food industry because of the following reasons: decentralized network, immutability of the encryption, consensus of its transaction, democratic and transparent nature of its governance. The potential impact of blockchain-based IoT solutions or traceability systems on the consumers can be a restoration and boosting of confidence in a specific food product especially after a recall among others because of its interoperability and integration across the food value-chain and the accessibility of such information to consumers in real-time.

References

[1] Qian, J., Ruiz-Garcia, L., Fan, B., Robla Villalba, J. I., McCarthy, U., Zhang, B., . . . Wu, W. (2020). Food traceability system from governmental, corporate, and consumer perspectives in the European Union and China: A comparative review. *Trends in Food Science and Technology*, *99*, July 2019 402–412. https://doi.org/10.1016/j.tifs.2020.03.025.

[2] Verraes, C., Van Boxstael, S., Van Meervenne, E., Van Coillie, E., Butaye, P., Catry, B., . . . Herman, L. (2013). Antimicrobial resistance in the food chain: A review. *International Journal of Environmental Research and Public Health*, 10, https://doi.org/10.3390/ijerph10072643.

[3] Xiang, Y., Li, M., Guo, X., Wu, Y., Ying, Y., Wen, Y., & Yang, H. (2018). Raman rapid detection of environmental hormone. *Sensors and Actuators, B: Chemical*, *262*, 44–49. https://doi.org/10.1016/j.snb.2018.01.196.

[4] Xu, M. L., Gao, Y., Han, X. X., & Zhao, B. (2017). August 16. Detection of Pesticide Residues in Food Using Surface-Enhanced Raman Spectroscopy: A Review. *Journal of Agricultural and Food Chemistry*, 65, 6719–6726. https://doi.org/10.1021/acs.jafc.7b02504.

[5] Mc Carthy, U., Uysal, I., Badia-Melis, R., Mercier, S., O'Donnell, C., & Ktenioudaki, A. (2018). Global food security – Issues, challenges and technological solutions. *Trends in Food Science and Technology*, *77*, 11–20. https://doi.org/10.1016/j.tifs.2018.05.002.

[6] Devleesschauwer, B., Haagsma, J. A., Mangen, M.-J.-J., Lake, R. J., & Havelaar, A. H. (2018). The Global Burden of Foodborne Disease. *Food Safety Economics*, 107–122. https://doi.org/10.1007/978-3-319-92138-9_7.

[7]   Grace, D., Wu, F., & Havelaar, A. H. (2020). MILK Symposium review: Foodborne diseases from milk and milk products in developing countries – Review of causes and health and economic implications. *Journal of Dairy Science, 103*(11), 9715–9729. https://doi.org/10.3168/jds.2020-18323.

[8]   *WHO estimates of the global burden of foodborne diseases*. (2015).

[9]   Bhaskar, S. V. (2017). Foodborne diseases – disease burden. *Food Safety in the 21st Century*, 1–12. https://doi.org/10.1016/B978-0-12-801773-9/00001-7.

[10] Hussain, M. A., & Dawson, C. O. (2013a). Economic impact of food safety outbreaks on food businesses. *Foods, 2*(4), 585–589. https://doi.org/10.3390/foods2040585.

[11] Petruzzi, L., Corbo, M. R., Sinigaglia, M., & Bevilacqua, A. (2017). Microbial Spoilage of Foods. *The Microbiological Quality of Food*, 1–21. https://doi.org/10.1016/B978-0-08-100502-6.00002-9.

[12] Farber, J. M., Zwietering, M., Wiedmann, M., Schaffner, D., Hedberg, C. W., Harrison, M. A., & Gummalla, S. (2021). May 1. Alternative approaches to the risk management of Listeria monocytogenes in low risk foods. *Food Control*. 123, https://doi.org/10.1016/j.foodcont.2020.107601.

[13] Bintsis, T. (2017). Foodborne pathogens. *AIMS Microbiology, 3*(3), 529–563. https://doi.org/10.3934/microbiol.2017.3.529.

[14] *Centers for Disease Control and Prevention (CDC). Surveillance for Foodborne Disease Outbreaks United States, 2015 : Annual Report*. (2017). Atlanta, Georgia.

[15] Focker, M., & van der Fels-Klerx, H. J. (2020). December 1. Economics applied to food safety. *Current Opinion in Food Science*, 36, 18–23. https://doi.org/10.1016/j.cofs.2020.10.018.

[16] McLinden, T., Sargeant, J. M., Thomas, M. K., Papadopoulos, A., & Fazil, A. (2014). Component costs of foodborne illness: A scoping review. *BMC Public Health, 14*(1), 1–12. https://doi.org/10.1186/1471-2458-14-509.

[17] Hoffmann, S., & Ahn, J. (2021). *USDA Economic Research Service*. Retrieved from https://www.ers.usda.gov/amber-waves/2021/april/economic-cost-of-major-foodborne-illnesses-increased-2-billion-from-2013-to-2018/

[18] Pozo, V. F., & Schroeder, T. C. (2016). Evaluating the costs of meat and poultry recalls to food firms using stock returns. *Food Policy, 59*, 66–77. https://doi.org/10.1016/j.foodpol.2015.12.007.

[19] Loader, R., & Hobbs, J. E. (1999). Strategic responses to food safety legislation. *Food Policy, 24*(6), 685–706. https://doi.org/10.1016/S0306-9192(99)00073-1.

[20] Ades, G., Henry, C. W., & Feldstein, F. (2011). The food safety challenge of the global food supply chain. Food Safety Magazine, 32–39.

[21] Ahmed, S. (2014). *Measuring the Benefit of Reducing Food-Borne Illness: Valuing Symptoms Rather than Pathogens*. April Ottawa: University of Ottawa.

[22] Olanya, O. M., Hoshide, A. K., Ijabadeniyi, O. A., Ukuku, D. O., Mukhopadhyay, S., Niemira, B. A., & Ayeni, O. (2019). Cost estimation of listeriosis (Listeria monocytogenes) occurrence in South Africa in 2017 and its food safety implications. *Food Control, 102*, 231–239. https://doi.org/10.1016/j.foodcont.2019.02.007.

[23] Ebel, E. D., Williams, M. S., Ward-Gokhale, L. A., & Kisselburgh, H. M. (2019). Assessing the maximum size of annual foodborne outbreaks in the United States: An analysis of 1973–2016 outbreaks. *Microbial Risk Analysis, 12*, 20–26. https://doi.org/10.1016/j.mran.2019.02.002.

[24] Soon, J. M., Brazier, A. K. M., & Wallace, C. A. (2020). March 1. Determining common contributory factors in food safety incidents – A review of global outbreaks and recalls 2008–2018. *Trends in Food Science and Technology*, 97, 76–87. https://doi.org/10.1016/j.tifs.2019.12.030.

[25] Ahsan, K., & Gunawan, I. (2014). Analysis of Product Recalls: Identification of Recall Initiators and Causes of Recall. *Operations and Supply Chain Management: An International Journal, 7*(3), 97–106. https://doi.org/10.31387/oscm0180115.

[26] Kramer, M. N., Coto, D., & Weidner, J. D. (2005). The science of recalls. *Meat Science, 71*(1), 158–163. https://doi.org/10.1016/j.meatsci.2005.04.001.

[27] United States Department of Agriculture Food Safety and Inspection Service (2021). https://www.fsis.usda.gov/recalls

[28] Southall, M. (2019). Industry Benefits. *Food Traceability*, 51–62. https://doi.org/10.1007/978-3-030-10902-8_4.

[29] Roberts, H., & Veil, S. R. (2016). Health literacy and crisis: Public relations in the 2010 egg recall. *Public Relations Review*, *42*(1), 214–218. https://doi.org/10.1016/J.PUBREV.2015.07.013.

[30] Hoelzer, K., Moreno Switt, A. I., Wiedmann, M., & Boor, K. J. (2018). Emerging needs and opportunities in foodborne disease detection and prevention: From tools to people. *Food Microbiology*, 75, 65–71. https://doi.org/10.1016/j.fm.2017.07.006.

[31] RASFF. (2020). The Rapid Alert System for Food and Feed (RASFF) Annual Report 2020. In *Office of the European Union,2021*. Retrieved from http://www.ssrn.com/abstract=1152122

[32] De Leo, F., Coluccia, B., Miglietta, P. P., & Serio, F. (2021). Food contact materials recalls and international trade relations: an analysis of the nexus between RASFF notifications and product origin. *Food Control*, *120*, 107518. https://doi.org/10.1016/J.FOODCONT.2020.107518.

[33] Uelze, L., Becker, N., Borowiak, M., Busch, U., Dangel, A., Deneke, C., & Malorny, B. (2021). Toward an Integrated Genome-Based Surveillance of Salmonella enterica in Germany. *Frontiers in Microbiology*, *12*. https://doi.org/10.3389/FMICB.2021.626941/FULL.

[34] Llarena, A., Ribeiro-Gonçalves, B. F., Nuno Silva, D., Halkilahti, J., Machado, M. P., Da Silva, M. S., & Rossi, M. (2018). INNUENDO: A cross-sectoral platform for the integration of genomics in the surveillance of food-borne pathogens. *EFSA Supporting Publications*, *15*(11), https://doi.org/10.2903/sp.efsa.2018.en-1498.

[35] Rantsiou, K., Kathariou, S., Winkler, A., Skandamis, P., Saint-Cyr, M. J., Rouzeau-Szynalski, K., & Amézquita, A. (2018). Next generation microbiological risk assessment: opportunities of whole genome sequencing (WGS) for foodborne pathogen surveillance, source tracking and risk assessment. *International Journal of Food Microbiology*, *287*, 3–9. https://doi.org/10.1016/J.IJFOODMICRO.2017.11.007.

[36] Valeeva, N. I., Meuwissen, M. P. M., & Huirne, R. B. M. (2004). Economics of food safety in chains: A review of general principles. *NJAS – Wageningen Journal of Life Sciences*, *51*(4), 369–390. https://doi.org/10.1016/S1573-5214(04)80003-4.

[37] Nganje, W. E., Burbidge, L. D., Denkyirah, E. K., & Ndembe, E. M. (2021). Predicting Food-Safety Risk and Determining Cost-Effective Risk-Reduction Strategies. *Journal of Risk and Financial Management*, *14*(9), 408. https://doi.org/10.3390/jrfm14090408.

[38] Kong, D., Shi, L., & Yang, Z. (2019). Product recalls, corporate social responsibility, and firm value: Evidence from the Chinese food industry. *Food Policy*, *83*, 60–69. https://doi.org/10.1016/j.foodpol.2018.11.005.

[39] Betancourt, R., & Gautschi, D. (1988). The economics of retail firms. *Managerial & Decision Economics*, *9*(2), 133–144. https://doi.org/10.1002/MDE.4090090208.

[40] Barton, A. (1999). Public and private sector accounting – The non-identical twins. *Australian Accounting Review*, *9*(18), 22–31. https://doi.org/10.1111/J.1835-2561.1999.TB00105.X.

[41] Roberts, T. (1987). Valuing Food Safety. *Risk Assessment and Management*, 453–465. https://doi.org/10.1007/978-1-4757-6443-7_44.

[42] Crutchfield, S., & Allshouse, J. (1998). *The economics of improving food safety* (pp. 53–64). pp. 53–64.

[43] Landefeld, J. S., & Seskin, E. P. (1982). The economic value of life: linking theory to practice. *American Journal of Public Health*, *72*(6), 555–566. https://doi.org/10.2105/AJPH.72.6.555.

[44] Viscusi, W. K. (2014). The Value of Individual and Societal Risks to Life and Health. *Handbook of the Economics of Risk and Uncertainty*, 1, 385–452. https://doi.org/10.1016/B978-0-444-53685-3.00007-6.

[45] Valeeva, N. I., Meuwissen, M. P. M., Bergevoet, R. H. M., Oude Lansink, A. G. J. M., & Huirne, R. B. M. (2005). Improving food safety at the dairy farm level: Farmers' and experts' perceptions. *Review of Agricultural Economics*, *27*(4), 574–592. https://doi.org/10.1111/j.1467-9353.2005.00265.x.

[46] Antle, J. M. (1999). Benefits and costs of food safety regulation. *Food Policy*, *24*(6), 605–623. https://doi.org/10.1016/S0306-9192(99)00068-8.

[47] Hussain, M. A., & Dawson, C. O. (2013b). Economic impact of food safety outbreaks on food businesses. *Foods*, *2*(4), 585–589. https://doi.org/10.3390/foods2040585.

[48] Focus Food (2021). https://www.focos-food.com/food-recalls-in-germany/

[49] Ivanek, R., Gröhn, Y. T., Tauer, L. W., & Wiedmann, M. (2005). The cost and benefit of Listeria monocytogenes food safety measures. Critical Reviews in food Scince and Nutrition, 44(7–8), 513–523.

[50] Dewey-Mattia, D., Manikonda, K., Hall, A. J., Wise, M. E., & Crowe, S. J. (2018). Surveillance for foodborne disease outbreaks – United States, 2009–2015. MMWR Surveillance Summaries, 67(10), 1.

[51] Tack, D.M., Marder, E.P., Griffin, P.M., Cieslak, P.R., Dunn, J., Hurd, S., Scallan, E., Lathrop, S., Muse, A., Ryan, P., Smith, K., Tobin-D'Angelo, M., Vugia, D.J., Holt, K.G., Wolpert, B.J., Tauxe, R., Geissler, A. L., 2019. Preliminary incidence and trends of infections with pathogens transmitted commonly through food – Foodborne Diseases Active Surveillance Network, 10 U.S. sites, 2015–2018. Am. J. Transplant. 19, 1859–1863. https://doi.org/10.1111/ajt.15412

[52] World Bank Annual Report 2018 (English). Washington, D.C. : World Bank Group. http://documents. worldbank.org/curated/en/838591538415475420/World-Bank-Annual-Report-2018

[53] Liao, X., Ma, Y., Daliri, E.B.M., Koseki, S., Wei, S., Liu, D., Ye, X., Chen, S., Ding, T., 2020. Interplay of antibiotic resistance and food-associated stress tolerance in foodborne pathogens. Trends Food Sci. Technol. https://doi.org/10.1016/j.tifs.2019.11.006

[54] Begley, M., Hill, C., 2015. Stress adaptation in foodborne pathogens. Annu. Rev. Food Sci. Technol. https://doi.org/10.1146/annurev-food-030713-092350

[55] Hoffmann, V., & Moser, C. (2017). You get what you pay for: the link between price and food safety in Kenya. Agricultural Economics, 48(4), 449–458.

[56] Stein, R. A., & Chirilă, M. (2017). Routes of transmission in the food chain. In Foodborne diseases (pp. 65–103). Academic Press.

Bhekisisa C. Dlamini and Adeoluwa I. Adetunji

Chapter 11
Hazard Analysis and Critical Control Point (HACCP) and Food Safety Management Systems

Abstract: The occurrence of foodborne diseases continues to be a burden to developed and developing economies in the world. Globalization, changing consumer habits, emerging pathogens and antimicrobial resistance are adding pressure to food safety management systems. HACCP is globally recognized as a system that effectively control food safety hazards. However, its success depends on several factors such as proper implementation and maintenance and whether it is tailored towards control of hazards that are specific to a food business. Considering this, the effectiveness of any food safety management system must be continuously evaluated. Firstly, this chapter will introduce HACCP and its implementation requirements. Further, the challenges to successful implementation of a HACCP system are discussed as well as recent developments to HACCP system implementation. Lastly, tools that can be used to evaluate the effectiveness of a food safety management system that is based on the HACCP system are discussed.

11.1 Introduction

Food Safety Management System (FSMS) can be defined as a system that consists of interrelated elements that collectively ensures the production of safe food or food that will not cause harm to human health. The elements of FSMS include food safety programs, plans document control and clear responsibilities of senior management toward the system. A FSMS must be specific to a company due to the differences in product production processes, requirements of local food control systems and sometimes various demands from target markets [1]. Numerous reports have shown that proper implementation of a FSMS is effective in addressing food safety issues in the world [2]. In developed countries such as those in Europe, most food companies have a sound FSMS [3, 4] in place when compared to food businesses in developing countries like those in Africa [5]. The ease of establishing a FSMS seem to be depended on the size of the company due to financial challenges, lack of technical skills and other

Bhekisisa C. Dlamini, Department of Biotechnology and Food Technology, University of Johannesburg, Johannesburg, Gauteng, South Africa
Adeoluwa I. Adetunji, Department of Research & Development, Labworld, Division of Philafrica Foods, Johannesburg, Johannesburg

https://doi.org/10.1515/9783110748345-011

challenges associated with the implementation and maintenance of such systems [6–8]. As a result, large companies find it easier to implement an effective FSMS while micro, small and medium enterprises face challenges, in particular with certified FSMS [7, 8].

HACCP, as defined by Motarjemi [11], is a risk-based system that entails the implementation of some basic good wholesome practices. That is, HACCP is a tool for assessing risks and establishing control systems that focuses on control measures for key hazards along the food chain. In most cases, FSMS's include the Hazard Analysis and Critical control Points (HACCP) program. This is because an effective FSMS is expected to control the occurrence of food safety hazards (biological, chemical, and physical) and subsequently improve the organizations performance [9]. Due to the importance of HACCP, food business within the EU member states is required by law to implement a FSMS based on HACCP principles [10]. FSMS's that incorporates the HACCP system include the International Standards Organization (ISO 22000), British Retail Consortium (BRC), Safe Quality Foods (SQF), and the International Food Standard (IFS). This chapter seeks to discuss the recent developments in the application of FSMS with more emphasis placed on the HACCP system. Further, other certified FSMS will be reviewed as well as recent developments in evaluating a FSMS.

11.2 Hazard analysis and critical control point (HACCP) system

The food value chain system, from producers through dealers, processors, and other contributors such as food distributors and retail shops, continue to experience changes. As a result, ensuring safety throughout the entire food value chain is extremely tough. Hence, an effective FSMS must be implemented to ensure customers' and consumers' trust in the value chain system. An FSMS based on HACCP principles can be defined as a science-based and structured approach used to identify specific hazards and control actions [11]. In other words, this is a process of identifying risk concerns in food production processes, as well as implementing preventative and monitoring measures to ensure that the plan of action is implemented. There are some prerequisite requirements that must be met in order to have an efficient HACCP system. The unique feature of the complete food value chain system determines these essential criteria. They include the following: good agriculture practice, good hygienic practice, good manufacturing practice, and good transport or storage practice [12].

11.2.1 Principles of hazard analysis and critical control point (HACCP) system

The principles of HACCP may be found in every aspect of food production and it has established and preventive FSMS. This system is scientific and practical technique for detecting, measuring, and monitoring particular risk factors which are significant in attaining product safety [12]. This approach is founded on seven fundamental concepts. Hazards may be identified and controlled before they endanger the safety of food items and the health of consumers, thanks to an FSMS based on these seven principles. According to the Codex Alimentarius Commission [12], the seven principles of the HACCP system are presented in Table 11.1.

The HACCP system has a monitoring component that is crucial in ensuring proper control of food hazards [11]. Monitoring is explained as a procedure of keeping track of critical processes as they happen. All activities that are deemed critical control points (CCPs) must be monitored, which means that monitoring operations that are not considered CCPs is optional and outside of food safety considerations, such as quality assurance, should be indicated as such. The results of the monitoring must demonstrate that the targeted preventative or control measures are being implemented and are successful. In high-risk food enterprises, monitoring must concentrate on high-risk factors and practices identified, because they can result in foodborne illness if not controlled. For example, the monitoring of temperature and prevention of cross-contamination in ready-to-eat food are important in ensuring food safety. Therefore, effective monitoring is carried out to ensure that food attributes fulfill food safety criteria and that production, processing, preparation, and storage methods follow the HACCP system's guidelines [11].

Table 11.1: The seven principles of the HACCP system (adapted from CXC 1-1969).

HACCP principle	Explanation	Activities involved
Principle 1: Conduct hazards analysis and identify control measures (step 6)	Involves identifying all possible hazards that can occur in raw materials, the environment and at each stage of processing. Hazards are categorized as biological, chemical, and physical hazards.	– All potential hazards associated with product are identified. – Significant hazards are listed including their source and reason for presence. – The known or unknown risk of the significant hazards is considered. – The availability of measures to control the hazards is considered.
Principle 2: Determine the critical control points (CCPs) (step 7)	Point, stage, procedure or operation at which control shall be applied to prevent, reduce to acceptable levels or eliminate a food hazard.	– The CCP decision tree can be used to identify CCPs. – CCP's must be identified for only significant hazards. – One CCP may control more than one hazard (e.g., cooking step in killing more than one pathogen).

Table 11.1 (continued)

HACCP principle	Explanation	Activities involved
Principle 3: Establish validated critical limits for each CCP (step 8)	Criterion used to determine whether a CCP is in control and differentiates acceptable from unacceptable. It can be measurable or observable.	– Establish critical limits for each CCP. – Validate the critical limit using scientific methods to ensure that they are able to effectively control the hazard. – Validation of critical limits can also be done based on existing literature, regulations, guidance from competent authorities or studies by competent third-party organizations.
Principle 4: Establishment of a monitoring system (step 9)	Scheduled measurement or observation that detects a deviation at a CCP. Monitoring must allow timely detection of a deviation of a critical limit for effective isolation of affected products.	– Establish a monitoring system, i.e., observable (pump settings, or correctness of labeling) or measurable (physical and chemical measurements). – Decide on whether monitoring will be continuous (e.g., temperature and time) or at intervals (e.g., moisture levels). – Ensure that monitoring personnel is properly trained and monitoring documents are signed and kept.
Principle 5: Establish corrective actions (step 10)	Action taken when monitoring indicates that a particular CCP is not under control. Corrective action must prevent or minimize reoccurrence of the deviation.	– Establish a corrective action that is specific to the CCP. – Ensure that the corrective action is able to bring the CCP under control after a deviation. – Use knowledgeable people or experts to decide on whether food product can be re-used or destroyed after a deviation. – Conduct root cause analysis to identify and correct the source of the deviation.

Table 11.1 (continued)

HACCP principle	Explanation	Activities involved
Principle 6: Validation of HACCP plan and verification procedures (step 11)	**Validation** of control measures involves obtaining evidence that a control measures or combination thereof, is capable of controlling hazards. Validation is performed during development of HACCP plan and after implementation. **Verification** involves using methods, procedures tests and other assessments to confirm effectiveness of HACCP system. Verification is performed after implementation of the HACCP plan and on an ongoing basis.	**Validation:** – Includes review of scientific literature; use of mathematical models, validation studies and/ or guidance developed by authoritative sources. – Evidence should be obtained during operation to confirm that control of hazards can be achieved constantly under production. **Verification:** – Includes observations, calibration, sampling and testing, records review and auditing (internal or external).
		– Must be carried out by person who is independent from responsibility of performing the monitoring and corrective actions. – Frequency must be sufficient to confirm effectiveness of HACCP
Principle 7: Establishment of documentation and record keeping (step 12)	Ensures establishment of documentation keeping of records that are appropriate to the nature and size of operation. This assists the business to verify that the HACCP controls are in place and properly maintained.	– Documentation examples: HACCP team composition; hazard analysis and scientific reports for hazards; CCP determination; Critical limit determination; and validation of control measures. – Record examples: CCP monitoring activities; deviations and associated corrective actions; verification procedures performed.

Twelve critical steps were described in the Codex Guidelines in the application of HACCP system. It is worth noting that the twelve critical steps include the seven principles of the HACCP system. In fact, principle 1 of the HACCP system is also the sixth step as shown in Table 11.1. Steps 1–5 are shown below, and the reader is directed to the CXC 1- 1969 document for further details on these steps.

Step 1: Assemble HACCP team and identify scope
Step 2: Describe product
Step 3: Identify intended use and users
Step 4: Construct flow diagram
Step 5: On-site confirmation of flow diagram

11.2.2 Challenges to successful HACCP system implementation and application

Several factors have been identified to impact effective implementation and application of HACCP system. Some of the factors include primarily financial, technical, managerial, organizational, and educational constraints [13–15]. As a result of these difficulties, many food enterprises fail to effectively implement a HACCP system. According to Milios et al. [16], all of these challenges stem from a lack of understanding of HACCP and its benefits, a lack of training, a lack of management commitment, variability in production lines, variability in consumer demand, and the uniqueness of food businesses based on their capacity. They also mentioned that the cost of developing a HACCP system, as well as the cost of implementing and maintaining a HACCP system, are key barriers to proper HACCP system implementation. Organizational elements of HACCP system, as it relates to extensive and overly complex documentation, have been identified as one of the primary impediments in the application of the HACCP principles, in addition to the difficulties to effective implementation of the HACCP system [13, 17–20]. In relation to the development of HACCP system, financial, competence-related, and technical hurdles are the most common issues impeding the proper implementation of the HACCP system particularly in the small-scale food businesses [17, 21].

According to the study by Dzwolak [22], typical flaws and gaps that have a detrimental impact on the effective implementation of the HACCP system with specific focus on small-scale food enterprises were identified. The following issues were noted in the author's findings, complications encountered during the development of the HACCP plans, as well as limitations during the on-site review of HACCP documentation. The most difficult aspects relating to the designing HACCP programs were hazard assessment and severity assessment. These issues further revealed that there was a high level of noncompliance in the areas of documentation, hazard identification and assessment, process flow diagrams, and system verification. In addition to these concerns, a number of small-scale food businesses halted or curtailed activity in the HACCP system area after its adoption, confirming observations reported in other studies and proving that the difficulty with maintaining the HACCP system still persists. The study also established that a number of mistakes and errors are committed by the HACCP teams operating in the small-scale food business sector having insufficient knowledge and abilities [22]. This clearly indicates the need to carefully look into the process of selecting and equipping the HACCP team in order to ensure effective development of a HACCP system.

All these factors, as mentioned above, have a significant impact on the development of a good HACCP plan. According to Wallace et al. [23]. developing a good HACCP strategy is critical to guaranteeing the HACCP system's effective implementation and application. This means that a thorough assessment of the HACCP plan can discover flaws, deficiencies, and inefficiencies, ensuring that the HACCP system is effectively implemented and leads to successful food safety management [17, 20, 24, 25]. Adequate motivation is essential to overcome these challenges that impact on proper development and deployment of the HACCP system in any food business. Some of the motivations include: the necessity for efficient processing operations; the need to reduce product recalls; the want to exceed consumers' expectations; the prevention of previously experienced food safety concerns; human capacity development; and effective management decision-making [16].

Hence, having a guidebook and/or manual for implementing a HACCP system would be advantageous [16]. This tool should be adaptable enough that multiple formats of guidance documents can be prepared for the implementation of the HACCP system in different types of food businesses. This will help with dealing with specific technical challenges that arise in various food enterprises throughout the food value chain. In addition, the implementation guide document can be used to train personnel in the development, implementation and maintenance of the HACCP system. Likewise, Dzwolak [22] emphasized the necessity of a flexible HACCP methodology to assist various types of small food businesses. For this system to work effectively, depending on the circumstances and capacities of the food industry operation, a more flexible approach to HACCP application may be required. With the implementation of a HACCP system, certain changes may be required relating to the need for modifications to processing parameters, processing steps, manufacturing technology, final product attributes, distribution method, and/ or intended use. Every HACCP system should be able to adapt to changes in plant operation, processing procedures, or technological innovations.

11.2.3 Recent development to HACCP system application

The effectiveness of the food safety system in protecting public health necessitated ongoing evaluation of the entire system, particularly in terms of how the HACCP study is being conducted and how the HACCP plan of actions is being established. In the context of a HACCP study, a thorough and accurate hazard analysis is critical to formulating a comprehensive HACCP action plan. Expertise with accurate understanding of hazards, as well as health risks associated with specific hazards, are required in hazard analysis for it to be effective. Because of the complexities of this procedure, one of the most recent innovations in HACCP system implementation relate to separating raw material/ingredient-related hazard analysis from process-step-related hazard analysis [26]. Separate assessment of ingredients-related risks versus process steps, according to these authors, allows the HACCP team to conduct a full hazard analysis and identify appropriate control strategies for preventing and treating these issues. This approach enables for adequate hazard analysis by distinguishing inherent dangers in ingredients from hazards that could be introduced, regulated, or amplified throughout the processing phases. This indicates that a clear distinction between the two permits appropriate implementation of relevant preventative measures, allowing for an effective HACCP system application.

Another recent HACCP development is a shift in the approach to preventative controls with the goal of fostering a risk-based approach to safety management in the food production value chain. This relates to the newer hazard analysis and risk-based preventive control (HARPC) plans. The traditional HACCP strategy is based on reacting to the possibility of contamination, whereas HARPC is based entirely on prevention [27]. In a nutshell, the typical HACCP system is based on the concept of "reasonably possible hazards" in the management of food safety hazards via prerequisite programs or CCPs. According to McEntire [28], numerous food types such as fresh fruits and vegetables lack true CCP, thereby application of traditional HACCP strategy based on CCP is insufficient. McEntire [28] further noted that moving from a reactive approach to a preventive strategy is critical in ensuring food safety by shifting away from testing food for safety and toward preventing contamination. As a result, the notion of HARPC is based on "predictable or reasonably foreseeable hazards." The differences between the two are compared in Table 11.2.

HACCP is not a stand-alone program; this system is built on the foundation of prerequisite programs (PRPs). PRPs provide the appropriate environment and operating conditions in the production of safe food items. However, it is necessary to have a hazard analysis plan that is proactive in assuring quality and eliminating costly safety issues throughout the food business value chain. This necessitated the conception of a newer HARPC in relation to recent developments in the HACCP system, which is based on the introduction of the idea of "operational prerequisite programs (OPRPs)," as stated in ISO 22000 standards [29].

OPRP concept according to Scanlin and Sayler [29] is aimed at bridging the gap between the HACCP and the HARPC. Prerequisite programs (PRPs) are required to function as control measures for preventing severe hazards, and OPRPs can be defined as such. As a result, PRPs and key control points are now considered preventive control measures in hazard prevention plans. OPRPs are identified through risk assessment, which includes determining what is measurable and how to quantify things like critical control limits (CCP). Further, Panghal et al. [30] defined OPRPs as PRPs identified by the hazard analysis as critical in controlling the probability of hazard infiltration, contamination, or proliferation of food safety hazards in the product(s) or processing environment. These are necessary to manage product-specific dangers that are not covered by CCP. Control measures are chosen for each identified serious risk in the hazard analysis. They are classified according to whether they are handled using OPRPs or through CCP control methods. The product, process, and hazard to be

Table 11.2: Traditional HACCP versus newer HARPC (adapted from Scanlin and Sayler [29]).

Traditional HACCP	Newer HARPC
Reasonably possible hazards	Reasonably expected hazards
Hazards control infers element of risk based	Hazards control is risk based
Hazards control measures explicit	Hazards control is not explicit
Requires a flowchart	Does not require a flowchart
Utilizes "decision tree"	Utilizes "risk-based matrix/table"
HACCP requires a team	HARPC requires a "Preventive Controls Qualified Individual" (PCQI)
HACCP requires annual review of the documentation	HARPC plan reanalyzed at least every three years depending changes to the processes

controlled all play a role in determining whether a control measure is linked to a CCP or an OPRP. OPRPs, according to Panghal et al. [30], have a better chance of limiting contamination and proliferation than PRPs, which are primarily concerned with maintaining a safe production environment. For effective food safety management, PRP, OPRP, and control measures at CCPs can stand alone or be integrated. OPRPs and CCPs are usually product-specific in controlling product-specific hazards determined by hazard analysis, whereas PRPs are applied to the entire manufacturing plant. As shown in Figure 11.1, absolute control measures apply to CCPs, whereas nonabsolute control measures apply to OPRPs [30].

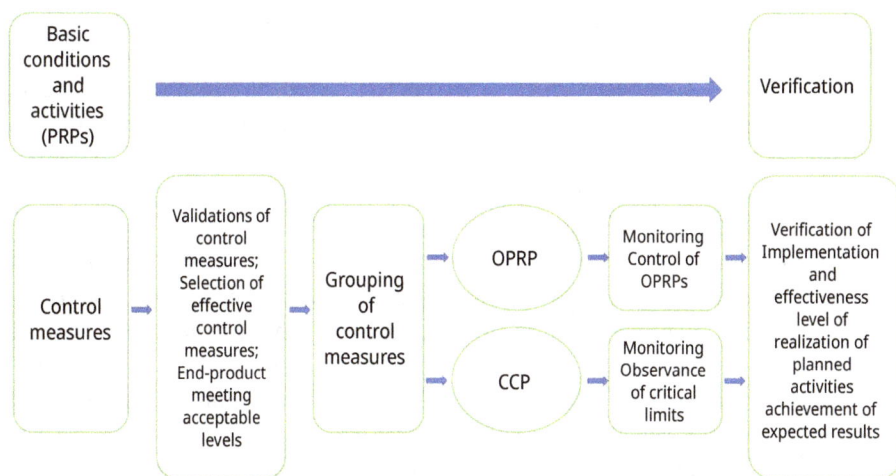

Figure 11.1: System approach to deciding and verifying CCP and OPRPs [30].

In addition to the differences highlighted in Table 11.1, HARPC plan does not have critical limit definition and the hazard sources are nonspecific [27]. In summary, HARPC approach requires having documented plans that identify hazards, detail the activities needed to reduce or eliminate those hazards, identify monitoring procedures and record the results, and specify what actions to take if problems develop [27]. Further noted that the HARPC concept is required to assess products and processes for the following:

–   Biological, chemical, physical, and radiological hazards
–   Natural poisons, pesticides, medication residues, decomposition, parasites, allergies, and unauthorized food and color additives are all things to be concerned about
–   Hazards that arise naturally or are accidentally added
–   Hazards that were purposefully introduced, such as terrorism

Adoption of HARPC concept can effectively reduce or eliminate probable exposure to hazards or sources of contamination. Further, proper identification, monitoring, verification, and execution of remedial control measures, as well as record-keeping are essential. Ultimately, an effective HARPC strategy is generally based on compliance with HACCP mandates, but a proven HACCP plan is the greatest precursor to a successful HARPC plan [27].

Furthermore, the implementation of the revised CODEX HACCP 2020 also contributes to recent advancement in the HACCP system. Management commitment to food safety, consolidation and expansion of definitions, revamping of the use of the Codex decision tree, strengthening of the training and competence section, inclusion of food safety culture, validation of critical limits as a key requirement, and requirement for a traceability/ product tracing system are among the changes incorporated in the revised version. Further, the 2020 CXC 1-1969 [31] document put more emphasis on the following:

–   The use of a science-based preventative approach to ensure food safety and suitability and the importance of effective GHP's in preventing the contamination of food.
–   The proper application of prerequisite programs before considering implementation of a HACCP system.
–   That the responsibility to identify hazards associated with raw materials, other ingredients, the production environment lies with the food business operator.
–   The application of GHPs may be sufficient to control hazards in some instances depending on the type of food, risk of significant hazards and processing procedures. In cases where GHPs cannot be applied alone, a combination of GHPs and other control measures at CCP should be considered.
–   The scientific validation of control measures that reduce hazards to acceptable levels.

- That the extent of monitoring control measures, their corrective actions, verification and documentation must consider the nature of food product and the size of the food business.
- The review of food hygiene systems to determine need for their update. The review must be done occasionally and when changes that can affect the impact of significant hazards and their control measures occur.
- The importance of appropriate communication about the food and food process to ensure food safety at all stages of the food chain.

Another key development is in the light of artificial intelligence (AI). This new development is vital in guaranteeing successful execution of the HACCP system with the introduction AI. Enforcing adherence to the implementation of the HACCP plan, according to Matt Brown (CEO, Wherefour, Petaluma), is a critical component in ensuring that the plan translates into compliance activity [32]. Matt further noted that this approach requires adoption of suitable software system or technology in strengthening compliance with regard to implementation of following correct procedure and prevent personnel from deviating from the plan. The additional advantage is that a software provides a better view of the entire process by keeping a permanent record of who did what and when, which may decrease the temptation to cut corners and/or skip crucial steps. McEntire [28] proposed an alternative HACCP concept and/or preventive controls strategy. This relate to failure mode and effects analysis (FMEA) concept, which is not new in the field of food safety. It adds a new dimension to HACCP by making it more deliberative and semi quantitative. This indicate that FMEA offers the ability of measuring how likely it is that the incident will be identified before the detrimental consequence occurs. In addition, it provides for a more detailed distinction and display of the risk spectrum. Although there are still a variety of ways to group and cluster different sorts of hazards, which may add subjectivity into the study, it does provide a way to visualize and explain areas where more attention and resources are needed [28]. In essence, FMEA approach is based on proactive investigation of what could go wrong or cause a failure. Therefore, critical thinking placed a demand for a root-cause analysis of failure thereby aiding the identification of effective preventive controls and the assessment of failure. In line with suggestion of McEntire [28], effective management of food safety necessitates the critical thinking of skilled professionals regarding any food production systems, as it relates to factors under their control as well as factors beyond their control, indicating that HACCP hazard analysis goes beyond simple yes/no exercise, but rather a more thoughtful process.

11.3 Evaluating the performance of an FSMS

Over the past decade, food processing companies have invested immensely in establishing and maintaining FSMS. As indicate above, more and more companies are adopting FSMS which means continued investment into the system. It is not surprising then, that there have been calls on whether FSMS investments yield positive returns in terms of improved food safety and reduction in food safety incidences. Traditionally, the performance of a FSMS is assessed through inspection or auditing. The former is normally done against local laws and regulations enforced by government authorities [33]. The major challenge with inspections is that they are generally carried out after a long period of time or following a food safety incident in most developing countries. This is due to shortage of qualified staff, lack of resources such as costs for analysis, transportation, and in some instances, inspectors may have a wider scope that includes nonfood items. Therefore, inspections do not always give a good indication of the status of a FSMS when the above challenges are considered. Despite this, very useful findings on the performance of FSMS can be obtained from inspections through recognizing noncompliances/nonconformities [34]. Also, it must be mentioned that inspections have evolved nowadays and now somewhat follow a similar process to audits [35].

Food safety audits can be divided into first-party audit, second-party audit and third-party audit. The latter is more important in the context of evaluating the performance of a FSMS because it is done by an independent body and may result in certification. The primary purpose for an audit is to verify that all requirements of a standard are implemented correctly and that there is proof (documents) to support their effectiveness [33]. In addition, audits can identify gaps in the system and this is important for continual improvement of the business [36]. Although audit outcomes give a good indication of the status of a FSMS, they are mainly specific to the standard or set of standards audited against. Also, the scope and frequency of audits may vary depending on the standard and the risks associated with the product being processed. It is worth mentioning that the quality of audit results may be influenced by the competence and integrity of the auditor.

Given the above, new tools for assessing the performance of FSMS have been designed lately. These are believed to be more robust and comprehensive because they are based on food safety performance indicators, include the influence of human behavior and in some instances, involve actual data from sampling. Luning et al. [37, 38] developed an FSMS diagnostic instrument (FSMS-DI) that is based on a systematic assessment of the core control activities and assurance activities of a company's FSMS (Figure 11.2). The tool is not depended on the existing quality assurance and or guidelines that feed into the FSMS of a company. To describe control and assurance activities, the FSMS-DI uses indicators at four levels (0, 1, 2, and 3). Further, context factors (organization of the company, type of product and processing activities) as well as the microbiological status of the FSMS are covered within the FSMS-DI tool. The effectiveness of the FSMS-DI tool has been validated in animal-based food [39, 40] and fresh produce industries in Europe [41]. Other food sectors that have used the tool are shown in Table 11.3.

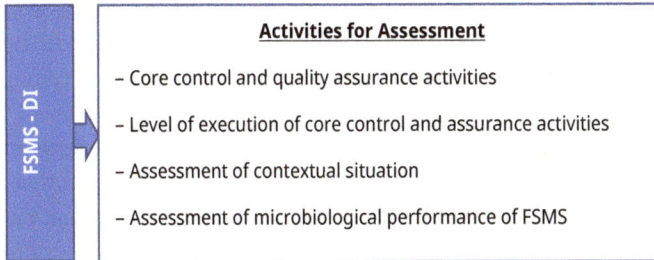

Figure 11.2: Assessment activities followed with the FSMS-DI tool.

Another tool that has been developed to assess the performance of FSMS is known as the Microbial Assessment Scheme (MAS) (Figure 11.3). This tool was designed by Jacxsens et al. [42], and it focuses on assessing the microbiological status of an existing FSMS. With the MAS tool, the microbiological safety of a company can be classified using three levels (1, 2 and 3). A company that achieves the highest level (3) is assumed to have a good safety performance (e.g., low microbial counts), while level 1 depicts a poor microbial status. The levels from selected external and internal food safety performance indicators are further summed up (18–21, in total) to get a microbiological safety level profile. Thereafter, assigned scores are used to obtain an overall indication of the actual food safety performance.

Table 11.3: Sectors of the food industry where the FSMS-DI tool has been applied.

Country	Sector	Food product	Performance outcome	Reference
Kenya	Dairy processors	Dairy products	Scale of production influences performance of FSMS	Njage et al. [7]
Serbia	Orchard farms and cold stores	Raspberries	Moderate to high risk FSMS	Rajkovic et al. [4]
South Africa	Primary production	Mushrooms	Average FSMS output	Dzingirayi et al. [43]
Kenya	Food processors	Fresh produce – export	Moderate FSMS output	Sawe et al. [44]
Brazil	Primary production	Organic lettuce	Moderate to high risk for microbial contamination	De Quadros Rodrigues et al. [45].

Since the MAS tool involves actual microbiological data, the following challenges have been identified when using the tool: (i) a need for profound knowledge on microbial behavior under different production processes, (ii) high costs and time required for analysis (iii) an expert is needed to properly relate results to the FSMS. To circumvent this, Jacxsens et al. [46], designed a tool that can diagnose the microbiological status of FSMS without performing actual microbiological analysis. The approach is based on the use of carefully selected and validated (in Europe) food safety performance indicators. For more information on MAS, the reader is referred to Jacxsens et al. [47].

The MAS protocol was successfully used to analyze the microbiological status of FSMS applied along the lamb production chain in Spain [40]. The investigation reported that the slaughterhouses contained the highest microbiological contamination. Recently, Cheah et al. [48], used the MAS tool in combination with the FSMS-DI protocol to evaluate the FSMS of powdered beverage manufactures. The combined tools were able to differentiate the FSMS performance of the manufactures based on the type of FSMS certification.

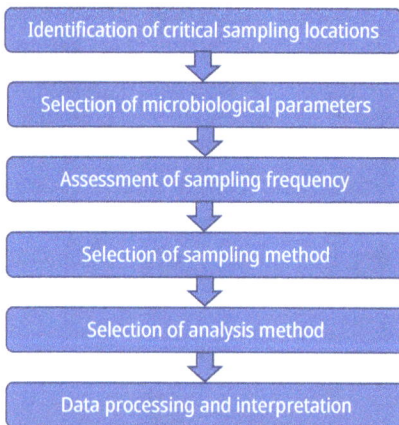

Figure 11.3: Procedure for using the microbiological assessment scheme (MAS) tool [42].

In Singapore, a retrospective approach based on reported foodborne disease outbreaks and food hygiene violations was used to evaluate the effectiveness of implemented FSMS [49]. Using a controlled interrupted time-series analysis, the investigation did not find significant reduction in food hygiene violations, while foodborne disease outbreaks were reduced following implementation of the FSMS's. The performance of FSMS implemented at pork slaughter plants in China was determined using a questionnaire and a checklist with a scoring system [50]. The checklist was adopted from the Global Food Safety Initiative Global Markets Programme. The findings identified various factors that affected the performance of the FSMS and weaknesses thereof. In addition, the authors suggested that unannounced assessment of FSMS give more reliable results than announced assessments.

11.4 Quality assurance standards

Quality assurance standards and guidelines are an important element of FSMS since they ensure food quality and safety. They are implemented to support existing food control legislation that is enforced by local authorities. The type of quality assurance standard that a business can implement depends on demands by food safety authorities, target market, and requirements of sector organizations. Over the past years, there has been an increase in the number of companies that adopt quality assurance standards like ISO 22000, British Retail Consortium (BRC), and International Featured Standards (IFS). This is because QAS with valid certification allow companies to have access to export markets. Although the rise in the adoption of QAS is mainly associated with improved food quality and safety, other authors have reported otherwise. All QAS are continuously revised and new versions or editions are published. This is to ensure that they are able to effectively control current and new hazards and that they are able to meet changing customer requirements.

11.4.1 British Retail Consortium (BRC) Global Food Safety Standard

The BRC Global Food Safety Standard was first published in 1998 to assist the food industry to comply with requirements of the food safety laws in the United Kingdom and European Union [51]. Later, the standard became popular worldwide as a benchmark for the production of good quality and safe food using systems that are in line with regulatory and statutory requirements. The BRC provides a clear outline of operational criteria requirements that organizations must follow to properly manage the quality and safety of food, thus meeting customer requirements. The standard is used worldwide by food manufacturers, retailers, suppliers and companies that export food. It is divided into seven sections that include senior management commitment and requirements for continual improvement; implementation of a food safety plan based on HACCP system; food safety and quality management system; site standards; product control; process control and personnel management plan. Organizations that implement the BRC standard are audited at intervals by a reputable third-party body to get certification. The certification gives customers confidence that the food manufacturer follows globally accepted best practice in the production of good quality and safe food.

11.4.2 ISO 22000 food safety management system

The ISO 22000 is an international recognized standard that is designed to ensure food quality and safety at every stage of the food chain and thus hailed as the "farm to form" standard [52]. The standard was first published in 2005 and it is applicable to all organizations, irrespective of size. It integrates the principles of HACCP outlined by the Codex Alimentarius Commission and some elements of the ISO 9001 standard. In addition, the standard provides a framework for the development and maintenance of an effective FSMS within the context of the organization and includes communication (internal and external) requirements and continual improvement. The standard can be applied alone or integrated with available FSMS in an organization. In general, the ISO 22000 standard is used voluntarily by many food organizations worldwide as a yardstick for best practice. In cases where a certificate is required, third-party audits are carried out against the standard followed by certification. Organizations that correctly implement the ISO 22000 are able to demonstrate their ability to produce food that is safe for human consumption. Although the ISO 22000 is recognized worldwide, it is not benchmarked by the GFSI. As an alternative, the Food Safety System Certification (FSSC 2000) is adopted by companies that require a GFSI recognized standard because it was designed using the ISO 22000 standard, ISO 9001, ISO/TS 22003, and ISO/TS 22002-1.

11.4.3 International Featured Standard (IFS)

The IFS Food is a GFSI benchmarked standard that uses a risk-based approach to develop guidelines for the manufacture of safe and quality food along the supply chain [53]. The standard was developed with suppliers, retailers and certifications bodies and it is applicable to food processing and food packaging companies, including those that package loose food products. It covers the complete food supply chain including distributors, agents and brokers. The IFS takes a nonprescriptive approach on required processes along the food chain and this allows companies to integrate processes that are specific to their own environment. Requirements for organizations that own brands are found within the IFS to ensure that the quality of branded products meets customer requirements and brands are protected against liabilities. The primary areas covered by the IFS are: governance and commitment; food safety and quality management system; resource management; operational processes; measurements, analysis and improvements; and food defense plan. The audit approach for certification is thorough, consistent, and transparent throughout the supply chain.

11.4.4 Safe Quality Food (SQF)

Safe Quality Food is a global food safety certification scheme that was developed in 1994 and first recognized by GFSI in 2004 [54]. It is managed by the Safe Quality Food Institute (SQFi). The standard provides rigorous, transparent and systematic procedures for managing food safety risks to ensure food safety and quality. Certification covers all sectors from farm to fork, that is, primary production, manufacturing, packaging, distribution and wholesale. The SQF standard is divided into three programs. The SFQ Fundamentals Program (previously Level 1) is targeted at low risk products produced by small and medium enterprises. It constitutes food safety fundamentals such as good agricultural practices and good manufacturing practices. This program is not recognized by the GFSI but provides a pathway to the other recognized programs. The second program is the SQF Food Safety Program. It has several codes and focuses on systematic implementation of the HACCP system in the control of food safety risks and hazards. This program is recognized by the GFSI and it is suitable for organizations that are targeting local and international food markets. The third program is known as the SQF Quality Program, and it is also recognized by the GFSI. SQF Quality is targeted at organizations that want to implement a robust quality program to an existing and effective food safety plan. Quality threats are identified and controlled using the Codex HACCP procedures in this program.

11.5 Conclusions

Proper implementation and maintenance of an effective FSMS is crucial in preventing food safety issues. A number of FSMS include the HACCP system because it follows a science-based preventative approach with validated control measures. Although effective, a number of barriers can prevent the implementation of HACCP system by small-scale companies. Therefore, more interventions are required to enable small-medium sized companies to adopt flexible HACCP systems without compromising the safety of food. Over the past few years, the HACCP approach has been updated and recent developments suggest ways in which it can be improved. These include fostering a risk-based approach to safety management and implementation of the FMEA concept. The proposed ideas appear sound but they must be thoroughly evaluated against the improved conventional systems to ensure that they don't become additional barriers to implementation of HACCP. Tools for effective evaluation of FSMS have been developed. Their relevance to different sectors of the food industry has also been proven. The outcome from these tools is important for continuous improvement and for directing resources to priority areas. Therefore, the use of the FSMS evaluation tools is encouraged because they can encourage governments, companies and all other stakeholders to invest in FSMS.

References

[1] Jacxsens, L., Luning, P. A., Marcelis, W. J., van Boekel, T., Rovira, J., Oses, S., et al. (2011). Tools for the performance assessment and improvement of food safety management systems. *Trends in Food Science and Technology, 22*, S80–S89.

[2] Kirezieva, K., Luning, P. A., Jacxsens, L., Allende, A., Johannessen, R. S., Tondo, E. C., et al. (2015). Factors affecting the status of food safety management systems in the global fresh produce chain. *Food Control, 52*, 85–97.

[3] Piira, N., Kosola, M., Hellsten, C., Fagerlund, A., & Lundén, J., (2021). Comparison of official food control results in Finland between food establishments with and without a certified food safety management system. *Food Control, 129*, 108230. https://doi.org/10.1016/j.foodcont.2021.108230

[4] Rajkovic, A., Smigic, N., Djekic, I., Popovic, D., Tomic, N., Krupezevic, N., et al. (2017). The performance of food safety management systems in the raspberries chain. *Food Control, 80*, 151–161.

[5] Tutu, B. O., & Anfu, P. O. (2019). Evaluation of the food safety and quality management systems of the cottage food manufacturing industry in Ghana. *Food Control, 101*, 24–28.

[6] Hou, M. A., Grazia, C., & Malorgio, G. (2015). Food safety standards and international supply chain organization: A case study of the Moroccan fruit and vegetable exports. *Food Control, 55*, 190–199.

[7] Njage, P. M. K., Opiyo, B., Wangoh, J., & Wambui, J. (2018). Scale of production and implementation of food safety programs influence the performance of current food safety management systems: Case of dairy processors. *Food Control, 85*, 85–97.

[8] Chaoniruthisai, P., Punnakitikashe, P., & Rajchamaha, K. (2018). Challenges and difficulties in the implementation of a food safety management system in Thailand: a survey of BRC certified food productions. *Food Control, 93*, 274–283.

[9] Kafetzopoulos, D. P., & Gotzamani, K. D. (2014). Critical factors, food quality management and organizational performance. *Food Control, 40*, 1–11.

[10] EC. (2004). Regulation (EC) No 852/2004 of the European Parliament and of the Council of 29 april 2004 on the hygiene of foodstuffs. *Official Journal of the European Union, L226*, 3.

[11] Motarjemi, Y. (2016). Chapter 8 – Food safety management: State of the art. In *Handbook of hygiene control in the food industry.* http://dx.doi.org/10.1016/B978-0-08-100155-4.00008-X

[12] CAC (Codex Alimentarius Commission). (2003). Hazard analysis and critical control point (HACCP) system and guidelines for its application. In *Food hygiene basic texts (CAC/RCP 1-1969) Rev. 4, Recommended international code of practice general principles of food hygiene.* Rome: Food and Agriculture Organization, 23.

[13] Chen, E., Flint, S., Perry, P., Perry, M., & Lau, R. (2015). Implementation of non-regulatory food safety management schemes in New Zealand: A survey of the food and beverage industry. *Food Control, 47*, 569–576.

[14] Dora, M., Kumar, M., Van Goubergen, D., Molnar, A., & Gellynck, X. (2013). Food quality management system: Reviewing assessment strategies and a feasibility study for European food small and medium-sized enterprises. *Food Control, 31*, 607–616.

[15] Milios, K., Pantelis, E. Z., Pantouvakis, A., Mataragas, M., & Eleftherios, H. D. (2013). Techno-managerial factors related to food safety management system in food businesses. *British Food Journal, 115*(9), 1381–1399.

[16] Milios, K., Drosinos, E. H., & Zoiopoulos, P. E. (2012). Factors influencing HACCP implementation in the food industry. *Journal of the Hellenic Veterinary Medical Society, 63*(4), 283–290.

[17] Dzwolak, W. (2014). HACCP in small food businesses – The Polish experience. *Food Control, 36*, 132–137.

[18] Dzwolak, W. (2017). Documenting HACCP in a small restaurant – A practical approach. *Quality Assurance and Safety of Crops & Foods, 9*(2), 187–194.

[19]   Toropilova, J., & Bystricky, P. (2015). Why HACCP might sometimes become weak or even fail. *Procedia Food Science, 5*, 296–299.

[20]   Wallace, C. A., Sperber, W. H., & Mortimore, S. E. (2018). *Food safety for the 21st century. Managing HACCP and food safety throughout the global supply chain* (2nd Edition). Oxford: Wiley.

[21]   Charalambous, M., Fryer, P. J., Panayides, S., & Smith, M. (2015). Implementation of Food Safety Management Systems in small food businesses in Cyprus. *Food Control, 57*, 70–76.

[22]   Dzwolak, W. (2019). Assessment of HACCP plans in standardized food safety management systems – The case of small-sized Polish food businesses. *Food Control, 106*, 1–16. https://doi.org/10.1016/j.food cont.2019.106716

[23]   Wallace, C. A., Halyoak, L., Powell, S. C., & Dykes, F. C. (2014). HACCP – The difficulty with Hazard analysis. *Food Control, 35*, 233–240.

[24]   Mortimore, S., & Wallace, C. (2013). *HACCP: A practical approach* (3rd Edition). New York, Heidelberg, Dordrecht, London: Springer.

[25]   Mortimore, S. E., & Wallace, C. A. (2015). *HACCP: A food industry briefing* (2nd Edition). Oxford, Hoboken: Wiley Blackwell.

[26]   Kottapalli, B., Dobmeier, N., & Ledenbach, L. H. (2020). Ingredients Hazard Analysis and Process-Step Hazard Analysis: Why Separate the Two? Food Safety Magazine, August 17, 2020.

[27]   Denis, S. (June 15, 2021). HACCP and HARPC: Key differences and definitions. *Trace Gains.* https://www.tracegains.com/ (accessed: 21 June 2021)

[28]   McEntire, J. (2021). Beyond HACCP and Preventive Controls: Promoting True Risk-Based Thinking Tied to Public Health Outcomes. Rethinking preventive approaches to food safety. Food Safety Magazine, June 20, 2021.

[29]   Scanlin, A., & Sayler, A. (2018). RBHAAPC, HACCP Plus, HARPC Minus or HACCPARBPCPHF: What's in a Name? Food Safety Magazine, March 20, 2018.

[30]   Panghal, A., Chhikara, N., Sindhu, N., & Jaglan, S. (2018). Role of Food Safety Management Systems in safe food production: A review. *Journal of Food Safety*, e12464. https://doi.org/10.1111/jfs.12464

[31]   CXC 2020 (Codex Alimentarius Commission). (2020). General Principles of Food Hygiene. CXC 1-1969 (CODEX, 2020). Food and Agriculture Organization, Rome. Available from: https://www.codexalimen tarius.org

[32]   Parker, L. (2020). Current best practices for establishing and refining food industry HACCP plans. Food Safety Magazine, November 13, 2020.

[33]   Motarjemi, Y., & Mortimore, S. (2014). Food safety assurance systems: Audits of food safety management systems. *Encyclopedia of Food Safety, 4*, 288–292.

[34]   Turku, M., Lepistö, O., & Lundén, J. (2018). Differences between official inspections and third party audits of food establishments. *Food Control, 85*, 459–465.

[35]   Kotsanopoulos, K., & Arvanitoyannis, I. S. (2017). The role of auditing, food safety, and food quality standards in the food industry: A review. *Comprehensive Reviews in Food Science and Food Safety, 16*, 760–775.

[36]   GFSR. (2016). Food safety audits. Available from: http://globalfoodsafetyresource.com/food-safety /food-safety-audits# (accessed 2017 January 2).

[37]   Luning, P. A., Bango, L., Kussaga, J., Rovira, J., & Marcelis, W. J. (2008). Comprehensive analysis and differentiated assessment of food safety control systems: A diagnostic instrument. *Trends in Food Science and Technology, 19*, 522–534.

[38]   Luning, P. A., Marcelis, W. J., Rovira, J., Van der Spiegel, M., Uyttendaele, M., & Jacxsens, L. (2009). Systematic assessment of core assurance activities in a company specific food safety management system. *Trends in Food Science and Technology, 20*, 300–312.

[39]   Sampers, I., Toyofuku, H., Luning, P. A., Uyttendaele, M., & Jacxsens, L. (2012). Semiquantitative study to evaluate the performance of a HACCP-based food safety management system in Japanese milk processing plants. *Food Control, 23*, 227–233.

[40] Osés, S. M., Luning, P. A., Jacxsens, L., Jaime, I., & Rovira, J. (2012). Food safety management systems performance in the lamb production chain. *Food Control*, *25*, 493–500.

[41] Kirezieva, K., Nanyunja, J., Jacxsens, L., van der Vorst, J. G. A. J., Uyttendaele, M., & Luning, P. A. (2013). Context factors affecting design and operation of food safety management systems in the fresh produce chain. *Trends in Food Science and Technology*, *32*(2), 108–127.

[42] Jacxsens, L., Kussaga, J., Luning, P. A., Van der Spiegel, M., Devlieghere, F., & Uyttendaele, M. (2009). A microbial assessment scheme to measure microbial performance of food safety management systems. *International Journal of Food Microbiology*, *134*, 113–125.

[43] Dzingirayi, G., & Korsten, L. (2016). Assessment of primary production of horticultural safety management systems of mushroom farms in South Africa. *Journal of Food Protection*, *79*(7), 1188–1196.

[44] Sawe, C. T., Onyango, C. M., & Njage, P. M. K. (2014). Current food safety management systems in fresh produce exporting industry are associated with lower performance due to context riskiness: Case study. *Food Control*, *40*, 335–343.

[45] de Quadros Rodrigues, R., Loiko, M. R., de Paula, C. M. D., Hessel, C. T., Jacxsens, L., Uyttendaele, M., et al. (2014). Microbiological contamination linked to implementation of good agricultural practices in the production of organic lettuce in Southern Brazil. *Food Control*, *42*, 152–164.

[46] Jacxsens, L., Uyttendaele, M., Devlieghere, F., Rovira, J., Osés Gomez, S., & Luning, P. (2010). Food safety performance indicators to benchmark food safety output of food safety management systems. *International Journal of Food Microbiology*, *141*, 180–187.

[47] Jacxsens, L., Devlieghere, F., & Uyttendaele, M. (2009). *Quality management systems in the food industry*. ISBN 978-90-5989-275-0.

[48] Cheah, H. Y., Merican, S. E., Rashid, N. K. M., Bakar, A. Z. A., Omar, S., & Maimunah Sanny, M. (2021). Assessing the Performance of Food Safety Management System Using Food Safety Management System Diagnostic Tools and Microbial Assessment Scheme: A Case of Powdered Beverage Manufacturers. *Malaysian Journal of Medical Science*, *28*(3), 129–142.

[49] Aik, J., Turner, R. M., Kirk, M. D., Heywood, A. E., & Newall, A. T. (2020). Evaluating food safety management systems in Singapore: A controlled interrupted time-series analysis of foodborne disease outbreak reports. *Food Control*, *117*, 107324.

[50] Xiong, C., Liu, C., & Zheng, L. (2017). Performance assessment of food safety management system in the pork slaughter plants of China. *Food Control*, *71*, 264–272.

[51] BRC. BRC (British Retail Consortium) Global Food Safety Standard. (2021). Available from: https://www.brc.org.uk (accessed 30 June 2021).

[52] ISO 2000 (International Organization for Standardization). (2018). Food Safety Management. Available from: https://www.iso.org(accessed 30 June 2021).

[53] IFS (International Featured Standards). (2020). IFS Standards. Available from: https://www.https://www.ifs-certification.com/index.php/en/standards (accessed 30 June 2021).

[54] SQF. (2021). SQF (Safe Quality Food). Available from: https://www.sqfi.com (accessed 30 June 2021).

Christiana Eleojo Aruwa* and Alaofin Sefunmi

Chapter 12
Food safety culture and enhancing food employee compliance

Abstract: Worldwide food borne diseases (FBDs) are associated with the consumption of unsafe foods which cause millions of deaths annually. Food contamination can occur at any point along the food supply and production chains and requires the monitoring of personnel involved, regulations implementation, and upward review of food business value systems with regard to ensuring safety of food products intended for the end user or consumer, among a host of other interconnected factors. In this chapter, authors provide an up to date overview on food safety culture (FSC) practices and requirements for all food industry actors, especially employees, as well as business owners. It expatiates on the concepts of food safety, FSC, FSC standards, as well as means to ensure and enforce employee and employer compliance. Likewise, it informs on the need to evolve FSC standards and compliance levels to stand the test of time. Such evolution should be built into and enforced in a food business's structure and vision (value statement and system). Finally, the chapter covers methods to monitor, control and prevent unwanted food safety-related incidences by strengthening employee-employer and other industry stakeholder partnerships.

12.1 Background and introduction

Globally, unsafe foods result in illnesses known as foodborne diseases (FBDs). FBDs arise from the ingestion of harmful microorganisms (bacteria, fungi, viruses, and parasites) and/or microbial toxins or chemicals found in and on contaminated foods [1, 2]. FBDs constitute a common public health concern. Food contamination may occur at any stage from harvesting and production to consumption. FBDs can lead to morbidity

Acknowledgement: Authors acknowledge the National Research Foundation (NRF) Innovation Postdoctoral Grant awarded to Dr. C. E. Aruwa, and tenable at the Department of Biotechnology and Food Science, Durban University of Technology, South Africa.

*Corresponding author: Christiana Eleojo Aruwa,** Faculty of Applied Sciences, Department of Biotechnology and Food Science, Durban University of Technology (DUT), South Africa,
e-mail: ChristianaA@dut.ac.za
Alaofin Sefunmi, Department of Plant Pathology, North Dakota State, University, PO Box. 6050, Fargo, United States of America.

https://doi.org/10.1515/9783110748345-012

and, occasionally, mortality. Incidences of FBDs are on the rise globally, and predominantly in developing countries. This may be associated with the total or partial neglect of the necessary food and personal hygiene principles and practices. Illnesses that are foodborne pose a threat to public health, public safety, as well as national and international economic development. As worldwide travel, trade, and immigration increase, there is also the increased probability of transfer of pathogens and contaminants across borders. Annually, more than 2 million deaths are recorded, and mostly children from developing areas, as a result of FBDs. The consumption of *Salmonella typhi* in foods which cause typhoid fever is reported to be responsible for about 600,000 deaths annually and globally. In America, unsafe foods cause millions of infections, more than 300,000 hospitalizations and about 5,000 deaths on an annual basis. One of six persons in the United States was affected by FBD in 2011, according to American Centres for Disease Control and Prevention (CDC) [1–3].

Some studies and surveillance programs have shown that implicated bacterial species which cause FBDs include *Salmonella, Campylobacter, Shigella, Cryptosporidium*, Shiga toxigenic *E. coli* (STEC non-O157), *Vibrio, Yersinia, Listeria*, and *Cyclospora* species. *Listeria, Cyclospora*, and *Vibrio* were most implicated and highest among the elderly populations (≥65 years of age), while the other foodborne pathogens were more prevalent in children below the age of 5 [1, 2]. Of the microorganisms listed, *Campylobacter* and *Salmonella* (most frequently isolated causative agent) are the most commonly implicated in FBDs, and their isolation and incidence in relation to FBDs continue to rise in places such as the United Kingdom, Norway, Finland, Denmark, Sweden, Iceland, Switzerland, and the Netherlands. Again, increasing world populations and demand for food contribute to the increased length and complexity of the food supply chains. Also, the expanding international food distribution network and incidence of FBDs at any point within the chain can potentially and adversely impact the health of populations and economies around the globe. Thus, international and multisector collaborations must be maintained, and complex connections between stakeholders and active participants along the food chain must be essentially protected and backed by relevant regulations to ensure public safety and health. The need to evaluate and enhance food safety (FS) measures, culture, and compliance levels at the local, national, and international scene cannot be overemphasized [1].

Unfortunately, although some expert opinions and articles on FBD cases are reported, many cases remain shrouded in uncertainty and are either unreported or underreported. This is worsened by the fact that most FBDs are not easily diagnosed due to the array of symptoms, and especially when they are also self-limiting. The inability to identify, diagnose, and report cases is also closely linked to lack of adequate testing facilities, personnel technical know-how, and surveillance strategies. Some foodborne illness symptoms include diarrhea-related dehydration, stomach upset, chills, fatigue, vertigo, mild to intense cramps, and mild fever. Some symptoms may eventually result in death. Fast foods and ready-to-eat meals are the primary causes of FBDs. Still home-cooked meals may also be implicated in foodborne illnesses [4]. Most food contaminations happen during food preparation and processing on site. Contaminations can be mitigated when guidelines by FS monitoring and regulatory authorities are adhered to, from the start of production till finish. In about ≤50% of FBD cases, reheated, undercooked, and improperly cooled foods were involved, while inadequate food cooking and preparation practices accounted for less than 20% of FBD cases [1, 4]. In relation to health, safe food is therefore considered as food that is clean (microbiologically, chemically, and physically), hygienically prepared and packaged, and which has not lost its nutritional value and is not stale or expired. Factors contributing to food contamination therefore threaten food safe. Thus, every resource that aid in the mitigation of food contamination at all points within the food chain must be exhausted [1].

12.2 Food safety

FS embodies all methods, practices, and principles surrounding food handling, preparation and storage, which are channeled for prevention of FBDs. An FBD outbreak is said to occur when 2 or more cases of similar infections, illnesses or disease is reported to be linked to the consumption of a common food source [5]. Routine FS protocols are then followed to prevent further health hazards and outbreaks, thus FS-related incidents could call for application of overlapping food defense strategies. Safety of foods and food products need to be assured among food growers, producers, the industry, market and the end users or consumers. From the industry to market practices, FS would cover food origins, hygiene and labeling practices, pesticide residues in foods, food additives, plus relevant food systems certifications, and food policies and guidelines that mitigate exports and imports across borders and regions. With regard to market and consumer safety, it is often the belief that products which get to markets are safe for purchase, safely prepared, and safely delivered to the target market [5]. Unsafe foods serve as vectors for the transmission of pathogens that cause FBDs or death in animals and humans. Foods also serve as growth and reproductive media for pathogens. Developed economies possess intricate food processing standards which are few, absent or present but not regularly or well enforced in developing regions. The lack of safe and portable water sources is also an integral contributor to diseases spread [6].

12.3 Food safety cultures (FSC)

The structure of FS culture is largely dependent on the demeanors, attitudes, beliefs, values, and competencies held by the groups and individuals that make up a food organizational system. These attributes determine the organizational style, commitment level, and program for ensuring end user safety and health [7]. Food safety culture

(FSC) refers to the particular culture of a food production facility that includes all characteristics highlighted above and determines what is done behind closed doors in the absence or presence of FS regulations. FSC also covers food business stakeholders like owners, employees, employers, and investors, and how they act and think on a daily basis to ensure FS. When both producers and customers can be proud that safe foods and food products are always distributed for sale in markets, then a sound and healthy FSC is being practiced [8].

The robustness and soundness of a food facility's FSC system is also largely a function of employee commitment, behavior, beliefs, and attitudes [9]. If organizations and stakeholders understand the adverse impacts of unsafe foods on heath and the economies of scale, then it would be easier to achieve a sound FSC. Thus, whatever the enterprise, multinational, national, small, or medium scale, and from the food facility manager or owner to the roadside vendor, there should be some degree of comprehension on food handling and FS practices. FS standards and regulations are well known, maintained and practiced in developed regions since its food production and trade industry is mainly involved in worldwide sourcing of raw materials, food mass production, processing, and distribution. Nonetheless, since this globalized food system increases the complexity of the food supply chain and delivery systems, it would require more stringent laws that circumvent regulatory oversights to ensure safe foods and safety practices. FBDs incidence studies have shown that good manufacturing practices (GMPs) failures are implicated in most cases and are scarcely associated with failures in FS systems, thus indicating the necessity to comprehend individual attitudes, behaviors, and factors that influence their actions [8].

Food safety can be improved with favorable behavioral changes among food business employees' compliance levels. Hence, FS equates with prevailing demeanors [9]. Since FBDs are closely associated with unsafe foods consumption, improvements in FBD surveillance and unsafe foods detection, and sensitization in FS responsibilities, can reduce FBD incidence trends. Data on points where a food contamination took place and where the food was improperly processed/handled are also essential for a robust FBD case investigation [3]. Central to FS are ensuring the use of adequate processing techniques, equipment, holding temperatures, and good employee personal hygiene. More importantly, FSC should integrate food and other applied sciences to create a sound behavior-based FS culture or management system. Food establishment with set FS goals should also have measures to assess levels of compliance. Lagging indicators are used in outcome-based measures by FS professionals to determine levels of progress being made against FBDs. Implementing effective FSC processes involve behavioral changes which constitute leading indicators. FS goals and measurements paired with appropriate consequences are critical to achieving efficient and improved employee performance [3].

12.3.1 Importance of food safety culture

Since foods are integral to survival, it is important that production practices are structured to ensure the safety of foods. Therefore, each individual relies on another in a food supply chain to make the "healthy" and right decisions. These decisions are integral components of a FSC and can either hinder or enable FS and hygiene practices [10]. The main aim of an FS professional is the creation of a sound FSC with protective attributes against FBD and outbreaks. When FS is relegated to the background by food production facilities and not substantially prioritized, it constitutes endangerment of end users of products and a platform for the occurrence of FBDs with subsequent harm to the food business [11, 12].

Food businesses which do not understand and value the basic principles of customer service and fundamental human right to safe foods may lose the business over time. Customers are great at spreading bad product reviews which inadvertently could result in destruction of brands. Food hygiene violations are also an offshoot of a poor FSC. Consumers can also contract FB illnesses from consumption of improperly stored and cross contaminated food products. Media monitoring systems are also now available which track FS violators and can tarnish the image of implicated brands [13]. It was estimated in 2012 that the average cost to a food company due to food product recalls could be as much as $10 M, minus the sales and brand loss which also accrue. Many recalls are associated with the identification of allergens undeclared in the labels and detection of food microbial pathogens as contaminants in food products. Thus, a wholesome FSC helps a food business facilitate the control and prevention of product deviations from set safety and quality regulatory standards. FSC can be used in certifying the effectiveness of food preparation, processing, handling, packaging, storage, and supply practices. Hence, it also impacts on recall severity and likelihood of a recall occurring [13].

In addition, research has also shown that influencing an individual's value system might in turn change their attitude toward an FS policy. Employee attitudes to affirmative action changed more when information was provided as a message associated with the value of equality as against similar message presented as a policy on affirmative action. Thus, associating FSC with personal belief systems is more efficient in bringing about behavioral changes in employees. Nonetheless, since the smallest signs of misdemeanor can induce widespread unwholesome behavior in other FS areas, measures must be in place within the FSC to properly and immediately address them [9].

12.4 Effect of unsafe foods

In 2003, about 30% of food-related outbreaks occurred in private homes in Europe [14], and millions of cases linked to ingestion of unsafe foods result in hospitalizations and deaths [2, 15]. As a result, several food products are recalled annually due to the presence of exogenous adulterants, allergens, pathogens, and/or their toxins. Unsafe foods are also produced from poor hygienic practices and may be linked to mistakes that happen during handling. Such scenarios can arise when food business or facility employees and handlers are not well trained, untrained or their FS and hygiene knowledge is not periodically updated, and the facility comes under regulatory auditing and inspections. In other words, food preparation processes and food business employees cannot be extricated from standard regulatory compliance practices and must also be prioritized. The knowledge base of food handlers on safety, hygiene and GMP is a constant determinant of FS and hence, must be periodically and upwardly reviewed and assessed [15, 16].

12.5 Causes of unsafe food

Food safety can be compromised with contaminants during harvesting, preparation, processing, production, packaging, storage, transportation, distribution, and sales. Contaminants can be physical, or biological, chemical, or physical [17]. Physical foreign contaminants include pest and/or their fecal droppings, plant stalks, plastic pieces, hair stands, glass, metal and stone or dirt pieces, fingernails, and jewelry pieces [7]. Chemical contaminants could be artificial or natural, and some examples are air, water, or soil chemical pollutants, veterinary drugs, pesticides, natural toxins, chemical migrants from food packages, food adulterants, and additives that are unregulated and cross-contaminants from processing [17, 18]. Biological or living contaminants are mostly implicated in FBDs and include rodents, microorganisms transmitted through blood, saliva, fecal and pest droppings, pests, and humans [19]. High protein and starch food environments, as well as food matrices with neutral pH, oxygen levels, water activity, and temperature between 5 and 60 °C for about ≤20 min favor bacterial growth and survival [20]. In a decade, the worst case of FBD was associated with the *Escherichia coli* O157:H7 strain outbreak which may have originated from Yuma, Arizona, the United States and lasted for about 2 months [21–23]. A death was recorded and about fourteen individuals developed kidney failure [21]. Foods can also be made unsafe due to negligent attitude to stipulated regulations.

12.6 Ensuring food safety

12.6.1 Setting food safety standards

The International Organization for Standardization (ISO) developed standards for FS like the ISO 22000 published in 2005, which is derived from the ISO 9000 standard. ISO 22000 stipulates regulations for FS management systems and incorporates interactive system management, communication, HACCP principles, and prerequisite programs. Food standards are built up and formed from varied FS concern aspects, as well as food-related sources over time, both geared toward providing and assuring finished foods and food products safety. Standards are reviewed every fifth year to make sure they remain relevant to ever progressive food businesses and the food industry [24].

12.6.2 How to make food safe

The acts and practices of preventing food contamination, separating cooked from raw foods to prevent cross-contamination, food processing for the adequate length, time, and temperature; appropriate storage of foods, and utilizing safe raw materials and water in production processes make up the five basic and central principles of food hygiene [25]. A complete adherence to these practices decreases the likelihood of occurrence of contamination incidents. Living and physical contaminants can be reduced with the use of well-sealed materials and containers in products packaging and storage to prevent the entry of air and moisture. Tools and surfaces must also be kept sanitary, debris-free and clean between batch preparations. Still, when all safety standards have been met, there remains a window of possibility for biological contaminants like microorganism to survive and grow during product storage. In other words, foods with short shelf lives should be kept at cold temperatures and consumed (usually between 1 and 7 days) before their expiry dates or consumed in 1–12 months if instantly stored and frozen [26, 27].

Perishable foods should be refrigerated within the first and second hour when temperatures are above 32.2 °C. Freezing and refrigeration temperatures also require regular checks with a thermometer. Refrigeration temperatures should be ≤4.4 °C and freezing temperature at ≤0 °F. Hot or cold geographical locations are also contributing factors to FS. Foods stored in exposed surroundings with pests and rodents must be inspected prior to consumption. All potential contaminations must be envisaged in determining if a food is safe or not since certain contaminations do not present visible signs on foods or change organoleptic properties of foods. Contaminated food products should be timely discarded, and the food environment assessed for additional contaminations [20]. The surrounding and storage conditions, storage technique employed, and type of food and its properties determine the shelf life (time within which the food is safe to consume). The likelihood of a food becoming unsafe to cause FBD increases the further the product is from its expiry or best before date [20].

12.7 Theoretical means of ensuring FSC compliance in food facilities

12.7.1 Shared values, beliefs, and norms

FSC must be imbibed and integrated into food establishments' norms and values. Sound culture and values need to then be passed down to new employees, stakeholders, and managements. The formal and informal transfers of wholesome FSC practices cannot be overemphasized. Documented company cultures still have to pass through human and individual translations over time before they become norms. When translations transform into beneficial or harmful demeanors they can still be shared and passed on to new entrants into a food company. Hence, while a food establishment's formal values may remain, the offshoot norms and behaviors can still be shared, and are largely unspoken, informal, unwritten, and undocumented [10].

12.7.2 Impacting mindsets and behavior

If food company belief and value systems and their needs thereof well comprehended by each member and employee, behaviors and mindsets are more easily affected and compliance to rules better achieved. A sustainable, strengthened and mature FSC is achievable if there is a basic comprehension that the food establishment's overall vision, mission and values affect the mindset of each company member. For instance, job description attached to every role must be clearly understood, and members must actively participate in defining and reviewing those roles. Again, employees must understand that specific roles should favorably contribute to the mission statement of the organization. The mindset with which employees view their employers or senior colleagues is a significant determinant of their commitment levels to ensuring FS. The food organization owners must therefore continuously liaise with its employees because they are important to the company's FSC [10].

12.7.3 Defining FSC throughout the food establishment

There is no one FSC that fits all food businesses; hence, FS must be clearly defined to every department and member within a company. Expectations and terms must also be plainly outlined and understood. In other words, roles and expectations will vary across the maintenance, production, purchasing and quality control and assurance departments. All departmental heads and members must then work to ensure functions align with company economies of scale, values, objectives and FS regulations. In case of smaller food business operators, owners must possess exemplary FS behaviors and significantly influence the business and employee FSC. Setting reliable direction and targets requires in-depth planning and thought for successful achievement of set FS goals. It also calls for an almost total comprehension of what the company considers a success in both the short and long run. Once company directions are set, they must be periodically reviewed to ensure that values evolve to maintain viability and sustainability. Food safety must be an integral aspect that is considered, discussed, and reviewed consistently in a food business's direction-setting activity.

12.7.4 Defining the business structure and purpose

A food business structure, mission, and vision are determined by the owner, its senior leadership team, stakeholders, or board of directors. The business purpose must reflect in the organization's media image communicated to its target market, and do not need to make specific mention of FS. However, the essential nature of their FS values must be demonstrated and communicated through their annual reports, websites, advertisements, agent or distributors, to all employees, and potential customers. Likewise, the business's core values surpass all business aspects and inform the FSC.

12.7.5 Commitment to leadership

The type and tone of a food establishment leadership also contribute to set the FSC direction by aligning with and supporting the overall company mission. From the owners to executives and employees in the production and other departments, all members are considered "leaders" in their own right since each one significantly affects the development of the organizational FSC. If resources (personnel, time, raw materials, finance, and equipment) are appropriately allocated, it portrays a leadership dedicated to FS. The leadership must also ensure the compliance of raw materials, foods and products, employee, and environment to set jurisdictional standards and policies. True compliance is however measured by the extent to which regulations are safely and positively translated into good production practices. An example is the clear translation of laws and procedures by unit heads and leaders into behavioral expectations for all employees. Accountable and credible leaders who commit to FS make up fundamental elements of a company's FSC [10].

12.7.6 Effective communication and messaging

The use of clear, effective, and consistent messaging techniques have been shown to favorably convey food companies expectations to staff and employees. Messaging is used to periodically update and remind staff of the establishment's FS practices and approaches. FS messaging aim to increase awareness, inform and educate both existing and new employees on the FSC of the organization. Each employee can then take responsibility for their role in protecting the company brands and consumers and ensuring FS. Food safety policy statements must also be always made accessible to all and regularly referred to in the establishment's communications. A type of messaging that explains the essence of policies put in place and how they must be complied with should be targeted toward all members, for example, permanent, contract and temporary employees and external stakeholders. Communication of an organization's FS expectations and practices using multiple media in multiple locations within the food establishment is also key to enhancing compliance. Use of specific, simple and relatable slogans, pictures, symbols, posters, and wherever the behavior is needed may ensure better compliance. Messages communicated will also have to be modified periodically to update employees on industrial FS needs. It is also important to introduce new posters, signs, symbols and other creative communication medium, and mix things up to continually to get employees attention. While words are good, they could be used minimally, as pictures could be more effective than words. Engaging employees to ask questions also makes them a part of the organizational FS solution [3].

12.7.7 The employees and people

Personal behaviors, habits, and actions are critical contributors to FSC and can increase or decrease potential risks to FBDs. "People" refers to each person that is an actor in a food industry, from farm to fork and includes marketers, distributors, salespersons, the customer service personnel, and other actors within a food chain. "People" also covers the recruitment processes, capacity-building and on-boarding, to employee empowerment and education. Incentivizing people and employee roles can also aid the establishment of a company's FSC, governing structure and metrics. A robust system which lays out consequences of bad practices and behaviors must also be set. Individual and group competencies in FS should also be focused on and elaborated to their most basic elements. This further provides every member with relevant tools to maintain a safe food environment. Such tools which upgrade skill sets may cover metrics, procedures, standards, knowledge, accountability, and standards, and the efficient maintenance of FS practices [16]. In addition, a food company's FSC is said to be "mature" when all stakeholders accede to and share the same FS goals, work synergistically to achieve company objectives and take responsibility for their roles in meeting and maintaining set standards. Therefore, under normal conditions, the typical members who constitute an FS group or team are not solely responsible for a company's FSC.

12.7.8 Food safety governance

Food safety should be an important part of a food company's FSC. Company standards must also align with international best practices. The components of FS governance include policies and standards, culture and behaviors, strategic direction, risk and issues management, and organizational structure and accountability. Results are best when the company's FSC is made formal and appropriately documented to show defined roles throughout the organization. Bigger food conglomerates may also make clear separations between their commercial and safety-oriented decision making processes. This serves to reduce potential conflicts of interest. A company's FS governing rules should also encourage whistleblowing to facilitate communication between employees and management, and employee education on protocols to follow in matters of ethical concern. Employees should also be convinced that their concerns would be seriously considered and appropriately investigated. Companies may also look into the creation of independent escalation routes which allow FS teams to expressly report to executive leaders instead of to senior operations personnel. A good FS governance should also aim to achieve the following:

a) **People empowerment**: FS governance should give people just enough authority, knowledge and capacity to act and positively impact the food company's ability to improve, adapt, maintain and sustain its FSC. Employees should be empowered enough to lead and drive positive change.

b) **Employee capacity building:** When an employee's capacity to work effectively is built by the company through periodic trainings and workshop, they may develop a greater sense of commitment toward the company's and mission to maintain and meet FS guidelines. Competency or capacity building is a developmental initiative that encompasses peculiar and technical FS activities alongside the acquisition of management and leadership skills (communications, negotiation, and problem-solving skills). Since capacity building is expected to enhance personnel behavior and performance, employee development should be monitored over a period of time. A food business typically adapts better to a changing food industry environment when its workers are empowered to assume challenging new roles [16]. It has also been suggested that education and training of employees should be based on ability to identify risks, as well as demeanors, tasks and topics of greater relevance which are more frequently associated with FBDs. Training curricular should also be made user friendly and simplified to improve employee comprehension [3].

12.8 Practical means to ensure FSC in food facilities

The behavior of food handler and employees determines the outcome of food and food products. Changes to FSC must begin at the senior levels of management since they have the last say over resources allocation. FSCs can be assured in food facilities through some of these practical ways. Foremost is the management's commitment to entrenching a sound and beneficial FSC within the company, and lead by example. Management's exemplary practices could be as basic as wearing appropriate personal protective equipment (PPE) and frequent hand washing whenever necessary. Second is establishing FS functions and incorporating them in employee job descriptions, from top to bottom. The use of SOPs and posters in work areas would also suffice as constant reminders of what is expected of them [9].

Third is the efficient communication of FSC to all staff and includes educating and informing food handlers on FS procedures to be followed and the consequences of poor FS practices. This is followed by the provision of resources for training of key personnel and may call for the assistance of FS consultants. The effective use of both physical and virtual or digital training avenues the extent to which a company's FSC can switch and evolve over time. Use of digital techniques can also extricate scenarios of potential document falsifications. Records that are time-stamped should also be used. This would make data comparison from varied sites and reviews easier for manager and supervisors from a single computer. The fifth step is to measure the level of success of a sound FSC in relation to employee objectives through recognition schemes. This culminates into the last step which is tailored toward the continued improvement of finished food products using the organization's certified, reviewed and updated hazard analysis and critical control point (HACCP) document. HACCP reviews are done annually or whenever circumstances call for it since regulations can change at any time [28].

12.9 The learning organization's contribution to FS compliance

Besides the frontline employees, training is also essential for mid- and senior-level supervisors and managers who also have FS-related training needs. Where management personnel were excluded from trainings, they show a lack of basic comprehension of FS risks. Thus, capacity building must be from top to bottom, through and across a food company. The extent of compliance to FSC depends on extent of compliance by peers and other businesses, trainings effectiveness, and other factors already discussed in this chapter. Besides FS training for "all," it is important to also work out the most efficient technique to use in training delivery and assessment for comprehension. Some of optimal methods for learning include mentoring, coaching, classroom delivery, self-directed study, and on-the-spot assessment and training. The content of trainings should be both practical and theoretical, and applicable and relevant to the food industry and role competencies. Trainers must be good communicators who possess the required and transferrable technical know-how. Trainings must be followed by periodic refresher courses. Training benefits include improvement in providing customized professional training and development, effective performance evaluation, clarified employee expectations and accountability, better recruitment procedure for selection of new staff, enhanced identification of capability gaps and skills, and improved succession planning. Other aspects of the learning organization which include programs and practices to impact FS compliance are discussed in the following paragraphs.

12.9.1 Food safety influencers

An FS influencer (FSI) is one who has some influence on FS behaviors or procedures and could adversely or positively impact a food company's FSC. Identifying FSIs and tailoring FS communications and trainings to incorporate FSIs could optimize employee compliance to FS regulations and programs. Some common influencing behaviors are the "carrot and stick" (I get punished or rewarded), "guilt and conscience" (I know it is the right thing to do), "herd effect" (everyone does it), "follow my leader" (my manager does it), and "I can be seen" (shine the light) behaviors. In a bid to gain better insight into the links that exist among employee behavior, FSI impact and consequences attached to behaviors, the "Antecedents, Behaviors, and Consequences," that is, the ABC model can be used. The antecedent takes place before behavioral traits begin to show and could be a communication, training, stated expectations, circumstances, policy and stimulus, past experience, or job aids. Many antecedents like skilled senior managers and leaders, built competencies, trust, confidence and openness, sufficient time, simplified protocols, tracking and trending, appropriate equipment and tools, and data measurement, could be used to align employee behaviors [10, 28].

12.9.2 Company incentives, rewards, and recognition programs

The food industry and company's recognition style and drive help to manage and guide desired FS behaviors. The approaches are structured to accommodate cultural differences within the organization. Some deterrents and incentives used to achieve consistent compliance are peer, corporate and employee recognition programs, praise, money awards, time compensations and holidays, demotions, promotions, learning from failures, team and individual awards, sharing best practices, and feedback systems (negative and positive).

12.9.3 Accountability programs

Defined responsibilities enhance employee accountability and adherence to FS system. When an employee can acknowledge and account for products, policies, actions, and decisions that fall within the purview of his/her specific role and position, they are also obligated to report and explain deviations from set regulatory standards. Employees must know whom and where to report FS-related issues that are beyond their responsibilities. Hence, accountability must be covered in a food company's structure and at all organizational levels.

12.9.4 Periodic performance assessment

Employee performance is assessed in line with FS policies, and expectations attached to roles. Good performance should be acknowledged, and improvements suggested and made where necessary. Assessments must aim to support the organization's continuous improvement, and not witch hunt employees. There exists a strong link between assessments made and behaviors which follow afterward. So, performance assessments and associated consequence or reinforcement systems should be carefully planned prior to implementation. FS performance measurements should also address product and process performance. Proactive and reactive measures achieve different objectives, so nature of performance measures should also be carefully considered. Proactive measures mitigate potential effects on the customer and move the risk a step away from them.

12.9.5 Proper FS documentations

Adequate and appropriate FS documentation enhances decision-making. Usually records from processes, trainings and production activities, as well as data on FS plans, protocols, expectations, are entered into books and forms and help to ascertain product consistency. Keeping updated documentations safeguards against sudden regulatory agency visits and eliminates sole reliance on employee knowledge. Food-safety-related documentations should be well kept, easy to comprehend, appropriate to the organization, up to date, not forged, and relevant to the industry, comprehensive. Food safety documents include standard operating procedures (SOPs), food industry safety forms, policies, and certifications, and FS manuals for varied schemes [28].

12.9.6 Adaptability schemes

Adaptability is defined as the extent to which a food organization can easily adjust to fluctuating conditions and influences within and outside the organization and industry. It also covers the company's response within its current state or its move to a new state. Adaptability is impacted by the company's FSC, and changes may or may not be expected, but the speed, nature and success of the company response depends on individual employee adaptabilities. Adaptability in a sound FSC is demonstrated through its skill in preparing for, anticipating and responding to unforeseen disruptions and changes to ensure the organization's ultimate survival. Important elements of adaptability include:

a) **Food safety current state and expectations:** If the food company's direction and expectations are stated clearly in its mission and vision, the business can better assess its current FSC against the expectations. The company can also better answer questions surrounding the level of use of FS practices in decision making at all levels by employees and in crisis situations if FS was truly a part of their vision statement.

b) **Crisis management and problem-solving:** This is linked with a business's level of anticipation and response to crisis given their laid down values and expectations. A good crisis management scheme would include a post-crisis review to enable sustained improvement and learning. Insights, near-misses, measures, and other occurrences determine the problem-solving capabilities of a business and the responses to identified challenges. An effective crisis management strategy must focus on determining root cause and enforcing control and preventive measures to combat them in the long run.

c) **Agility:** Agility implies being able to think and arrive at relevant conclusions rapidly. This concerns assessing threats and opportunities and adjusting strategies accordingly. A company's strategy can either enhance or hinder its agility and adaptability. Oversight of performance against expectations will inform agility.

12.9.7 Hazard and risk awareness strategies

This strategy separates FSC from the general culture of a food establishment. Hazard and risk (HAR) identification and mitigation are central to the efficient sustenance of FSC. It is essential that the technical and scientific fundamentals of HAR assessments are understood and accessible to all. To increase awareness and comprehension on HARs, food organizations must update current knowledge on the industry, that is, FS regulation changes, analytical advancements, and major technological changes. This will broaden awareness and understanding of potential risks and hazards. Evolutions in FSC would involve advancements that require baseline knowledge of HAR by employees, which is achievable by worker training and education. The focus of such evolutions goes beyond to meeting minimum regulatory specifications. It is important that all members of a food company are alert to their roles in potential HAR recognition. All require training in HARs peculiar to their role, but most importantly the technician, production floor staff and equipment operators. Trainings should be typically led by persons who spearhead the development of FSC values and norms and facilitated by teams in charge of FS risk evaluation and hazard analysis. Trainers and facilitators must communicate risks adequately and should be up to date on measurements of likelihood of occurrence and severity of effect. Levels of HAR awareness can be verified through audit drives, near-misses review, and behavioral observation and assessment [10].

12.9.8 Engaging employees

This embodies management's willingness to commit finances and formal trainings to boost employee FS practices. FS hazards continue to emerge and must be communicated to employees on a regular basis to inspire trust in the company's FSC. Employees can then better understand why and how accidents happen and how to prevent them, to safeguard theirs and consumer health. Production plant workers comprehend this all too well since they are monitored periodically to achieve related metrics. Still impacts of FS mishaps may not instantly affect them. Hence, employees must be actively engaged to fully comprehend the importance of control and preventive scheme and the likely consequences which may accrue when strategies fail or are deviated from. In the same vein, support must however be available to them to fully implement remedial actions like product isolation, production line shut down, and report of near-miss events. These must be backed by company and management resources to address resultant actions. Likewise, the work environment should encourage the report of improper behavior, FS potential hazards, and sharing of experiences.

12.10 Additional ways to enhance food employee compliance

Food business managements are burdened with the task of ensuring employees comply with FS standards and values. Management can also achieve compliance by monitoring employee practices and behaviors while following guidelines. Monitoring strategies can be active or reactive. Active monitoring (AM) ensures that FS guidelines are appropriately adhered to in respective workstations prior to events such as illness, incidents and accidents occurring. It aims to detect standardized compliance and maintain and recognize good performance. It also identifies deviations from and noncompliance with set standards in order to identify the cause and proffer appropriate control measures to resolve shortfalls.

AM can be done through systematic and safety inspections. Systematic inspections focus on one or more of the 4 Ps which constitute the workplace or premises, production plant, personnel behavior and work ethics, plant or equipment, using different inspection schemes at regular intervals. For example, daily checks on machinery and monthly housekeeping inspections. Safety inspections involve comparing actual values and records to company and legal standards. It is a routine check to determine if general standards are acceptable or not, and if control measures are required [29, 30]. Safety sampling is another AM technique that checks compliance with specific workplace guideline by assessing samples only. The rule of thumb is that larger sample sizes give the truest and most significant representation of the workplace. Benchmarking, safety tours, or surveys also fall under AM [31].

Reactive monitoring (RM) utilizes illness, incident and accident records as safety, health and performance indicators to show lagging areas. Its weakness is twofold. In the first instance, it is not proactive, focuses on setbacks and assesses failures. Secondly, since an event had already occurred certain factors need to be adjusted afterward and not before. On the other hand, AM praises and rewards identified good performances and are proactive in nature. Nonetheless, RM remains a useful tool if used together with some AM techniques. RM methods include determining events root causes and resolving the event. Resolutions can be reached through employee trainings around the cause of the event and AM exercises. One RM method encourages the investigation, recording and reporting of events or incidents. Examples are environmental monitoring studies which highlight illness prevalence, potential safety and health risks and proffer solutions [32, 33]. However, the second RM method focuses on data collection and use of statistical analysis to identify potential event trends [34].

In light of the forgoing, it is clear that FSC and employee compliance levels are closely linked. While several practical and theoretical measures exist to ensure workers comply with general FS standards and organizational FSC, the practical and on-site applications are far from being realized, especially in food businesses in developing regions of the world. This is further worsened by the fact that most food businesses in developing economies fall under the unregulated sectors of the food supply chain and are therefore outside standard regulatory lines of enforcement and implementation. Nevertheless, given the potential consequences of unsafe foods in form of adverse health effects and economic losses, all governments and food industry stakeholders must work in consonance to continuously develop wholesome FSCs that invariably and positively impact on employee compliance.

References

[1] Ucar, A., Yilmaz, M. V., & Cakiroglu, F. P. (2016). Food safety – Problem and Solution. *Significance, Prevention Control Food Relative Distribution*, 3.

[2] CDC. (2007). Preliminary foodnet data on the incidence of infection with pathogens transmitted commonly through food – 10 States, United States, 2006. *Morbidity and Mortality Weekly Report, 56*, 336–339.

[3] Yiannas, F. (2009). Food Safety Culture: Creating a behaviour-based food safety management system. Food Microbiology and Food Safety. Arkansas USA: Springer Science and Business Media.

[4]   Aruwa, C. E., & Akinyosoye, F. A. (2015). Microbiological assessment of ready-to-eat foods (RTEs) for the presence *Bacillus* species. *Journal of Advances in Biology & Biotechnology*, *14*, 145–152.

[5]   Texas food establishment Rules. (2015). Texas DSHS website: Texas Department of State Health services, 5.

[6]   Shiklomanov, I. A. (2000). Appraisal and assessment of world water resources. *International Water Resources Association*, 11–32.

[7]   ACSNI Human Factors Study Group. (1993). Third report – Organising for safety HSE Books.

[8]   Greenlee, S. (2019). The importance of food safety cultures. Creating a safe food culture: The role of attitude and behaviour.

[9]   Yiannas, F. (2015). *Food safety = behavior. food microbiology and food safety*. New York USA: Springer Science and Business Media.

[10]   GSFI. (2018). A culture of food safety. A position paper from the global Food safety Initiative, 1.0.

[11]   Queensland Health. (2019). Foodborne disease outbreaks, (Accessed July 8, 2021, at https://www. health.qld.gov.au.Updated 26 July 2019).

[12]   Scallan, E., Hoekstra, R. M., Angulo, F. J., Tauxe, R. V., Widdowson, M., & Roy, S. L. (2011). Foodborne illness acquired in United State-major pathogens. *Emerging Infectious Diseases*, *11*, 7–15.

[13]   Food Marketing Institute. (2012). Food politics. Food navigator reports.

[14]   World Health Organization – WHO. (16 December 2003). Several foodborne diseases are increasing in Europe. World Health Organization. Accessed June 27, 2021. Archived from the original on 16 April 2005.

[15]   World Health Organization – WHO. (2021). Food safety and foodborne illness. World Health Organization. Retrieved 10 December 2010. Accessed June 22.

[16]   Aruwa, C. E., Akindusoye, A. J., & Awala, S. I. (2017). Socio-demographic characteristics and food hygiene level assessment of food handlers in cafeterias around a Federal University in Nigeria. *Journal of Scientific Research & Reports*, *14*, 1–9.

[17]   Campden BRI. (2018). Physical contaminants in food, identification and prevention at Campden BRI, (Accessed June 17, 2021, at https://www.campdenbri.co.uk. Retrieved 10 June 2018.

[18]   Food Safety Magazine. (2018). Modern analysis of chemical contaminants in food, (Accessed July 13, 2021, at https://www.foodsafetymagazine.com.)

[19]   Australian Institute of Food Safety. (2018a). What are the different types of food contamination?, (Accessed July 21, 2021, at https://www.foodsafety.com.au/faq/what-are-the-different-types-of-food-contamination.)

[20]   Australian Institute of Food Safety. (2018b). Food safety and the different types of food contamination, (Accessed July 19, 2021, at https://www.foodsafety.com.au/blog/food-safety-and-the-different-types-of-food-contamination.)

[21]   Food Safety News. (2018). North Dakota confirms *E. coli* outbreak case; 26 states hit. Food Safety News, Accessed July 16, 2021.

[22]   Hoffman, J. (2018). Romaine riddle: Why the *E. coli* outbreak eludes food investigators. Accessed June 18, 2021.

[23]   The Mercury News. (2018). FDA may never find source of Romaine *E. coli* outbreak, Accessed July 10, 2021.

[24]   International Standards Organization – (ISO). (2017). ISO 22000 Revision. International Organization Standards ISO, Accessed July 2, 2021.

[25]   World Health Organization – WHO. (2010). Prevention of foodborne disease: Five keys to safer food. World Health Organization, Accessed July 17, 2021.

[26]   Zeratsky, K. (2018). How long can you safely keep leftovers in the refrigerator? Mayo Clinic. Katherine Zeratsky, R.D., L.D. November 21.

[27]   FoodSafety.gov. (2021). Storage times for the refrigerator and freezer. https://en.wikipedia.org/wiki/Food_safety–cite_note-13 Accessed July 4.

[28] Aruwa, C. E. (2020). Food regulations and governance. In *Food science and technology: trends and future prospects, part VI, food business: entrepreneurship and regulation* (pp. 485–510). Germany: DeGruyter.

[29] Aruwa, C. E., & Ogundare, O. (2017). Microbiological quality assessment of pupuru and plantain flours in an urban market in Akure, Ondo State, Southwestern Nigeria. *OA Library Jo, 4*, e3783.

[30] Aruwa, C. E., & Farotimi, A. R. (2019). Microbiological and proximate analyses of lebanese bread (Pita) from Akure metropolis. *South Asian Journal of Research in Microbiology, 3*, 1–11.

[31] Zutshi, A. (2019). How to develop well-informed and assured teams to embed food safety compliance in the UK catering and hospitality industry. *Perspectives in Public Health, 139*, 294–295.

[32] Dada, E. O., & Aruwa, C. E. (2016). Asymptomatic bacteriuria prevalence among primary school children in the Federal University of Technology, Akure (FUTA), Ondo State, Nigeria. *Journal of Applied Life Sciences International, 4*, 1–8.

[33] Dada, E. O., & Aruwa, C. E. (2014). Microorganisms associated with urine contaminated soils around lecture theatres in Federal University of Technology, Akure, Nigeria. *International Journal of Applied Microbiology and Biotechnology Research, 2*, 79–85.

[34] United States Food and Drug Administration. (2015). Basics for Handling Food Safely. USDA Food Safety and Inspection Service. Accessed July 16, 2021.

Jane Misihairabgwi*, Maria Angula, Elizabeth Sivhute, Jose Lorenzo
and Ahmad Cheikhyoussef

Chapter 13
Natural food toxicants and health implications

Abstract: Despite increased awareness of the importance of consuming safe food, concerns regarding food contamination have largely been directed at exogenous chemicals in the food chain such as artificial or synthetic additives, agrochemicals, environmental contaminants, or adulterants. Comparatively, little to no attention has been directed to toxic and potentially hazardous naturally occurring food constituents and chemicals which naturally occur as contaminants in foods. Such compounds, referred to as natural food toxicants, have been associated with adverse human health complications. Acute and chronic toxicity effects of some inherent plant, animal, and fungal constituents, as well as naturally occurring toxic contaminants have been researched for some time and current research continues to avail updated information on the properties, toxic effects, mechanism of toxicity and causal association between some natural constituents and contaminants and their negative health effects. Additionally, advances have been made in developing methods to rapidly detect, identify, and quantify natural food toxicants to inform regulation and mitigation measures. This chapter summarizes and discusses the major toxic food plant and animal constituents, naturally occurring toxic food contaminants and toxicants produced on food processing that pose health risks to consumers.

*Corresponding author: Jane Misihairabgwi**, Department of Human, Biological and Translational Medical Sciences, School of Medicine, Faculty of Health Sciences and Veterinary Medicine, University of Namibia, Private Bag 13301, Windhoek, Namibia, e-mail: jmisihairabgwi@unam.na
Maria Angula, Department of Human, Biological and Translational Medical Sciences, School of Medicine, Faculty of Health Sciences and Veterinary Medicine, University of Namibia, Private Bag 13301, Windhoek, Namibia
Elizabeth Sivhute, Department of Food Science and Systems, School of Agriculture, University of Namibia, Windhoek, Namibia
Jose Lorenzo, Centr Centro Tecnológico de la Carne de Galicia, Adva. Galicia n° 4, Parque Tecnológico de Galicia, San Cibrao das Viñas, 32900 Ourense, Spain; Área de Tecnología de los Alimentos, Facultad de Ciencias de Ourense, Universidad de Vigo, 32004 Ourense, Spain
Ahmad Cheikhyoussef, Science and Technology Division, Multidisciplinary Research, Centre for Research Services, University of Namibia, Private Bag 13301, Windhoek, Namibia

https://doi.org/10.1515/9783110748345-013

13.1 Introduction

The World Health Organization (WHO) defines natural toxins (NTs) as secondary metabolites produced by living organisms with poisonous and harmful impacts on human or animal health when consumed [1]. A large variety of plants, marine and aquatic animals, fungi, and algae produce toxins naturally to serve specific functions, which include chemical defense against insects, microorganisms, and predators [1–4]. The NTs, which exhibit diverse chemical structures, nature, and toxicity, generally do not harm the producing organisms but can harm other organisms and the environment [1, 2, 5, 6].

According to Taylor and Helfle [7], natural toxicants in foods may be classified as either food constituents which inherently occur in plant, animal, or fungal foods, or naturally occurring contaminants that are produced in foods through microbial and insect infestation. Contaminants of biological origin, which remain in foods after destruction of the biological source, exhibit natural occurrence in the foods and can therefore be considered as natural food toxicants.

Both acute and chronic health effects can result from exposure to natural food toxicants, with chronic toxicity being more likely due to poor dietary diversity [7, 8]. Clinically, acute toxicity symptoms generally include mild gastrointestinal and neurological disorders, respiratory paralysis and fatality, while chronic toxicity is associated with carcinogenic and mutagenic effects [4].

Toxicity from inherent toxins in foods is largely dependent on the amounts of the foods consumed, individual susceptibility to the toxins, and food processing procedures employed. Serious health risks arise when the foods are consumed in sufficient quantities to cause toxicity, are inadequately processed, or when various processing procedures result in production of toxic products [7]. Commonly employed food processing procedures such as chopping, mixing, boiling, and frying, which improve the organoleptic properties of foods, may result in chemical reactions which enhance toxicity. Novel food processing procedures and culinary practices without consideration of natural food toxicant behavior poses health risks to consumers. Regulation of natural food toxicants requires up to date knowledge of their origin, properties, mechanism of action, detection, and analysis to ensure consumer protection. This chapter presents an up-to-date discussion of both the major inherent NTs in food and naturally occurring toxic contaminants, focusing mainly on their properties, mechanism of toxicity, health effects, processing effects, mitigation, and advances in detection and analysis.

13.2 Natural toxins in plant foods

Numerous edible plants contain diverse NTs, with the toxin variation being dependent on factors including the plant species, growth conditions, and the geographical factors in the environment where the plant grows [1, 6]. Of over 300,000 plant species worldwide, only hundreds are edible and the edible plant parts, which may contain NTs, include foliage, fruits, roots, and tubers [4–6]. The toxins may be produced to serve as defense against microbial threats, insects, and predators, or may be present owing to natural selection and/or new breeding methods associated with enhanced protection [6]. When some of the food plants are consumed in excess or without appropriate processing, acute poisoning from NTs is a huge concern. Toxic effects range in severity from mild vital organ damage symptoms, carcinogenicity, and death, occasionally [4]. While poisoning in humans may also arise from wild plants not intended for human consumption due to wild food picking, or to improper identification of edible plants, this chapter focuses on natural toxicants in foods meant for human consumption. Widely distributed natural plant constituents possessing potent toxicity and currently of major concern regarding inflicting toxicity in humans include cyanogenic glycosides, alkaloids, furocoumarins and lectins, which will be discussed in this chapter.

13.2.1 Cyanogenic glycosides

Cyanogenic glycosides are natural plant toxins which plants synthesize for protection from herbivores and pathogens [9]. The nitrile-containing plant secondary metabolites, which consist of sugar moieties, are found in more than 2,500 plant species, including cassava and lima beans, which are widely consumed in large quantities in some countries [10]. Approximately 50 cyanogenic glycosides are known, and of these, linamarin, a β-glycosylated acetone cyanohydrin present in the leaf and root portions of cassava plants, is of major importance regarding toxicity arising from plant food consumption [11]. Amygdalin, which mainly occurs in inedible plant parts such as apple, peach, and apricot seeds, has recently become a toxin of concern due to its occurrence in processed fruit juices [5].

Cyanogenic glycosides in cassava exhibit potential toxicity arising from products of enzymatic degradation by linamarase (β-D-glucosidase), an enzyme found in cassava plant tissues. Linamarase catalyzes cleavage of the β-linkage of linamarin, forming acetone cyanohydrin and glucose. Acetone cyanohydrin subsequently liberates hydrogen cyanide (HCN) which exhibits higher toxicity [9, 12]. HCN is therefore produced during food processing procedures such as cutting and chopping of cassava leaves and roots, presenting toxicity on consumption of inadequately processed products. Consumption of fresh cassava can also cause cyanide toxicity following decomposition of linamarin by action of β-D-glucosidase enzymes of gut flora and cyanide absorption in the gastrointestinal tract [11–14].

There are reports of acute cyanide intoxication and adverse long term health issues, which include goiter, cretinism, tropical neuropathy, and *konzo* (spastic paraparesis) in Africa and South America following consumption of cassava [15–17]. Deaths and various serious health complications have occurred after ingesting fruit pits, which contain higher levels of cyanide than in cassava [7, 10]. Overconsumption of cyanogenic plants causes cyanide poisoning, manifesting with symptoms including tachypnea, dizziness, vomiting, headache, and delirium [17, 18]. Cyanide inhibits cytochrome c oxidase, an important enzyme required in the mitochondrial respiratory chain [16, 17].

Efforts to detoxify cassava roots include employment of processing treatments like peeling, steeping, wet pounding, sun drying, fermenting, and boiling, as well as application of chemical treatments [19, 20]. The toxic cyanide in cassava can be removed more effectively by soaking in water or cutting into small pieces before cooking. Many species of cassava and fruit seeds are inedible without prior processing due to high levels of toxic cyanogenic glycosides [7, 20]. Bitter apricot seeds which are toxic when eaten raw, are safe for consumption after boiling and have been used in the preparation of some Chinese soups [20].

To ensure consumer safety, it is vital that foods are assessed for their cyanogenic potential and cyanide contents. Based on the cyanohydrin levels in cassava pulp, Kobawila [21] classified cassava toxicity as high (>100 mg HCN/kg), moderate (50–100 mg HCN/kg) and nontoxic (<50 mg HCN/kg pulp). According to FAO/WHO, in 2017 [22], a maximum of 10 mg HCN/kg/dry matter was recommended as safe. Kaiser and coworkers [5] advocate for plant breeders to consider inherent toxins and desirable characteristics of plants to ensure food safety.

Advances in sample extraction and analytical techniques have been applied to enable risk assessment from cyanogenic glycosides. A method for ultrasonic assisted sample extraction has been developed by Zhong and coworkers [23], while the analytical procedures high-performance liquid chromatography (HPLC) and liquid chromatography tandem mass spectrometry (LC–MS/MS) have recently been widely applied in separating and quantifying individual cyanogenic glycosides [24].

13.2.2 Alkaloids

The term "alkaloids" is collectively used in the description of a group of chemicals characteristically exhibiting alkali-like properties and structurally possessing one or more nitrogen atoms in a heterocyclic ring [25]. Owing to the presence of alkaloids in many plant species, humans are consistently exposed to alkaloids upon plant food consumption [7]. Despite alkaloid levels in the edible plants being usually not high enough to bring forth acute toxicity, the chronic toxicity arising from consistent exposure dietary exposure to low levels requires investigation. Of great concern regarding human health are the glycoalkaloids, which include solanine and chaconine in potatoes, and the pyrrolizidine alkaloids (PAs) [25].

13.2.2.1 Glycoalkaloids

Glycoalkaloids are steroidal glycosides containing nitrogen, which are widely produced as secondary metabolites in edible plants of the genus *Solanum*, which encompasses tomato *Solanum lycopersicum* (tomato), *Solanum tuberosum* (potato), and *Solanum melongena* (aubergine) [8, 25]. The principal glycoalkaloids in potatoes, solanine, and chaconine, normally occur at low levels in all potato tubers, being concentrated in potato skin. Green or blighted potatoes, and potato sprouts can contain high glycoalkaloid levels [7].

Glycoalkaloid toxicity in humans, including solanine poisoning attributed to consumption of blighted, or sprouted potatoes, has been reported as early as two decades ago [7, 25]. The toxicity arises from the membrane-disruptive and acetylcholinesterase inhibitory properties of the glycoalkaloids. Toxicity symptoms posed by solanine glycoalkaloids include dyspnea, drowsiness, vomiting, and diarrhea [8, 26].

A potato tuber usually contains a total glycoalkaloid content of only 20–130 mg/kg tuber, which would not result in acute toxicity [7]. However, elevated glycoalkaloid levels are found in green tubers, sprouts, leaves, and tubers of certain wild potato species [7, 25]. Toxicity in humans has resulted from consumption of parts of potatoes containing 200–800 mg/kg tuber glycoalkaloid content [7, 25]. Given the increasing tendency towards consumption of potato peel, which is perceived as a rich source of dietary fiber and vitamins, exposure to glycoalkaloids may be on the rise. Toxicity can occur from cooked potato parts since glycoalkaloids are neither destroyed by heat nor decreased by washing or soaking. Via the oral route, glycoalkaloid toxicity in humans occurs at a level of 2–5 mg/kg body weight [7].

13.2.2.2 Pyrrolizidine alkaloids

PAs are toxic plant secondary metabolites, produced primarily by plants of the Boraginaceae, Asteraceae, and Fabaceae families, as a shield from herbivores [7, 27]. They frequently occur in edible plant species and have recently been detected in salads, honey, milk, meat, and eggs [28]. Structurally, pyrrolizidines are bicyclic, constituting two fused five pyrrolidine rings, sharing a common nitrogen atom. PAs may also occur in plants as their corresponding *N*-oxides (PANOs) [27].

PA/PANOs naturally occur in plants of the families of Asteraceae, Boraginaceae, Apocynaceae, and Fabaceae [29]. More than 660 known PA/PANOs are estimated to occur in diverse compositions in 3% of all flowering plants, with the toxins being mainly concentrated in seeds, and lower amounts being found in leaves, stems, and roots [30]. About half of more than 350 different PAs described to date are assumed to be hepatotoxic [31]. Variation in PA contents in plants is influenced by species, plant part, stage of growth, and environmental factors including soil and climatic conditions [32].

PA/PANOs primarily occur as contaminants in culinary herbs and spices, honey, or grain-derived products, and humans get exposed to them through consumption of these contaminated food products [33, 34]. In a study carried out in Europe, out of 305 herb and spice samples tested, 178 (58.4%) contained one or more PA or PANO [33]. Additionally, the same study recorded the highest quantity of 24.6 mg/kg in oregano spice. Contamination of honey by PAs is common, with retail honey produced using echium and other PA producing plants reportedly containing up to 2,000–4,000 µg of dehydro-PAs/kg [33, 35].

Following ingestion, PAs are hydrolyzed to toxic necines and necic acids in the liver, or oxidized to form PANOs or pyrrolic products, which are biological alkylating agents that lead to tissue damage and induce genetic mutations [3, 36, 37].

There have been extensive reports of PA intoxication via contaminated teas, cereals, salads, and cereals, with symptoms of acute toxicity including hepatotoxicity, vomiting, and diarrhea [27, 31]. Even though PA levels are frequently not sufficient to elicit acute poisoning, the quantities taken frequently exceed maximum allowable regulatory limits and may lead to chronic diseases [27]. Chronic intake of low doses of PAs from herbal products is associated with hepatotoxicity, predominantly manifesting with cirrhosis and cancer [27].

Various processing effects on PAs have been reported [35, 39]. Kaltner and coworkers [35] reported decreased levels of PANO in honey following chemical derivatization or dimerization due to the action of bee digestive enzymes. However, possible detoxification of the PAs could not be concluded [35]. As much as foreign seeds, among which are some PA containing seeds, can be selected for discard during grain sorting, chronic poisoning from PAs remains possible from heavily contaminated grain where sorting does not eliminate PA occurrence in the cleaned grains [39].

Passage of low amounts of PAs into milk has been reported when cows and goats were fed PA containing plants, while no PAs were quantified in milk products such as yoghurt, cheese, or infant formula [34, 40]. De Nijs et al. [41] found out that, pasteurization as well as fermentation procedures during production of cheese and yoghurt led to a decrease of PA/PANOs.

Information regarding the maximum regulatory limits for PAs in food worldwide is sparse, although discussions have been held on possible regulatory levels of PAs in several foods such as teas, honey, herbal infusions, honey, and food supplements [42]. For herbal infusions such as rooibos, chamomile and peppermint, the proposed maximum limit has been set at 400 µg/kg with 200 µg/kg being set for some other herbal infusions [43]. A PA limit of 4 µg/kg has been set by the EU [43].

Challenges lie in determining the toxicity dosage for various PA types, given that even within the same types of PAs, variation exists in the levels inducing toxicity. Ma and coworkers [28] advocate for a systematic assessment system for determining toxic levels of PAs in foods and medicinal herbs, predicting PA potency, and establishing fast and efficient detection methods. Conventionally, the spectrophotometric detection methods used for dehydro-PAs and PANOs are limited to the quantification of total PA contents but not individual PAs. Recently, liquid chromatography tandem mass spectrometry (LC-MS/MS) has become the principal method for the simultaneous detection and quantification of PAs and PANOs while NMR is employed for structural identification [44, 45], providing important baseline information for further toxicological investigations. Qie et al. [45] have recently developed a simple and effective method for analysis of several toxic alkaloids in forage grass using the Quick, Easy, Cheap, Effective, Rugged, and Safe (QuEChERS) method and LC-MS/MS.

13.2.3 Furocoumarins

Furocoumarins, which are produced as plant self-defense chemicals against insect predation, fungal invasion, bacterial attack, and physical damage, occur naturally in several edible plants in the Umbelliferae and Rutaceae families, and possess phototoxic properties [1, 46, 47]. Lemons, limes, and grapefruits are examples of fruits of the Rutaceae family, which contain furocoumarins, while the juice, oil and flesh of oranges contains relatively low concentrations of furocoumarins [46, 47]. Carrots, parsnips, celery, and parsley are examples of umbelliferous plants containing furocoumarins [1, 46–48]. Due to increased consumption of a wide range of fruit and vegetable parts in diverse fruit and vegetable smoothies and salads, furocoumarins have received much attention because of their strong photoactivity and ability to intercalate DNA under UV light resulting in genetic mutations, carcinogenicity and phytophotodermatitis [47].

Furocoumarins' natural toxicity is based on direct contact and photoactivation. Numerous experimental studies have revealed the phototoxicity and photocarcinogenicity of furocoumarins. The compounds are photoactive and can intercalate DNA and induce mutation [47]. This happens after their irradiation with UV light and undergoing photoactivation, making them highly reactive. In the reactive state, mechanisms by which furocoumarins induce toxicity include DNA-adduct formation, protein denaturation, saturated fatty acid cycloadduct formation and formation of cellular damage inducing reactive oxygen species. Cross-linkage is associated with cutaneous phototoxicity or carcinogenicity [49]. These reactions have also been reported after consumption of large quantities of certain vegetables with high levels of furocoumarins [47]. Psoralen and angelicin are examples of furocoumarins exhibiting photoactivity, with psoralen forming both monoadducts and interstrand DNA cross links under UV irradiation, due to its linear structure. Angelicin, on the other hand, owing to its steric structure, only forms monoadducts, which can be rapidly repaired by the cells, resulting in its lower phototoxicity relative to psoralen [49, 50, 52].

Production of citrus fruit juice using citrus fruit peels can lead to dietary exposure to furocoumarins, owing to the presence of bergamottin (bergamot) in citrus fruit peels, which inhibits some cytochrome P450 isozymes, preventing metabolism of some drugs and consequent high levels of the drugs in the bloodstream [51, 52]. Bergamot is also phototoxic, leading to skin toxicity upon exposure to sunlight [53]. Nevertheless, the levels of furocoumarin toxins diminish by cooking, microwaving, or boiling. The study of Ypshihiro and Kiminori [54] revealed that heat treatment of grape juice reduces the concentrations of the furocoumarins bergamottin and 6',7'-dihydroxybergamottin, hence eliminating the potential for drug interactions.

Due to lack of information regarding the actual amounts of furocoumarins in foods, currently no specific regulations focused on the presence of furocoumarins in food exist.

Historically, furocoumarin analysis was done via thin-layer chromatography (TLC) and HPLC [47]. Recently, a QuEChERS method has been developed and is used to extract furocoumarins from food samples by acetonitrile extraction and analysis via ultraperformance liquid chromatography, coupled with tandem mass spectrometry (UPLC-MS/MS) [55]. Zhao et al. [56] recently developed a method for to simultaneously detect and quantify coumarins, furocoumarins, flavonoids, and phenolic acids.

13.2.4 Lectins

Lectins are proteins that bind to specific carbohydrate moieties through specific existing domains. Their functions include recognition, interaction with, and reversibly binding to specific sugar moieties [57]. The proteins were originally detected in castor beans and are now currently known to be present in many other bean types, including green, red and white kidney beans as well as in grain products [7].

Exposure to high levels of lectins, mostly from legume seeds, causes gastroenteritis by destroying gastrointestinal tract epithelia, interfering with cell division, causing red blood cell agglutination, and damage to the kidney, liver, and heart [57, 61]. Acute toxicity from lectin exposure manifests with symptoms such as abdominal pain, vomiting, and diarrhea [57].

Culinary practices play an important role in determining exposure to dietary lectins. Consumption of foods high in lectins, without appropriate processing procedures such as cooking, baking or extrusion poses health risks [7]. To prevent exposure to the high concentrations of lectins in legumes such as beans, it is advisable to soak them in fresh water before cooking then and boil them thoroughly at high temperature [1]. Recent culinary trends such as the inclusion of partially cooked beans in salads, mainly red kidney beans, poses risks for toxicity and consumer awareness is necessary in this regard.

13.3 Natural toxins in animal foods

In animal foods, NTs may be metabolic products or chemicals passed along the food chain [1, 3]. Although it is rare to get food poisoning after consumption of terrestrial animals, avoiding the toxin containing tissues and poisonous glands is important when these animals are used as food [1]. Poisoning due to marine toxins, which accumulate in fish and other seafoods, occurs in many parts of the world [3].

13.3.1 Aquatic biotoxins

Natural toxins occur in fish and shellfish, with most of the toxins being produced by marine algae (phytoplankton) [8]. The phytoplankton associated toxins are referred to as phycotoxins. "Red tides" are toxic red-brown pigmented phytoplankton associated with poisoning, although incidences of poisoning have also been reported in the absence of the red tides [8].

13.3.2 Marine algal toxins

Marine toxins are naturally occurring compounds that are produced by aquatic microorganisms [58]. The algal toxins are behind a wide range of human health conditions linked to consumption of contaminated seafood, of the skin when it gets in contact with contaminated water, or the effects of aerosolized toxins on respiration [59]. Such health conditions are inclusive of ciguatera and paralytic shellfish poisoning (PSP) effects, which are acute neurologic diseases, and chronic dementia resulting from exposure to domoic acid (DA) [60]. Identified marine biotoxin classes are saxitoxin (STX), DA, ciguatoxin (CTX), brevetoxin (BTX), tetrodotoxin (TTX), okadaic acid (OA), azaspiracid (AZA), and palytoxin (PLTX) [58, 61]. Changes in climatic conditions and increase in availability of nutrients are reportedly among the main factors pushing expansion of marine biotoxins to new areas [61]. However, other factors that led to the increased scientific interest in these biotoxins could be the extreme biological invasions and the development of sensitive analytical methods such as nontargeted high-resolution mass spectrometry [62–64], whole-genome sequencing [65], short sequence tags [66], chemical processes such as cationization [67] and analytical instrumentation (LC-MS/MS) to detect very low thresholds of these biotoxins [61]. Although several challenges been encountered owing to the absence of certified standards and sensitivity dearth of the high resolution MS, significant improvements have been reported [68]. Some of the reported toxic diseases associated with marine algal toxins are listed in Table 13.1.

Table 13.1: Toxic diseases associated with marine algal toxins [58, 59].

Diseases	Causative organism	Primary vector	Toxin	Symptomatology	Pharmacologic target
PSP	*Alexandrium* spp. *Gymnodinium* spp. *Pyrodinium* spp.	Shellfish	Saxitoxins	– Gastrointestinal problems – Paralytic phenomena – Recovery or death	Voltage-gated sodium channel Site 1
NSP	*Karenia brevis*	Shellfish	Brevetoxins	– Gastrointestinal and neurological problems – Respiratory problems – Recovery or death	Voltage-gated sodium channel Site 5
CFP	*Gambierdiscus toxicus*	Reef fish	Ciguatoxins	– Gastrointestinal problems – Cardiovascular or neurological problems	Voltage-gated sodium channel Site 5
ASP	*Pseudo-nitzschia* spp.	Shellfish	Domoic acid	– Gastrointestinal and neurological problems – Cardiac or respiratory problems – Recovery or death	Glutamate receptors
DSP	*Dinophysis* spp. *Prorocentrum* spp.	Shellfish	Okadaic acid Pectenotoxin Yessotoxin	– Gastrointestinal problems – Gastrointestinal problems – Gastrointestinal problems	Ser/thr protein phosphatases
	Alexandrium spp. *Karenia* spp.		Cyclic imine	– Lack of observations in humans	
	Palythoa spp. *Ostreopsis* spp.		Palytoxin	– Gastrointestinal problems – Muscle and cutaneous problems	

13.3.2.1 Saxitoxins

Saxitoxins, otherwise called paralytic shellfish toxins (PSTs), are neurotoxic heterocyclic guanidines mainly produced by dinoflagellates from three genera: *Alexandrium*, *Gymnodium*, and *Pyrodinium* [58]. PSTs cause human intoxications following consumption of contaminated molluscan shellfish, resulting in neurotoxicity. Saxitoxin infection leads to PSP syndrome [58]. The mechanism of toxicity is via saxitoxin binding to voltage-gated sodium channels, blocking neuronal activity, with the peripheral nervous system in humans being the primary site of action primary site of action [59].

Toxicity symptoms begin 30 min to several hours following consumption of contaminated shellfish [58]. Symptoms of mild intoxication include tingling sensation or numbness of the perioral area, headache, dizziness, nausea, and to a lesser extent, gastrointestinal illness [58]. Severe PSP has, in some cases, resulted in ataxia, incoherence, and weakness [59]. These symptoms may be accompanied by respiratory difficulty which may lead to respiratory arrest and death, unless respiratory support is provided [59]. Death may occur as early as 3–4 h after ingestion of contaminated seafood, with no antidotes being available [58]. PSP may, however, be prevented by large-scale, proactive monitoring programs, and rapid closures of harvest in areas containing dinoflagellate algal blooms [59].

13.3.2.2 Domoic acid

The naturally occurring excitotoxin, DA, originally isolated from the marine red alga *Chondria armata*, is also produced by the toxic diatoms *Nitzschia*, *Pseudo-nitzschia*, and *Amphora* [58], which are found in polar, temperate, subtropical, and tropical regions. The diatoms are responsible for many toxic blooms worldwide affecting coastal environments in Europe, America, or Oceania [58]. Amnesic shellfish poisoning (ASP) results from consumption of shellfish such as oysters, clams, mussels, and scallops. Sea fish such as sardines, other sea animals such as squid, lobster, and crab, and anchovies are also vectors for DA contamination [59]. ASP is symptomized by gastrointestinal disturbances manifesting as nausea, vomiting and diarrhea, and neurologic effects such as seizures, disorientation, and permanent loss of short-term memory [59]. Ingestion of 1–5 mg/kg domoic acid has reportedly resulted in human toxicity [46].

13.3.2.3 Ciguatoxins

Ciguatoxins, which are cyclic polyether compounds, are a major class of marine algal toxins [61]. They are produced by epibenthic dinoflagellates of the genera *Gambierdiscus* and *Fukuyoa*, characteristically growing attached to macroalgae and coral surfaces in shallow waters [58]. Following the food web, ciguatoxins, therefore, accumulate in herbivorous fish and, subsequently, in carnivorous fish [58]. The toxins are classified into three groups: Pacific, Caribbean, and Indian ciguatoxins, with slight structural differences, and usually occur in tropical and subtropical waters of the Caribbean Sea, Pacific Islands, and Indian Ocean [58]. Humans are exposed to ciguatoxins mainly through consumption of finfish and mollusks which cause ciguatera fish poisoning (CFP) [46, 61]. Intoxication due to ingestion of ciguateric fish in humans is the most frequently reported seafood-related poisoning, with general symptoms of CFP including gastrointestinal disturbances (nausea, vomiting and diarrheal) within 2–6 h after exposure, followed by neurologic symptoms, including numbness of the perioral area and extremities, muscle and joint aches, tachycardia, hypertension, blurred vision, and paralysis. Rarely, CFP causes death [59]. Despite being produced by organisms that live beneath the surface, ciguatoxin is not routinely monitored for concentration in seafood, and intoxication can only be prevented by completely abstaining from ingesting tropical reef fish. Currently, no FDA regulations exist to limit the levels of ciguatoxins in fish [46]. Caribbean ciguatoxin-1 (C-CTX-1) is mainly responsible for the CTX contamination in fish from Maderia, a Portuguese area, and the Selvagem Islands, with other C-CTX1 analogues and metabolites having also been reported [61, 69].

13.3.2.4 Brevetoxins

Brevetoxins (BTXs or PbTxs), which are lipophilic, cyclic polyether compounds, were initially produced by the dinoflagellate *Karenia brevis* [58]. Owing to their rapid metabolism by many animals, including shellfish, there exist a variety of different toxin profiles responsible for toxicity in animals. BTXs can accumulate along the marine food web, causing massive mortality of fish, birds, and marine mammals in impacted areas. In humans, the ingestion of BXT-contaminated molluskan shellfish, such as clams, oysters, and scallops, can trigger neurotoxic shellfish poisoning (NSP) in humans, with characteristic symptoms being paresthesia, respiratory distress, and ataxia [59]. Additionally, muscle, liver, and stomach contents of some planktivorous fish can lead to NSP [58]. Because BTXs are not diminished by rinsing or cleaning, similar to many marine toxins, and the toxins cannot be detected by taste, smell, or changes in morphology of seafood, consumers of noncommercial seafood are at increased risk of poisoning [58]. In many cases, NSP is associated with recreationally harvested shellfish collected during or post "red tide" blooms [58]. However, there is no documentation of NSP fatal intoxication in humans, unlike PSP [59].

13.3.2.5 Okadaic acid group

Okadaic acid (OA) and dinophysistoxins (DTXs) are lipophilic heat-stable toxins that are produced by marine dinoflagellates, *Dinophysis* or *Prorocentrum* [58]. The toxins inhibit serine/threonine phosphatases, critical components of signaling cascades that regulate numerous metabolic processes, ion balance, neurotransmission and cell cycles [46]. Okadaic acid and dinophysistoxins accumulate in filter-feeding shellfish (mussels, oysters, scallops, clams, and cockles). Consumption of contaminated seafood may result in the gastrointestinal syndrome called diarrhetic shellfish poisoning (DSP) in humans [59]. Acute DSP manifests with mild symptoms, comparative to other poisoning situations linked to shellfish consumption. Symptoms may include diarrhea, abdominal cramps, nausea, chills, or vomiting, occurring 30 min to several hours after ingestion of DSP toxins and generally resolving within 2–3 days, with or without medical treatment [46]. Chronic exposure to low levels of DSP toxins may, however, be more serious, as they have been shown to cause tumors. DSP has been reported in the United States, South America, Europe, Canada, New Zealand, and Japan, among other areas in the world [58]. The recommended FDA level of okadaic acid is 0.2 ppm plus 35-methyl okadaic acid (DXT 1) [46].

13.3.2.6 Azaspiracids

Dinoflagellates of the genera *Azadinium* and *Amphidoma* produce the polyether compounds, azaspiracids, while bioconversion of phytoplanktonic parental compounds in shellfish results in the formation of numerous azaspiracid analogs [58]. Bioconversion of phytoplanktonic parental compounds in shellfish contributes to the high number of azaspiracid analogs. The toxins have a global distribution, being found in shellfish from many coastal regions in areas including northwest Africa, eastern Canada, and western Europe. Azaspiracids accumulate in other marine organisms following the trophic web, mostly in filter feeding mollusks including mussels, oysters, scallops, and clams. Azaspiracid shellfish poisoning (AZP) may arise from consumption of shellfish contaminated with azaspiracids, although the biological target of this class of toxins is still unknown [58]. Symptoms in humans include severe gastrointestinal illness.

13.3.2.7 Palytoxin

PLTXs are marine polyether biotoxins produced by soft corals of the genera *Palythoa*, *Zoanthus*, and *Parazoanthus*, by planktonic and benthic dinoflagellates of the genus *Ostreopsis*, and by cyanobacteria of the genus *Trichodesmium* [58].

13.3.3 Nonalgal toxins

13.3.3.1 Gempylotoxin

Gempylotoxins are naturally occurring nonalgae toxins produced by Escolar (*Lepidocybium flavobrunneum*), and Oilfish or Cocco (*Ruvettus pretiosus*), a marine fish of the snake mackerel family [46]. These toxins contain a strong purgative oil, that when consumed can cause Gempylidae fish poisoning, gempylotoxism, or keriorrhea [46]. The toxin consists of wax esters (C_{32}, C_{34}, C_{36}, and C_{38} fatty acid esters), the primary component of which is $C_{34}H_{66}O_2$; these constitute a substantive portion of the lipid present in these fish (14–25% by weight). Escolar oil contains >90% wax esters. Ingestion of fish containing wax esters in large amounts, coupled with their indigestibility and low melting point, results in diarrhea [46]. No tolerances have been established, and the FDA recommends avoidance of these fish [46].

13.3.3.2 Tetramine in whelks

Tetramine is a toxin found in the salivary glands of *Buccinum*, *Busycon*, or *Neptunia* spp., a type of whelk or sea snail that is distributed in temperate and tropic waters and has been a source of food for humans [46]. Whelks are associated with a heat-stable neurotoxin, tetramine, which upon ingestion can cause eyeball pain, headache, dizziness, abdominal pain, ataxia, tingling in the fingers, nausea and diarrhea [46]. Although the FDA recommends removal of the salivary gland to avoid possible intoxication, tetramine is present in other tissues, albeit at lesser concentrations [46].

13.3.3.3 Trimethylamine oxide

Trimethylamine oxide is a toxin found in meat of the Greenland shark (*Somniosus microcephalus*), dogfish family, and the pacific sleeper shark (*Somniosus pacificus*) [46]. Trimethylamine oxide, which is broken down to trimethylamine in the gut, probably by enteric bacteria, is toxic to both humans and dogs [46].

13.3.3.4 Tetrodotoxins

Tetrodotoxin (TTX) is a potent neurotoxin produced by *Vibrio* and *Pseudomonas* bacteria, among others [58, 70]. TTX poisoning was traditionally associated with consumption of pufferfish in Japan, but was also found in other marine animals, including goby fish, gastropods, crabs, or bivalves in countries such as Australia, Bangladesh, Brazil, China, Israel, Morocco, Singapore, Taiwan, and the USA [58, 70]. Recently, TTXs have been found in gastropods and marine bivalves along the European Coasts of France [71], Spain [72], Italy [73], UK [74], Portugal [75], and the Netherlands [76]. TTX and analogues bind to site 1 of the voltage-gated sodium channels (Nav) and prevent access of monovalent cations to the sodium channel pore. TTX affects both action potential generation and impulse transmission, resulting in a blockade of the nerve conduction and in muscle paralysis [58]. Additionally, the toxin has been identified in the skin of certain frogs and as the poisonous principal in the venom of the blue-ringed octopuses. The signs of intoxication induced by tetrodotoxin in experimental animals are comparable with those caused by the PSP compounds. However, for the same degree of neuromuscular paralysis, a systemic, lasting arterial hypotension is produced by tetrodotoxin, which is also a highly potent hypothermic agent. The mode of action of tetrodotoxin is very similar to that of saxitoxin [61, 77].

13.4 Mycotoxins

Mycotoxins, which are diverse fungal secondary metabolites, fall under the naturally occurring food contaminants that negatively impact human health [78]. Fungi of the *Aspergillus*, *Penicillium*, and *Fusarium* genera are principal producers of mycotoxins, with aflatoxins (AFs), fumonisins (FBs), ochratoxin A (OTA), zearalenone (ZEN), ergot alkaloids (EAs), patulin (PAT), *Alternaria* toxins (ATs), and deoxynivalenol (DON) mainly being of food concern [78–80]. Under favorable conditions of temperature and humidity, fungal infestation and subsequent mycotoxin contamination may occur on numerous agricultural products including cereals, nuts, dried fruits and spices at various food and feed chain phases, affecting quality and safety of the foods [81–83]. Following an extensive global survey, Gruber-Dorninger and coworkers [84] reported that mycotoxin contamination in grains and animal feed was common. Mycotoxins may be ingested via direct consumption of contaminated food or indirectly via consumption of animal food products including meat, milk, and eggs [79]. Due to reported carcinogenic, mutagenic, teratogenic, neurotoxic, and immunosuppressive effects of mycotoxins, both acute and chronic mycotoxin exposures are of global health concern [85–88].

To protect consumer health, countries in Europe have set and continuously revise the maximum limits for mycotoxins in foods, continually monitoring and improving methods employed in agricultural production and improved agricultural production to protect human health and ensure safe food trading [88, 89]. European Commission (EC) regulated mycotoxins include AFs (AFB$_1$, B$_2$, G$_1$, G$_2$, and M$_1$), FBs (FB$_1$ and B$_2$), OTA, DON, and ZEN. However, emerging mycotoxins, which are insufficiently toxicologically characterized, are neither regulated via any legislation, nor regularly monitored in food [89]. The majority of emerging mycotoxins, which have been reported in grains, nuts and fruits, are produced by fungi of the genera *Aspergillus*, *Fusarium*, *Penicillium*, *Alternaria*, and *Claviceps* [90, 91]. *Alternaria* mycotoxins, chiefly alternariol (AOH), alternariol monomethyl ether (AME), and the altertoxins (ATX I, II, and III), are important emerging mycotoxins, due to their genotoxic and mutagenic effects reported in different cell lines [91–93]. Other emerging mycotoxins of *Fusarium* fungal origin include moniliformin, beauvericin, and enniatins.

13.4.1 Aflatoxins

AFs, the most studied mycotoxins in terms of abundance, toxicity, and human impact, are a group of structurally related, toxic, secondary metabolites produced chiefly by *A. flavus* and *A. parasiticus* species [94, 95]. *A. flavus* strains produce only AFs B$_1$ (AFB$_1$) and B$_2$ (AFB$_2$), while *A. parasiticus* and *A. nomius* strains can produce AFB$_1$, AFB$_2$, G$_1$ (AFG$_1$), and G$_2$ (AFG$_2$) [96, 97]. The hydroxylated metabolites of AFs B$_1$ and B$_2$, which are AFs M$_1$ and M$_2$, respectively, and can occur in milk or milk products obtained from livestock exposed to contaminated feed [96]. AFB_1, a potent carcinogen, with direct correlation to liver cancer and many other adverse effects in many animal species, is the most toxic among the AFs, and has been a causal factor for reported periodic acute aflatoxicosis [87, 98–100].

Dietary exposure to AFs is via ingestion of contaminated food following fungal infection pre- or postharvest, or during processing. AFs commonly occur in spices, dried fruits, peanuts and cereals, including corn and millet, posing serious health consequences worldwide, especially in African and Asian countries [82, 101–103].

To maintain the health of the population, strict regulations for AFs in food have been set and implemented in various countries. Globally, about 99 countries have set legislation for mycotoxins, with the limit of AF in food ranging between 1 and 20 ppb [104]. The Codex Alimentarius have set regulatory standards at 15 ppb for total AFs in peanuts and 0.5 ppb for AF M1 in milk, whereas the US Food and Drug Administration (FDA) has set the limit at 20 ppb for AFs in peanuts [104]. The strictest regulations for AFs are imposed by the European Union, with AFB_1 being set at 2 ppb and total AFs at 4 ppb in any directly consumed food [102].

13.4.2 Fumonisins

FBs, which were discovered in South Africa in 1988, are a group of mycotoxins produced by *Fusarium* species, primarily *F. proliferatum* and *F. verticillioides*, which are commonly occurring maize pathogens [105]. Four structurally distinct groups of the FBs, designated A, B, C, and P, are documented [106]. FB toxicity has been linked to esophageal carcinoma, interference with cellular folate uptake, and increased occurrence of neural tube defects [107]. Among FBs, FB_1, the most abundant and most studied, is also the most commonly occurring, accounting for about 70% of the total FB contamination in food and feed, and the most toxic [105]. Although FBs predominantly contaminate maize and maize products globally, they have also been detected in other foods such as the detection of FB_1 in sorghum malt and beer [103, 108] and FB_2 in coffee beans, wine, and beer [109].

To protect consumer health and ensure safe food trade, setting and compliance to harmonized international standards in crucial. The Codex Alimentarius Commission laid down standards for allowable maximum fumonisin levels in food, with levels of 4,000 and 2,000 ppb being laid down for raw maize grain and maize flour and meal, respectively. However, advisory limits set by the US FDA include 2,000 ppb for degermed dry milled corn products including corn meal and flour with fat content of 2.25% dry weight basis; 4,000 ppb for dry milled corn bran and cleaned corn intended for mass production; and 3,000 ppb for cleaned corn intended for popcorn [104]. Regulations set by the EU are lower, being 2,000, 1,000, and 200 ppb for unprocessed maize, maize meal or flour, and maize-based infant and young child foods respectively. All other maize based foods destined for direct consumption are regulated at 400 ppb [104].

13.4.3 Ochratoxin

Ochratoxins, which are secondary fungal metabolites produced by *Penicillium* in temperate climate zones and *Aspergillus* in tropical climate zones, were discovered in 1965 in South Africa [97]. OTA, OTB, and OTC are the three major ochratoxin types, with OTA, the most potent and carcinogenic, being mainly produced by *Aspergillus ochraceus*, *Aspergillus ostanius*, and *Penicillium verrucosum* [11]. Natural occurrence of ochratoxins has been documented in diverse agricultural product, including cereals such as such as corn, wheat, and barley, in legumes such as beans, in coffee, and notably in dried vine fruits, grape juice and wine [110]. Coffees, wines, and stored grains are identified as major contributors of dietary OTA intake. In addition to being a carcinogenic agent, OTA has been reported to cause nephrotoxicity and development of renal tumors in a variety of animal species although human health effects of OTA are less well-defined. Moreover, OTA has been linked to the development of the diseases Balkan endemic nephropathy and chronic interstitial nephropathy, as well as other renal diseases in several studies [112, 113]. To harmonize practices and minimize chances of OTA exposure, many countries worldwide have set a maximum regulatory level of 5–20 ppb. EU maximum limits are set for various food commodities such as wine and grape juice (2 ppb), uncooked cereal grains (5 ppb), processed cereal products (3 ppb), roasted coffee (5 ppb), instant coffee and raisins (10 ppb) [104].

13.4.4 Zearalenone

ZEN, which is classified as a group 3 carcinogen by IARC, is structurally related to naturally occurring estrogens and thus indicated as a mycotoxin responsible for inducing estrogenic effects in humans and animals [114, 115]. Fungi of the *Fusarium* genera, including *Fusarium graminearum*, *Fusarium cerealis*, and *Fusarium culmorum* are regular contaminants of cereals, mainly maize, wheat, and barley, making them prone to ZEN contamination and increasing the likelihood of ZEN occurrence in grain-based products meant for human consumption such as baked goods, pasta, breakfast cereals and bread [114]. ZEN can also be detected in milk from cows that consumed contaminated feed, posing health risks for consumers of the milk.

Limited studies reported high levels of ZEN in different food commodities, examples being beans (157 ppb), peanuts (186 ppb), and maize from Cameroon, and Nigerian rice (1,169 ppb). There are no regulations on ZEN in African countries or US FDA. According to the European Commission [115, 116] the maximum levels of ZEN are regulated in unprocessed maize (200 ppb) and for other unprocessed cereals (100 ppb).

Additionally, the limits are set for cereal snacks (50 ppm), breakfast cereals and processed cereal-based foods (50 ppm), and baby foods (20 ppb) [116].

13.4.5 Deoxynivalenol

DON, a group B trichothecene mycotoxin having a tetracyclic 12, 13 epoxytrichothecene skeleton, is also referred to as vomitoxin, and is produced by fungal pathogens such as *Fusarium graminearum* and *Fusarium culmorum*, especially at temperatures ranging from 25 to 30 °C and moisture content of 23–25% [117]. Food commodities commonly reported to be associated with DON are the *Fusarium* head blight infection prone cereals such as maize, wheat, rice, rye, oat, and barley. DON contaminated foods have been reported to cause gastroenteritis occurrences, manifesting with such symptoms as vomiting, (hence the name "vomitoxin" for DON), feed refusal, reduced weight gain, anorexia, and impaired immune function in many animal studies [118, 119]. Even though human gastroenteritis occurrences have never been conclusively linked to DON intoxication, some global outbreaks suggest a possible linkage. However, there is limited empirical evidence on human health effects of DON [120].

Regulation limits have been set for some countries, regarding DON in food. DON maximum levels have been set at 500 and 750 ppb for cereal products, and flour used as raw material in food products, respectively by the EU [104] while the US FDA has set a guideline for DON in processed human food at 1 ppb. For Codex Alimentarius Commission, the maximum limit of DON is set at 200 ppb for cereal-based foods intended for consumption by infants and young children, and at 1,000 ppb for cereal products derived from wheat, maize or barley, such as flour, meal, semolina, or flakes [121].

13.4.6 Patulin

PAT, a polyketide mycotoxin mainly produced by *Penicillium patulum* and *Penicillium expansum* species [122], was recognized for its application as an antimicrobial and anticancer compound, as well as a drug to treat common cold arising from infection by *Penicillium patulum* (now *Penicillium griseofulvum*). However, recent research has demonstrated toxicities in humans arising from PAT exposure, which include vomiting, ulceration, and hemorrhage [122]. The IARC has classified PAT as a group 3 carcinogen. Common sources of PAT in human diets are apples and apple-based products, while other fruits such as grapes, peaches and pears are also reported to be prone to PAT contamination [122]. A number of studies have reported PAT in apple juice, for instance in South Africa, PAT contamination of apple juice was found at a level of 5–45 ppb [123].

Levels of PAT in apple juice are regulated at 50 ppb in South Africa, which is the same as the EU established 50 ppb maximum recommended concentration in apple and its products [104].

Some recommended mitigation strategies for reducing mycotoxin contamination of foods include awareness creation to promote ideal agricultural pre- and postharvest practices, use of improved crop varieties and dietary diversity [83]. Recently, identification of genes associated with resistance to mycotoxin accumulation or fungal infection is underway using genetic mapping [5]. In maize, genomic selection is a valuable tool to simultaneously select for alleles associated with imparting resistance to mycotoxigenic fungi [5].

The detection and quantification of mycotoxins in food and feed is crucial to guarantee food safety, therefore the development of robust analytical methods is of paramount importance. Competitive immunoassays, chromatography and methods based on sensors are widely used for the detection of mycotoxins [124]. Challenges were experienced in detection and quantification of masked mycotoxins, which are mycotoxin derivatives modified by plant defense mechanisms such as glycosylation and which are undetectable using the conventional analytical techniques. The regulated mycotoxins DON, ZEN, FUM, and OTA, as well as the emerging mycotoxins AME and AOH are among the mycotoxins producing masked mycotoxins. LC-MS/MS has recently been effectively used in the detection of multiple mycotoxins and masked mycotoxins with reliable precision and accuracy [125]. LC-MS/MS methods, however, are unsuitable for rapid, real-time analysis of foods, requiring skilled personnel. Polymerase chain reaction, fluorescence/near-infrared spectroscopy-based methods, and hyperspectral imaging are emerging rapid robust methods for easy detection of mycotoxins [126].

QuEChERS sample preparation methods are used nowadays for various food matrices enabling detection and quantification of multiple mycotoxins present in food samples [127].

13.5 Toxicants produced during food processing

Food processing technologies such as smoking, pasteurization, irradiation, freezing, and canning often result in changes of the food nutrients and other chemical components [128]. The changes not only cause a reduction in nutritional value but may lead to formation of some toxic substances, examples being polycyclic aromatic hydrocarbons (PAHs), protein or amino acid pyrolysates, and nitrosamines. The Maillard reaction, which involves reaction of reducing sugars with amino acids to form complex mixtures, plays the most important role in the formation of various chemicals during food processing, including toxic ones [128].

13.5.1 Polycyclic aromatic hydrocarbons

Incomplete combustion of fossil fuels, including wood, coal, and oil, results in the formation of toxic PAHs [46]. PAHs, which are widely occurring environmental contaminants, being mainly found in dust, soil, water, also enter the food chain via environmental contamination or food processing in numerous foods [129]. Structurally, PAHs are comprised of at least three fused benzene rings and carbon and hydrogen molecules in a cyclic arrangement [130]. High temperature processing of carbohydrates in foods in the absence of oxygen, as well as grilling of meat, fish, or other foods under high heat or in direct contact with flames promotes production of PAHs [129]. Charcoal grilling of meat has generally been associated with the highest PAH levels, followed by smoking, roasting, and steaming [46]. Vegetable oil is among the most abundant food sources of PAHs, possibly due to endogenous production processes [129].

13.5.2 Heterocyclic aromatic amines

The two major classes of heterocyclic aromatic amines (HAAs), namely, pyrolytic HAAs and aminoimidazoarenes (AIAs) exist, with pyrolytic HAAs being formed from the pyrolysis of amino acids or proteins at high temperature while AIAs are formed from creatine, free amino acids, and monosaccharides, via the Maillard reaction [46]. Numerous protein-rich foods derived from animals, such as cooked poultry, meat, and fish, as well as gravies made from the meat stocks and sauces made from pan scrapings following frying of meat contain HAAs. HAA formation and quantities arising from processing are influenced by the temperature used for cooking, and the length of cooking time, with higher HAAs being associated with higher cooking temperatures and longer cooking times. The technique used in cooking as well as the equipment also affects HAA concentrations, with higher concentrations being associated with grilling and panfrying comparative to roasting. Exposure to HAAs in cooked meats is highly variable [46].

Another emerging toxin which is widely used in the production of consumer products is bisphenol A (BPA), an environmental endocrine disrupting chemical [131] widely used in the synthesis of plastics from polycarbonate [28]. Through the use of animal experiments, harmful impacts of BPA exposure on systems and organs have been reported [28], which include reproductive system [132–135], development [136–138], immune system [139–141], and nervous system [142–145] disorders. BPA is a widely occurring environmental toxin, with a *p*-cresol related structure which accumulates in chronic kidney disease (CKD) via promotion of mitochondrial dysfunction, subsequently resulting in depletion of energy, oxidative stress, and apoptosis [146]. Recently, Mielech and coworkers [147] reported that BPA is associated with heart disease, and can cause more frequent heart attacks, and ischemic heart disease.

According to IARC, acrylamide is classified as a carcinogenic compound possessing potent neurotoxic, carcinogenic, and genotoxic effects [147]. Being part of the most intensively researched food-borne contaminants, acrylamide mainly occurs in most thermally treated plant-based foods, regardless of the whether the food has been prepared at home or in a factory setting [128]. Asparagine, the main precursor of acrylamide, is an amino acid that rapidly reacts with reducing sugars and carbonyls via the Maillard reaction cascade to produce acrylamide [148, 149]. Acrylamide can disrupt mitochondrial function, leading to cell apoptosis, enzymatic mechanisms disruption, hormonal balance destabilization, muscle dysfunction, and infertility [150]. Among the main sources of acrylamide in the human diet are baked, fried or roasted foods such as potato products, coffee, and cereal grains [151]. In the case of children's food, acrylamide mainly comes from heat-treated potato-based products such as french fries, potato chips, and potato pancakes [152].

13.5.3 Nitrosamines

Nitrosamines are compounds formed by chemical reactions of nitrates, nitrites, and other proteins/secondary amines, sharing a general structure, in which the amine moiety may be derived from any organic secondary amine. N-Nitrosodimethylamine (NDMA) is one of the most frequently occurring nitrosamines in human dietary foods [153]. NDMA is a potent carcinogen, capable of inducing malignant tumors in a variety of tissues, including the lungs, stomach and liver [154]. Other common nitrosamines documented in meat products include N-nitrosopirrolidine (NPYR), N-nitrosopiperidine (NPIP), N-nitrosodiethylamine (NDEA), and N-nitrosomorpholine (NMOR) [155]. The production of nitrosamines depends on several variables including the temperature and acidity of the food, as well as storage conditions. High concentrations of residual nitrite in foods are linked to longer times of storage. Conditions such as heating, acidification, or oxidation of nitrites lead to the formation of nitrous acid, and the resulting nitrosonium cation ($N = O^+$) formed reacts with dimethylamine to generate nitrosamines [155].

Nitrosamines are widespread, occurring ubiquitously in the human environment, in air, water, drugs and food. Occurrence of nitrosamines and/or their precursors in diverse processed foods, such as processed meats, cheese, vegetable oil and alcoholic beverages, is unintentional, via food processing by-products [156]. On consumption of nitrosamine-precursor containing foods, nitrosamine formation can occur in the mouth or stomach. Furthermore, naturally occurring or added nitrites or nitrates in food may combine with amines forming nitrosamines in the acidic environment of the mouth or stomach. Industries processing fish and meat products commonly use nitrates and nitrites in the prevention of toxin production by *Clostridium botulinum*, and for color, preservation, and flavor of the products. However, they can be potential sources of nitrosamines in meat products. Products such as bacon, minced beef, and cured meats, are rich in amines due to their high protein content, and they are also reported to have high levels of nitrosamines [155, 157]. Over the past three decades, there has been a decline in nitrosamine levels in foods, paralleled by decreased use of nitrite in foods. Levels of nitrosamines in foods have been decreasing during the past three decades, in parallel with a decreased use of the nitrite in food. Following a survey of the contents of nitrosamines in processed meat products in the European Union, low mean contents of below 2.7 mg/kg for NDMA and below 0.9 mg/kg for NDEA were reported in meat products [158].

Numerous nitrosamines, such as NDMA, which are carcinogenic to animals, are reasonably anticipated to be human carcinogens. According to IARC, NDMA and NDEA are classified as probably carcinogenic to humans, and NDBA, NPIP, NPYR as possibly carcinogenic to humans [159].

Nitrosamine toxicity occurs when the compound undergoes α-hydroxylation, catalyzed by cytochrome P450 enzymes, followed by dealkylation of the primary nitrosamine. The unstable primary nitrosamine further decomposes to diazonium, a DNA alkylating agent, which results in DNA damage can lead to cancer [160, 161].

N-Nitrosamines can induce tumors in a variety of organs, such as the esophagus, kidney, liver, and lungs, depending on the species. Nitrites resulting from nitrate metabolism as well as those deriving from food can react in the gastrointestinal tract with precursors of *N*-nitroso compounds, including amines and amides, resulting in the formation of *N*-nitroso compounds, the carcinogenic nitrosamines being an example [162]. High levels of nitrosamines in the diet and consequent high exposure to nitrates and nitrites, can lead to toxicity in the form of methemoglobinemia [163].

Food processing procedures including drying, kilning, salting, smoking, or curing are reported to promote nitrosamine formation. The reaction of alkyl amines with nitrites used during curing or drying processes may result in the formation of carcinogenic nitrosamines. Although nitrites undergo rapid depletion during processing procedures, the levels remaining in the final product depend on the initial amounts added and the processing conditions employed [164]. Formation of nitrosamines is dependent on several factors including nitrite concentration, acidity or alkalinity, temperature, storage conditions, and the presence of catalysts or inhibitors.

An increase in temperature during food processing may favor formation of nitrosamines. In a study investigating nitrosamine production in bacon under various cooking conditions such as broiling, baking, frying and microwave cooking [165], results indicated that, lower operating temperatures and indirect heating procedures reduced the occurrence of nitrosamines. Substances such as ascorbic acid, erythorbic acid, and tocopherol can also inhibit the nitrosation reaction [166]. The risk of formation of nitrosamines such as NMDA can be mitigated by several methods under development and regulation, an example being the addition of ascorbic acid to processed meats [167]. Occurrence of NDMA in water is however, a fairly recent discovery, more research is needed in the processing methods to be employed.

There is limited information regarding the regulation limit of nitrosamines in food. The USDA has set limits of nitrite in finished meat products to a level not exceeding 200 ppb while the WHO has set the maximum level of nitrites in processed meat at 10 ppb [168].

For currently cured foods such as bacon and cooked ham, research is continually directed at seeking for sodium nitrate/nitrite substitutes to be used as preservatives. For instance, some processed meats producers have resorted to using natural nitrate sources such as celery juice or spinach extracts, instead of sodium nitrate/nitrite. However, the nitrates present in celery juice or spinach extract are reduced to nitrites by the addition of starter bacterial cultures and as a result contribute to nitrosamine formation [169]. Therefore, further research is needed to seek alternatives to mitigate the production of nitrosamines in food.

13.5.4 Trans fats

Unsaturated fatty acids with one or more double bonds in the trans configuration are referred to as trans-fatty acids (TFAs) [170]. Besides being present as industrially produced trans fats (iTFAs), TFAs are found naturally in foods as well, being commonly known as ruminant trans fats (rTFAs) [170].

Industrial TFAs (iTFAs) are manufactured through industrial processes by partial hydrogenation of vegetable oils, and heating oil at very high temperature, elaidic acid being the prime iTFA [170]. Industrially produced trans-fats (iTFAs) are bought from the oil factories by food producers because the fat is inexpensive, solid at room temperature, has some technical advantages for food processing, and prolongs products' shelf life compared to nonhydrogenated oils. iTFAs may be used at varying concentrations in the production of biscuits, pastries, margarine, spreads, confectionery creams, and fillings and in fried foods served at fast food restaurants.

Fat from ruminants, such as the fat in dairy products, may contain up to 5% TF (R-TF), which is much lower than the up to 50% TF that may be present in partially hydrogenated fat found in human food. Vaccenic acid is the major fatty acid in R-TF [171]. Abundance of literature focuses on the effects of industrially produced trans fats in different diseases, overlooking rTFAs. However, typically we consume 2–9% of our total fatty acid content as rTFAs [170]. Sheep and cattle meats as well as dairy products [cheese, milk, and butter] contain rTFAs [170]. The major contributor within the category of rTFAs is vaccenic acid. Vaccenic acid (18:1 trans-11), a precursor to another TFA, conjugated linoleic acid, constitutes 50–80% of all ruminant derived trans fats [170, 172, 173]. While trans fats may not be considered of any nutritional value, they find application in commercial procedures in food industries to yield semi fat foods and increase the shelf life of products like margarine, crackers, deep-fried fast foods, pancakes, and can also be found in restaurants [175]. TFA intake varies in different settings worldwide, depending on dietary habits and varying quantities of iTFA in processed foods [176].

Trans fatty acids have been identified as an important cause of cardiovascular diseases, strokes, and heart attacks [170]. Trans fat consumption promotes systemic inflammation, increasing the risk of coronary heart disease (CHD) by increasing the level of low-density lipoprotein, decreasing the level of high-density lipoprotein, and raising triglycerides in the bloodstream [177, 178]. It has been estimated that a 2% rise in energy consumption from TFA is linked with a 23% increased risk of CHD [157]. Studies have been directed at understanding the impact of trans fats in humans with recent findings having now identified an important cardioprotective role for the ruminant trans fats, necessitating enquiry into the mechanisms responsible for, not only the toxic actions of trans fats, but also their potential as beneficial compounds within the diet [170].

With the advent of the SARS-CoV-2, patients with noncommunicable diseases (NCDs) have exacerbated unfavorable outcomes. The relationship between SARS-CoV-2 and Trans Fatty Acid (TFA) is the increased risk factors and worse effects of developing cardiovascular diseases [179]. TFA is linked with an increased risk of cardiovascular diseases [179]. NCDs share risk factors by triggering the severity of SARS-CoV-2. Addressing TFA elimination in the food systems will lessen the possible complications of SARS-CoV-2 and NCDs particularly, cardiovascular diseases [179].

Consciousness of the well-established health complications of trans fats is limited among consumers. Many products with almost 0.5 g trans-fat, if consumed over the course of a day, may approximate, or exceed the 2 g maximum as recommended by the American Heart Association [177]. The WHO has intensified efforts to minimize the intake of industrial trans-fat worldwide, with dietary trans-fat elimination being a priority [171, 180]. The six strategic actions to eliminate industrially produced trans fats namely the "REPLACE" package including Review, Promote, Legislate, Assess, Create, and Enforce has been outlined with the goal of global trans-fat-free by 2023 [180]. According to the Food and Drug Administration (FDA), partially hydrogenated oils (PHOs), as the main source of artificial trans fat in the global food supply, is no longer "generally recognized as safe," or GRAS [179]. Consequently, the deleterious effects of TFA accompanied by a lack of awareness of food products containing TFA-like fat-based spreads, shortenings, and some baked goods increase the modifiable risk factors. It may not be stated quantitatively in the nutrition labels, but it can be found qualitatively in the packet ingredient list as "partially hydrogenated," a hidden source of TFA [179].

Gas chromatography (GC) has been commonly used for the determination of fatty acid profiles in food products [180]. Although the official GC methods are precise and reliable for the analysis of fatty acid, to improve the analysis time and coelution problems, experimental optimization processes were frequently used. Recently, Pojjana-pornpun and coworkers [180] developed a program to simulate fatty acid peak profiles on a gas chromatograph coupled with a 100 m CP-Sil 88 column. The approach was effectively applied to analyze trans fat alternatives, including shortenings, margarines, butter blends and nondairy creamers. These products are composed of natural fats and oils as well as fully hydrogenated vegetable oils and contain 0.0–6.3 g TFA/100 g fat, like that found in nature.

References

[1]   WHO. (2018). Natural Toxins in food (Factsheet), Available online: (Accessed June 02, 2021 at https://www.who.int/news-room/factsheets/detail/natural-toxins-in-food).

[2]   Fletcher, M. T., & Netzel, G., (2020). Food safety and natural toxins. *Toxins*, *12*, 236.

[3]   Gadallah, M. (2019). Natural toxins in food from animal origin. Available from: http:///:doi:10.13140/RG.2.2.30473.08803

[4]   Salem, M. A., Serag, A., El-Seedi, H. R., Hamdan, D. I., Ezzat, H. M., & Zayed, A., (2021). In A. G. Mtewa & Chukwuebuka E. (Eds.), *Identification and analysis of toxic phytochemicals* (pp. 443–479). Elsevier: Phytochemistry, the Military and Health.

[5]   Kaiser, N., Douches, D., Dhingra, A., et al. (2020). The role of conventional plant breeding in ensuring safe levels of naturally occurring toxins in food crops. *Trends in Food Science and Technology*, *100*, 51–66.

[6]   Osman, A. G., Ali, Z., Chittiboyina, A. G., & Khan, I. A., (2021). An update on plant toxins posing human health risks. In J. G. Morris & Vugia D. J. (Eds.), *Foodborne Infections and Intoxications 5th end* (pp. 479–491). Academic Press.

[7]   Taylor, S. L., & Hefle, S. L., (2017). Naturally Occurring Toxicants in Foods. In C. E. R. Dodd, Aldsworth T., Stein R. A., Cliver D. O. & Riemann H. P. (Eds.), *Foodborne Diseases* (3rd ed, pp. 327–344). London: Academic Press.

[8]   Hwang, D. F., & Chen, T. Y., (2016). Toxins in Food: Naturally Occurring. In C. Benjamin, Paul M. F. & Fidel T. (Eds.), *Encyclopedia of Food and Health* (pp. 326–330). Academic Press.

[9]   Mosayyebi, B., Imani, M., Mohammadi, L., et al. (2020). An update on the toxicity of cyanogenic glycosides bioactive compounds: Possible clinical application in targeted cancer therapy. *Materials Chemistry and Physics*, *246*, 122–841.

[10] Panter, K. E., (2018). Cyanogenic Glycoside-Containing Plants. In R. C. Gupta, *Veterinary toxicology: Basic and clinical principles* (3rd ed, pp. 935–940). Houston, TX: Elsevier.

[11] Kudtak, B., Wieczerzak, M., & Namiesnik, J., (2017). Cyanogenic compounds and estrogen disruptors. In A. Witczak & Sikorski Z. E. (Eds.), *Toxins and other harmful compounds in foods* (pp. 243–251). Boca Raton FL: CRC Press.

[12] Cressey, P., & Reeve, J., (2019). Metabolism of cyanogenic glycosides: a review. *Food and Chemical Toxicology: An International Journal Published for the British Industrial Biological Research Association*, *125*, 225–232. https://doi.org/10.1016/j.fct.2019.01.002.

[13] Rivadeneyra-Dominguez, E., Vazquez-Luna, A., Rodriguez-Landa, J. F., & Diaz-Sobac, R., (2013). Neurotoxic effect of linamarin in rats associated with cassava [Manihot esculenta Crantz] consumption. *Food and Chemical Toxicology*, *59*, 230–235.

[14] Shim, Y. Y., Olivia, C. M., Liu, J., Boonen, R., Shen, J., & Reaney, M. J., (2016). Secoisolariciresinol diglucoside and cyanogenic glycosides in gluten-free bread fortified with flaxseed meal. *Journal of Agricultural and Food Chemistry*, *64*(50), 9551–9558.

[15] Ayankunbi, M. A., (1991). Effect of methods of preparation on the nutrient composition of some cassava products – Garri [eba], "Lafun" and "Fufu". *Food Chemistry*, *41*(3), 349–354.

[16] Adamolekun, B., (2020). Etiology of Konzo, epidemic spastic paraparesis associated with cyanogenic glycosides in cassava: role of thiamine deficiency. *Journal of Neurological Sciences*, *296*, 30–33.

[17] Kashala-Abotnes, E., Okitundu, D., Mumba, D., Boivin, M. J., Tylleskär, T., & Tshala-Katumbay, D., (2019). Konzo: a distinct neurological disease associated with food (cassava) cyanogenic poisoning. *Brain Research Bulletin*, *145*, 87–91.

[18] Jackson, J., Chiwona-Karltun, L., & Gordon, A., (2020). Food safety and quality considerations for cassava, a major staple containing a natural toxicant. In G. André (Ed.), *Food safety and quality systems in developing countries* (pp. 343–366). Academic Press.

[19] Samson, S. O., & Akomolafe, O. M., (2017). Fermentation: a means of treating and improving the nutrition content of cassava (Manihot esculenta C.) peels and reducing its cyanide content. *Journal of General and Applied Microbiology*, *8*(3), 17–25.

[20] Panghal, A., Munezero, C., & Sharma, C. N., (2019). Cassava toxicity, detoxification and its food applications: a review. *Toxin Reviews*, *40*, 1–16.

[21] Kobawila, S. C., (2002). Reduction of the cyanide content during fermentation of cassava roots and leaves to produce bikedi and ntoba mbodi, two food products from Congo. *African Journal of Biotechnology*, *4*, 689–696.

[22] Food and Agriculture Organization of the United Nations (FAO) and World Health Organization (WHO) (2017 April). *Codex Alimentarius Commission. Eleventh session of the Codex committee on contaminants in foods*. Rio de Janeiro (Brazil): Joint FAO/WHO Standards Program.

[23] Zhong, Y., Xu, T., Ji, S., et al. Effect of ultrasonic pretreatment on eliminating cyanogenic glycosides and hydrogen cyanide in cassava. *Ultrasonics Sonochemistry*, *78*, 105742.

[24] Zhao, M., Bergaentzlé, M., Flieller, A., & Marchioni, E., (2019). Development and validation of an ultra-high performance liquid chromatography-high resolution mass spectrometry method for simultaneous quantification of cyanogenic glycosides and secoisolariciresinol diglucoside in flaxseed (Linum usitatissimum L.). *Journal of Chromatography A*, *1601*, 214–223.

[25] Al Sinani, S. S. S., & Eltayeb, E. A., (2017). The steroidal glycoalkaloids solamargine and solasonine in Solanum plants. *South African Journal of Botany*, *112*, 253–269.

[26] Milner, S. E., Brunton, N. P., Jones, P. W., O'Brien, N. M., Collins, S. G., & Maguire, A. R., (2011). Bioactivities of glycoalkaloids and their aglycones from Solanum species. *Journal of Agricultural and Food Chemistry*, 2011 *59*, 3454–3484.

[27] Moreira, R., Pereira, D. M., Valentão, P., & Andrade, P. B., (2018). Pyrrolizidine Alkaloids: Chemistry, Pharmacology, Toxicology and Food Safety. *International Journal of Molecular Sciences*, *19*(6), 1668.

[28] Ma, C., Liu, Y., Zhu, L., et al. (2018). Determination and regulation of hepatotoxic pyrrolizidine alkaloids in food: A critical review of recent research. *Food and Chemical Toxicology*, *119*, 50–60.

[29] Boppre, M., (2011). The ecological context of pyrrolizidine alkaloids in food, feed and forage: An overview. *Food Additives and Contaminants - Part A Chemistry*, *28*, 260–281.

[30] Wiedenfeld, H., (2011). Plants containing pyrrolizidine alkaloids: Toxicity and problems. *Food Additives and Contaminants - Part A Chemistry*, *28*, 282–292.

[31] He, Y., Lian, W., & Ding, L., (2021). Lung injury induced by pyrrolizidine alkaloids depends on metabolism by hepatic cytochrome P450s and blood transport of reactive metabolites. *Archives of Toxicology*, *95*, 103–116.

[32] Kirk, H., Vrieling, K., Van Der Meijden, E., & Klinkhamer, P. G., (2010). Species by environment interactions affect pyrrolizidine alkaloid expression in Senecio jacobaea, Senecio aquaticus, and their hybrids. *Journal of Chemical Ecology*, *36*(4), 378–387.

[33] Kaltner, F., Rychlik, M., Gareis, M., & Gottschalk, C., (2020). Occurrence and risk assessment of pyrrolizidine alkaloids in spices and culinary herbs from various geographical origins. *Toxins*, *12*(155), 1–15.

[34] Mulder, P. P. J., Lopez, S. P., Castelari, M., Bodi, D., Ronczka, S., & Preib-Weigert, A., (2018). These A. Occurrence of pyrrolizidine alkaloids in animal- and plant-derived food: Results of a survey across Europe. *Food Additives and Contaminants - Part A Chemistry*, *35*, 118–133.

[35] Kaltner, F., Rychlik, M., Gareis, M., & Gottschalk, C., (2018). Influence of Storage on the Stability of Toxic Pyrrolizidine Alkaloids and Their N-Oxides in Peppermint Tea, Hay, and Honey. *Journal of Agricultural and Food Chemistry*, *66*(20), 5221–5228.

[36] Xia, Q., Chou, M. W., Edgar, J. A., Doerge, D. R., & Fu, P. P., (2006). Formation of DHP-derived DNA adducts from metabolic activation of the prototype heliotridine-type pyrrolizidine alkaloid, lasiocarpine. *Cancer Letters*, *231*, 138–145.

[37]   Chen, T., Mei, N., & Fu, P. P., (2010). Genotoxicity of pyrrolizidine alkaloids. *Journal of Applied Toxicology: JAT, 30*, 183–196.

[38]   Chou, M. W., Wang, Y. P., Yang, Y. C., et al. (2003). Riddelliine N-oxide is a phytochemical and mammalian metabolite with genotoxic activity that is comparable to the parent pyrrolizidine alkaloid riddelliine. *Toxicology Letter, 145*, 239–247.

[39]   Molyneux, R. J., Gardner, D. L., Colegate, S. M., & Edgar, J. A., (2011). Pyrrolizidine alkaloid toxicity in livestock: a paradigm for human poisoning? *Food Additives and Contaminants, 28*(3), 293–307.

[40]   Chung, S. W., & Lam, A. C., (2017). Investigation of pyrrolizidine alkaloids including their respective N-oxides in selected food products available in Hong Kong by liquid chromatography electrospray ionisation mass spectrometry. *Food Additives and Contaminants - Part A Chemistry, 34*(7), 1184–1192.

[41]   De Nijs, M., Mulder, P. P. J., Klijnstra, M. D., Driehuis, F., & Rlap, H., (2017). Fate of pyrrolizidine alkaloids during processing of milk of cows treated with ragwort. *Food Additives and Contaminants - Part A Chemistry, 34*(12), 2212–2219.

[42]   European Commission. (2020). Rapid Alert System for Food and Feed (RASSF) Portal. https://webgate.ec.europa.eu/rasff.

[43]   European Commission [EC] (2003). Regulation No 1425/2003. Official J European Union L 203, 1–3.

[44]   Avula, B., Sagi, S., Wang, Y. H., Zweigenbaum, J., Wang, M., & Khan, I. A., (2015). Characterization and screening of pyrrolizidine alkaloids and N-oxides from botanicals and dietary supplements using UHPLC-high resolution mass spectrometry. *Food Chemistry, 178*, 136–148.

[45]   Qie, M., Li, S., Guo, C., Yang, S., & Zhao, Y., (2021). Study of the occurrence of toxic alkaloids in forage grass by liquid chromatography tandem mass spectrometry. *Journal of Chromatography A, 1654*, 462463.

[46]   Dolan, L. C., Matulka, R. A., & Burdock, G. A., (2010). Naturally occurring food toxins. *Toxins, 2*, 2289–2332. http://dx.doi.org/10.3390/toxins2092289.

[47]   Melough, M. M., Cho, E., & Chun, O. K., (2018). Furocoumarins: A review of biolochemical activities, dietary sources and intake, and potential health risks. *Food and Chemical Toxicology, 113*, 99–107.

[48]   Mahendra, C. K., Tan, L. T. H., Lee, W. L., et al. (2020). Angelicin – a furocoumarin compound with vast biological potential. *Frontiers in Pharmacology, 11*, 1–34.

[49]   Ranganath, Y. S., Babu, V. H., Sandeep, G., & Parameshwar, R., (2011). Synthesis and evaluation of some novel furocoumarin derivatives for radical scavenging profile and cytotoxic studies. *Journal of Chemical and Pharmaceutical Research, 3*(4), 62–68.

[50]   Munakata, R., Olry, A., Karamat, F., et al. (2016). Molecular evolution of parsnip (Pastinaca sativa) membrane-bound prenyltransferases for linear and/or angular furanocoumarin biosynthesis. *New Phytologist, 211*, 332–344.

[51]   Girennavar, B., Poulose, S. M., Jayaprakasha, G. K., Bhat, N. G., & Patil, B. S., (2006). Furocoumarins from grapefruit juice and their effect on human CYP3A4 and CYP1B1 isoenzymes. *Bioorganic and Medicinal Chemistry, 4*, 2606–2612.

[52]   Bailey, D. G., Malcom, J., Arnold, O., & Spence, J. D., (1998). Grapefruit juice-drug interactions. *British Journal of Clinical Pharmacology, 46*, 101–110.

[53]   Placzek, M., Fromel, W., Eberlein, B., Gilbertz, K. P., & Przybilla, B., (2007). Evaluation of phototoxic properties of fragrances. *Acta Dermato-Venereologica, 87*, 312–316.

[54]   Yoshihiro, U., & Kiminori, M., (2006). The Use of Heat Treatment to Eliminate Drug Interactions Due to Grapefruit Juice. *Biological and Pharmaceutical Bulletin, 29*, 2274–2278.

[55]   Lee, S. G., Kim, K., Vance, T. M., Perkins, C., Provatas, A., Wu, S., Qureshi, A., Cho, E., & Chun, O. K., (2016). Development of a comprehensive analytical method for furanocoumarins in grapefruit and their metabolites in plasma and urine using UPLCMS/MS: a preliminary study. *International Journal of Food Sciences and Nutrition, 7486*, 1–7.

[56] Zhao, X. J., Guo, P. M., Pang WH, Z., et al. (2020). A rapid UHPLC-QqQ-MS/MS method for the simultaneous qualitation and quantitation of coumarins, furocoumarins, flavonoids, phenolic acids in pummelo fruits. *Fluorine Chemistry*, *325*, 126835.

[57] Chettri, D., Boro, M., Sarkar, L., & Verma, A. K., (2021). Lectins: Biological significance to biotechnological application. *Carbohydrate Research*, *506*, 108367.

[58] Vilarino, N., Louzao, M. C., Abal, P., et al. (2018). Human Poisoning from Marine Toxins: Unknowns for Optimal Consumer Protection. *Toxins*, *10*(8), 324.

[59] Dolah FM, V. (2000). Marine Algal Toxins: Origins, health effects, and their increased occurrence. *108*, 133–141.

[60] Baden, D. G., Rein, K. S., & Gawley, R. E., (1998). Marine Toxins: How They are Studied and What They Can Tell Us. In K. E. Cooksey (Ed.), *Molecular Approaches to the Study of the Ocean*. Dordrecht: Springer. Available from https://doi.org/10.1007/978-94-011-4928-0_26.

[61] Gerssen, A., & Gago-Martínez, A., (2019). Emerging Marine Biotoxins. *Toxins*, *11*(6), 314.

[62] Estevez, P., Castro, D., Pequeño-Valtierra, A., Leao, J. M., Vilariño, O., Diogène, J., & Gago-Martínez, A., (2019). An Attempt to Characterize the Ciguatoxin Profile in *Seriola fasciata* Causing Ciguatera Fish Poisoning in Macaronesia. *Toxins*, *11*, 4, 221.

[63] Patria, F. P., Pekar, H., & Zuberovic-Muratovic, A., (2020). Multi-Toxin Quantitative Analysis of Paralytic Shellfish Toxins and Tetrodotoxins in Bivalve Mollusks with Ultra-Performance Hydrophilic Interaction LC-MS/MS-An In-House Validation Study. *Toxins*, *12*(7), 452.

[64] Turner, A. D., Dhanji-Rapkova, M., Fong, S. Y. T., et al. (2020). Ultrahigh-Performance Hydrophilic Interaction Liquid Chromatography with Tandem Mass Spectrometry Method for the Determination of Paralytic Shellfish Toxins and Tetrodotoxin in Mussels, Oysters, Clams, Cockles, and Scallops: Collaborative Study. *Journal of AOAC International*, *103*(2), 533–562.

[65] Zhang, S., Li, J., Qin, Q., et al. (2018). Whole-Genome Sequencing of Chinese Yellow Catfish Provides a Valuable Genetic Resource for High-Throughput Identification of Toxin Genes. *Toxins*, *10*(12), 488.

[66] Wilson, D., & Daly, N. L., (2018). Nuclear magnetic resonance seq (NMRseq): A new approach to peptide sequence tags. *Toxins*, *10*(11), 437.

[67] Klijnstra, M. D., & Gerssen, A., (2018). A sensitive LC-MS/MS method for palytoxin using lithium cationization. *Toxins*, *10*(12), 537.

[68] Dom, I., Bire, R., Hort, V., Lavison-Bompard, G., Nicolas, M., & Guérin, T., (2018). Extended Targeted and Non-Targeted Strategies for the Analysis of Marine Toxins in Mussels and Oysters by (LC-HRMS). *Toxins*, *10*(9), 375. http://doi:10.3390/toxins10090375.

[69] Costa, P. R., Giraldez, J., Rodrigues, S. M., et al. (2021). High Levels of Tetrodotoxin (TTX) in Trumpet Shell Charonia lampas from the Portuguese Coast. *Toxins*, *13*(4). 250. http://doi:10.3390/toxins13040250.

[70] Biessy, L., Boundy, M. J., Smith, K. F., Harwood, D. T., Hawes, I., & Wood, S. A., (2019). Tetrodotoxin in marine bivalves and edible gastropods: A mini-review. *Chemos*, *236*, 124404. http://doi:10.1016/j.chemosphere.2019.124404.

[71] Hort, V., Arnich, N., Guérin, T., Lavison-Bompard, G., & Nicolas, M., (2020). First Detection of Tetrodotoxin in Bivalves and Gastropods from the French Mainland Coasts. *Toxins*, *12*(9), 599.

[72] Leao, J. M., Lozano-Leon, A., Giradez, J., Vilarino, O., & Gago-Martinez, A., (2018). Preliminary results on the evaluation of the occurrence of tetrodotoxin associated to Marine Vibrio spp. in Bivalves from the Galician Rias [Northwest of Spain]. *Marine Drugs*, *16*(3), 81.

[73] Bacchiocchi, S., Campacci, D., Siracusa, M., et al. (2021). Tetrodotoxins [TTXs] and *Vibrio alginolyticus* in Mussels from Central Adriatic Sea [Italy]: Are They Closely Related? *Marine Drugs*, 19 6, 304.

[74] Turner, A. D., Dhanji-Rapkova, M., Coates, L., et al. (2017). Detection of Tetrodotoxin Shellfish Poisoning (TSP) Toxins and Causative Factors in Bivalve Molluscs from the UK. *Marine Drugs*, *15*(9), 277.

[75] Costa, P. R., Giraldez, J., Rodrigues, S. M., et al. (2021). High Levels of Tetrodotoxin (TTX) in Trumpet Shell Charonia lampas from the Portuguese Coast. *Toxins, 13*(4). 250. http://doi:10.3390/toxins13040250.

[76] Gerssen, A., Bovee, T. H. F., Klijnstra, M. D., Poelman, M., Portier, L., & RLAP, H., (2018). First Report on the Occurrence of Tetrodotoxins in Bivalve Mollusks in The Netherlands. *Toxins, 10*(11), 450.

[77] Roue, M., Darius, H. T., Ung, A., Viallon, J., Sibat, M., Hess, P., Amzil, Z., & Chinain, M., (2018). Tissue Distribution and Elimination of Ciguatoxins in Tridacna maxima [Tridacnidae, Bivalvia] Fed Gambierdiscus polynesiensis. *Toxins, 10*(5), 189.

[78] Agriopoulou, S., Stamatelopoulou, E., & Varzakas, T., (2020). Advances in Occurrence, Importance, and Mycotoxin Control Strategies: Prevention and Detoxification in Foods. *Foods, 9*, 137.

[79] Kebede, H., Liu, X., Jin, J., & Xing, F., (2020). Current status of major mycotoxins contamination in food and feed in Africa. *Food Control, 110*, 106975.

[80] Luo, S., Du, H., Kebede, H., Liu, Y., & Xing, F., (2021). Contamination status of major mycotoxins in agricultural products and food stuff in Europe. *Food Control, 127*, 108120.

[81] Kluczkovski, A. M., (2019). Fungal and mycotoxin problems in the nut industry. *Current Opinion in Food Science, 29*, 56–63.

[82] Gonçalves, B. L., Coppa, C. C., De-Neeff, D. V., Corassin, C. H., & De Oliveira, C. A. F., (2018). Mycotoxins in fruits and fruit-based products: Occurrence and methods for decontamination. *Toxin Reviews, 38*, 263–272.

[83] Misihairabgwi, J. M., Ezekiel, C. N., Sulyok, M., Shephard, G. S., & Krska, R., (2019). Mycotoxin contamination of foods in Southern Africa: A 10-year review (2007–2016). *Critical Reviews in Food Science and Nutrition, 591*, 43–58.

[84] Gruber-Dorninger, C., Jenkins, T., & Schatzmayr, G., (2019). Global Mycotoxin Occurrence in Feed: A Ten-Year Survey. *Toxins, 11*, 375.

[85] International Agency for Research on Cancer (1993). IARC Monographs on the Evaluation of Carcinogenic Risks to Humans; International Agency for Research on Cancer: Lyon. *France, 56.*

[86] IARC (International Agency for Research on Cancer) (2002). *Traditional herbal medicines, some mycotoxins, naphthalene, and styrene. Monographs on the evaluation of the carcinogenic risk of chemicals to humans* (Vol. 82, pp. 82–171). Lyon, France: IARC Press.

[87] 2012). *IARC (International Agency for Research on Cancer) Mycotoxins and Human Health* (Vol. 158, pp. 87–104). Lyon, France: IARC Press.

[88] Braun, D., Eiser, M., Puntscher, H., Marko, D., & Warth, B., (2021). Natural contaminants in infant food: The case of regulated and emerging mycotoxins. *Food Control, 123*, 107676.

[89] Luo, S., Du, H., Kebede, H., Liu, Y., & Xing, F., (2021). Contamination status of major mycotoxins in agricultural products and food stuff in Europe. *Food Control, 127*, 108120.

[90] Fraeyman, S., Croubels, S., Devreese, M., & Antonissen, G., (2017). Emerging Fusarium and Alternaria mycotoxins: Occurrence, toxicity and toxicokinetics. *Toxins, 9*, 228.

[91] Jarolim, K., Del Favero, G., Ellmer, D., Stark, T. D., Hofmann, T., Sulyok, M., et al. (2017). Dual effectiveness of Alternaria but not Fusarium mycotoxins against human topoisomerase II and bacterial gyrase. *Archives of Toxicology, 91*, 4. 2007–2016.

[92] Aichinger, G., Kruger, F., Puntscher, H., Preindl, K., Warth, B., & Marko, D., (2019). Naturally occurring mixtures of Alternaria toxins: Anti-estrogenic and genotoxic effects in vitro. *Archives of Toxicology, 93*(10), 3021–3031.

[93] Mujahid, C., Savoy, M. C., Baslé, Q., et al. (2020). Levels of Alternaria Toxins in Selected Food Commodities Including Green Coffee. *Toxins, 12*(9), 595.

[94] Liu, Y., & Wu, F., (2010). Global burden of aflatoxin-induced hepatocellular carcinoma: A risk assessment. *Environmental Health Perspectives, 118*, 818–824.

[95] Pickova, D., Ostry, V., & Malir, F., (2021). A recent overview of producers and important dietary sources of aflatoxins. *Toxins, 13*(3), 186.

[96] Bennett, J. W., & Klich, M., (2003). Mycotoxins. *Clinical Microbiological Reviews, 16*, 497–516.

[97] Anukul, N., Vangnai, K., & Mahakarnchanakul, W., (2013). Significance of regulation limits in mycotoxin contamination in Asia and risk management programs at the national level. *Journal of Food and Drug Analysis, 21*, 227–241.

[98] Martins, M. L., Martins, H. M., & Bernardo, F., (2001). Aflatoxins in spices marketed in Portugal. *Food Additives and Contaminants, 18*, 315–319.

[99] Probst, C., Bandyopadhyay, P. J., & Cotty, P. J., (2014). Diversity of aflatoxin-producing fungi and their impact on food safety in sub-Saharan Africa. *International Journal of Food Microbiology, 174*, 113–122.

[100] Ashiq, S., (2015). Natural occurrence of mycotoxins in food and feed: Pakistan perspective. *Comprehensive Reviews in Food Science and Food Safety, 14*, 159–175.

[101] Anukul, N., Vangnai, K., & Mahakarnchanakul, W., (2013). Significance of regulation limits in mycotoxin contamination in Asia and risk management programs at the national level. *Journal of Food and Drug Analysis, 21*, 227–241.

[102] Mahato, D. K., Lee, K. E., Kamle, M., et al. (2019). Aflatoxins in Food and Feed: An Overview on Prevalence, Detection and Control Strategies. *Frontiers in Microbiology, 10*, 1–10.

[103] Nafuka, S. N., Misihairabgwi, J. M., Bock, R., Ishola, A., Sulyok, M., & Krska, R., (2019). Variation of Fungal Metabolites in Sorghum Malts Used to Prepare Namibian Traditional Fermented Beverages Omalodu and Otombo. *Toxins, 11*(165), 1–18.

[104] Food and Agricultural Organization-FAO (2004). Worldwide regulations for mycotoxins in food and feed in 2003. *FAO Food and Nutrition Paper No, 81*, Rome, Italy.

[105] Marasas, W. F. O., (2001). Discovery and occurrence of the fumonisins: a historical perspective. *Environmental Health Perspectives, 109*, 239–243.

[106] Rheeder, J. P., Marasas, W. F. O., & Vismer, H. F., (2002). Production of fumonisin analog by Fusarium species. *Applied and Environmental Microbiology, 68*, 2101–2105.

[107] Mismer, S. A., Suarez, I., Felkner, M., et al. (2006). Exposure to fumonisins and the occurrence of neural tube defects along the Texas Mexico border. *Environmental Health Perspectives, 114*, 237–241.

[108] Roger, D. D., (2011). Deoxynivalenol (DON) and fumonisins B1 (FB1) in artisanal sorghum opaque beer brewed in North Cameroon. *African Journal of Microbiology Research, 5*, 1565–1567.

[109] Scott, P., (2012). Recent research on fumonisins: A review. *Food Additives and Contaminants: Part A, 29*, 242–248.

[110] Magnolia, C. E., Astoreca, A. L., Chiacchiera, S. M., & Dalcero, A. M., (2007). Occurrence of ochratoxin A and ochratoxigenic mycoflora in corn and corn-based foods and feeds in some South American countries. *Mycop, 163*, 249–260.

[111] Ostry, V., Malir, F., Toman, J., & Grosse, Y., (2017). Mycotoxins as human carcinogens-the IARC Monographs classification. *Myc Res, 33*, 65–73.

[112] Abid, S., Hassen, W., Achour, A., et al. (2003). Ochratoxin A and human chronic nephropathy in Tunisia: is the situation endemic? *Human and Experimental Toxicology, 22*, 77–84.

[113] Bui-Klimke, T. R., & Wu, F., (2015). Ochratoxin A and human health risk: a review of the evidence. *Critical Reviews in Food Science and Nutrition, 55*(13), 1860–1869.

[114] Mally, A., Solfrizzo, M., & Degen, G. H., (2016). Biomonitoring of the mycotoxin Zearalenone: Current state of the art and application to human exposure assessment. *Archives of Toxicology, 90*, 1281–1292.

[115] Richard, J. L., (2007). Some major mycotoxins and their mycotoxicoses-An overview. *International Journal of Food Microbiology, 119*, 3–10.

[116] European Commission-EC. (2006). Commission Regulation (EC) No 1881/ 2006 of 19 December as amended, on setting maximum levels of certain contaminants in foodstuffs. http://eurlex.europa. eu/LexUriServ/LexUriServ.do?uriDOJ:L:2006:364:000 5:0024:EN: PDF.

[117] Mostrom, M., (2016). *Mycotoxins: Toxicology. Encyclopedia of food and health* (pp. 43–48). Fargo, USA: North Dakota State University.

[118] Pestka, J. J., (2010). Deoxynivalenol: Mechanisms of action, human exposure, and toxicological relevance. *Archives of Toxicology, 84*, 663–679.

[119] Wu, F., Groopman, J. D., & Pestka, J. J., (2014). Public Health Impacts of Foodborne Mycotoxins. *Annual Review of Food Science and Technology, 5*, 351–372.

[120] Chen, C., Turna, S. N., & Wu, F., (2019). Risk assessment of dietary deoxynivalenol exposure in wheat products worldwide: Are new codex DON guidelines adequately protective? *Trends in Food Science and Technology, 89*, 11–25.

[121] Codex Alimentarius Commission (2015). *38th Session of the Codex Alimentarius Commission*. Geneva, Switzerland: http://www.fao.org/fao-who-codexalimentarius/roster/detail/en/c/297672/.

[122] Puel, O., Galtier, P., & Oswald, I. P., (2010). Biosynthesis and toxicological effects of patulin. *Toxins, 2*, 613–631.

[123] Leggott, N. L., & Shephard, G. S., (2001). Patulin in South African commercial apple products. *Food Control, 12*, 73–76.

[124] Mahato, D. K., Lee, K. E., Kamle, M., et al. (2019). Aflatoxins in Food and Feed: An Overview on Prevalence, Detection and Control Strategies. *Frontiers in Microbiology, 10*, 1–10.

[125] Iqbal, S. Z., (2021). Mycotoxins in food, recent development in food analysis and future challenges; a review. *Current Opinion in Food Science, 42*, 237–247.

[126] Tao, S., Wang, L., Zhu, Z., Liu, Y., Wu, L., Yuan, C., Zhang, G., & Wang, Z., (2018). Adverse effects of bisphenol A on Sertoli cell blood-testis barrier in rare minnow Gobiocypris rarus. *Ecotoxicology and Environmental, 171*, 475–483.

[127] Sulyok, M., Beed, F., S. boni, A., Abass, A. M., & Krska, R., (2015). Quantitation of multiple mycotoxins and cyanogenic glucosides in cassava samples from Tanzania and Rwanda by LC-MS/ MS-based multi-toxin method. *Food Additives andvContaminants: Part A, 32*(4), 488–502.

[128] Stadler, R. H., (2019). In M. Granvogl & MacMahon S. (Ed.), *Food Process Contaminants* (pp. 1–13). Food-Borne Toxicants: Formation, Analysis, and Toxicology Washington, DC: American Chemical Society.

[129] Abdel-Shafy, H. I., & Mansour, M. S. M., (2016). A review on polycyclic aromatic hydrocarbons: Source, environmental impact, effect on human health and remediation. *Egyptian Journal of Petroleum, 25*(1), 107–123.

[130] Sampaio, G. R., Guizellini, G. M., Da silva, S. A., et al. (2021). Polycyclic Aromatic Hydrocarbons in Foods: Biological Effects, Legislation, Occurrence, Analytical Methods, and Strategies to Reduce Their Formation. *International Journal of Molecular Sciences, 22*, 11, 6010.

[131] Sonavane, M., & Gassman, N. R., (2019). Bisphenol A co-exposure effects: a key factor in understanding BPA's complex mechanism and health outcomes. *Critical Reviews in Toxicology, 49*, 371–386.

[132] Ji, H., Miao, M., Liang, H., et al. (2018). Exposure of environmental Bisphenol A in relation to routine sperm parameters and sperm movement characteristics among fertile men. *Scientific Reports, 8*(1), 17548.

[133] Mustieles, V., Ocon-Hernandez, O., Mínguez-Alarcon, L., et al. (2018). Bisphenol A and reproductive hormones and cortisol in peripubertal boys: The INMA-Granada cohort. *Science of the Total Environment, 618*, 1046–1053.

[134] Ozel, Ş., Tokmak, A., Aykut, O., Aktulay, A., Hançerlioğulları, N., & Engin Ustun, Y., (2019). Serum levels of phthalates and bisphenol-A in patients with primary ovarian insufficiency. *Gynaecological Endocrinology, 35*, 364–367.

[135] Pollack, A. Z., Mumford, S. L., Krall, J. R., Carmichael, A. E., Sjaarda, L. A., Perkins, N. J., Kannan, K., & Schisterman, E. F., (2018). Exposure to bisphenol A, chlorophenols, benzophenones, and parabens

in relation to reproductive hormones in healthy women: A chemical mixture approach. *Environment International*, *120*, 137–144.

[136] Mammadov, E., Uncu, M., & Dalkan, C., (2018). High Prenatal Exposure to Bisphenol A Reduces Anogenital Distance in Healthy Male Newborns. *Journal of Clinical Research in Pediatric Endocrinology*, *10*, 25–29.

[137] Jensen, T. K., Mustieles, V., Bleses, D., et al. (2019). Prenatal bisphenol a exposure is associated with language development but not with ADHD-related behavior in toddlers from the Odense Child Coh. *Environmental Research*, *170*, 398–405.

[138] Wang, Z., Liang, H., Tu, X., et al. (2019). Bisphenol A and pubertal height growth in school-aged children. *Journal of Exposure Science and Environmental Epidemiology*, *29*, 109–117.

[139] Xu, J., Huang, G., & Guo, T. L., (2016). Developmental Bisphenol A Exposure Modulates Immune-Related Diseases. *Toxins*, *4*(4), 23.

[140] Tao, S., Wang, L., Zhu, Z., et al. (2019). Adverse effects of bisphenol A on Sertoli cell blood-testis barrier in rare minnow Gobiocypris rarus. *Ecotoxicology and Environmental Safety*, *171*, 475–483.

[141] Berger, K., Eskenazi, B., Balmes, J., Kogut, K., Holland, N., Calafat, A. M., & Harley, K. G., (2019). Prenatal high molecular weight phthalates and bisphenol A, and childhood respiratory and allergic outcomes. *Pediatric Allergy and Immunology*, *30*(1), 36–46.

[142] Li, J., Wang, Y., Fang, F., Chen, D., Gao, Y., Liu, J., Gao, R., Wang, J., & Xiao, H., (2016). Bisphenol A disrupts glucose transport and neurophysiological role of IR/IRS/AKT/GSK3β axis in the brain of male mice. *Environmental Toxicology Pharmacology*, *43*, 7–12.

[143] Kharrazian, D., & Vojdani, A., (2017). Correlation between antibodies to bisphenol A, its target enzyme protein disulfide isomerase and antibodies to neuron-specific antigens. *Journal of Applied Toxicology*, *37*, 479–484.

[144] Xin, F., Fischer, E., Krapp, C., et al. (2018). Mice exposed to bisphenol A exhibit depressive-like behavior with neurotransmitter and neuroactive steroid dysfunction. *Hormones and Behavior*, *102*, 93–104.

[145] Pan, R., Wang, C., Shi, R., et al. (2019). Prenatal Bisphenol A exposure and early childhood neurodevelopment in Shandong, China. *International Journal of Hygiene and Environmental Health*, *222*(5), 896–902.

[146] Bosch-Panadero, E., Mas, S., Civantos, E., et al. (2018). Bisphenol A is an exogenous toxin that promotes mitochondrial injury and death in tubular cells. *Environmental Toxicology: An International Journal*, *33*(3), 325–332.

[147] Mielech, A., Puscion-Jakubik, A., & Socha, K., (2021). Assessment of the Risk of Contamination of Food for Infants and Toddlers. *Nutrition*, *13*, 2358.

[148] Stadler, R. H., Blank, I., Varga, N., Robert, F., Hau, J., Guy, P. A., Robert, M. C., & Riediker, S., (2002). Acrylamide from Maillard reaction products. *Nature*, *419*(6906), 449–450.

[149] Mottram, D. S., Wedzicha, B. L., & Dodson, A. T., (2002). Acrylamide is formed in the Maillard reaction. *Nature*, *419*, 448–449.

[150] Koszucka, A., Nowak, A., Nowak, I., & Motyl, I., (2020). Acrylamide in human diet, its metabolism, toxicity, inactivation and the associated European Union legal regulations in food industry. *Critical Reviews in Food Science and Nutrition*, *60*(10), 1677–1692.

[151] Lineback, D. R., Coughlin, J. R., & Stadler, R. H., (2012). Acrylamide in foods: a review of the science and future considerations. *Annual Reviews of Food Science and Technology*, *3*, 15–35.

[152] Semla, M., Goc, Z., Martiniaková, M., Omelka, R., & Formicki, G., (2017). Acrylamide: a common food toxin related to physiological functions and health. *Physics Research*, *66*(2), 205–217.

[153] Jakszyn, P., Agudo, A., Berenguer, A., Ibanez, R., Amiano, P., Pera, G., et al. (2006). Intake and food sources of nitrites and N-nitrosodimethylamine in Spain. *Public Health Nutrition*, *9*, 785–791.

[154] Anderson, L. M., Souliotis, V. L., Chhabra, S. K., Moskal, T. J., Harbaugh, S. D., & Kyrtopoulos, S. A., (1996). N-nitrosodimethylamine-derived O[6]-methylguanine in DNA of monkey gastrointestinal and

urogenital organs and enhancement by ethanol. *International Journal of Cancer Research, 66,* 130–134.

[155] Flores, M., Mora, L., Reig, M., & Toldra, F., (2019). Risk assessment of chemical substances of safety concern generated in processed meats. *Food Science and Human Wellness, 8,* 244–251.

[156] Yurchenko, S., & Molder, U., (2006). Volatile N-nitrosamines in various fish products. *Fluorine Chemistry, 96*(2), 325–333.

[157] Ho, I., & Bratzler, L., (1970). Effect of sodium nitrite on flavor of cured pork. *Journal of Food Science, 35,* 668–670.

[158] EFSA (2017). Re-evaluation of potassium nitrite (E 249) and sodium nitrite (E 250) as food additives. *EFSA Journal,15,* 4786.

[159] IARC (International Agency for Research on Cancer) (1987). Overall evaluation of carcinogenicity: an updating of IARC monographs on the Evaluation of the Carcinogenic Risk of Chemicals to Humans. *Int Ag for Res on Cancr, 8*(7), 1–42.

[160] Hecht, S. S., (2003). Tobacco Carcinogens, Their Biomarkers and Tobacco-Induced Cancer. *Nature Reviews Cancer, 3,* 733–744.

[161] Williams, D. L. H., (2004). *Nitrosation Reactions and the Chemistry of Nitric Oxide*(1st ed). Amsterdam, BV: Elsevier.

[162] DellaValle, C. T., Xiao, Q., Yang, G., et al. (2014). Dietary nitrate and nitrite intake and risk of colorectal cancer in the Shanghai Women's Health Study. *International Journal of Cancer, 134,* 2917–2926.

[163] Chan, T. Y., (2011). Vegetable-borne nitrate and nitrite and the risk of methaemoglobinaemia. *Toxicology Letter, 200,* 107–108.

[164] Barbieri, G., Bergamaschi, M., & Franceschini, M., (2013). Kinetics of nitrite evaluated in a meat product. *Meat Science, 93,* 282–286.

[165] Pensabene, J. W., Fiddler, W., Gates, R. A., Fagan, J. C., & Wasserman, A. E., (1974). Effect of frying and other cooking conditions on nitrosopyrrolidine formation in bacon. *Journal of Food Science, 39,* 314–316.

[166] Herrmann, S. S., Duedahl-Olesen, L., & Granby, K., (2015). Occurrence of volatile and non-volatile N-nitrosamines in processed meat products and the role of heat treatment. *Food Control, 48,* 163–169.

[167] Mitch, W. A., Sharp, J. O., Trussel, R. R., Valentin, E. R. L., Alvarez-Cohen, L., & Sedlak, D. L., (2003). N-nitrosodimethyl amine (NDMA) as a drinking water contaminant: A review. *Environmental Engineering Science, 20*(5), 389–403.

[168] Epley, R. J., Addis, P. B., & Warthesen, J. J., Nitrite in meat. In *Animal Science.* AG-FS-0974-A Minnesota Extension Service, University of Minnesota.

[169] Islam, M. A., Amin, M. N., Siddiqui, S. A., Hossain, M. P., Sultana, F., & Kabir, M. R., (2019). Transfatty acids and lipid profile: A serious risk factor to cardiovascular disease, cancer and diabetes. *Diabetes and Metabolic Syndrome, 13,* 1643–1647.

[170] Ganguly, R., & Pierce, G. N., (2015). The toxicity of dietary trans fats. *Food and Chemical Toxicology, 78,* 170–176.

[171] Stender, S., (2020). Trans fat in foods in Iran, South-Eastern Europe, Caucasia and Central Asia: a market basket investigation. *Food Policy, 96,* 0306–9192.

[172] Lock, A. L., Corl, B. A., Barbano, D. M., Bauman, D. E., & Ip, C., (2004). The anti-carcinogenic effect of trans-11 18:1 is dependent on its conversion to cis-9, trans-11 CLA by delta 9- desaturase in rats. *Journal of Nutrition, 134,* 2698–2704.

[173] Stender, S., Astrup, A., & Dyerberg, J., (2008). Ruminant and industrially produced trans fatty acids: health aspects. *Food and Nutrition Research, 52*(1), 1651.

[174] Alonso, L., Fontecha, J., Lozada, L., Fraga, M. J., & Juarez, M., (2019). Fatty acid composition of caprine milk: major, branched-chain, and trans fatty acids. *Journal of Dairy Science, 82*(5), 878–884.

[175] Mozaffarian, D., Martijn, B. K., Alberto, A., Stampfer, M. J., & Walter, C. W., (2006). Trans fatty acids and cardiovascular disease. *New England Journal of Medicine*, *354*, 1601–1613.

[176] Downs, S. M., Loeh, S., & Wu, J. H., (2016). Trans fatty acids: a summary of the evidence relating consumption to cardiovascular outcomes and the efficacy of prevention policy to reduce levels in the food supply. In A. Bendich & Deckelbaum R. (Eds.), *Preventive nutrition* (5th ed). Switzerland: Springer.

[177] Islam, M. A., Amin, M. N., Siddiqui, S. A., Hossain, M. P., Sultana, F., & Kabir, M. R., (2019). Transfatty acids and lipid profile: A serious risk factor to cardiovascular disease, cancer and diabetes. *Diabetes and Metabolic Syndrome*, *13*, 1643–1647.

[178] Oh, K., Hu, F. B., Manson, J. E., Stampfer, M. J., & Willett, W. C., (2005). Dietary fat intake and risk of coronary heart disease in women: 20 years of follow-up of the Nurses. *American Journal of Epidemiology*, *161*, 672–679.

[179] Ablao, C. J. N., Sagum, R. S., Maddela, A. K. M., & Macapagal, J., (2020). Road to trans-fat free Philippines: An emerging milestone amidst COVID-19 pandemic. *Lancet Regional Health – Western Pacific*, *1*, 2666–6065.

[180] Pojjanapornpun, S., Nakornsadet, A., Sombutsuwan, P., Chumsantea, S., Krisnangkura, K., & Aryusuk, K. (2020). Simulation of fatty acid peak profile on a 100-m CP-Sil 88 column and application for analysis of commercial trans-fat alternatives in Thailand. *Journal of Food Composition and Analysis*, *91*, 103525.

Jamiu Olaseni Aribisala and Saheed Sabiu*

Chapter 14
Synthetic toxicants and their health implications

Abstract: Synthetic toxicants are environmentally hazardous chemical agents that can contaminate abiotic systems (water, soil, and air), creating negative health impacts in biotic ecosystems (animals, food, plants, and microorganisms). While the point sources release encompasses synthetic toxicants from a single, confined, and identifiable channel, the nonpoint toxicants are from many diffused sources and are usually caused by atmospheric deposition. The release of synthetic toxicants into the environment has been linked to several environmental disasters with accompanying health implications relating to cancer, food intoxication, gastrointestinal infections, neurological and kidney damage, congenital defects, and reproductive issues. This chapter examined selected toxicological disasters over the years, their impacts on the environment, and their associated health implications. Also, background information on the type of synthetic toxicants and underlying mechanisms of action, food adulteration, and legislation against them in several countries were discussed. As synthetic toxicants have continuously been released in higher concentrations into the ecosystem, threatening humans' health, it is hoped that the recognition of these threats would drive toxicological studies into continuous monitoring and assessment of their deleterious impact.

14.1 Introduction

Environmental toxicants and pollutants are eco-harmful physical and chemical agents that can contaminate the abiotic systems (water, soil, and air), consequently causing adverse health effects in the biotic ecosystems (animals, food, plants, and microbes) [1]. The emerging contaminants, endocrine disruptors, carcinogens, mutagens, neurotoxins, and several other environmental toxicants emanate from a variety of point and nonpoint sources. While the point sources encompass toxicants from a single,

Acknowledgment: The TWAS-NRF bursary awarded to Aribisala O. Jamiu for his PhD program tenable at Durban University of Technology is thankfully acknowledged.

*Corresponding author: Saheed Sabiu,** Department of Biotechnology and Food Science, Faculty of Applied Sciences, Durban University of Technology, P.O. Box 1334, Durban 4000, South Africa, e-mail: sabius@dut.ac.za
Jamiu Olaseni Aribisala, Department of Biotechnology and Food Science, Faculty of Applied Sciences, Durban University of Technology, P.O. Box 1334, Durban, 4000, South Africa

https://doi.org/10.1515/9783110748345-014

confined, and identifiable channel, the nonpoint toxicant are from many diffused sources and are usually caused by atmospheric deposition as well as oil and toxic chemical spillage [2, 3]. Unlike the synthetic toxicants, the natural toxicants such as essential oil, collagen, and enzymes are toxic compounds that are naturally produced by living organisms [4]. These toxins are harmless to the organisms themselves but could be dangerous to humans and animals if consumed [5]. The synthetic toxicants, on the other hand, tend to buildup in fatty tissues and organs of animals and contribute to the so-called "Chemical Body Burden" and have been found to reduce the body's ability to fight diseases [2, 5, 6]. For instance, parabens (found in over 90% of skin care products) have been linked to breast tumors in fatty tissues of the breasts, while sulfates accumulate in the brain, heart, liver, and lungs, creating a variety of health problems [6].

Globally, the synthetic toxicants remain one of the primary causes of food contamination and have been implicated in outbreaks of foodborne diseases [7]. Synthetic toxicants come from diverse sources, including soil, disinfection byproducts, healthcare products, air, water, and packaging materials [8] and could easily contaminate food products. Although, the incidences of food contamination are traceable back to more than 8,000 years, the current advent and rise in globalization and agribusinesses have worsened the situation globally [7, 8]. For example, about 1527 foodborne disease outbreaks attributable to synthetic toxicants occurred in the United States from 2009 to 2010, causing 23 mortality and 29,444 illnesses [9]. Similarly, in 2010, around 500 Nigerian children succumbed to acute lead toxicity death following consumption of toxicants-contaminated food harvested from lead-polluted environment and soil [10]. Generally, due to increased industrial activities, synthetic toxicants have also become more prevalent in recent years, and such activities have largely resulted in atmospheric/air/environmental contamination [2, 3, 10]. Besides environmental contamination, foods contaminated with synthetic toxicants have far-reaching implications on human health and well-being as could be seen with heavy metals and pesticides ingestion, culminating in gastrointestinal diseases [10]. Hence, a continuous account and understanding of the links between the synthetic pollutants and human health is imperative to know the risk assessment indicators with a view to avoid potential health risks. Here, we present an insight on synthetic toxicants, their implications on human health and well-being, and common regulations governing their controls within the ecosystem.

14.2 Environmental exposure to toxicants

The contamination of the ecosystem due to human actions remains a complex and persistent global challenge confronting scientists, stakeholders, and regulatory agencies [3]. The world has recently seen a quick and unequal growth in technology and industrialization, as well as unfair and unsustainable use of nonrenewable resources

[2]. The aftermath of this growth has culminated in contaminants, posing danger to the environment's wholesomeness [3]. Residue of toxicant deposited in the environment during transportation and distribution of product made of synthetic toxicant have negative impacts that range from short to long term on the immediate environment [4]. Consequently, environmental pollution has become a key global public health concern as the accompanying environmental contaminants, especially the trace elements, have posed numerous risks to ecological, plant, and human health [11]. In humans, exposure to elevated amount of these contaminants may have deleterious health implications as synthetic toxicants are extremely persistent and can accumulate in human tissues, leading to a variety of diseases [12]. Also, plants and other live forms in the environment are not left out as synthetic toxicants could pollute their tissues once a contact is established [13, 14]. As a result, the negative consequences of the widespread presence of toxicants have emerged as a major technical and societal impediment to the production of pollutant-free foods [5].

Generally, human's exposure to contaminants could be through several mechanisms such as inhalation and absorption from the environment, food, water, and soil [14]. From the environment, exposure could be through inhalation of chemicals unintentionally discharged into the atmosphere from industrial activities, while ingestion of heavy metals such as lead contaminated foods and water is a possible means of human exposure to synthetic contaminants through food and water. From the soil, contaminants can buildup in edible tissues of plants which are subsequently consumed by animals and/or by humans [15]. Studies have traced the potential mechanisms of pollutant transmission and adverse effects from soil to plants, and subsequently, to humans [15]. These observations have consequently sparked several interests in evaluating the risks associated with toxicants from marketable edibles provided by rural and urban agriculture [14, 15], which will undoubtedly help in predicting the likely paths and quantities of toxicant exposure to humans through these sources. Such risk assessments could help to avoid health-related risks as recently being reported [6, 14].

The increased environmental consciousness in industry has resulted in the birth of yet another significant area known as occupational medicine [6]. This area entails a thorough comprehension of illnesses that arise due to human exposure to synthetic occupational and industrial toxicants, and the methods for controlling and preventing the illnesses [6]. For example, the ability of synthetic toxicants to impede the body's immune system in fighting diseases have been demonstrated to result in several illnesses such as cancer, food intoxication, gastrointestinal infections, neurological and kidney damage, congenital defects, and reproductive issues [10]. Also, occurrence of environmental disasters has been studied to have consequential environmental impacts and accompanying health implications due to release of varying types of toxicants [14]. Selected disasters over the years and their impacts on the environment are presented in Table 14.1.

Table 14.1: Selected toxicological disasters and their health implications.

Year	Place	Descriptions and health implications	References
2018	China	Carbon nine is escaping into the atmosphere. Some fishermen were taken to the hospital. Contaminants were released into the sea, killing many farmed fish. The sale of aquatic items was halted following the incident, and the adjoining salt field was shut down.	[16]
2012	China	The Guangxi Longjiang River and water supplies have been poisoned with toxic cadmium. Between January 15 and February 2, more than 40,000 kg of fish were found dead within the city limits of Hechi, according to authorities.	[6, 17]
2010	Nigeria	After eating food polluted with lead-infected soil and dust, an approximate of 400–500 kids succumbed to acute lead toxicity.	[18]
2007	Australia	Thousands of bird's deaths gave awareness to toxicity threat, due to the inappropriate transportation of lead ore. The federal government paid for a multimillion-dollar clean-up after excessive levels of lead were found in several adults and children, as well as in water tanks.	[19]
2003	Iraq	The largest synthetic emission of SO_2 experienced to date due to fire outbreak over a 3-week period in June 2003. A maximum of 21,000 tons of SO_2 per day were released. People were hospitalized, and most of the vegetation destroyed.	[20]
1990	Bronx, New York	At the Happy Land Social Club, toxic fumes containing carbon monoxide and cyanide killed 87 people.	[21]
1986	Chernobyl, Soviet Union	The atomic control plant's ionizing radiation killed 32 people right away and harmed 5 million others.	[22]
1984	Bhopal, India	Three thousand people died and 200,000 were injured because of an industrial emission of methyl isocyanate vapor.	[23]
1982	Iraq and Iran	Thousands of people died because of mustard gas used during chemical warfare.	[24]
1978	Love Canal, New York	Toxic wastes sparked widespread alarm.	[25]
1976	Seveso, Italy	After an unintentional discharge of dioxin into the atmosphere, the prevalence of chloracne has risen.	[26]
1970s	California	The nematocide 1,2-dibromo-3-chloropropane implicated in sterility of pesticide personnel.	[27]
1971	Iraq	Four hundred and fifty-nine people died because of methylmercury-contaminated grain.	[28]
1952	London, England	Due to photochemical haze approximately 4,000 people died.	[29]

Table 14.1 (continued)

Year	Place	Descriptions and health implications	References
1950	Minamata Bay, Japan	Eating of lead to organic mercury contaminated fish lead to food poisoning.	[30]
1948	Donora, Pennsylvania	Smog caused 20 deaths and thousands of people to become ill because of the contaminated air.	[31]
1939–1954	Japan	Itai-Itai disease occurred due to water contaminated with cadmium.	[32]
1930	Meuse Valley, Belgium	Smog resulted in 64 deaths and illnesses.	
1920s	Worldwide	Asbestos exposure puts loads of people in danger, leading to a high rate of asbestos-related sickness and cancer.	[33]
1900s	Worldwide	Bladder cancer rate increase among workers in dye industry due to exposure to β-naphthylamine.	[34]
1700s	England	As the first occupation-related cancer to occur, there has been a significant increase in scrotal cancer in smokestacks caused by polycyclic aromatic hydrocarbon exposure	[20]
79 A.D	Pompeii	Thousands of people died because of volcanic gas, particularly nitrogen and sulfur oxides, released by Mount Vesuvius' eruption.	[14, 34]

14.2.1 Types of synthetic toxicants and their underlying toxicity mechanisms

The synthetic toxicants can either be organic or inorganic. The organic toxicants include pesticides, polychlorinated biphenyls (PCBs), polycyclic aromatic hydrocarbons (PAHs), azo dyes, endocrine-disrupting chemicals, chlorinated phenols, and so on, whereas CO, SO_2, and several heavy metals including Hg, As, Pb, Cr, and Cd constitute the inorganic synthetic toxicants [35].

14.2.1.1 Synthetic organic and inorganic toxicants

Synthetic organic toxicants include man-made substances such as solvents, pesticides and fertilizers, dyes, plasticizers from the industries, detergents, surfactants, prescription products, and explosives used in industrial operations and war [36]. These agents may end up in the environment as a result of production, usage, spillage, or inadequate management. For instance, polychlorinated dibenzo-*para*-dioxins and polychlorinated dibenzofurans are the end products of production of certain chlorinated pesticides, as well as the anaerobic digestion of chlorophenolic wastes [37]. Although most of the organic toxicants polluting the environment are man-made, natural events also contribute to atmospheric pollution [35]. For example, bushfires are a major source of PAHs that are naturally present in small quantity in the atmosphere; however, there has been significant increase in their environmental concentrations due to increased fossil fuel applications recently [35, 38]. Even though oil spills from the unintentional source (such as leakages from oil tankers) are not particularly threatening, they can have other temporary adverse effects on the habitats [6, 39].

In addition to their inherent toxicity, toxic acids are formed when carbon monoxide, sulfur dioxide and other inorganic toxicants produced during volcanic eruptions react with water vapor in the air. At a concentration beyond 2%, CO_2 becomes hazardous [40], and human activities such as the use of fertilizer, coal and petroleum fuel combustion have increased the amount of CO_2 and other toxicants in the environment [1]. Some selected synthetic organic and inorganic toxicants and their mechanisms of toxicity are presented in Table 14.2.

Table 14.2: Some synthetic organic and inorganic toxicants and their toxicity mechanisms.

Organic compounds	Toxicity mechanisms	References
2,4-D and triclopyr	Disruptors of auxin: They inhibit the development of plant by imitating auxins (hormones of plant).	[5]
Acids, anilines, diquat, formalin, paraquat, and solid bases	Tissues disruptors: These toxicants induce the tissues of living organisms through oxidation or reduction process.	[5]
Benzene, dioxins, and PAH	Carcinogenic and mutagenic: These toxicants form adducts with DNA, resulting in mutation of gene which contribute to the development of cancer.	[41]
Bromoxynil and triazines	Photosynthesis inhibitors: They inhibit chloroplasts in plant cells and algae.	[41]
Carbamate insecticides, organochlorines, organophosphorus, pyrethroids, pyrethrum, and synthetic	Neurotoxic: Some of these toxicants block the neuronal cells receptors while others act on voltage-dependent axonal sodium channels, affecting the brain impulse and triggering seizures.	[42]

Table 14.2 (continued)

Organic compounds	Toxicity mechanisms	References
Chloroacetamides, dinitroanilines, dichlobenil, imidazoles, morpholines thiocarbamates, and triazoles	Germination inhibitors: These toxicants impair cell development through disruption of protein, lipid, and fatty acid synthesis. Some toxicants also halt ergosterol synthesis, a key constituent of fungus cell layers while others interrupt the formation of microtubules, thus preventing cell development.	[41, 43]
Eliquis, lixiana, and pradaxa	Anticoagulants: They include toxicants that prevent blood from clotting through vitamin K regeneration inhibition, resulting in defective hepatocytes. In some cases, it may lead to hemorrhaging. They are mostly used against rodents, vertebrates, and birds.	[44]
Fungicides, organometallic compounds, and strobilurin	Respiratory inhibitors: They interfere with the mitochondria of bacteria specifically by inhibiting complex III involved in electron transfer process during ATP generation.	[45]
Herbicides, glyphosate, insecticides, and sulfonamides	Inhibitors of biosynthetic processes: They target specific processes such as inhibition of amino acids/proteins production in fungi and plant cells as well as preventing arthropods from producing chitin, preventing them from mounting and developing.	[45]
Phthalates, organochlorines, and organometallic compounds	Endocrine system disruptors: They act in the same way as hormones do in animals, controlling complex physiological processes such as: activation of the complex P450 for detoxification as well as forming metabolites that interact with thyroxine thereby, lowering retinol levels in the blood and causing thyroid cancer.	[43]
Inorganic compounds		
Arsenic	Compete with phosphorus in several phosphorylation reactions, causing disruption in ATP production.	[40]
Cadmium	Replaces some metals such as zinc and calcium in certain metabolic processes, and can cause cancer if inhaled.	[40]

Table 14.2 (continued)

Organic compounds	Toxicity mechanisms	References
Carbon monoxide	Attach with hemoglobin thereby preventing it from transporting oxygen in the blood.	[1]
Copper	Buildup in some cells (e.g., algae and fungi spores) and stop them from germinating. Copper ions disrupt sodium control in fish, and can cause damage to the liver in mammals.	[46]
Lead	Interferes with calcium ions during nerve conduction and inhibits several enzymes involved in hemoglobin synthesis.	[47]
Mercury	They are toxic due to their ability to enter tissues and reach the nervous system.	[47]
Zinc	In free state, they could be corrosive with wide range of biocidal activity.	[46]

14.2.1.2 Mechanism of action of toxicants

Toxicants trigger toxicity following transportation to target sites in plants and animals after been absorbed via appropriate organs (leaves and roots) in plants and skin, digestive and respiratory systems in animals [48]. At the target site, the toxicant or its biotransformed product is absorbed while the remaining unabsorbed portions are excreted and metabolized to harmless products, or retained in lipid tissues [40, 46, 48]. Toxicants that have been absorbed interact with endogenous target molecules, stimulating cellular dysfunction and possibly disrupting repair mechanisms at all functional levels [40]. Events culminating in toxicities are interconnected and entail a series of processes that begin with exposure and progress through delivery, interactions with target cells, and changes in the biological environment, culminating in toxicity and dysfunctional repair (Figure 14.1).

14.2.2 Metabolism of synthetic toxicants

A toxicant's ability to be metabolized and excreted is one of the most significant factors determining its survival and eventual toxicological effects [37]. Many families of metabolic enzymes are involved in the metabolism of toxicants, several of which have a wide range of substrate specificity [37, 49]. The flavin-containing monooxygenases (FMOs), cytochrome P450 monooxygenases (CYPs), amine oxidases, cyclooxygenases, reductases, hy-

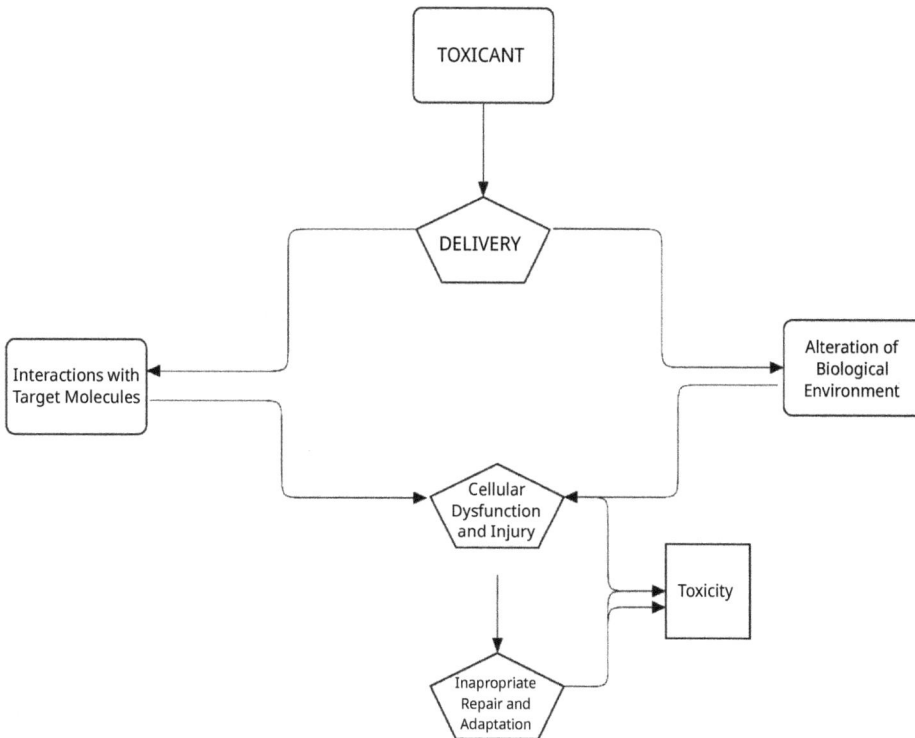

Figure 14.1: A typical illustration of toxicity mechanism of synthetic toxicants (self-creation).

drolases, sulfotransferases, methyltransferases, glutathione transferases, and several other conjugating enzymes are representatives of the most significant enzymes implicated in the metabolism of toxicants [49]. The liver, an organ where several crucial functional proteins are synthesized, is also where most toxicants' metabolism occurs, as it can mediate chemical transformations. The lipophilic nature of many toxicants makes it easy for them to be transported by lipoproteins for further processing [37, 49].

Usually, toxicants go through one or two phases of metabolism (Figure 14.2) in the liver and other metabolizing organs. A polar reactive group is added to the molecule in phase I, making it a suitable substrate for phase II enzymes [2]. The CYPs, FMOs, and hydrolases are commonly involved in phase I metabolism [50, 51], where bulky substituents like sugars, sulfates, or amino acids, are conjugated to significantly boost the toxicant's water solubility trait for further detoxification and subsequent excretion [2, 37, 51]. Due to the products' normal properties of being electrophiles capable of reacting with nucleophilic substituents on macromolecules, monooxygenations during phase I are likely to produce reactive intermediates more often than phase II metabolic processes [52].

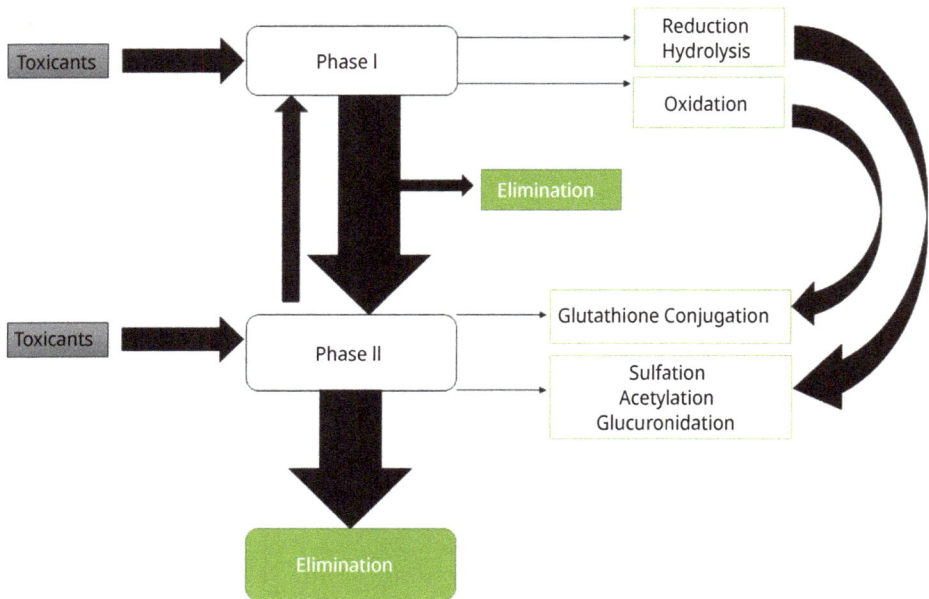

Figure 14.2: A typical representation of phases I and II of toxicant metabolisms (self-creation). Toxicants are rendered more polar in phase I through functionalization (reduction, oxidation, hydrolysis, etc.) in preparation for phase II conjugation reactions (glucuronidation, acylation, sulfation, etc.) for subsequent elimination. Sometimes, direct elimination may also occur after phase I without necessarily progressing to phase II.

14.3 Synthetic toxicants and food adulteration

By combining various forms of hazardous chemicals and poisonous synthetic colorants, as well as mixing rotting and perishable foods with healthy product to market, food is polluted or adulterated. Various chemicals and techniques have been used to preserve food during storage since the dawn of civilization [53]. When such preservatives or additives are not hazardous to human health, then using chemicals as preservatives is good and ethically sound [53]. However, problem arises when hazardous synthetic substances are added to food in the name of preservation to either preserve its freshness or extend the shelf life [54]. The food adulterated by these practices does not meet the Food and Agriculture Organization's (FAO) or World Health Organization's (WHO) definition of healthy food, and several factors contributing to food adulteration such as financial interest, carelessness, lack of knowledge and awareness about sanitation, packaging, storage, and transportation have been identified over the years [55]. Consequently, consumers are deceived and are exposed to various chronic and acute diseases. Food adulterations are more prominent in developing countries as there are no regulations against them but in developed countries, globalization is making them less common

[55]. Adulteration of foods and their products may be either accidental or deliberate. The accidental adulteration involves occurrence of substandard foods due to natural phenomenal such as drought, inadequate rainfall, resulting in short storage conditions [44]. Intentional adulteration, on the other hand, is done with the purpose of defrauding or misleading customers [44]. This is more severe, as it affects the citizens' well-being, including the health of innocent individuals who consume these tainted foods [55]. Adulteration occurs at all stages of the food chain, from preparation to consumption [55]. Most food producers, processors, restaurant owners, retail suppliers, retailers, and so on are all culprits in this illegal activity of food adulteration in one way or another [44].

14.3.1 Sources of food adulteration

Adulteration of food embroils addition of excessive, preventable, and/or harmful substances into food through unintentional or intentional means. The substance added could be injurious or of low quality which could occur in food naturally or originate from the environment, during food manufacturing, through food additives or through migrants from packaging materials [56].

14.3.1.1 Natural occurring food adulteration

Native toxins in crops and animals are sources of food adulteration which occur naturally and when consumed in huge quantities could lead to serious illness. This type of adulteration includes the presence of certain organic or inorganic chemicals or native radicals in foods like toxic varieties of crops and vegetables, sea foods, and mushrooms [57]. For instance, consumption of phalloidin which is a toxin present in amanita mushroom has been shown to cause liver and kidney damage [57]. Neurotoxin in *Lathyrus sativus* (grass pea) is another good example of naturally occurring toxin present in food which has been shown to cause crippling disease known as lathyrism when consumed in huge amount [58]. Some foods such as stone fruits, cassava, bamboo roots, almonds, and sorghum have been shown to contain high concentration of cyanogenic glycoside whose toxicity has been linked to mental disorientation, vomiting, headache, low blood pressure, cyanosis, rapid breathing, dizziness, stomach pains, and diarrhea [57]. Similarly, lectins are poisons found in many varieties of beans and if eaten uncooked can cause severe vomiting, stomach pain, and diarrhea. However, food processing such as cooking and fermentation has been shown to eliminate lectins from beans [57].

14.3.1.2 Environmental food adulteration

Environmental pollutants are impurities found in water, air, and soil that are either introduced into food by humans or occur naturally. These pollutants can be identified using biosensor assays [59]. Because of industrial contamination, it is possible for foods to be contaminated with heavy metals (Pb, Hg, Cd, etc.) and PCBs, and this was the case with thousands of tons of heavy metals that were smelted in the industrial district of Huludao in Northeast China [14]. These pollutants could find their way into food through several means such as plant absorption of these toxicants from the environment [60]. Since plants are at the bottom of the food chain, they are able to quickly consume toxins from the soil, which contaminates not only fruits and vegetables but also fish when these pollutants find their way into water bodies through water run-off and industrial linkage [61, 62]. Polycyclic aromatic hydrocarbons (PAHs) which are well known for their carcinogenic and mutagenic properties when consumed at high concentration can find their way into the food chain through incomplete fossil fuels, burning of oil and coal as well as via food processing through the burning of woods for meat and fish smoking [63]. Many pesticides used as plant defense agents penetrate into the food chain, affecting human's health in adverse ways, including itching, brain damage, and causing cancer [61]. Pesticides are applied in the amount of 3 billion kg per year around the world, posing a significant hazard [14, 60]. Since the maximum residue levels (MRLs) of pesticides, on the other hand, determine their toxicity, in order to reduce consumer exposure to pesticide residues in food, laws in many countries have limited their usage [60]. However, in many developing nations, such legislation does not exist or is poorly implemented. For instance, in Israel, at least two types of pesticides are often being identified in honeybee with the lipophilic pesticides must often found in beeswax which can impede the well-being of children when consumed [64]. Furthermore, veterinary drug residues in farm animals, similar to pesticide residues, can remain in the meat, posing threat to human through drug residue exposure, antibiotic resistance transmission, and allergy risk [62]. Also, contamination of foods in the environment can occur because of some natural phenomenon native to the environment from where the foods are harvested or collected [65]. For instance, algal toxins produce by some species of algae during blooming have been found in the ocean contaminating drinking water and shellfish, such as mussels, scallops, and oysters, which when consumed by human and other animals can cause diarrhea, vomiting, tingling, paralysis, and other severe symptoms [65]. Furthermore, some weeds mostly belonging to the family Boraginaceae, Asteraceae, and Fabaceae have been shown to produce pyrrolizidine alkaloids (PAs) toxins [66]. Many of these weeds spread through fields and contaminate food crops with PAs and when found around water bodies can also contaminate streams and seepage water in groundwater wells [66]. The PAs have been shown to be severely toxic with DNA-damaging potential which might lead to cancers when consumed by humans. The PAs are stable during processing and have been found in herbal teas, spices and honey as well as some other foods such cereals and cereal goods [66].

Due to several compound being linked to and many crops bearing PAs, the FAO/WHO Codex Committee on Contaminants in Food is currently developing management techniques to prevent PA-containing plants from entering the food chain [67].

14.3.1.3 Adulteration from production to preparation of food

Unwanted contaminants that are created during processing of foods include unfit substances which result from canning, roasting, baking, boiling, hydrolysis, and fermentation [68, 69]. During the manufacturing process, heat treatment has been shown to be a source of pollutants [70]. Both in households and in industries, a high cooking temperature is a popular food-processing method. The use of a high temperature in combination with external factors such as food packaging materials can compromise the quality and safety of food attributable to toxicant formation [7]. Toxicants like acrylamide, nitrosamines, chloropropanol, furane, and PAH have been shown to be formed when foods are subjected to either heating, roasting, grilling, baking, canning, fermentation, or hydrolysis [70]. Fryers produce a wide variety of toxic substances during food preparation [7]. Furthermore, microwave cooking can introduce pollutants into food, since most foods are cooked in undesirable containers or film-wrapped, which are prone to toxicant production and subsequent release [71]. Food safety and quality issues have been shown to be caused by microwaving foods with packaging materials such as paperboard, composites, and plastics as toxic compounds or residues from such packaging materials can be transferred into the food during cooking [72].

During food processing, even though packaging material serve to protect food during long-distance transport by covering food product, most of the materials used for packaging food rarely undergo testing to know their quality and compositions so hence can contaminate foods they come in contact with which can present a number of hazards [72]. Packaging materials are enhanced by adding additives, such as stabilizers, antioxidants, and plasticizers. Any direct or indirect interaction between food and packaging material can, however, result in the transfer of these substances from the packaging material into the food, a process known as migration, and constitute potential adulteration to the food sample [73]. Similarly, using metallic cans in packaging results in contamination of the food by ion migration [73].

Corrosion is prevented by coating cans with varnishes such as epoxy resins, which provide protection from corrosion and can easily find their ways into food when used for packaging. Minor by-products from epoxy resin such as bisphenol A diglycidyl ether (BADGE), bisphenol A, or cyclo-di-BADGE have been shown to have negative impact on human acting as endocrine disruptor [73] and can also cause hormonal imbalances in the body [71].

Food additives and preservatives are substances that are purposefully applied to foods during processing to give them specific desirable properties. They are used for a variety of purposes, such as preservation, aesthetics, flavor masking, and sweetening [62, 73]. Examples are colorants, sweeteners, preservatives, and anti-caking agents. However, food additives may also be a source of contaminants. Some of the safety issues regarding food additives have been associated with cytotoxicity, genotoxicity, and induction or potential for induction of mutagenicity [74]. For underserved financial gains, traders mix various harmful chemicals with food to keep it fresh for a longer period [69, 75], while ignoring the health hazards they can cause to the human body, as well as using cheap fillers to reduce the cost of products and increase profits.

Educating consumers about food safety at home is critical in curbing dangers associated with food adulteration at household level which can lead to increased incidence of foodborne disease outbreaks [76]. Food adulteration due to handling errors during preparation of food at household ranges from cross-contamination, improper food storage, insufficient cooking or reheating temperatures and handling by diseased food handlers. However, research has shown that following appropriate hygiene measures can significantly minimize the prevalence of food adulteration during food preparations [76].

Other source of food contaminations includes contamination during food transportation with diesel and petrol exhausts [69]. Contamination from equipment use in food processing as cleaning agents can leave chemical residues that can easily be introduced to the food during processing and preparation [73]. Majority of issues relating to contamination in this way have been attributed to nonconformity of food products. For instance, in India, 67% of nonconforming milk samples were observed to contain 45% of skimmed milk powder, 8% of detergent and 27% of glucose [77]. Table 14.3 presents a list of common foods, their frequently reported adulterants and possible health implications.

Table 14.3: Common foods, their adulterants, and possible health implications.

Food item	Adulterants	Common health implications	References
Asafoetida	Chalk powder, soap, and stone	Diarrhea, collapse, drooling, hoarseness, and low blood pressure	[78]
Barley, wheat, and rice	Damaged grain, dust, stone, and pebble	Malnutrition and gastrointestinal infections	[78]
Bone meal and meat	Rock phosphate, sand, formalin, leather meal, and blood meal	Cancer, nausea, coughing, shortness of breath, pulmonary edema, headache, and dizziness	[79–81]
Black pepper, edible oils, and fats	Mineral oil (petroleum fractions and white oil)	Anemia	[82]

Table 14.3 (continued)

Food item	Adulterants	Common health implications	References
Candies, jams, and juice	Non-permitted colorants and yellow aniline dye (like metanil yellow and other artificial food dyes)	Diarrhea, collapse, low blood pressure, drooling and hoarseness hyperactivity, leaning impairment, depression, and irritability	[83]
Curry powder	Coloring agents in food and dust	Hyperactivity, leaning impairment, depression, and irritability	[83, 84]
Coffee	Maize, soybean, rye, and triticale	Malnutrition	[85]
Coriander powder and chilli	Salts soluble in water, rhodamine B dye, synthetic colors soluble in water, dung powder, other common salts, and redbrick powder	Hyperactivity, leaning impairment, depression and irritability low blood pressure, drooling, and hoarseness	[85, 86]
Chili powder, green chili cloves, and turmeric powder	Non-permitted colorants, yellow aniline dye, like metanil yellow	Diarrhea, collapse, hyperactivity, leaning impairment, depression and irritability low blood pressure, drooling, and hoarseness	[84, 86]
Fresh vegetables, fish, fruits, and sea foods	Oxytocin, malachite green, copper sulfate, wax, saccharin, and formalin	Organ damage, cancer developmental abnormalities, coughing, dizziness, headache, and death	[80, 87]
Fruit juices and soft drinks	Cadmium	Kidney, bone and lung diseases, Immune system weakening, malnutrition, gastrointestinal diseases, disease impairing psychosocial functions	[37, 54]
Honey	Inverted syrup and starch syrup	Heart diseases, diabetes, obesity	[37]
Ice cream and sweetmeats	Metanil yellow, ethyl acetate, nitrate, pepper oil, and gum	Liver and kidney damage and dizziness	[88]
Jaggery	Chalk powder, metanil yellow, and washing soda	Diarrhea, collapse, low blood pressure, drooling, and hoarseness	[88]
Milk	Hydrogen peroxide, formalin, chalk powder, starch, urea, soap powder, and starch	Cancer, nausea, coughing, shortness of breath, pulmonary edema, headache, and dizziness	[80, 89]
Salt	White powder and urea	Diarrhea, collapse, low blood pressure, drooling, and hoarseness	[89]

Table 14.3 (continued)

Food item	Adulterants	Common health implications	References
Sugar	Washing soda and chalk powder	Diarrhea, collapse, low blood pressure, drooling, and hoarseness	[89]
Tomato sauces	Colors and pumpkin pulp that are nonedible	Hyperactivity, leaning impairment, depression, and irritability	[83, 84]
Vinegar	Mineral acid	Acute marginal deficiencies, anemia, and impair muscle functions	[37, 54]

14.3.2 Health implications of food contaminants

More than 48 million Americans get sick each year through food contamination [81]. Chemically contaminated food seriously harms health and has symptoms that range from mild stomach upset to serious health complications. Chemicals in food and water such as arsenic, PCBs, and lead, have been shown to be associated with neurodevelopmental disorders such as attention deficit disorder, autism and cerebral palsy [81]. It has been shown that, even at much lower concentration than those that can impair development of adult's brain, some of these chemicals can induce brain damage and perpetual disabilities in developing fetuses [89]. Additionally, heavy metal such as lead and cadmium are quickly absorbed into food chain resulting in reducing immune defenses, malnutrition, gastrointestinal, disease impairing psychosocial functions and can cause intrauterine growth retardation (Table 14.3) [37, 54]. The consumption of PCBs in food may negatively impact the development of children's nervous systems and immune systems [37, 81]. Toxic pesticides in food pose serious health risks in causing neurological damage, kidney damage, congenital defects, and reproductive issues. Excessive exposure to these pesticides has been shown to cause cancer [37, 81]. Because of pesticide residues built up in the body's tissues, metabolic dysfunction may also arise [89].

14.4 Controlling food contamination through preventative measures

A robust surveillance and response system is necessary to limit the introduction and misuse of chemicals in foods. Food regulations require the FDA to set minimum quantities of chemicals permitted in food, such as the maximum concentration of pesticide allowed in food [74]. In developing and underdeveloped countries, including those in Africa, there is still no legislation governing how much of these pollutants can be found in food. Especially concerning are non-regulated chemicals [90], and further research is needed to determine what pollutants are undetectable in human's foods. Individual customers are also important because they can have a big impact on how they manage their health [90] since they are able find information via the internet and help to reduce the rate of food contamination. Also, reports by news media and journalists on diseases, threats, and causes are critical to public health [90] and consumers should maintain a healthy level of skepticism about food safety problems published in the news and refrain from eating potentially tainted food products until scientific evidence supports action. Lastly, food manufacturers must acknowledge the need to be more honest and upfront about their food products to protect the public from foodborne illnesses [81].

14.4.1 Legislations against food adulteration

It is important for the government to take corrective actions against food counterfeiting and ensure the safety of foods. This has been achieved over the years through improved cooperation between law enforcement and food-related agencies, wherein an imposable penalty becomes operational under the Food Act [95]. In order to combat food adulteration, a number of laws and regulations were adopted around the world. A few of these are herein discussed.

In Nigeria, there are legislative provisions in response to the food safety challenge. Of these, the two principal penal statutes which include the Criminal Code and the Penal Code contain provisions that prohibit and prescribe punishment for any persons who sells or is in possession of things that are noxious or unfit for food or drink with the intent to sell them as food or drink; sells adulterated or adulterates with the intent to sell noxious food or drinks [91]. And the punishment for the two offenses under the Criminal Code is 1 year in prison, whereas the punishments under the Penal Code are terms of imprisonment that may extend to 2 years for the offence [92, 91].

The South African Foodstuffs, Cosmetics, and Disinfectants Act of 1972 was enacted to prohibit any form of adulterations or contaminations regarding foods, food products and drinks, and person or persons who violates a provision of these or allows such a violation to occur is guilty of an offense and subject to a penalty influenced by the seriousness of the offense [93].

To control the abuse of formalin, the Bangladesh's cabinet on Formalin Control Act passed a legislation that carries a maximum sentence of life imprisonment or a fine of 20,000 taka or both, depending on the offence [94]. Statutory requirements require traders to have a license in order to import, store, sell, and market formalin. It is illegal to store formalin in homes, offices, companies, or cars without a formalin license. The legislation stipulates a maximum sentence of 7 years in prison or a fine of 500,000 taka, or both [95]. A fine of 20,000,000 taka is the maximum penalty for possession of formalin processing equipment [94]. Bangladesh's comprehensive law on pure food allows police to arrest anyone regardless of the approval of the court. Authorities offer mobile courts to deal with immediate issues when possible [95].

Enacted in 2006, the Food Safety and Standards Act (FSSA) of India certifies and regulates the safety of foodstuffs and ensures that food for human consumption is available in an ecologically safe, sustainable, and dignified manner. Food manufacturers, warehouses, transporters, and sellers must be registered with Food Safety and Standards Authority India (FSSAI) and must comply with FSSAI regulations [96]. Adulterated food products cannot be distributed, imported, exported, sold, or stored under the law. This crime carries a minimum sentence of 6 months in jail, with a maximum sentence of 3 years, as well as a fine of 1,000 rupees [96]. Importing, manufacturing, storing, selling, or distributing adulterated food or ingredients that are detrimental to one's health is illegal and it attracts 1 year in jail, and a minimum fine of 2,000 rupees. In the selling or delivery of food, toxic or other health-damaging products are discovered, resulting in death or serious injury. The penalty is a minimum of 5,000 rupees and a 3-year jail term that can be extended up to life [96].

In the United States, the Food and Drug Administration (FDA) was created, consisting of qualified staff with the authority to enact laws in order to control the quality of food and medication available on the market [97, 98]. This is also similar to other countries such as China, UAE and Pakistan. Those who are involved in the adulteration of food are subjected to severe punishment influenced by the seriousness of the offense [98].

Food safety procedures and principles are set by the European Union (EU) legislation (EC Regulation 178/2002) in all European member countries. It supports the establishment of the European Food Safety Authority (EFSA) and defines food safety principles [99, 75]. Food law of the EU protects consumers' interests and gives them information about the food they can consume so that they can make informed decisions about what to buy. The goal of the law is to avoid consumer fraud and deception, as well as food adulteration and other activities that may lead to consumer misinformation. There is no distinction between food safety criteria and food standards criteria in the provisions of Article 8 [75, 99]. Consequently, Europe developed the rapid warning system for food and feed (RASFF) in order to detect anomalous conditions in its countries. For EFSA, Member State (MS), EU, and international organizations to exchange information on emerging risks, the emerging risk exchange network (EREN) has been established. Through the system, which is comprised of experts from around the world, national databases [e.g., Rapid Alert System for Food and Feed portal (https://ec.europa.eu/food/safety/rasff-food-and-feed-safety-alerts_en) and Schengen Information System (https://edps.europa.eu/data-protection/european-it-systems/schengen-information-system_en#do)] are accessible and shared, resulting in enhanced knowledge exchange [75, 99].

14.5 Conclusion

In recent years, food contaminants have become a serious issue, posing health risks to consumers. Synthetic toxicants occur in water, air, soil, and foods. Their influence on humans has resulted from intoxication and residues ingested through plant and animal products obtained from the ecosystem. Advances in analytical techniques and toxicological assessments have lent scientific credence to understanding the deleterious impacts of toxicants on animals and humans. The synthetic toxicants such as drugs, nonpesticidal, and pesticidal substances abound in the ecosystem and many of them, particularly the environmentally persistent ones and those implicated as carcinogens, have been banned or withdrawn from use. The recognition of their environmental- and health-impacting effects has necessitated the development and introduction of regulations and laws to monitor or control their applications. Synthetic toxicants have continued to threaten humans' health as many are continuously released in higher concentrations into the ecosystem. It is hoped that the recognition of these threats on both the environment and humans' well-being and health would drive toxicological studies into continuous monitoring and assessment of the synthetic toxicants.

References

[1] Alengebawy, A. Abdelkhalek, S. T., Qureshi, S. R., & Wang, M.-Q. (2021) Heavy metals and pesticides toxicity in agricultural soil and plants: ecological risks and human health implications. *Toxics, 9*, 42. https://doi.org/10.3390/toxics9030042

[2] Cai, L. M., Wang, Q. S., Wen, H. H., Luo, J., & Wang, S. (2019). Heavy metals in agricultural soils from a typical township in Guangdong Province, China: Occurrences and spatial distribution. *Ecotoxicology and Environmental Safety, 168*, 184–191.

[3] Sarwar, T., Shahid, M., Khalid, S., Shah, A. H., Ahmad, N., Naeem, M. A., Ul haq, Z., Murtaza, B., & Bakhat, H. F. (2019) Quantification and risk assessment of heavy metal build-up in soil–plant system after irrigation with untreated city wastewater in Vehari, Pakistan. *Environ Geochem Health, 42*(12), 4281–4297

[4] Khalid, S., Shahid, M., Murtaza, B., Bibi, I., Natasha, N. M., & Niazi, N. (2020) A critical review of different factors governing the fate of pesticides in soil under biochar application. *Science of the Total Environment, 711*, 134–645

[5] Masindi, V., & Muedi, K. L. (2018). Environmental contamination by heavy metals. In H. E.- D. M. Saleh, & Aglan R. F., (Eds.), *Heavy metals* (pp. 115–132). london, UK: *IntechOpen*.

[6] Wang, X., Liu, W., Li, Z., Teng, Y., Christie, P., & Luo, Y. (2020) Effects of long-term fertilizer applications on peanut yield and quality and plant and soil heavy metal accumulation. *Pedosphere, 30*, 555–562

[7] Faille, C., Cunault, C., Dubois, T., & Bénézech, T. (2017). Hygienic design of food processing linesto mitigate the risk of bacterial food contamination with respect to environmental concerns. *Innovative Food Science and Emerging Technologies, 46*, 65–73

[8] Robertson, L., Sprong, H., Ortega, Y., Van der giessen, J., & Fayer, R., (2014). Impacts of globalisation on foodborne parasites. *Trends Parasitol, 30*, 37–52.

[9] CDC, (2013). Surveillance for foodborne disease outbreaks – United States, 2009-2010. *Annals of Emergency Medicine*, *62*, 91–93.

[10] Song, Q., Zheng, Y. J., Xue, Y., Sheng, W. G., & Zhao, M. R. (2017). An evolutionary deep neural network for predicting morbidity of gastrointestinal infections by food contamination. *Neurocomputing*, *226*, 16–22.

[11] Natasha, S. M., Saleem, M., Anwar, H., Khalid, S., Tariq, T. Z., Murtaza, B., Amjad, M., & Naeem, M. A. (2020b) A multivariate analysis of comparative effects of heavy metals on cellular biomarkers of phytoremediation using *Brassica olerac*ea. *International Journal of Phytoremediation*, *22*, 617–627

[12] Shahid, M., Natasha, D. C., Niazi, N., Xiong, T., Farooq, A., & Khalid, S. (2020) Ecotoxicology of heavy metal (loid) enriched particulate matter: foliar accumulation by plants and health impacts. *Rev Environ Contam Toxicol*, *253*, 65–113

[13] Hurley, P. M. (1998) Mode of carcinogenic action of pesticides inducing thyroid follicular cell tumors in rodents. *Environmental Health Perspectives*, *106*(8), 437–445.

[14] Shahid, M., Niazi, N. K., Dumat, C., Naidu, R., Khalid, S., Rahman, M. M., & Bibi, I., (2018). A meta-analysis of the distribution, sources and health risks of arsenic-contaminated groundwater in Pakistan. *Environmental Pollution*, *242*, 307–319

[15] Natasha, S. M., Farooq, A. B. U., Rabbani, F., Khalid, S., & Dumat, C. (2020a). Risk assessment and bio physiochemical responses of spinach to foliar application of lead oxide nanoparticles: a multivariate analysis. *Chemosphere*, *245*,125605

[16] Wang, X. N. F. (2018) Fujian donggang petrochemical carbon nine leak official: atmospheric indicators have returned to norma. *Beijing News Real-time News*, (2018–11–08)

[17] Burkitt, L., (1 February 2012). Cadmium spill threatens water supply. In *WSJ*. Retrieved 25 May 2015.

[18] Dooyema, C. A., Neri, A., Lo, Y. C., Durant, J., Dargan, P. I., Swarthout, T., Biya, O., Gidado, S. O., Haladu, S., Sani-Gwarzo, N., Nguku, P. M., Akpan, H., Idris, S., Bashir, A. M., & Brown, M. J., (2012). Outbreak of fatal childhood lead poisoning related to artisanal gold mining in northwestern Nigeria, 2010. *Environ Health Perspect*, *120*(4), 601–607.

[19] Eliza, B., & Emily, S. (2020) Search suspended after shark attacks Esperance man Andrew Sharpe, *ABC News*, Retrieved (11 October 2020).

[20] Björnham, O., Grahn, H., Von schoenberg, P., Liljedahl, B., Waleij, A., & Brännström, N. (2017). The 2016 Al-Mishraq sulphur plant fire: Source and health risk area estimation. *Atmospheric Environment*, *169*, 287–296.

[21] Koleniak, M., & Peyser, A. (1990). Fire Hardened Vets Shaken by the Piles of Bodies. *New York Post*, Retrieved (March 26, 1990).

[22] Güttinger, H., & Stumm, W. (1992). An Analysis of the Rhine Pollution caused by the Sandoz Chemical Accident, 1986. *Interdisciplinary Science Reviews*, *17* (2), 127–136.

[23] Eckerman, I. (2013). Bhopal gas catastrophe 1984: Causes and consequences. In *Reference module in earth systems and environmental sciences* (pp. 272–287). Elsevier.

[24] Dunn, P. (1986) The chemical war: Journey to iran, NBC Defense and Technology International, Retrieved (11 April 1986), p. 28

[25] Depalma, A. (2004). Love canal declared clean, ending toxic horror. *The New York Times*. ISSN 0362-4331. Archived from the original on February 17, 2017. (Retrieved February 16, 2017).

[26] Robbe, F. (2016). *Seveso 1976. Oltre la diossina*. Castel Bolognese: Itaca Books.

[27] Thrupp, L. A. (1991). sterilization of workers from pesticide exposure: the causes and consequences of dbcp-induced damage in Costa Rica and beyond. *International Journal of Health Services 21* (4),731–757.

[28] Skerfving, S. B., & Copplestone, J. F. (1976). Poisoning caused by the consumption of organomercury-dressed seed in Iraq. *The Bulletin of the World Health Organization*, *54*(1), 101–112.

[29] Brimblecombe, P. (1976). Attitudes and Responses Towards Air Pollution in Medieval England. *Journal of the Air Pollution Control Association*, *26*(10), 941–945.

[30] Nabi, S. (2014). Methylmercury and Minamata Disease. In *Toxic effects of mercury* (pp. 187–199). New Delhi: Springer.

[31] Hamill, S. D. (2008). Unveiling a Museum, a Pennsylvania Town Remembers the Smog That Killed 20. *The New York Times*. Retrieved (November 2, 2008).

[32] Friberg, L., Piscator, M., & Nordberg, G. (1971). The Itai-itai disease. In L. Friberg, M. Piscator & G. Nordberg (Eds.) *Cadmium in the environment* (pp. 111–114). Ohio: *CRC press*.

[33] Goldberg, M., & Luce, D. (2009). The health impact of nonoccupational exposure to asbestos: what do we know? *European Journal of Cancer Prevention*, *18*(6), 489–503

[34] Aksoy, M., (1989). Hematotoxicity and Carcinogenicity of Benzene. *Environmental Health Perspectives*, *82*, 193–197.

[35] Fan, Y., Li, Y., Li, H., & Cheng, F., (2018). Evaluating heavy metal accumulation and potential risks in soil-plant systems applied with magnesium slag-based fertilizer. *Chemosphere*, *197*, 382–388.

[36] Hernando, M. D., Mezcua, M., Fernández-Alba, A. R., & Barceló, D. (2006). Environmental risk assessment of pharmaceutical residues in wastewater effluents, surface waters and sediments. Talanta, *69*(2), 334–342.

[37] Kaur, R., Mavi, G. K., Raghav, S., & Khan, I. (2019) Pesticides classification and its impact on environment. *International Journal of Current Microbiology and Applied Sciences*, 8, 1889–1897.

[38] Schafer, K. (2002). Persistent toxic chemicals in the US food supply. *Journal of Epidemiology and Community Health 56*, 813–817.

[39] Saravi, S. S. S., & Shokrzadeh, M. (2011). Role of pesticides in human life in the modern age: A review. In M. Stoytcheva (Ed.), *Pesticides in the modern world-risks and benefits* (pp. 3–12). Rijeka, Croatia: *IntechOpen*, 2011.

[40] Barman, H., Das, S. K., & Roy, A. (2018). Zinc in Soil Environment for Plant Health and Management Strategy. *Universal Journal of Agricultural Research*, *6*,149–154.

[41] Njinga, R. L., Moyo, M. N., & Abdulmaliq, S. Y. (2013). Analysis of essential elements for plants growth using instrumental neutron activation analysis. *International Journal of Agronomy*, 1–9.

[42] Łukowski, A., & Dec, D. (2018). Influence of Zn, Cd, and Cu fractions on enzymatic activity of arable soils. *Environmental Monitoring and Assessment*, *190*, 278–298.

[43] Frenk, S., Ben-Moshe, T., Dror, I., Berkowitz, B., & Minz, D. (2013) Effect of metal oxide nanoparticles on microbial community structure and function in two different soil types. *PLoS ONE*, *8*, e84441

[44] Zeng, F., Wei, W., Li, M., Huang, R., Yang, F., & Duan, Y. (2015) Heavy metal contamination in rice-producing soils of Hunan province, China and potential health risks. *International Journal of Environmental Research and Public Health*, *12*, 15584–15593.

[45] Fuller, C., Bonner, J., & Page, C. (2004). Comparative toxicity of oil, dispersant, and oil plus dispersant to several marine species. *Environmental Toxicology and Chemistry*, *23*(12): 2941–2949.

[46] Kamunde, C. N., & Woods, C. M. (2004). Environmental chemistry, physiological homeostasis, toxicology, and environmental regulation of copper, and essential element in freshwater fish. *Australasian Journal of Ecotoxicology, 10*, 1–20.

[47] Wijngaarden, R. V., Brock, T. M., & Brink, P. D. (2005). Threshold levels for effects of insecticides in freshwater ecosystems: a review. *Ecotoxicology*, 14(3), 355–380

[48] Connell, D. W. (2005) Concepts of Environmental Chemistry (pp 446–589). Boca Raton, FL: *Taylor and Francis*.

[49] Hodgson, E., & Goldstein, J. A. (2001) Metabolism of toxicants: Phase I reactions and pharmacogenetics. In E. Hodgson & Smart R. C. (Eds.). *Introduction to biochemical toxicology* (3rd ed, pp. 67–113). New York: Wiley.

[50] Lan, M. M., Liu, C., Liu, S. J., Qiu, R. L., & Tang, Y. T. (2020). Phytostabilization of cd and pb in highly polluted farmland soils using ramie and amendments. *International Journal of Environmental Research and Public Health*, *17*, 16–61.

[51] Tukey, R. H., & Strassburg, C. P. (2000). Human UDP-glucuronosyltransferases: Metabolism, expression, and disease. *Annual Review of Pharmacology and Toxicology, 40*, 581–616.

[52] Benedetti, M. S. (2001) Biotransformation of xenobiotics by amine oxidases. *Fundamental and Clinical Pharmacology, 15*, 75–84.

[53] Marsh, K., & Bugusu, B. (2007). Food packaging–roles, materials, and environmental issues. *Journal of Food Science, 72*, R39–R55.

[54] Khan, S., Cao, Q., Zheng, Y. Huang, Y., & Zhu, Y. (2008). Health risks of heavy metals in contaminated soils and food crops irrigated with wastewater in Beijing, China. *Environmental Pollution, 152*, 686–692.

[55] Kantiani, L., Llorca, M., Sanchís, J., & Farré, M. (2010). Emerging food contaminants: a review. *Analytical and Bioanalytical Chemistry, 398*, 2413–2427.

[56] Mastovska, K., (2013). Modern Analysis of Chemical Contaminants in Food. *Food Safety Magazine. 10*, 57–72.

[57] Dickey, R. W., & Plakas, S. M. (2010). Ciguatera: a public health perspective. *Toxicon, 56*(2), 123–136

[58] Ghosh, B., Mitra, J., Chakraborty, S., Bhattacharyya, J., Chakraborty, A., Sen, S. K., et al. (2015) Simple detection methods for antinutritive factor β-ODAP Present in *Lathyrus sativus* L. by high pressure liquid chromatography and thin layer chromatography. *PLoS ONE, 10*(11), e0140649.

[59] Baeumner, A. (2003). Biosensors for environmental pollutants and food contaminants. *Analytical and Bioanalytical Chemistry, 377*, 434–445.

[60] Zheng, N., Wang, Q., Zhang, X., Zheng, D., Zhang, Z., & Zhang, S. (2007). Population health risk due to dietary intake of heavy metals in the industrial area of Huludao city, China. *Science of the Total Environment, 387*, 96–104.

[61] Krishna, A., & Govil, P. (2006). Soil Contamination due to heavy metals from an industrial area of Surat, Gujarat, and Western India. *Environmental Monitoring and Assessment. 124*, 263–275.

[62] Peralta-Videa, J., Lopez, M., Narayan, M., Saupe, G., & Gardea-Torresdey, J. (2009). The biochemistry of environmental heavy metal uptake by plants: implications for the food chain. *The International Journal of Biochemistry & Cell Biology, 41*, 1665–1677.

[63] Ledda, C., Loreto, C., Bracci, M., *et al.* (2018). Mutagenic and DNA repair activity in traffic policemen: a case-crossover study. *Journal of Occupational Medicine and Toxicology, 13*, 24–49.

[64] Bommuraj, V., Chen, Y. K., Sperling, H., Barel, S, R., & Shimshoni, J. A. (2019) Pesticide and trace element residues in honey and beeswax combs from Israel in association with human risk assessment and honey adulteration. *Food Chemistry, 299*, 125–123.

[65] Mattarozzi, M., Cavazza, A., Calfapietra, A., Cangini, M., Silvia, P., Bianchi, F., & Careri, M. (2019) Analytical screening of marine algal toxins for seafood safety assessment in a protected Mediterranean shallow water environment. *Food Additives & Contaminants, 36*(4), 612–624,

[66] Kisielius, V., Hama, J. R., Skrbic, N., *et al.* (2020). The invasive butterbur contaminates stream and seepage water in groundwater wells with toxic pyrrolizidine alkaloids. *Scientific Reports, 10*, 84–97

[67] Malysheva, S. V., Mulder, P. P. J., & Masquelier, J. (2020). Development and Validation of a UHPLC-ESI-MS/MS Method for Quantification of Oleandrin and Other Cardiac Glycosides and Evaluation of Their Levels in Herbs and Spices from the Belgian Market. *Toxins, 12*, 243

[68] C., N., Aznar, M., & Carrizo, D. (2016). Food contamination during food process. *Trends in Food Science and Technology, 48*, 63–68.

[69] Schrenk, D. (2004). Chemical food contaminants. *Bundesgesundheitsblatt Gesundheitsforschung Gesundheitsschutz 47*, 841–877.

[70] Vardhan, K. H., Kumar, P. S., & Panda, R. C. (2019). A review on heavy metal pollution, toxicity and remedial measures: Current trends and future perspectives. *Journal of Molecular Liquids, 290*, 111–197.

[71] Azevedo, F. A., Carvalho, L. R., Grinberg, L. T., Farfel, J. M., Ferretti, R. E., Leite, R. E., Jacob, F. W., Lent, R., & Herculano-Houzel, S. (2009). Equal numbers of neuronal and nonneuronal cells make the

human brain an isometrically scaled-up primate brain. *Journal of Comparative Neurology, 513*(5), 532–541.

[72] Dougherty, C., Holtz, S., Reinert, J., Panyacosit, L., Axelrad, D., & Woodruff, T. (2000). Dietary exposures to food contaminants across the United States. *Environmental Research, 84*, 170–185.

[73] Cabado, A., Aldea, S., Porro, C., Ojea, G., Lago, J., & Sobrado, C. (2008). Migration of BADGE (bisphenol A diglycidyl-ether) and BFDGE (bisphenol F diglycidyl-ether) in canned seafood. *Food and Chemical Toxicology, 46*, 1674–1680.

[74] Bajwa, U., & Sandhu, K. (2011). Effect of handling and processing on pesticide residues in food-areview. *Journal of Food Science and Technology, 51*, 201–220.

[75] Randles, P. (2012) Emerging Risks. In *Paper given to the 57th Meeting of the Advisory Committee on Animal Feedstuffs on 7th March 2012*. Available at: http://www.food.gov.uk/multimedia/pdfs/commit-tee/acaf1201.pdf

[76] Jian, C., & Goleţ, I. (2014) Knowledge of food safety and hygiene and personal hygiene practices among meat handlers operating in western Romania. *Food Control, 42*, 214–219.

[77] Manning, L., & Soon, J. M. (2014) Developing systems to control food adulteration. *Food Policy, 49*, 23–32.

[78] Perry, M. J. (2008) Agricultural Health and Safety, International Encyclopedia of Public Health, *Academic Press*, 69–76,

[79] Block, G., Matanoski, G. M., Seltser, R., & Mitchell, T. (1988) Cancer morbidity and mortality in phosphate workers. *Cancer Research, 48*, 7298–7303.

[80] Rahman, M., Sultan, M., Rahman, M., & Rashid, M. (2015). Food Adulteration, a serious public health concern in Bangladesh. *Bangladesh Pharmaceutical Journal, 18*(1), 1–7.

[81] Schantz, S., Gardiner, J., Gasior, D., McCaffrey, R., Sweeney, A., & Humphrey, H. (2004). Much ado about something: the weight of evidence for PCB effects on neuropsychological function. *Psychology Schools, 41*, 669–679.

[82] Sinha, N., Deshmukh, P. R., & Garg, B. S. (2008). Epidemiological correlates of nutritional anemia among children (6-35 months) in rural Wardha, Central India. *Indian Journal of Medical Sciences, 62* (2), 45–54.

[83] Spears, K. (1988). Developments in food colourings: the natural alternatives. *Trends in Biotechnology, 6*, 283–288.

[84] Griffiths, J. C. (2005). Coloring Food and beverages. *Food Technology, 59*(5), 38–44.

[85] Aherne, A. S., & Obrien, N. M. (2002). Dietary flavonols: Chemistry, food content and metabolism. *Nutrition, 18*, 75–81.

[86] Sahar, S. A. S., & Manal, M. S. (2012). The effects of using color foods of children on immunity properties and liver, kidney on rats. *Food and Nutrition Sciences, 3*, 897–904

[87] Hallagan, J. B., Allen, D. C., & Barzelleca, J. F. (1995). The safety and regulatory status of food, drug and cosmetic color additives exempt from certification. *Food and Chemical Toxicology, 33*(6), 515–528.

[88] Abdel-Rahim, E. A., Ahmed, E. A., El-Desoky, G. E., & Ramadan, M. E. (1987). Biochemical role of some natural and synthetic colorants on liver function of rats. *Mania Journal of Agricultural Research, 9*(3), 11–17.

[89] Kher, S., De Jonge, J., Wentholt, M., Deliza, R., De andrade, J., & Cnossen, H. (2011). Consumer perceptions of risks of chemical and microbiological contaminants associated with food chains: a cross-national study. *International Journal of Consumer Studies, 37*, 73–83.

[90] Villanueva, C., Kogevinas, M., Cordier, S., Templeton, M., Vermeulen, R., & Nuckols, J. (2014). Assessing exposure and health consequences of chemicals in drinking water: current state of knowledge and research needs. *Environmental Health Perspectives, 122*, 213–221.

[91] Ojebode, M. (2017) "Nigerian Regulatory Agencies and Food Poisoning" Financial Nigeria (3 May 2017) accessed (17 October 2017).

[92] NAFDAC (2018). Food Safety and Applied Nutrition (FSAN), accessed (1 May 2018)

[93] Neser, J. J. (2001) 'Mandatory minimum sentences in the South African context' (2001), Vol. 3 No. 3 Crime Research in South Africa, available at http://www.crisa.org.za/volume3/vvs.html.

[94] Kamruzzaman, M. D. (2016). Formalin crime in Bangladesh, A Case Study. *European Journal of Clinical and Biomedical Sciences*, *2*(5). Available: http://article.sciencepublishing-group.com/html/10.11648.j.ejcbs.20160205.12.html#paper-content-1-2

[95] Lakshmi, V. (2012). Food Adulteration ijsit (www.ijsit.com), Volume 1, 2, November-December 2012. 106. pp. 106-113. Available from http://www.ijsit.com/admin/ijsit_files/FOOD%20 ADULTERATION_1.2.4. pdf Accessed (23 April 2017).

[96] Food and Safety Standard Authority India (2006). policy document. Available from: http://www.fssai.gov.in/Portals/0/Pdf/Article_on_fruits.pdf

[97] Identification and Prevention of Adulteration Guidance Document (2020). ASTA: The Voice of the U.S. Spice Industry in the Global Market. Available at: https://www.astaspice.org/food-safety/identifi-cation-prevention-adulteration-guidance-document/

[98] Jeffrey, M., & Pilcher, (2017). *Food in World History* (p. 59). New York: Routledge, 2006, from https://en.wikipedia.org/wiki/Adulterant Accessed (23 April 2017)

[99] Louise, M., (2014). Developing system to control food adulteration. Food Policy, *49*(1), 23–32. availavle at: http://clok.uclan.ac.uk/14025/1/

Olabisi Omowumi Adebisi and Christiana Eleojo Aruwa*

Chapter 15
Food allergies: causes and control

Abstract: Although food are materials that are typically intended to provide certain nutrients to ensure wholesome health, growth and wellbeing when ingested and assimilated by body tissues and cells, food immunological reactions in form of food allergies (FAs) do occur in response to certain food ingredients. Such food constituents are called food allergens. Food allergy is not a new food-related concept, but studies have continued to show new scientific insights to better understand FA types and causal factors. Likewise, such scientific insights could further aid the development of preventative and control strategies to better mitigate emerging spikes in FA trends. In addition, food allergies affects all age groups, and there has been an increase in adult onset FA prevalence. In light of the foregoing, this chapter aims to provide a comprehensive overview on FA types, causes, control, symptoms, epidemiology and their diagnosis. The chapter also shows FA-related study gaps, such as the need to clearly demonstrate and prove the food allergen-carcinogen link.

15.1 Background

Food is any edible substance that provides the body with essential nutrients when consumed, digested, and assimilated by the body cells and tissues [1]. These nutrients have the ability to maintain and sustain life and stimulate growth and development. Foods in varied diet combinations also protect the body from disease and regulates body functions. Foods generally contain different nutrients like carbohydrates, fats, proteins, vitamins, minerals, or trace elements for maintenance of healthy and daily activities. Health and nutrition are closely linked, and nutrition depends on the food intake. So, food is an important single factor for health and fitness [1]. Most foods consumed by humans are of plant, animal or microbial origin and can be classified based on nutritional composition, method of preparation, origin and sources, physicochemical proper-

Acknowledgment: Authors thank the South African National Research Foundation (NRF) for the Innovation Postdoctoral Fellowship Grant awarded to Dr. C. E. Aruwa, and tenable at the Department of Biotechnology and Food Science, Durban University of Technology, South Africa.

***Corresponding author: Christiana Eleojo Aruwa,** Department of Biotechnology and Food Science, Faculty of Applied Sciences, Durban University of Technology (DUT), Durban, South Africa, e-mail: ChristianaA@dut.ac.za
Olabisi Omowumi Adebisi, Molecular Biology & Interdisciplinary Life Sciences (MBIL), New Mexico University, Las Cruces, PO Box 30001, New Mexico, United States.

https://doi.org/10.1515/9783110748345-015

ties and purpose of the food besides nutritional benefits [2]. Despite the nutritional benefits associated with food consumption, some foods are known to elicit some adverse effects referred to as food allergies (FAs) [3, 4].

One is said to have an FA when they develop abnormal hyperactive immune reaction(s) to apparently harmless or generally innocuous substances in foods or food additives. It is an adverse reproducibly reaction that arises even in close proximity to the suspected food or food component with similar configuration [4]. FA is sometimes confused with food intolerance (FI). FI is a health state linked to an inability to digest or assimilate a particular food component. For example, lactose in milk and gluten in wheat are not easily digested due absence of needed enzymes [3]. It is an abnormal reaction which has nothing to do with immunological responses and is not life-threatening in most cases. Histamine toxicity, food poisoning, and food additive sensitivity are other food-related maladies that mimic FAs. Different types of FAs are also governed by various immune modulators [4].

FAs are quite common and experienced by a significant number of adults and children worldwide. The number of patients with FAs is also on the rise. While any food may cause an allergy. Some foods and food ingredients are more allergenic than others. When a food initiates or induces an unwanted immune response, an FA is said to have happened. Even minute exposure to the allergen can bring about allergic immune responses. Immediate reactions occur within minutes of exposure or ingestion, while delayed reactions take hours and longer. FA symptoms may be mild to life-threatening and include low blood pressure, tongue, face or mouth swelling, difficulty in breathing, hives, vomiting, diarrhea and itchy rashes [5]. In every thirteen children it is believed that FAs occurs in at least one child. On the other hand, its frequency in adults may be overlooked. In adults, FA affects one in ten; that is, its occurrence is more frequent in children, and 50% of adult FAs begin in adulthood [6, 7]. To give a general insight and overview on FAs, this chapter, therefore, discusses FA causes and control, alongside its epidemiology, diagnostic methods, therapies, and FA onset in adults.

15.2 Types of food allergies and their triggers

Immunological reactions arise when food-related allergens are inhaled, ingested, or via atopic contact [8, 9]. This reaction usually occurs when an individual's immune system mistakes a certain food protein for dangerous unfamiliar molecule and proceeds to attack it [1, 2]. A vast array of protein types with varying function *in vivo* are allergenic. These include binding and structural proteins, biological enzymes, and their inhibitors [9–11]. The amount of allergens needed to cause an adverse reaction and the resulting symptoms vary from person to person [10]. For instance, in a susceptible individual the consumption of a little quantity of the suspected food can give rise to an allergic reaction. In other cases, minute quantities of the allergen does trigger deadly responses in extremely at-risk persons [10, 11]. Susceptibility reactions to allergens also take place on the skin (especially in cases where the skin integrity has been tampered with in conditions such as eczema), gastrointestinal (GI), and respiratory tracts. In some cases, the presence of proteinaceous food allergens in vaccine preparations can trigger reactions to such vaccines [12, 13].

15.2.1 Food allergy classifications

FA classifications depend on the nature of allergen and immunological response or mechanism of action involved, and the type of clinical manifestation presented. Based on immunological response, the following types are known:

1. **The Ig E antibody food allergies:** The immune system utilizes antibodies which are blood proteins in recognizing and fighting infections. In this type of FA, the antibody Ig E is released in response to a food allergen. This is the most common type of FA and symptoms occur shortly after eating and may involve anaphylaxis (Table 15.1) [14, 15].
2. **The non-Ig E antibody FA:** In this type of FA, the Ig E antibodies are not expressed; instead, other immune cells and products are utilized in fighting the threat or allergen (Table 15.1). Symptoms within this category occur some hours after consumption of the suspected food [4, 14].
3. **Ig E and/or non-Ig E-mediated:** This is a hybrid of the two categories (Table 15.1) discussed earlier [15].

   On the basis of the nature of food allergen, there are two classes, that is, class 1 and 2 FAs.
1. **Class 1 FAs:** These types of FAs are caused by egg white, peanut, and cow milk allergens. These allergens show resistance to gastric digestion, are unaffected by acid and heat, trigger allergic reactions through the GIT, and cause systemic reactions [11, 15].
2. **Class 2 FAs:** Unlike class 1, this category of allergens is prone to gastric digestion, are common in adults, and reactions arise from allergen inhalation. Here allergens have close similarity (cross-reaction/reactivity) to proteins found in birch tree pollen. The FA symptoms in this class include anaphylaxis and oral allergy syndrome. Most class 2 allergens are hard to isolate and identify and are also very unstable [11].

In the third major category based on clinical manifestations of FAs, clinical signs are observed in almost every organ of the body. The FA reactions can therefore be in the form of digestive or nondigestive manifestations [15]. Digestive-related FAs may present as gastrointestinal disorders, oral allergy syndrome, eosinophilic esophagitis, eosinophilic gastroenterocolitis, dietary protein enterocolitis or proctitis. Nondigestive-related FAs may manifest as cutaneous, respiratory, or systemic symptoms [16].

Table 15.1: Food allergy classifications: symptoms, mechanisms, and food sources.

Ig E-mediated food allergy

Disorder	Symptoms	Mechanism of action	Food source
Oral allergy syndrome	Primarily mild pruritus and angioedema	Ig E-mediated and occurs as a result of primary sensitization to pollen proteins which are homologous to food proteins in certain fruits/vegetables	Raw fruits and vegetables
Urticaria/angioedema, anaphylaxis	Cutaneous, gastrointestinal, respiratory symptoms	Ig E-mediated and occurs as a result of cross-linking of Ig E which leads to release of other immune mediators	Milk, egg, wheat, soy, peanut, tree nuts, and fish
Food-dependent exercise-induced anaphylaxis	Food triggers anaphylaxis only if ingestion is followed by exercise	Enhanced mast cell activities and altered intestinal permeability when food ingestion is followed by exercise	Wheat, shellfish, and celery

Non-Ig E-mediated food allergy

Disorder	Symptoms	Mechanism of action	Food source
Food-protein induced enterocolitis	Delayed emesis, diarrhea few hours after ingestion, hypotension may occur in severe cases	Increased tumor necrosis factor (TNF)a, and decreased transforming growth factor (TGF) beta	Grains, soy, and milk
Food-protein-induced proctocolitis	Blood-stained infant stools	Eosinophilic infiltration in the colon	Breastmilk, soy, and milk

Mixed or hybrid food allergy

Disorder	Symptoms	Mechanism of action	Food source
Atopic dermatitis	Pruritus, chronic skin inflammation	Ig E mediates skin mast cells activation; late phase inflammatory cells (T cells and eosinophils) infiltration	Egg, wheat, milk, and soy
Eosinophilic gastroenteropathies	Vomiting, dysphagia, abdominal complaints, nausea.	Eosinophils infiltrate GIT, eosinophil mediators (interleukin 5, eotaxin) mediate	

Source: Sicherer and Sampson [4], Nowak-Węgrzyn et al. [14], Olivier [15].

15.2.2 Food allergy triggers

FAs are caused by suspected foods or food components (food allergens) which elicit immunological reactions. Food allergens are proteins or derivatives of protein, and their consumption can evince immunological reactions [11, 17]. In most cases, these allergens are water-soluble glycoproteins (about 10–60 kD). This solubility in water contributes to their easy transport through the mucosal surface of the intestinal tract. The unique attributes of proteins like the presence of stable bonds in their quaternary and tertiary forms make them affective allergens. The same unique feature prevents their breakdown in digestive processes. In addition, genetic engineering has been implicated in increasing the allergenicity of food. Generally, almost 200 foods are known to provoke FAs. Of these, close to 10 are major allergens that account for more than 85% of almost all FAs. These foods are milk, egg, fish, shellfishes (crustaceans and mollusks), soya bean, peanuts, tree nuts, cereals, and sesame and mustard seeds [17].

a. **Milk allergens:** Milk, especially cow milk, is highly nutritious and a major part of the human diet [18], and it contains about 200 protein types. Of these, whey and caseins make up the largest proportions, and the ratios vary in different lactating ruminant animals/mammals. The α-s1-casein and β-lactoglobulin are the major allergens in milk [19]. Again, significant sequence similarities also exist in ewe, goat, and cow milk proteins. So, Ig E cross-reactions are common in these milks. Cross-reactions between soy and cow milk invariably knock out soy milk as a cow milk alternative for allergic patients [18, 19]. Fortunately, hydrolysis and irradiation techniques are known to modify the allergen protein structure of milk, but homogenization, pasteurization, and sterilization do not.

b. **Egg allergens:** Eggs are highly nutritious, full of vitamins, and have essential amino acids and in adequate amounts. Egg's gelation, emulsification, and foaming properties also enable it to be an excellent choice in several food production processes. Some major egg white protein allergens are ovomucin, conalbumin, ovalbumin, and ovomucoid, and minor albumen allergens include flavoprotein, ovoglobulins, and avidin. Egg yolk proteins include transferrin, lipovitellin, γ-globulin, phosvitin, and lipovitellenin. The presence of ovalbumin and ovomucoid makes egg white a more effective allergen compared to egg yolk [20].

c. **Fish allergens:** Fish is one of the most common food allergens. Some implicated fishes include mackerel, flounder, hake, cod, halibut, and haddock, among others. Parvalbumin is the primary fish allergen that triggers allergic reaction in sensitive individuals [21].

d. **Shellfish allergens:** Shellfish allergies have been reported for sea foods like shrimps, clams, squids, octopus, crayfish, crabs, mussels, lobsters, and prawns, among others [17].

e. **Allergens from soya bean:** Soya bean belongs to the family *Fabaceae*, is considered to be one of the best plant-based nutritional foods that provides a well-balanced mix of omega fatty acids and amino acid [22]. Regardless of its health benefits, it contains allergenic components like P34 (soya vacuolar protein) β-conglycinin, glycinin, soya bean hull proteins, the pathogenesis-related (PR) soya bean protein, among others [14, 19]. Significant cross-reactions also exist between other legumes (lentils and peanut) and soya bean [23]. Similar cross-reactivity has been demonstrated between soymilk and cow milk.

f. **Peanut allergens:** The peanut plant belongs to the family Leguminosae [24]. Its potent allergens include conglutin, vicilin (Ara h 1), profilin, legumin, and pathogenesis-related protein, Ara h 8. Conglutin and vicilin are the major allergens, while others are minor [25].

g. **Tree nut allergens:** Allergens have been reported for tree nuts like pine and macadamia nuts, almonds, walnut, hazelnut, pistachio, cashew, Brazil nuts, and pecans. Allergens vary depending on nut type [17].

h. **Cereal allergens:** Cereal allergens like globulin, albumins, and glutenin cause reactions following their inhalation and/or ingestion. Food components like certain vinegars, syrups, spices, additives, sauces, starches, seasonings, and beverages are potential hidden cereal sources. Cereal additives are also used in some medications, toy stuffing, pillows, and cosmetics [25, 26].

i. **Sesame allergens:** The sesame herb plant belongs in the family Pedaliaceae. It is used extensively due to its high content of nutrients. Sesame seeds, either crushed or whole, are potent allergens, and share similar biochemical and structural similarities with poppy seed, kiwi, rye, and various tree nuts, and thus may lead to cross-reactivity. It is a hidden food allergen since it is used in many food product varieties. Reactions triggered include skin inflammation and allergic dermatitis [17].

j. **Mustard seed allergens:** The mustard plant is from the family Brassicaceae. It is often used as a food condiment. The oriental and yellow varieties make up common mustard condiment mixtures utilized in food products like sauces, soups, and salad dressings. Sin a 1 and Bra j 1 allergens in yellow and oriental variants, respectively, are the most potent. Both share similar structures, and cross-reaction with cabbage, Brazil nut, and broccoli allergens has been reported [27].

k. **Other food allergens:** Besides those discussed above, other less common allergens have been found in corn, pea, fruits, vegetables, and chickpeas. Certain concentrations of sulfites (≥10 mg/kg) have been shown to elicit allergic responses [27, 28].

15.3 Factors influencing the activity of allergens

Factors that determine the activities of food allergens are subdivided into host and allergen factor categories [29, 30]. Host factors are related to the host and its environment and include host family history, age of the host, host's lifestyle, exercise, infection, host genetic makeup, presence of underlining health issues, medication history, and gut integrity. Allergen factors include entry route, time of exposure, quantity of the allergen exposed to, allergen type, allergen epitopes conformation, cross-reactivity with other allergen, and food preparation method [31, 32].

15.4 Mechanism of action of food allergens

FAs begin with sensitization where the allergen is marked as a foreign molecule and the body starts to mount immunological defense responses [33]. Thereafter, signs and symptoms may manifest as immediate or delayed responses. Immediate reactions are dominated by Ig E antibodies which are found in the blood and bind dendritic cells on Fc epsilon receptors, basophiles, macrophages, and mast cell surfaces [15]. Some of these immune cells express inflammatory mediators like cytokines. If a sensitized individual comes across an allergen they had encountered before, Ig E antibodies on basophils and mast cells bind the allergen [34, 35], inducing a reaction cascade which leads to degranulation of the immune cells. The cell granules then release mediators like heparin and histamine into the blood which combine with body cell receptors and cause mucus secretion, muscle contraction, and vasodilation [35]. Histamine expression is complete in about a half hour and leads to immediate hypersensitivity (allergy) symptoms. The later expression of cytokine mediators results in a more prolonged second wave of symptoms [4, 15, 35]. In cases involving the cardiovascular system, hypotension, loss of consciousness, and dysrhythmias may result. Even though the most common allergies affect the GIT, circulatory, pulmonary, cutaneous, and neurological systems, all body organs may be implicated. The severity and duration of FAs usually depend on individual susceptibility level, quantity of the allergen ingested, epitope recognition pattern, and intensity of Ig E binding [35]. In delayed allergy types, immune T cells induce symptoms that manifest after some hours, and even to days, upon ingestion of the food allergen. Identification of delayed allergic responses is harder compared to immediate reactions. These reactions are mediated by tissue-bound immune cells, and the acuteness of the reaction may be less than that of the immediate reactions [4, 35].

15.5 Epidemiology of food allergies

The prevalence of FA reactions is increasing worldwide. Studies have shown that less than 10% of children and adults experience FAs, but the numbers are increasing steadily [36, 37]. Some theories have been proposed to explain these increases [38–40]. Table 15.2 shows the prevalence rate of FAs for different allergens, with the most common allergens duly represented and milk taking the next highest prevalence range after shellfishes.

Table 15.2: Prevalence of food allergies.

Allergen	Prevalence rate (%)
Peanut	0.1–0.6
Tree nuts	0.4–0.6
Fish	0.1–0.5
Shellfish and crustacean	0.5–2.8
Egg	0.2–1.5
Milk	0.3–2.5
Wheat	0.1–0.5
Soy	0.1–0.5
Sesame	0.1–0.3

Source: Osborne et al. [36] and Sicherer [37].

15.6 Symptoms of food allergy

FA symptoms can be immediate or delayed after ingestion of a suspected food and are not predictive [4]. Allergic responses range from skin hives to diarrhea, nausea, bloating, gas and stomach cramps, vomiting, or cardiovascular and respiratory reactions. Symptoms vary with individuals and cannot be predicted. FA reactions may be severe, mild, or anaphylactic (can potentially lead to death) [5]. Reaction to the same food allergen may also vary for the same individual. Given the increase in FAs and its potential life-threatening reactions, there is need for added emphasis on truthful and clear information on ingredients used in foods and food product labels [6]. Table 15.3 gives an overview of the possible symptoms associated with FAs and the part of the body affected.

15.7 Food allergy diagnosis

Diagnosis of FAs is mostly carried out using oral food challenge, diet elimination, medical history and physical examination, skin prick, and blood tests. A patient's medical history essentially assesses past symptoms associated with the consumption of particular foods and the possible mechanism involved. Some FA diagnostic tests are discussed [41].

1. **Diet elimination test:** This involves elimination of the suspect food from the individual's diet for weeks and before it is again reintroduced one at a time to check for symptoms related to the particular food. The test drawback is that it cannot distinguish between FA, food sensitivity, or FI. The absence of a change in symptoms following diet elimination indicates that the food or component is not responsible for the disorder [41].

Table 15.3: Food allergy symptoms, body part affected, and response elicited.

Affected body part	Immediate symptoms	Delayed symptoms
Skin	Itching, hives, redness, measles-like red bumps, and swelling under the skin	Same as immediate symptoms plus rash similar to eczema and flushing
Eyes	Itching, tearing, redness, and swelling around the eyes	Same as immediate symptoms
Upper respiratory	Congestion of the nasal passage, runny nose, sneezing, voice box swells, and dry cough	Same as immediate symptoms
Lower respiratory	Cough, tight chest, shortness of breath, and wheezing	Cough, shortness of breath, wheezing
GI (oral)	Lip swells, itchy tongue or palate, and inability to swallow food	Same as immediate symptoms
GIT (lower)	Nausea, colicky abdominal pain, reflux, vomiting, and diarrhea	Same as immediate symptoms in addition to bloody stool, irritability, food refusal with weight loss in young children
Cardiovascular	Fast heartbeat (sometimes slow in anaphylaxis), loss of consciousness low blood pressure, fainting spell, and dizziness	Same as immediate symptoms
Other	Uterus contractions, light-headedness, and hoarse voice	Same as immediate symptoms

Source: Burk et al. [1], Sicherer and Sampson [2], and Sicherer and Sampson [4].

2. **Skin prick test:** In the Ig E-mediated response/disorder, this test is among the fastest for detecting sensitization to particular food allergens. It involves the use of a needle board (Multitest) or a bifurcated needle with which allergens are introduced to the body and directly placed on the skin. The formation of hives at the inoculating spot gives a positive result and an indication of the presence of allergen-specific antibodies against the introduced allergen. The demerit of this test is that it can only detect the presence or absence of an allergy but cannot predict the severity and duration of the reaction that may or may not occur if such foods are consumed. On the other hand, diagnosis of non-Ig E-related allergies is not possible with this test and requires more definitive tests like the quantitative Ig E tests which are integral to diagnosing FAs [42].

3. **Skin patch test:** This test is used in determining the ability of an allergen to cause skin inflammation in delayed FAs [43].

4. **Blood test:** This test quantifies the amount of specific food allergen antibodies Ig E produced against a particular food allergen. Radioimmunoassay (RIA), radioabsorbent test (RAST), and enzyme-linked immunosorbent assay (ELISA) methods

can be used. In carrying out the RIA test, a known quantity of the antigen is made radioactive by labeling it with a gamma-radioactive isotope of iodine. Increasing Ig E concentrations correspond to increased probability of an allergic reaction taking place. Quite a number of allergens can be screened using the same sample in this test, but non-Ig E-related FAs cannot be identified [44, 45].

5. **Oral food challenge test:** This test is for FAs that are IgE-mediated; in this test, the allergen is well packaged in form of a pill and is given to the patient to ingest directly. This test has to be done under good supervision to monitor the signs and symptoms and the possibility of anaphylaxis reaction. In this test, false-positive results are possible, even though they rarely occur. Hence, negative outcomes must be ruled out using the actual suspected food in a supervised and open feeding program [45].

15.8 Control and management of food allergies

FA control measures are mainly focused on eliminating the suspected or identified allergen [46, 47]. There is also a need for extensive education to understand food labels and restaurant meals/menus [48]. Patients must pay careful attention to food labels, foods from food vendors/restaurants, and other gatherings. Patients should also avoid cross-contamination of foods with allergen(s) in the process of preparing meals [49, 50]. Individuals at risk of anaphylaxis must be encouraged to obtain medical identification, educated on how to promptly identify preliminary symptoms, and taught to properly self-administer epinephrine auto-injections and antihistamines, which must always be available [51, 52]. Since no treatments are presently available to give a permanent cure for FAs, several therapies that are either allergen-specific or allergen-nonspecific have been researched and investigated to curtail the impact of FAs [53–56].

In certain treatments, specific allergens may be targeted in control measures to reduce or change the reactions to them. Among the treatments targeting particular allergens are subcutaneous immunotherapy (SCIT), sublingual immunotherapy (SLIT), peptide immunotherapy, oral immunotherapy (OIT), plasmid DNA immunotherapy (PDI), recombinant vaccines, and immunostimulatory sequences (ISS)-conjugated protein immunotherapy. Each has its own mechanism of action, health benefits, and concerns. The SCIT, OIT, and SLIT show similar mechanism of action and are proven therapies that act by inducing desensitization or tolerance to an allergen following steady exposure to the allergen. SCIT is efficacious for oral allergy syndrome therapy, but patients with peanut allergies, however, have a high risk of developing anaphylactic reactions. While SLIT has been reported to improve clinical tolerance to hazelnut allergy, the OIT effectively manages milk and egg allergies [57, 58]. The plasmid DNA immunotherapy acts by allergen gene immunization in order to enhance endogenous allergen production, resulting in possible induction of tolerance. The health concern linked to PDI has to do with safety in view of strain-dependent effects. Recombinant vaccines act by mutating Ig E binding sites such that proteins stimulate T cells to multiply, but with significantly decreased number of Ig E binding sites and capacity [58].

On the other hand, nonspecific allergen therapies target regulation of the general reaction and mostly benefit multiple FA sufferers. Some examples include cytokine/anti-cytokine (which block pro-allergic cytokines), anti-Ig E, and Chinese herbal medicine (which inhibit T helper 2-type immune cell response) therapies. The anti-Ig E treatment acts by decreasing free Ig E in circulation, inhibiting both early- and late-phase allergic responses, suppressing inflammation, and providing better control for allergic diseases. Its anti-Ig E therapy could be useful in combination with immunotherapy and is known to have provided an improved threshold against peanut-induced reactions. Chinese herbal medicine is administered orally, well tolerated, and generally safe, provides long-term protection from peanut anaphylaxis, and could be efficacious for managing multiple FAs. Although the cytokine/anti-cytokine therapy reportedly showed adverse systemic effects, the anti-Ig E and herbal medicine approaches have not shown adverse health concerns [58].

Other recommended food allergy control measures include the partial or exclusive breastfeeding for neonates first six months of life to help maintain GIT integrity, probiotics consumption [56], and the gradual introduction of other allergenic foods one at a time over a period of about 4–6 months after birth [57]. In addition, using modified allergens (allergoids) obtained by gene recombination technology, and deactivating conformational allergen epitopes through enzymatic polymerization, cooking, and some other food processing techniques could also suffice for FA control [57].

Reversing allergenic effects need not be a herculean task, as many have suggested that a simple avoidance of the food or food product causing the allergy would suffice. Nonetheless, it has been reported that some people can naturally overcome FAs if the allergies are correctly managed. While experiencing FAs, both adults and children greatly depend on other individuals for their safety. So, both FA patients and nonpatients should have a healthy and general comprehension of FAs. Surprisingly, in many cases sufferers are also not properly equipped to deal with their symptoms. Thus, a proper diagnosis is critical to better inform and equip on the most appropriate daily control measures. In most cases of self-diagnosis, diet is unnecessarily limited, and one may be prevented from getting access to adequate and proper care [7]. In cases of life-threatening symptoms, epinephrine autoinjectors can be used to save life. In certain patients where complaints lead to allergy tests, the tests may still show that one has no FAs. Under these circumstances, some patients have been advised to keep a journal close while documenting reactions to foods over a period. The journal can be helpful in showing allergenic patterns which might otherwise go unnoticed in tests.

15.9 Onset of food allergies in adults

It has been reported that the allergy onset in adults is increasing and has become a most confusing trend. This trend has to do with foods that have been consumed by an adult for years which all of a sudden begin to cause allergenic reactions [7]. Thus, FA can be developed to familiar foods over time and as one grows. Only fairly recently have more research data been generated on adult onset of FAs. Results from a wide survey involving thousands of participants showed that between 2015 and 2016 in the United States, a tenth of adults had FAs and more than half of these showed severe allergic reactions. Again, about 40% of respondents said that at least one of their allergic reactions usually required emergency care, but only 0.05% had a confirmed allergy diagnosis from a doctor. Also, less than 25% with FAs reported a recent prescription of epinephrine. The study highlighted a more worrisome trend than what was previously thought on emerging health issues in adults developing FAs to previously harmless foods. A confusing observation during the study was that only close to 20% of adult respondents thought they had an FA. Researchers linked such thoughts to the show of symptoms which are similar to those of other food-related conditions; hence, adverse food reactions and symptoms can be unclear, some of which are self-limiting, manageable, or life-threatening. Besides FAs, adverse food-related symptoms may also be attributed to FI, oral allergy syndrome, and sensitivities, among others [6]. The shellfish allergy is common in adults. This allergy is said to affect about 7.2 million Americans and adults scarcely outgrow the allergy. Allergy onset in adults also occurs at a high rate in response to foods like wheat, fin fish, peanut, egg, milk, sesame, tree nut, and soy, and affects half a million to almost five million adults in the United States. These onsets were also reported more in women than in men. No clear cause is known for adult-onset FAs, but possible microbial causes, hormonal changes such as those experienced in pregnancy, environmental (change in location), and genetic triggers are still being studied to give a clearer insight [6].

15.10 Food allergen–carcinogen relationship

Although cancer-causing substances seem not to have any direct link to allergies, some allergens may be carcinogenic. For example, individuals with celiac disease are more susceptible to bowel cancer if gluten is ingested from certain foods or food products. People like these are therefore advised to stick to gluten-free products. Fortunately, the list of gluten-free foods and products have continued to expand with time. Still, these examples hardly show the allergy–carcinogen link since they are considered under FI. Further research could provide better clear-cut links.

15.11 Closing remarks

Summarily, FAs are here to stay, and with increasing prevalence across the globe, it has become integral that more research be carried out to better understand its underlying and predisposing factors. These studies, alongside adequate monitoring, testing, and surveillance, are essential to appropriately tailor control, prevention, and treatment approaches to achieve the best results. Achieving a complete reversal of FAs would immensely pave the way to reduce current spiking trends, completely manage FAs, and nip future incidences in the bud before they evolve to become more problematic.

References

[1]   Burks, A. W., Tang, M., Sicherer, S., et al. (2012). Food allergy. *Journal of Allergy and Clinical Immunology, 129*, 906–920.

[2]   Sicherer, S. H., & Sampson, H. A. (2018). Food allergy: A review and update on epidemiology, pathogenesis, diagnosis, prevention, and management. *Journal of Allergy and Clinical Immunology, 141*, 41–58.

[3]   Sicherer, S. H., Allen, K., Lack, G., Taylor, S. L., Donovan, S. M., & Oria, M. (2017). Critical issues in food allergy: A National Academies Consensus report. *Paediatrics, 140*, e20170194.

[4]   Sicherer, S. H., & Sampson, H. A. (2010). Food allergy. *Journal of Allergy and Clinical Immunology, 125*, S116–25.

[5]   West, H. (2017). The 8 most common food allergies (Accessed August 29, 2021, at https://www.health line.com/nutrition/common-food-allergies.)

[6]   Gupta, R. S., Warren, C. M., Smith, B. M., et al. (2019). Prevalence and severity of food allergies among US adults. *JAMA Network Open, 2*, e185630.

[7]   Clopton, J. (2021). Adult-onset food allergies increasing, confusing, (Accessed August 27, 2021, at https://www.webmd.com/allergies/news/20190111/adult-onset-food-allergies-increasing-confusing.)

[8]   Blázquez, A. B., & Berin, M. C. (2017). Microbiome and food allergy. *Translational Research, 179*, 199–203.

[9]   James, J. M., & Crespo, J. F. (2007). Allergic reactions to foods by inhalation. *Current Allergy and Asthma Reports, 7*(3), 167–174.

[10]   Pham, M. N., & Bunyavanich, S. (2018). Prenatal diet and the development of childhood allergic diseases: Food for thought. *Current Allergy and Asthma Reports, 18*, 58.

[11]   Valenta, R., Hochwallner, H., Linhart, B., & Pahr, S. (2015). Food allergies: The basics. *Gastroenterology, 148*, 1120–1131.

[12]   Dunn, S. E., Vicini, J. L., & Glenn, K. C. (2017). The allergenicity of genetically modified foods from genetically engineered crops: A narrative and systematic review. *Annals of Allergy, Asthma and Immunology, 119*, 214–222.

[13]   Fong, A. T., Katelaris, C. H., & Wainstein, B. (2017). Bullying and quality of life in children and adolescents with food allergy. *Journal of Paediatrics and Child Health, 53*, 630–635.

[14]   Nowak-Węgrzyn, A., Katz, Y., Mehr, S. S., & Koletzko, S. (2015). Non-IgE-mediated gastrointestinal food allergy. *Journal of Allergy and Clinical Immunology, 135*, 111424.

[15]   Olivier, C. E. (2013). Food allergy. *Journal of Allergy and Therapy, S3*(2), 004.

[16]   Stiemsma, L. T., & Turvey, S. E. (2017). Asthma and the microbiome – Defining the critical window in early life. *Allergy, Asthma and Clinical Immunology, 13*, 3.

[17] Boye, J. I., Danquah, A. O., Thang, C. L., & Zhao, X. (2012). Food allergens. *Journal of Food Biochemistry and Food Processing, 2012,* 798–819.

[18] Skripak, J. M., Matsui, E. C., Mudd, K., & Wood, R. A. (2007). The natural history of IgE-mediated cow's milk allergy. *Journal of Allergy and Clinical Immunology, 120,* 1172–1177.

[19] Gil, F., Amezqueta, A., Martinez, D., et al. (2017). Association between caesarean delivery and isolated doses of formula feeding in cow milk allergy. *International Archives of Allergy and Immunology, 173,* 147–152.

[20] Hasan, S. A., Wells, R. D., & Davis, C. M. (2013). Egg hypersensitivity in review. *Allergy and Asthma Proceedings, 34,* 26–32.

[21] Lopata, A. L., Kleine-Tebbe, J., & Kamath, S. D. (2016). Allergens and molecular diagnostics of shellfish allergy: Part 22 of the series molecular allergology. *Allergy Journal International, 25,* 210218.

[22] Selb, R., Wal, J. M., & Moreno, F. J. (2017). Assessment of endogenous allergenicity of genetically modified plants exemplified by soybean – Where do we stand? *Food and Chemical Toxicology, 101,* 139–148.

[23] Vitaliti, G., Pavone, P., & Spataro, G. (2015). Legumes steam allergy in childhood: Update of the reported cases. *Allergologia et Immunopathologia J (Madr), 43,* 196–202.

[24] Sicherer, S. H., Furlong, T. J., Maes, H. H., Desnick, R. J., Sampson, H. A., & Gelb, B. D. (2000). Genetics of peanut allergy: A twin study. *Journal of Allergy and Clinical Immunology, 106,* 53–56.

[25] Du Toit, G., Roberts, G., Sayre, P. H., et al. (2015). Randomized trial of peanut consumption in infants at risk for peanut allergy. *New England Journal of Medicine, 372,* 803–813.

[26] Fern´andez-Rivas, M. (2008). Allergies to fruits and vegetables. *Pediatric Allergy and Immunology, 19,* 675–681.

[27] Clayton, E. (2012). Adverse effects of vaccines: Evidence and causality. Institute of Medicine, 65. ISBN 978-0-309-21435-3.

[28] Shroba, J., Rath, N., & Barnes, C. (2019). Possible role of environmental factors in the development of food allergies. *Clinical Reviews in Allergy and Immunology, 57,* 303–311.

[29] Plunkett, C. H., & Nagler, C. R. (2017). The influence of the microbiome on allergic sensitization to food. *Journal of Immunology, 198,* 581–589.

[30] Ravid, N. L., Annunziato, R. A., & Ambrose, M. A. (2015). Mental health and quality-of-life concerns related to the burden of food allergy. *Psychiatric Clinics of North America, 38,* 77–89.

[31] Roduit, C., Frei, R., Depner, M., et al. (2014). Increased food diversity in the first year of life is inversely associated with allergic diseases. *Journal of Allergy and Clinical Immunology, 133,* 1056–1064.

[32] Nicolia, A., Manzo, A., & Veronesi, F. (2014). An overview of the last 10 years of genetically engineered crop safety research. *Critical Reviews in Biotechnology, 34,* 77.

[33] Doll, R. J., Joseph, N. I., McGarry, D., Jhaveri, D., Sher, T., & Hostoffer, R. (2019). Epidemiology of allergic diseases. In M. Mahmoudi (Ed.), *Allergy and asthma.* Cham: Springer. https://doi.org/10.1007/978-3-030-05147-1_2

[34] Loh, W., & Tang, M. L. K. (2018). The epidemiology of food allergy in the global context. *International Journal of Environmental Research and Public Health, 15,* 2043.

[35] Nwaru, B. I., Hickstein, L., Panesar, S. S., et al. (2014). The epidemiology of food allergy in Europe: A systematic review and meta-analysis. *Allergy, 69,* 62–75.

[36] Osborne, N. J., Ukoumunne, O. C., Wake, M., & Allen, K. J. (2012). Prevalence of eczema and food allergy is associated with latitude in Australia. *Journal of Allergy and Clinical Immunology, 129,* 865–867.

[37] Sicherer, S. H. (2011). Epidemiology of food allergy. *Journal of Allergy and Clinical Immunology, 127,* 594–602.

[38] Boyce, J. A., Assa'ad, A., & Burks, A. W. (2010). Guidelines for the diagnosis and management of food allergy in the United States: Summary of the NIAID-sponsored expert panel report. *Journal of Allergy and Clinical Immunology, 126,* 1105–1118.

[39] Budimir, J., Mravak-Stipetić, M., Bulat, V., Ferček, I., Japundžić, I., & Lugović-Mihić, L. (2019). Allergic reactions in oral and perioral diseases-what do allergy skin test results show? *Oral Surgery, Oral Medicine, Oral Pathology, and Oral Radiology, 127*, 40–48.

[40] Rokaite, R., Labanauskas, L., & Vaideliene, L. (2014). Role of the skin patch test in diagnosing food allergy in children with atopic dermatitis. *Medicina (Kaunas), 40*, 1081–1087.

[41] Chafen, J. J. (2010). Diagnosing and managing common food allergies: A systematic review. *JAMA, 12*, 1848–1856.

[42] Chang, K. L., & Guarderas, J. C. (2018). Allergy testing: Common questions and answers. *American Family Physician, 98*, 34–39.

[43] Hill, D. J., Heine, R. G., & Hosking, C. S. (2004). The diagnostic value of skin prick testing in children with food allergy. *Pediatric Allergy and Immunology, 15*, 435–441.

[44] Allen, K. J., & Koplin, J. J. (2016). Prospects for prevention of food allergy. *Journal of Allergy and Clinical Immunology Practises, 4*, 215–220.

[45] Dunn, G. A., Chan, C. H., & Crevel, R. (2015). Precautionary allergen labelling: Perspectives from key stakeholder groups. *Allergy, 70*, 1039–1051.

[46] Allen, K. J., Turner, P. J., & Pawankar, R. (2014). Precautionary labelling of foods for allergen content: are we ready for a global framework? *World Allergy Organization Journal, 7*, 1–14.

[47] Roses, J. B. (2011). Food allergen law and the Food Allergen Labelling and Consumer Protection Act of 2004: Falling short of true protection for food allergy sufferers. *Food and Drug Law Journal, 66*, 225–242.

[48] Urisu, A., Ebisawa, M., & Ito, K. (2014). Japanese guideline for food allergy. *Allergology International, 63*, 399–419.

[49] de Silva, D., Geromi, M., & Panesar, S. S. (2014). Acute and long-term management of food allergy: Systematic review. *Allergy, 69*, 159–167.

[50] Turnbull, J. L., Adams, H. N., & Gorard, D. A. (2015). Review article: The diagnosis and management of food allergy and food intolerances. *Alimentary Pharmacology and Therapeutics (Review), 41*, 25.

[51] Deschildre, A., & Lejeune, S. (2018). How to cope with food allergy symptoms? *Current Opinion in Allergy and Clinical Immunology, 18*, 234–242.

[52] Dinakar, C., & Warady, B. (2016). Food allergy care: "It takes a team". *Missouri Medicine, 113*, 314–319.

[53] Nowak-Węgrzyn, A., & Sampson, H. A. (2011). Future therapies for food allergies. *Journal of Allergy and Clinical Immunology, 127*, 558873.

[54] Chiu, C. Y., Liao, S. L., Su, K. W., et al. (2016). Exclusive or partial breastfeeding for 6 months is associated with reduced milk sensitization and risk of eczema in early childhood: The PATCH Birth Cohort Study. *Medicine, 95*, 3–4.

[55] Chan, E. S., Abrams, E. M., & Hildebrand, K. J. (2018). Early introduction of foods to prevent food allergy. *Allergy, Asthma and Clinical Immunology, 14*, 57.

[56] Cuello-Garcia, C. A., Brożek, J. L., & Fiocchi, A. (2015). Probiotics for the prevention of allergy: A systematic review and meta-analysis of randomized controlled trials. *Journal of Allergy and Clinical Immunology, 136*, 952–961.

[57] Taylor, S. L., & Baumert, J. L. (2015). Worldwide food allergy labelling and detection of allergens in processed foods. Food allergy: Molecular basis and clinical practice. *Chemical Immunology and Allergy, 101*, 227–234.

[58] Wang, J., & Sampson, H. A. (2019). Food allergy: Recent advances in pathophysiology and treatment. *Allergy, Asthma and Immunology Research, 1*, 19–29.

Mduduzi Mokoena

Chapter 16
Risk assessment and risk analysis
for chemical hazards

Abstract: Food products are sensitive to contamination by various agents including microorganims and chemicals. Most chemical contaminants are heatstable and remain in the product after thermal processing of the food product. Such chemicals and their residues pose as serious health hazards to consumers. Their detection in food products requires health authorities to take serious sanctions resulting in the recall of the products from the shelves in the community outlets. This chapter will discuss risk assessment, risk management, and risk communication as the three major components of risk analysis and how they are implemented to ensure safety of food products to consumers.

16.1 Introduction

Chemical hazards are responsible for the very high number of recalls of food products globally, due to safety concerns. These entities can arise from the food processing plant itself, as well as natural and environmental sources. Sometimes they are natural constituents of the food or additives during the formulation of the food product. At a national level, the department of health can issue food safety alert in order to inform consumers about potential health risk associated with specific food products that could still be at supermarkets or in their homes. This food safety alert may equally be directed to health authorities to initiate recall of the products that pose a threat to public health.

Risk assessment, risk management, and risk communication are the three major components of risk analysis. The purpose of risk analysis is to ensure that proper decisions are taken regarding food products so as to safeguard public well-being and sustenance. This also enables governments to marshal different stakeholders to deal decisively with food risks when they arise, thereby ensuring food safety by following scientifically informed strategies to lower incidence of food-borne disease outbreaks. Risk assessment forms an integral and scientific part of risk analysis and also forms the basis of management strategies for the protection of human health, especially during times of uncertainty.

Mduduzi Mokoena, Department of Pathology, University of Limpopo, South Africa

https://doi.org/10.1515/9783110748345-016

16.2 Risk assessment

Chemical hazards are responsible for the high number of products that are globally withdrawn from the public. Hazard analysis and critical control point (HACCP) principle dictates that hazardous substances must be identified during the early stages of product development, and relevant measures be implemented to mitigate their impact on the quality and safety of the product. The source of hazardous substances could arise from the food product itself or from the additives that are part of the product formulation process. Some of them could be natural component of the food, whereas others could be naturally occurring in the environment.

The risk analysis paradigm formalizes a framework that ensures that the three components of food risk analysis function optimally, with proper communication among them to achieve order and scientific objectivity. The scientific part of this process, the risk assessment as far as food safety goes, is carried out by a FAO/WHO joint scientific committee. It is risk managers who eventually make a determination whether there is a possibility of risk, and develop policies and schedules, as well as providing essentials for successful completion of the tasks.

Pollutants of a chemical nature could arise from household products that are used for cleaning, disinfection, odor control, as well as weed-killers and insecticides. Of great concern is the fact that these contaminants also find their way to food and water, rendering these commodities unfit for consumption, with serious implications on human health and well-being [1]. Ingestion of chemically contaminated food can manifest as mild gastric discomfort to death as a result of liver, kidney, and nervous system failure.

Pesticides: Pesticides in food cause nervous system, genitourinary and reproductive systems due to their toxicity and carcinogenicity [2]. They can lead to the disruption of metabolic pathways that normally break down inimical compounds [3]. Some of these pesticides can result in mental and nervous system impairment in addition to endocrine system disruption [4, 5].

Animal drug residues: These are particularly problematic in foods from animal sources. Some of these residues can accumulate during the product development process, exceeding levels that can be tolerated by the human systems Depending on their chemical properties, some drug residues can become concentrated during the food manufacturing process. Extreme care must be taken, especially drugs whose tolerance levels are yet to be determined as these are hazardous substances [4, 5].

Heavy metals: These are of great concern as they may emanate from the agricultural processes, implements, and cutlery used in food preparation and consumption. Heavy metal poisoning can have severe health effects, including mental retardation in children, disorders associated with the skin, as well as heart disease and lifestyle diseases such as type II diabetes mellitus [4–6].

Environmental contaminants: These contaminants come into contact with food by virtue of being part of the environment where food is grown and processed. Should such contaminants be identified through hazard analysis, contaminant requires a preventive control; the control measure will depend on how the contaminant could get into the food product. In some cases, elevated concentrations of these contaminants emanate from their unintended incorporation in animal diet [4, 5], and these need to be avoided to ensure food safety.

Biological toxins: These include secondary metabolites of molds and fungi called mycotoxins, biogenic nitrogenous compounds, and some plant and algal toxins. Mycotoxins include aflatoxins, fumonisins, deoxynivalenol, ochratoxin, and patulin; are notorious for spoiling agricultural products in the field and in storage facilities. Some mycotoxins have been classified as carcinogens or potential carcinogens by the WHO. Proper monitoring and controls should be implemented throughout the supply chain to mitigate mycotoxin proliferation due to annual variance of growing and harvesting conditions [5].

Nitrogenous carcinogens such as histamines arise from the microbial catabolism of foods from animal sources. These may result in disorders such as headaches, nausea, and abdominal discomfort. Thus if aged cheeses used for product development pose a histamine hazard, preventive control should be implemented [4, 5].

Assessment of chemical hazard risk for food safety requires consideration of various factors that necessitate gathering of a huge amount of data, which is time-consuming and requires expertise; thus, a consultant is normally enlisted when developing and reviewing HACCP plans. Usage of modeling theories can expedite analysis of data, and to enable population-wide predictions [7, 8]. After the risk assessment task has been completed, further analysis may be warranted, with subsequent research activities. However, the report of the risk assessment team serves as the scientific testament for decision-making, including reopening the investigations when there is quest for more information.

16.3 Risk management

One of the goals of risk management is to safeguard public health by applying relevant strategies [9]. With the globalization of the economies, risk management must also be interpreted in the context of international trade. In regard to food safety matters, the WTO rules are set out to allow member countries to take appropriate measures to safeguard public health without creating unfair trade barriers. Hence, the country's policies must be within the parameters of justice and fairness, as disputes may be lodged against a country that applies stringent measures that are beyond what is reasonable to protect consumer health [10].

For risk management to play its meaningful role, it must be evaluated, taking into account scientific merits of the risks, as well as socioeconomic factors of public interest. This will allow authorities to make informed decisions regarding the potential of the risk to have adverse outcomes on the population, and, if so, to put measures in place to circumvent their impact. That said, all relevant parties must be involved to ensure that decisions are taken and implemented with the best interest for the public [10].

Risk management is a complex process, as it requires input from individual with diverse expertise. Thus, risk management protocols must be amenable to revision, and all activities associated with risk management process must be attended to. While risk managers have the overall responsibility for developing the risk profile, risk assessors and other stakeholders should also actively participate. For instance, a good risk profile will ensure that the initial risk management goals and questions to be answered during the risk assessment are explicit in order to avoid misunderstanding. This will require extensive interaction between risk assessors and risk managers, as well as dialogue with relevant external stakeholders. A good example of a risk profile is one that identifies the risk, how widespread it is, and gives an appraisal of its ramifications to consumer perceptions, and how the interventions will benefit the public. Hence it is a summary of all the information present at the moment. Risk managers and risk assessors must work cooperatively to avoid any misunderstandings, while answering all pertinent questions. Finally, the risk profile must be filed in such a way that it is clear to risk managers, and areas of priority outlined so that it is easy for them to decide whether a full-scale product recall is essential [10].

16.4 Risk communication

Risk communication is all about the sharing of the risk-specific information among all relevant stakeholders and authorities, including consumers. Hence risk communication plays a central role in bringing on board all necessary components of risk analysis to ensure food safety, taking into account all ethical considerations to protect the consumers as far as possible [10]. Many people are exposed to food safety and risks on a daily basis in all countries of the world. Incidents of food safety risks correspond to the quality of the preventive attempts put in place throughout the food chain at a local level [10].

In order to expedite prudent decision-making by all concerned, it is critical to convey messages about food risks efficiently and reliably, in a language that is easily understood, and with utmost accuracy, without any ambiguity. This goes a long way in building trust that is necessary in the food sector to avoid sanctions on agricultural produce that are often the result of public mistrust.

Much has been learnt by food regulators in the area of food risk communication over the years, including the fact that the communication does not occur precisely when the risk assessment is still underway. On the contrary, risk communication must be deliberately planned, taking into cognizance the implementation and management aspects that must be put in place, and a mechanism to process the results of risk analysis appropriately. In this regard, one person from the risk management team can be entrusted with the task of handling all the risk communication activities. Alternatively, communication experts may be called upon to plan and deploy the communication on behalf of the risk analysis team. Be that as it may, the tasks and activities embedded in the risk communication strategy must be clearly spelled out from the beginning, to ensure maximum public participation and coherent messages that are easy to understand [10].

The role of stakeholders cannot be overemphasized, as they ensure clarity of the messages by bridging the cultural ethos, values, and nuances associated with the local language. This facilitates rapport with affected consumers who then appreciate the measures that are going to be implemented to address the risk. A cordial, transparent sharing of information expedites the outcomes of the risk analysis process, as the ideas and opinions are allowed to flow freely. This collaboration with stakeholders benefits the risk management team in that outcomes of the process are well received by stakeholders, with less prospects to be resisted [8, 10].

Since various groups of people are involved in different areas of the food continuum, risk communication teams must identify all these stakeholders early in the process. Indeed, some stakeholders will readily communicate their interest in the risk analysis by virtue of their proximity to the industry, but a more protracted effort needs to be spurred to enlist those who may try to avoid involvement due to self-interest protection or ignorance. In some instances, incentives may have to be provided. While this may be tedious, nevertheless, the outcomes are worth it [10].

Many countries have developed policies on how and when public participation can be harnessed. Public gatherings are some of the ways to involve diverse stakeholders to some extent, while modern communication technologies tend to be more amenable to transparent two-way communication strategy. For example, web-based discussions, and radio and television phone-in programs can be used to share information with the community and to solicit views and feedback [8].

Perception of risk has analytical and emotional dimensions. The risk communication team must therefore take the emotional aspects of the public into cognizance while carefully communicating the scientific dimensions of the food safety risk. Given the fact that the public tends to be more interested in the nonscientific aspect associated with the food risk, the risk communication must carefully address such concerns to enable consumers to take informed decision regarding imminent food risk.

In the final analysis, how acceptable the measures to mitigate risk depends largely on public perception of the risk. If the public perceives the risk to be great, there will be a greater desire to have the risk reduced or avoided. But there are some cases where the public perceives risk reduction measures as more damaging than when there is no intervention. Thus, it is crucial for the risk management team to have thorough information regarding public perception about the risk [10].

16.4.1 Strategies for risk communication

Although risk communication takes place under various settings, it must be noted that that for it to be effective, the communication strategy should address the concerns of the stakeholders. Strategies employed during food crisis scenarios are remarkably different from those used to engender ongoing rapport with stakeholders [10]. Ideally, the strategy should inform the data collection and analysis concerning food risk, taking into account the views of the various stakeholders; formulation of coherent communications tailored to specific sections of the community; taking the stakeholders into confidence by giving a proper appraisal of the imminent risk and measures to circumvent it; assessing the efficacy of the risk communication outcomes as far as their clarity and impact. Sending the communication out at the earliest possible moment is critical to avert disruptions of the food supplies and their dire economic consequences. In the current technological era, untimely communication from an organization is sure to breed public mistrust and adversely affect its communication goal [10].

16.5 Conclusion

Chemical hazards in food are inimical to consumers, and measures must be put in place to identify and eliminate these hazards in food destined for human and animal consumption. Proper risk analysis, risk management, and risk communication form the core of such measures. Hence, organizations involved in food production and food product development must implement HACCP protocols aimed at timeous identification of health risks and to mitigate them. When breaches occur, and chemical hazards have made their way into food, timeous communication to recall affected food products is essential to ensure public health and future trust of the entity.

References

[1]   Rather, I. A., Koh, W. Y., Paek, W. K., & Lim, J. (2017). The sources of chemical contaminants if food and their health implications. *Frontiers in Pharmacology*, doi: doi.org/10.3389/fphar.2017.00830.

[2]   Bassil, K., Vakil, C., Sanborn, M., Cole, D., Kaur, J., & Kerr, K. (2007). Cancer health effects of pesticides. *Canadian Family Physician*, *53*, 1704–1711.

[3]   Androutsopoulos, V., Hernandez, A., Liesivuori, J., & Tsatsakis, A. (2013). A mechanistic overview of health associated effects of low levels of organochlorine and organophosphorous pesticides. *Toxicology, 307*, 89–94. doi: 10.1016/j.tox.2012.09.011.

[4]   Al-dabbas, M. (2020). Food chemical hazards. In F. A. Al-Rub (Ed.), *Food Safety Hazards*. doi: 10.29011/ 978-1-951814-03-8-005.

[5]   Thompson, L. A., & Darwish, W. S. (2019). Environmental chemical contaminants in food: review of a global problem. https://doi.org/10.1155/2019/2345283

[6]   Jaishankar, M., Tseten, T., Anbalagan, N., Mathew, B. B., & Beeregowda, K. N. (2014). Toxicity, mechanism and health effects of some heavy metals. *Interdisciplinary Toxicology, 7*(2), 60–72.

[7]   Jager, T., & Klok, C. (2010). Extrapolating toxic effects on individuals to the population level: The role of dynamic energy budgets. *Philosophical Transactions of the Royal Society of London B: Biological Sciences, 365*(1557), 3531–3540.

[8]   Safefood 360°. (2013). Chemical hazards in food safety. White Paper.

[9]   FAO/WHO. (1997). Risk Management and Food Safety. Report of a Joint FAO/WHO Consultation. Rome, Italy.

[10] FAO/WHO. (2016). Risk communication applied to food safety. Handbook. Food and Agriculture Organization of the United Nations World Health Organization.

George Ooko Abong and Jumbale Mwarome
Chapter 17
Risk assessment and risk analysis for biological hazards: a systematic literature review

Abstract: Risk assessment and risk analysis of biological hazards varies from country to country. Developed and developing countries have contrasting ways of hazard identification. These differences are associated with technological capacities, coupled with support structures in the food industries. While the public health effects of exposure to these hazards are universal, certain factors such as the dose, and duration of subjection determine the severity of the health of individuals. Food consumers are the primary victims of food safety compromises. When food products such as meat products, fruits, vegetables, and milk products are contaminated with biological agents, the effects on the health of consumers may be fatal. It is for this reason that modern technologies such as Risk Simulation Softwares are used to quantify the level of risks of biological hazards. However, the application of new technologies should be done within safe boundaries to protect the health of food processors, and food handlers. Furthermore, Hazard and Critical Control Points (HACCP) should guide upper control limits for biological hazards. The current review gives the general perspective of risk assessment and risk analysis for biological hazards.

17.1 Introduction

Food safety is a global public health concern since it affects people of all groups regardless of their age, gender, race, or economic status. The public health concerns with regard to food safety are points of alarm especially with the ever-increasing cases of antimicrobials in food. Foodborne diseases are a constraint in the food chain in both the developed and the developing countries. According to a report by the World Health Organization (WHO), 600 million cases of foodborne illnesses and 420,000 death cases were reported in 2010 [30]. Results from the study established that biological agents which caused diarrheal diseases accounted for the highest percentage of the infections. *Salmonella* was reported to be the leading cause of these infections followed by *Taenia solium*, hepatitis A virus, and aflatoxin [1]. It is the role of regulatory bodies to use risk assessment and risk analysis to bring the levels of food hazards to achievable and real-

George Ooko Abong, Jumbale Mwarome, Department of Food Science, Nutrition and Technology, University of Nairobi

https://doi.org/10.1515/9783110748345-017

istic levels. In order to address this issue, assessment of individual potential risks is important in alleviating the detrimental effects caused by specific hazards in food. Microorganisms can enter the food systems at any stage in the food chain. The ability of some microorganisms to adapt to various environmental conditions can enhance their capacity to multiply, adapt, and, consequently, increase their numbers in these environments. This scenario is a significant roadblock toward measures put in place to achieve safety of food [2]. The control of this menace can be achieved through different approaches. It is important, however, to understand the strategies that aim at eliminating the risks involved.

In food production, the safety of food is assured through proper application of the hazard analysis critical control point (HACCP) as well as the good manufacturing practices. The role of risk assessment and risk analysis comes into play in predicting the point at which a biological food hazard arises. The main objective of risk assessment and risk analysis of hazards in food is to safeguard the interests of consumers by eliminating the possibility of occurrence of harmful components of food. Risk assessment has been applied in assessing microbial resistance in food which is considered a human health threat worldwide by the WHO.

17.2 Biological hazards in the food chain

Biological hazards refer to a number of zoonotic and non-zoonotic agents that compromise the quality and safety of food. The agents include cell cultures, microorganisms, and other human endoparasites [3]. They can be organisms or products of organisms that can enter the food chain and cause food safety concerns. The presence of harmful microorganisms in food not only alters the eating quality of the food but also negatively affects the market acceptability of the product. Wide-spectrum antimicrobial resistance exhibited by foodborne microbes is a new area of concern in the food systems. The resistance may be due to the development of new nature of structures of certain bacteria. These structural changes in these microorganisms are caused by, but not limited to, genetic alterations through processes such as transduction, conjugation, and transformation [4]. Antimicrobial resistance can as well be attributed to processing treatments such as irradiation processes, acid treatments, and preservation. The common microorganisms that pose a public health threat in the food chain and of biological importance include *Salmonella* spp., *Escherichia coli*, *Listeria monocytogenes*, *Campylobacter jejuni*, *Bacillus cereus*, *Bacillus anthracis*, *Clostridium botulinum*, *Vibrio vulnificus*, *Yersinia enterocolitica*, *Vibrio parahaemolyticus*, and *Clostridium perfrigens*. Common viruses of concern include Norwalk virus and hepatitis A virus. In the recent past, foodborne parasites have become a problem especially in the tropical regions and in areas with poor sanitation. It is due to this nature of occurrence that these foodborne agents are considered of less concern to the

USA. However, in other European countries, foodborne illnesses affect 23 million people every year with 5,000 deaths reported. These cases are associated with *Salmonella-* and *Campylobacter*-related exposures [5]. According to a report by Doyle and Wi [6], the human body harbors about 70 species of protozoa, and more than 300 species of parasitic worms. However, the main parasites of concern in the tropical countries include *Toxoplasma gondi, Trichinella spiralis, Cryptosporidium, Taenia solium, Giardia,* and *Anisakis* spp. [7]. These parasites occur in different types of foods. Common food habitats for these parasites include raw vegetables, pork, beef, fish, raw fruits, and water [6]. The occurrence of biological hazards in food may be as a result of two different ways: agents intentionally introduced into the food products in such processes as production or research biotechnology and microbiological research or agents that may be unintentionally introduced into the food as a result of a human activity leading to the survival and growth of the microorganism. The presence of foodborne pathogens is readily detectable in the developed countries due to the existence of effective analytical and surveillance methods of analysis and detection. Risk communication through government advisories and media coverage influences consumer purchasing behaviors. Bitsch and Rombach [8] established that consumers in developed countries trust official advisories. Furthermore, there are other players in the food chain that help in alleviating the compromises in the food safety aspect of food. Through the Panel of Biological Hazards (BIOHAZ) of the European Food Safety Authority, biological hazards assessment and independent advice are given for both the food and feed chains [9].

However, in developing countries, these hazards remain unknown due to the insufficiency in methods of analysis. The situation becomes worse if the food has to be transported over long distances [7]. According to the Food and Agriculture Organization [31], the occurrence of biological hazards in food is through two main mechanisms that consequently lead to the development of foodborne illnesses. The first mechanism is where microorganisms produce toxins inside the human body, causing the body to develop symptoms which may range from mild to severe intoxications (food infection). The second mechanism is where the human body shows pathological responses toward an ingested toxin from a microorganism (food intoxication; FAO, 2010).

17.3 Risk assessment of biological hazards in food

Biological hazards not only affect consumers of food, they also pose a potential health risk to the food processors and other food handlers in the food chain through exposure to pathogenic microorganisms. The existence of biological hazards may cause environmental and economical losses. The assessment of risks resulting from food is vital in ensuring management planning for the sake of risk prevention, preparedness, and capacity development in case these hazards are experienced. Practical ways of conducting quantitative and qualitative risk assessment involve providing an estimate of the likelihood and extremity of illness in a given group as a result of consuming a contaminated food [3]. For example, *Quantitative Risk Model* [10] is a book that describes how the qualitative risk model is applied in poultry food chain. This model was implemented by the New Zealand Food safety Authority to manage the development of risks caused by *Campylobacter*. In the risk assessment design, the possibility of a carcass being contaminated with the bacteria and the exposure is used to derive dose–response information that predicts the number of people infected. One strategic approach to mitigate and prevent biological hazards is through quantitative risk assessment. Approaches by the WHO to risk analysis are strategically aimed at initiating an evidence-based approach by identifying the levels of risks and rank them according to priority. This has helped many countries to define their readiness and preparedness to handle food safety risks. Quantitative risk assessment has been shown to yield positive results in mitigating constraints of antimicrobial resistance in food. The main focus in this approach is to find a pathway in the treatment of foodborne bacteria in food that exhibit aversive responses toward antimicrobial agents [11].

Generally, risk assessors use four components in addressing the safety of food. These components include hazard identification, risk analysis, assessment of exposure, and risk characterization [12]. Because these components are equally important in assessing the degree of risk, a breach in one step can lead to a gap in the whole process. Although knowledge gaps are common in food safety risk analysis, Caffrey et al. [11] established that the component of hazard characterization has been an area of concern. This translates to lack of enough evidence to support the effects of foodborne diseases during risk assessment in the case of antimicrobial resistance.

Qualitative risk analysis involves definition and analysis of a potential food safety hazard to a certain population or a community, and using the results to predict the probable adverse effects [13]. Qualitative hazard analysis is useful when the data is scanty, for example, when ranking new emerging hazards. They are also advantageous in the way they help in giving evidence-based opinions based on their rankings. This is achieved through generation of descriptive materials based on the responses from participants [14]. Despite their distinctive methods of assessment, their focal point is their concept of conceptualizing the presence of hazard and the extremity of consequences to human health [15]. Although both quantitative and qualitative hazards assessment approaches are applied in evaluating the severity of risks, a detailed search on the subject literature by Caffrey et al. [11] showed that most of the articles (25/38) which were put under study had their focus on quantitative risk assessment. These articles were comprehensively obtained by identifying and describing literature for study of antimicrobial resistance, with special focus on products intended for treatment of bacterial infections.

17.4 Application of risk analysis of biological hazards in food safety

The principles of risk analysis are important in hazard prevention by applying the principles of HACCP steps. The science-based components are important in making informed management decisions and interventions for the sake of safety of food. Risk analysis has three main components that guide food industries and government regulators in ensuring safe food supply. These components of risk analysis include hazard assessment, risk management, and risk communication [16]. Each one of these components is important in collecting and interpreting data, making informed management decisions based on scientific evidence, and communicating the findings and decisions as required. Through the microbiological guidelines in the risk analysis components, manufacturing processes are monitored by regulatory agencies and conformity is ensured. Every country has its own food legislation protocols that are to be followed by all players in the food businesses. It is through these protocols that food businesses are required to implement Food Safety Management Systems which incorporate the prerequisite activities and HACCP principles which are management equipment for food safety guarantee [17]. The need to understand and rank risks brings into context the need for prerequisite activities. Prerequisite programs guide production and manufacturing activities in food industries and are a pillar in food safety and management systems.

17.5 Risk assessment of biological agents in meat and meat products

Stakeholders in the meat value chain need transparent, evidence-based, and conscientious approaches to estimate and collate the risk of biological hazards in meat and meat products. Ready-to-eat foods are associated with the development of listeriosis. The safety of meat and meat products can be compromised at any stage from the production, processing, storage all the way to the retail stage. Meat is considered inappropriate for human consumption when the microbial load reaches 10^7 colony-forming unit (CFU) per gram [18]. At this state, the sensory attribute and shelf life of the meat have deteriorated.

The control of foodborne illnesses should thus be initiated from the slaughter houses. The meat industry, especially in developing countries, is characterized by authorized, unmonitored, ill-equipped, roadside stands, and slaughterhouses where proper processing standards are not taken care of [18]. Routine inspection of slaughter houses should be the first intervention in reducing the possibility of food safety hazards occurrence. However, the inspection may not be effective enough to biological hazards since there are yet to be quantitative specifications levels for meat. In developed countries, the prevalence of the above-mentioned condition is quite high due to the contamination of foods with *Listeria monocytogenes* bacteria [19]. Long-term refrigeration of raw meat leads to the growth of *Listeria monocytogenes* and consequently reaches unacceptable levels [19]. Once consumed, the bacteria pose a public health risk. Qualitative microbial risk assessment (QMRA) can be used to establish the dose–response correlation, as well as exposure assessment, to determine the risk characterization out-turn for the specific microorganism. In this QMRA model, Monte Carlo simulations are utilized together with the appropriate software such as Risk Simulation Software® (https://www.palisade.com/risk/). The output from this model is a function of the possibility of postproduction and storage conditions of the meat products, and the growth of *Listeria monocytogenes* [20].

Salmonella is another common bacterium in meat and meat products that has been reported to cause foodborne illnesses. Raw-cured fermented sausages are one of the main target food products for this bacterium. One recent case of food poisoning occurred as a result of *Listeria* outbreak in a meat production plant in Polokwane, South Africa. More than 200 people died, with more than 1,000 cases reported within one month [21]. Other bacterial hazards of concern in meat and meat products, but are less frequently reported, are *Bacillus cereus*, *Staphylococcus aureus*, *Clostridium botulinum*, and *Clostridium perfrigens* [22]. Risk assessors can use different approaches in quantifying and analyzing the cases of human salmonellosis. The different approaches include analytical epidemiology, outbreak data, microbial subtyping, structured expert opinion, and comparative exposure assessment [9]. Each of these approaches has their own ¬rengths and weaknesses. Comparative exposure assessment has exhibited impressive ¬ired resolution in linking the proportion of salmonellosis cases with their causative ¬. However, the main bottleneck in utilizing this method is the fact that it has ¬en used in for risk assessment in human salmonellosis.

¬nalysis and risk assessment of biological ¬s in fruits and vegetables

¬f risks associated with fruits and vegetables may not match ¬roducts, the contamination of these products by microorgan- to the health of consumers. Raw fruits and vegetables nor- ¬croflora at the farm level. However, these products may ¬sms at the point of consumption. These pathogenic ¬oduct handling from the point of production to the ¬its and vegetables contributes the development of ¬ampylobacter jejuni*, *Escherichia coli*, *Listeria* *Vibrio* spp., and *Staphylococcus aureus*. All the above-mentioned bacteria have been associated with foodborne outbreaks [24]. Viruses can as well find their route to their human hosts through consumption of unhygienically produced and processed fruits and vegetables. The main viruses of concern include Rotavirus, and Norwalk virus. Parasitic protozoa which may be transmitted through person to person contact or via fecal contamination are a concern in safety of these food products. They include *Giardia lamblia, Cyclospora cayetanensis,* and *Cryptosporidium parvum* [3].

The global fruits and vegetables chain has been reporting cases of foodborne outbreaks, hence the need to do risk analysis for the fresh products. A lot of fruits have to be transported over long distances across countries and continents in search for market. For this reason, a food safety issue in one region of the world may cause public health concern in another region. Sporadic disease outbreaks have been reported in Australia, New Zealand, Canada, Japan, the United States and, the European Union [25]. The three main bacterial hazards associated with these outbreaks are norovirus, *Salmonella enteritica,* and *Staphylococcus aureus,* with their severity ranging from 42.2%, 19.9%, and 7.9%, respectively [25]. The risks factors to occurrence of pathogenic microorganisms in food include cross-contamination, irrigation water, infected food handlers, storage time and temperature abuse, and contaminated ingredients.

Risk assessment in fruits and vegetables can be made effective through describing CODEX principles and finding the association between the specific biological hazards and the association with the specific product. The risk of each pathogen is analyzed quantitatively, and separately. However, lack of enough data and information is one big drawback in quantitative analysis especially information regarding contamination at the growing stage of vegetables [26].

17.7 Risk assessment of biological hazards in milk and milk products

Industrial biotechnology is a modern trend in the food chain. An example of its application was used in modification of milk production using a synthetically modified milk hormone, rbST [27]. This technology has been proven to be beneficial in food production. However, its application can cause public health susceptibility due to the exposure of people, especially workers in the these industries, to biological agents [28]. Fermentation of milk is a common practice among many communities in Africa. In most instances, the preceding treatment before fermentation is sterilization and pasteurization. Qualitative risk assessment is theoretically useful in determining the risk in final products for milk and milk products [2]. The shortcoming of this approach, however, lies in the practicality of its application. The uncertainty and variability that comes with biological risk assessment is attributed to biological aspects of the biological hazards such as the microorganism's kinetics and dose responses.

The occurrence of biological hazards in the food chain is determined by the frequency of exposure. A study conducted by Adil et al. [29] indicated that the occurrence of the risk is directly related to the number of times the biological hazards occur in the marketing and production chain of the milk. The study which sorted to ascertain the seroprevalence of bovine brucellosis in milk also established that the total risk of brucella in milk is directly proportional to the exposure among consumers, farm workers, at constant intervals.

17.8 Conclusion

Biological agents play a significant role in the food chain. Quality assurance can be obtained through systematic approaches toward biological hazards quantification and ranking. Risk analysis is useful in evaluating the possibility of a hazard occurring and can be useful in mitigating adverse effects from potential biological hazards. Processing and retail of meat and meat products can be a vital step in the meat value chain and can lead to the entry of pathogenic microorganisms. For the case of fruits and vegetables, risk assessment should have its main focus in the retail stage. Handling of these commodities plays a key role in the quality of the final products that reach the consumer. Governments and other regulatory bodies should come up with microbial standards for slaughterhouses and foster conformity to those standards. Storage conditions of food products need to be reevaluated to monitor the growth of antimicrobial-resistant agents.

References

[1]   Todd, E. (2020). Food-borne disease prevention and risk assessment. *International Journal of Environmental Research and Public Health*, *17*(14), 1–13. https://doi.org/10.3390/ijerph17145129
[2]   Havelaar, A. H., Brul, S., de Jong, A., de Jonge, R., Zwietering, M. H., & ter Kuile, B. H. (2010). Future challenges to microbial food safety. *International Journal of Food Microbiology*, *139*(SUPPL. 1), S79–S94. https://doi.org/10.1016/j.ijfoodmicro.2009.10.015
[3]   Burzoni, S., Duquenne, P., Mater, G., & Ferrari, L. (2020). Workplace biological risk assessment: Review of existing and description of a comprehensive approach. *Atmosphere*, *11*(7). https://doi.org/10.3390/atmos11070741
[4]   Ali, M. E. (2017). Role of microorganisms in food contamination, processing and safety. *Journal Food Microbiology*, *1*(1), 9–10. http://www.alliedacademies.org/journal-food-microbiology/
[5]   Lee, H., & Yoon, Y. (2021). Etiological agents implicated in foodborne illness world wide. *Food Science of Animal Resources*, *41*(1), 1–7. https://doi.org/10.5851/KOSFA.2020.E75
[6]   Doyle, M. E., & Wi, M. (2006). Foodborne Parasites. *Foodborne Parasites*, October. https://doi.org/10.1007/0-387-31197-1
[7]   Schirone, M., Visciano, P., Tofalo, R., & Suzzi, G. (2017). Editorial: Biological hazards in food. *Frontiers in Microbiology*, *7*(JAN), 2016–2018. https://doi.org/10.3389/fmicb.2016.02154

[8] Bitsch, V., & Rombach, M. (2014). Risk Communication and Market Effects during Foodborne Illnesses: A Comparative Case Study of Bacterial Outbreaks in the U.S. and in Germany. *International Food and Agribusiness Management Review, 17*(3), 97–114.

[9] Romero-Barrios, P., Hempen, M., Messens, W., Stella, P., & Hugas, M. (2013). Quantitative microbiological risk assessment (QMRA) of food-borne zoonoses at the European level. *Food Control, 29*(2), 343–349. https://doi.org/10.1016/j.foodcont.2012.05.043

[10] *Quantitative risk model.* (2007).

[11] Caffrey, N., Invik, J., Waldner, C. L., Ramsay, D., & Checkley, S. L. (2019). Risk assessments evaluating foodborne antimicrobial resistance in humans: A scoping review. *Microbial Risk Analysis, 11*, 31–46. https://doi.org/10.1016/j.mran.2018.08.002

[12] Stohlgren, T. J., & Schnase, J. L. (2006). Risk analysis for biological hazards: What we need to know about invasive species. *Risk Analysis, 26*(1), 163–173. https://doi.org/10.1111/j.1539-6924.2006.00707.x

[13] Gizaw, Z. (2019). Public health risks related to food safety issues in the food market: A systematic literature review. *Environmental Health and Preventive Medicine, 24*(1), 1–21. https://doi.org/10.1186/s12199-019-0825-5

[14] Chen, Y., Dennis, S. B., Hartnett, E., Paoli, G., Pouillot, R., Ruthman, T., & Wilson, M. (2013). FDA-iRISK-A comparative risk assessment system for evaluating and ranking food-hazard Pairs: Case studies on microbial hazards. *Journal of Food Protection, 76*(3), 376–385. https://doi.org/10.4315/0362-028X.JFP-12-372

[15] Van der Fels-Klerx, H. J., Van Asselt, E. D., Raley, M., Poulsen, M., Korsgaard, H., Bredsdorff, L., Nauta, M., D'agostino, M., Coles, D., Marvin, H. J. P., & Frewer, L. J. (2018). Critical review of methods for risk ranking of food-related hazards, based on risks for human health. *Critical Reviews in Food Science and Nutrition, 58*(2), 178–193. https://doi.org/10.1080/10408398.2016.1141165

[16] Harris, K. B., Cross, H. R., Acuff, G. R., & Webb, N. B. (1995). Risk analysis, HACCP and microbial criteria in meat and poultry systems. *HACCP in Meat, Poultry, and Fish Processing*, 134–155. https://doi.org/10.1007/978-1-4615-2149-5_7

[17] Panel, E., Biohaz, H., Andreoletti, O., Baggesen, D. L., Bolton, D., Butaye, P., Cook, P., Davies, R., Nesbakken, T., Prieto, M., Ricci, A., Ru, G., Sanaa, M., Simmons, M., & Sofos, J. (2013). Scientific Opinion on the maintenance of the list of QPS biological agents intentionally added to food and feed (2013 update). *EFSA Journal, 11*(11), 1–107. https://doi.org/10.2903/j.efsa.2013.3449

[18] Das, A. K., Nanda, P. K., Das, A., & Biswas, S. (2019). Hazards and safety issues of meat and meat products. In *Food Safety and Human Health*. Elsevier Inc. https://doi.org/10.1016/B978-0-12-816333-7.00006-0

[19] Hadjicharalambous, C., Grispoldi, L., & Goga, B. C. (2019). Quantitative risk assessment of Listeria monocytogenes in a traditional RTE product. *EFSA Journal, 17*(S2). https://doi.org/10.2903/j.efsa.2019.e170906

[20] Harrison, R. L. (2009). Introduction to Monte Carlo simulation. *AIP Conference Proceedings, 1204* (January), 17–21. https://doi.org/10.1063/1.3295638

[21] Tchatchouang, C. D. K., Fri, J., De Santi, M., Brandi, G., Schiavano, G. F., Amagliani, G., & Ateba, C. N. (2020). Listeriosis outbreak in South Africa: A comparative analysis with previously reported cases worldwide. *Microorganisms, 8*(1). https://doi.org/10.3390/microorganisms8010135

[22] Omer, M. K., Álvarez-Ordoñez, A., Prieto, M., Skjerve, E., Asehun, T., & Alvseike, O. A. (2018). A Systematic Review of Bacterial Foodborne Outbreaks Related to Red Meat and Meat Products. *Foodborne Pathogens and Disease, 15*(10), 598–611. https://doi.org/10.1089/fpd.2017.2393

[23] James, J. (2005). Overview of Microbial Hazards in Fresh Fruit and Vegetables Operations. *Microbial Hazard Identification in Fresh Fruit and Vegetables*, 1–36. https://doi.org/10.1002/0470007761.ch1

[24] Rosenquist, H., Boysen, L., Krogh, A. L., Jensen, A. N., & Nauta, M. (2013). Campylobacter contamination and the relative risk of illness from organic broiler meat in comparison with conventional broiler meat. *International Journal of Food Microbiology, 162*(3), 226–230. https://doi.org/10.1016/j.ijfoodmicro.2013.01.022

[25] Li, M., Baker, C. A., Danyluk, M. D., Belanger, P., Boelaert, F., Cressey, P., Gheorghe, M., Polkinghorne, B., Toyofuku, H., & Havelaar, A. H. (2018). Identification of biological hazards in produce consumed in industrialized countries: A review. *Journal of Food Protection*, *81*(7), 1171–1186. https://doi.org/10.4315/0362-028X.JFP-17-465

[26] Bassett, J., & McClure, P. (2008). A risk assessment approach for fresh fruits. *Journal of Applied Microbiology*, *104*(4), 925–943. https://doi.org/10.1111/j.1365-2672.2007.03697.x

[27] Photographs, W. (1993). *Milk and Biotechnology: Maintaining Safe, Adequate Milk Supplies*. *17*(2), 27–31.

[28] Chen, C., & Reniers, G. (2020). Risk assessment of processes and products in industrial biotechnology. *Advances in Biochemical Engineering/Biotechnology*, *173*, 255–279. https://doi.org/10.1007/10_2018_74

[29] Adil, F., Mohammed, O., Mohammed, A., & Salman, A. (2020). Risk Assessment of Brucellosis in Dairy Cows in Bahri North Locality, Sudan. *EAS Journal of Nutrition and Food Sciences*, *1873*(3), 129–140. https://doi.org/10.36349/easjnfs.2020.v02i03.006

[30] World Health Organization (WHO). 2015. Estimating the burden of foodborne diseases. https://www.who.int/activities/estimating-the-burden-of-foodborne-diseases. Accessed on 31/08/2023.

[31] Food and Agriculture Organization (FAO). 2010. Foodborne diseases. https://www.fao.org/3/AB524E/AB524E.htm. Accessed on 31/08/2023.

Titilayo A. Ajayeoba*, Ebunoluwa J. Atolagbe, Olagboye O. Awosika,
Aina A. Daniel, Ohijeagbon O. Rebecca and Oluwatosin A. Ijabadeniyi

Chapter 18
Emerging food risk: factors that affect food security and human health

Abstract: Food products come in contact with various hazards from harvest, through production, storage, and distribution to the consumers' table. Such exposures may be biological (microbial, human, and animal origin) or chemical (food additive), while other challenges may be changing climate, pandemics, unfriendly environmental operation, national and international trade protocols, conflicts, and lack of consumers' awareness of the food handling. The food composition and preparation are gradually becoming an issue because they easily result in malnutrition or obesity when it is not properly monitored. Other key drivers stimulating food insecurity are the lack of trained personnel, and poor policy and standards on different food produce and on those involved in national and international trade. These risks may influence the quality and quantity of the produce, compromising the health of processors and consumers. The consequence exposes the populace to various threats, affecting the nation's economy and trigger communicable and noncommunicable diseases.

18.1 Introduction

Global food production and demand patterns have changed dramatically. Different incidences have influenced production, and increased prices have resulted in an unequal distribution and food access. Investigating and analyzing food access globally has become a priority, and it is an integral part of food security. Changes in income and pricing can disrupt the diet's proper balancing [1]. Furthermore, recent economic crises, along with significant price instability, have devastating effects on worldwide

*Corresponding author: **Titilayo A. Ajayeoba**, Department of Microbiology, Faculty of Science, Adeleke University, Ede, Osun State, Nigeria; Department of Biotechnology and Food Technology, Durban University of Technology, South Africa, e-mail: ajayeoba.titilayo@adelekeuniversity.edu.ng
Ebunoluwa J. Atolagbe, Olagboye O. Awosika, Department of Public Health Sciences, Faculty of Basic Medical Sciences, Adeleke University, Ede, Osun State, Nigeria
Aina A. Daniel, Department of Microbiology, School of Science and Technology Sciences, Babcock University, Illishan-Remo, Ogun State, Nigeria
Ohijeagbon O. Rebecca, Department of Food Sciences, Faculty of Food Science and Consumer Services, Ladoke Akintola University of Technology, Ogbomoso, Oyo State, Nigeria
Oluwatosin A. Ijabadeniyi, Department of Biotechnology and Food Technology, Durban University of Technology, South Africa

https://doi.org/10.1515/9783110748345-018

trade, global production, and subsequently of consumers' diets. Thus, the incidence of illnesses connected to eating an imbalanced diet has increased [2]. Though regional variances and dynamics determine balanced diets and sustainability, other recognized drivers include increasing wages, urbanization, different approaches to trade, and the varying food environment [1]. With the complexity of food systems, consumers are becoming more concerned about the quality and safety of food items in recent years, particularly with regard to food authenticity and food fraud. However, the extent of the problem and the dire repercussions of food fraud have received a lot of public attention. Fraud or mislabeling is still a common occurrence, particularly in animal-derived foods, which are generally considered the most contaminated commodities [3].

Food trade is increasing throughout the world, and the challenges with overarching food safety issues in Africa are becoming a burden globally. Food diseases affect human health, the affected country's economy, the workforce and industry [4]. These become necessary when considering the current consumer and retailer constant request to develop food products that have no additives, low salt, closer to neutral pH and are suitable for 'ambient storage' [5]. Foodborne illness and diseases mostly affect the vulnerable populations, especially the aged and immunocompromised; thus, outbreaks are likely to become more significant from a public health perspective if regulations are not in place. Therefore, this chapter evaluates the various factors that can trigger food insecurity and pose a challenge on public health.

18.2 Food composition and dietary pattern as a possible risk to human health

Diet is a collection of food compounds eaten by humans. The method of preparing food is crucial and ultimately affects their health. Certain food contains concentrations of bioactive compounds, particularly fruits and vegetables, and these types of diet are the most cardioprotective [6], but the preparation processes which mostly involve washing, peeling, cutting, and slicing of fresh-cut products may enhance rapid contamination and degradation. Since cardiovascular disease is still one of the leading causes of mortality worldwide, methods to control or prevent it are of paramount concern. Furthermore, some food processing methods can have unintended consequences, putting human health at risk. Since fresh-cut fruits and vegetables are perishable items that are more prone to contamination, they might enhance the growth of harmful pathogens. As a result, appropriate measures must be implemented to maintain quality and safety, as well as to avoid any potential outbreaks caused by the consumption of fresh-cut fruits and vegetables [7].

Microbial contamination of food can be from an external source or could be part of the normal flora if proper measures are not taken for preservation and storage [8]. Other nonmicrobial contamination may occur between production and consumption. The use of chlorine as a disinfectant is a chemical contaminant, and it is not considered safe when certain dosages are exceeded. To fulfill current safety standards, numerous developing disinfection technologies such as chemical treatments, such as chlorine dioxide, ozone, electrolyzed water, essential oils, high-pressure carbon dioxide, and organic acids, have been enhanced as alternatives to existing decontamination methods. UV light, pulsed light, ionizing radiation, high hydrostatic pressure, cold plasma, and high-intensity ultrasound are examples of nonthermal physical treatments that have proved to improve microbiological safety and retain optimal produce quality. However, relying solely on these disinfection technologies may not be sufficient to ensure food safety and high product quality [9].

Diet becomes a challenge and may constitute an emerging risk if the combination of food product meant for consumption is leading to poor nutrition. Food insecurity and improper physical and mental health status are associated with nutritional status of a location. Imbalanced diet becomes a norm in low-income earners, and these sects are prone to greater food insecurity while they try to keep up with social challenges. Inability to afford the right food combination with limited ability to take care of oneself contributes to an increased self-report of depressive symptoms, resulting in less favorable nutritional status [10]. In contrast, consumers with average or high income are also prone to wrong food combinations or imbalanced diets. The composition of a healthy diet varies and depends on the individual needs, which may include age requirement, gender, extent of physical activity, general health condition, cultural context, regional availability of foods, and dietary customs [11]. In addition, malnutrition becomes evident as a result of one or more of the following factors: inadequate food intake, wrong food choices, poor nutrient absorption, or a combination of these factors.

Foodborne infections and intoxications may be significantly influenced by increased environmental temperature which ultimately affects food temperature and leads to nutrient imbalance and microbial spoilage. Toxic chemicals released into the environment by industrial processes and agricultural practices may enter the human food chain and negatively impact food quality and nutrient absorption in the body because of the complexes formed within the food matrix, though the effects on health are thought to be minimal when small quantities of these complexes are present in food. Nevertheless, the concern with the presence of pesticides in food material is because they are known to suppress the immune system as evident in experimental animals [12].

When diet is compromised, food illness is inevitable. The role of governments and international agencies is important and should prioritize food safety to prevent foodborne illness, particularly among young children, and highlight effective food safety interventions that can be implemented in low- and middle-income countries

[13] because some microorganisms have represented an emerging threat to food safety. Norovirus ranks among the top causes of diarrheal disease along with *Campylobacter*, *Salmonella*, and *Taenia solium* [14].

18.3 Public health food drivers and hazards

Several factors are drivers that influence food production, distribution, and production. Environmental change via infectious disease emergence, biodiversity conservation, policymakers, and societal disruption also has direct human health outcomes, which is not directly integrated into planning for sustainable development [15]. Also, behavioral practices of food-handlers vary in different countries around world as the knowledge, attitude, and practice of food-handlers play a vital role in food safety and food service delivery. There are significant differences in food safety knowledge attitude and practice between consumers in different geographical locations [16]. Identifying certain food hazards are useful indicators to infer public health challenges. These hazards include the environment surrounding the location where food production is carried out, the food production chain that starts the farm to fork, and consumers [17]. Changes in the safety of the aforementioned pointers are signals that may require follow-up action to identify weak spots in the food production system. Furthermore, continual surveillance is required to monitor emerging and re-emerging hazards that are associated with generic food manufacturing practices [17].

If there needs to be a reduction in foodborne illness across different geographical locations, a high-level advocacy for research is necessary to identify, determine, and validate specific food safety interventions. This process requires a better knowledge of the disease and the burden on the environment. This include (i) identifying the localized burden of foodborne disease among vulnerable individuals and populations; (ii) disease burden estimation for high-risk foods; and (iii) increased human toxicity and disease from chemical contamination in food [18]. Such change may likely cause socioeconomic challenges and reduced resistance to diseases, including foodborne diseases. The increase in disease and uneven distribution resistant patterns may lead to serious problems of food security and safety, environmental degradation, large-scale migration from rural to urban areas and from poor to richer countries, and significant changes in ecosystems [4].

The threat leading to foodborne disease is significantly increased by biological and chemical contamination of areas where food is produced, processed, and consumed. These can be influenced by various factors which include price inflation of food commodities [19], populace growth, unplanned migration from rural to urban areas, and consequent slum formation that increase pollution. Poverty and inequity are the principal factors contributing to poor health. Although food irradiation is an affordable method of processing food to a safe state, consumers have concerns with the process of irradiating raw food [12].

One of the strategies necessary for foodborne disease control is the need to assess reliable information on the causative agents, the outcome on the populace, and the effects of food safety interventions put in place to prevent foodborne illnesses. This information will enable the development and implementation of effective surveillance systems in targets and measure progress in reducing foodborne diseases. An economic analysis of the costs of foodborne disease and the benefits of interventions would support these efforts [18].

The use of operational approaches from experts seems helpful, and it is an important tool in identifying and addressing information gaps associated with emerging food and feed risks. This can facilitate comparative analysis and harmonize mitigation strategies to address emerging risks and their drivers [20].

18.3.1 Social and environmental challenges along the agri-food supply chain

Along the agri-food supply chains, a number of social and environmental issues have arisen. Income, price variations and inflation of food commodities, gender differences, consumer demand for safe and quality foods, nutritional safety and health issues, access to market information and consumers' want, exposure to global opportunities for technology transfer, lack of updated information and intelligence are all socioeconomic issues that affect local populations or consumers. In terms of environmental concerns connecting food waste disposal (on-farm/off-farm industrial wastes, household waste), cross-contamination and deterioration due to microbial activities may arise if proper food safety measures are not followed [4].

Contaminants of emerging concern (CeC) in food and food products are considered as evolving hazards that pose growing challenges to food manufacturing process, supply chain, and the dietary exposure to these growing pollutants. Although CeC can either be primary or secondary [21], several elements including different types of emerging packaging material, evolving microorganisms due to changing climates, emerging health challenges and pandemics are gradually gaining attention because of the possible risk they can pose along the food chain. Some CeC can have biological effects on nontarget species. CeC dispersion in food products can be influenced by certain physicochemical features resulting from environmental factors (e.g., water solubility, solid–liquid distribution, polarity, and vapor pressure). Biological (microorganisms and wildlife) and abiotic (ultraviolet) activities can convert contaminants of rising concern [22]. Heavy metal pollution has also disrupted the ecosystem, posing major health risks, as well as a variety of new and growing challenges of food security on the soil-food crop, which have now become a global concern [23]. Some of these metal compounds are not completely removed by conventional treatment processes; once they enter the receiving environment, they degrade slowly in water, and form complexes with food materials and pose a risk to human health after consumption [24].

18.4 Biological contaminants

18.4.1 Zoonotic diseases

Most infections associated with human diseases pose a serious threat to health and have connections with animal as the source. In many cases, the altered food habit, climate variations, hostile environmental human operations, and changes in food habit may influence the occurrence and re-emergence of many zoonotic diseases because humans, animals, and their environment are interconnected [25]. These associations have also influenced pathogen diversity and the emergence of antibiotic resistance. The spread of pathogens from animals to humans may occur through direct or indirect contact. When food is involved between humans and animals, food and food products may serve as the medium for the spread of zoonotic pathogens, leading to a wide range of foodborne illnesses. Several types of insects including flies, cockroaches, beetles, caterpillars, and others naturally infest raw food that may be consumed by humans alongside food materials or prepared food products. Though they reduce the nutritional quality of the food, many of these edible insects pose health risks to humans by causing allergies and other disease conditions [25, 26]. Examples of other major emerging zoonoses that are posing a challenge to food security include avian influenza, feline cowpox, rotavirus infection, norovirus infection, Ebola, hantavirus infection, methicillin-resistant *Staphylococcus aureus* infection, rabies, brucellosis, tuberculosis (*Mycobacterium bovis*), and *Schistosoma* and the most recent coronavirus disease (COVID-19) [25].

For several helminthic and protozoan diseases, transmission to humans commonly depends on contaminated food and/or water. Consumption of raw or lightly cooked meat of infected food animals is the major factor in etiology of taeniasis (cysticercosis), echinococcosis, toxoplasmosis, and trichinellosis. Likewise, consumption of raw or undercooked freshwater fish (or products) in the form of exotic cuisines is gaining popularity worldwide, and may contribute to the spread of nematodes. Meat of wild boar may carry risk of trichinosis and echinococcosis. For waterborne parasitosis, the protozoa *Giardia*, *Cryptosporidium*, and *Toxoplasma* are the most significant causes [27].

18.4.2 Natural contaminants

Biogenic amines are linked to the growth of deteriorating decarboxylating microbes which represent a food quality index, serving as an indirect indicator of excessive microbial proliferation, nonfermented foods, low sanitary quality, and limited food freshness. Biogenic amines are linked to a variety of pathogenic syndromes, posing a health concern to consumers since they can cause headaches, heart palpitations, vomiting, diarrhea, allergies, and hypertensive crises. Some bioactive substances and natural poisons, in addition to pathogenic bacteria and viruses, can be responsible for food poisoning [28].

Ethyl carbamate (EtC) is found in a variety of fermented foods and alcoholic beverages, including cheese, bread, yoghurt, wine, whisky, and soy sauce. The interaction of urea with ethanol, thermal degradation of urea (for example, in baking) to ammonia and cyanic acid, and cyanide anion are three of the most prevalent means of producing EtC in acidic medium. EtC precursors include diethyl carbonate and azodicarbonamide, which are two examples of food additives [29]. While EtC may prevent the growth of bacteria and carcinoma in rat, inappropriate consumption and undefined quantity have been linked to the development of benign and malignant tumors, particularly lung cancer, and they can also exert a toxic effect on developing brains of fetus [30]. Therefore, necessary attention and protection of natural products is essential against EtC-induced toxicity.

18.4.3 Environmental contaminants

Heavy metal pollution has also disrupted the ecosystem, posing major health risks, as well as a variety of new and growing challenges of food security on the soil-food crop, which have now become a global concern. The primary sources of heavy metals in food are the soil environment, livestock manure, irrigation with polluted water, metallopesticides, phosphate-based fertilizers, and untreated sewage. In addition to natural sources, some metal substances may be involved in direct uptake by plants from soil to plant tissues or direct atmospheric deposition onto plant surface may also pose a challenge as it can be transferred to the food chain [23].

There are various ways by which transfer of high-technology rare earth elements (HtREEs) into the environment by hydrological and wind-driven processes takes place. Humans can be exposed to HtREEs by digestion and body absorption of contaminated water and food, inhalation from contaminated material/product, and/or direct intake when a medical activity is administered. External toxicological attributes include stunted growth in plants, reduced nutritional quality in food and food products, genetic disorder and neurological abnormally in animals, trophic bioaccumulation, increased pathogenicity and virulence in soil organisms. In humans, HtREEs cause severe damage to kidneys, resulting in neurological ailment, oxidative stress, pneumoconiosis, cytotoxic effect, and sterility in male [31].

Toxicological risk assessment in different sources of drinking water shows that the human health is impacted by the effect of these chemical contaminants in the water cycle; thus, monitored production and purification steps are essential for future monitoring [32].

18.4.4 Chemical contaminants

Colors are used in a variety of foods to impact the perception of food quality and stimulate hunger, but their usage as food additives is highly regulated in the developed world because many synthetic chemicals have been proved to be toxic and dangerous to consumers. Inconsistent regulations in the use of "approved" synthetic colors used in food products across countries have been linked to hyperactive behavior in children. Also, the replacement tendencies of artificial colors with natural ones may not be completely safe, and the illegal addition of known and emerging industrial dyes to foods is a challenge associated with food colors with potential health hazard to consumers [33].

18.5 Effects of food fraud in the global food system

Food fraud has become well known due to the challenges it presents along the food supply chains, technological advancements, globalization of food chain, and economic challenges faced by different countries, among others. Furthermore, food fraud is carried out by dishonest producers or suppliers who are mostly driven by economically motivated adulteration with the aim of making excessive profit from unsuspecting or vulnerable consumers [34]. Some integrity challenges which include vague definitions that do not communicate the proper idea, unspecific guidance in the production or processing of certain foods, issues of imitation, unguided supply chain, and other factors as shown in Figure 18.1 are major setbacks that have increased food risk across the globe [34, 35].

Consumers expect that the government will work together with experts in the food discipline to ensure manufacturers, producers, distributors, and retailers have guided principles that govern safe and quality production of food. Despite the fact that there are several agencies with different food policies in place, implementation and management, which involve education, awareness, and training of key stakeholders, are crucial to avoid food fraud [36]. Emerging challenges faced by some countries, particularly developing and underdeveloped nations, are low educational exposure, poor awareness, and lack of training on policies that guide food production and safety. In places where policies are in place, continual surveillance and creation of task force, where necessary, are important to maintain food integrity.

Agri-food is becoming a priority to the industrial sector because the state of the raw material/produce is important to the final product. The effect of agronomic practices and environment on the yield and quality of farm produce is very essential as the procedure may differ with geographical location due to climate change and experience of the farmer [37]. In addition, resilience of the supply chain and possible reduction of food fraud may require a good knowledge of the soil characteristics and crop development.

Food vulnerability may be caused by various elements which can include technical opportunities, periodic opportunities, motivations governed by tradition, culture, behavior, perception, level of awareness, economic challenges as well as preventive, management strategies and control measures in place [38]. Although food fraud does not always pose a risk to food safety, the authenticity and quality may be compromised when the production, manufacturing, and distribution do not follow appropriate channels and laid-down rules.

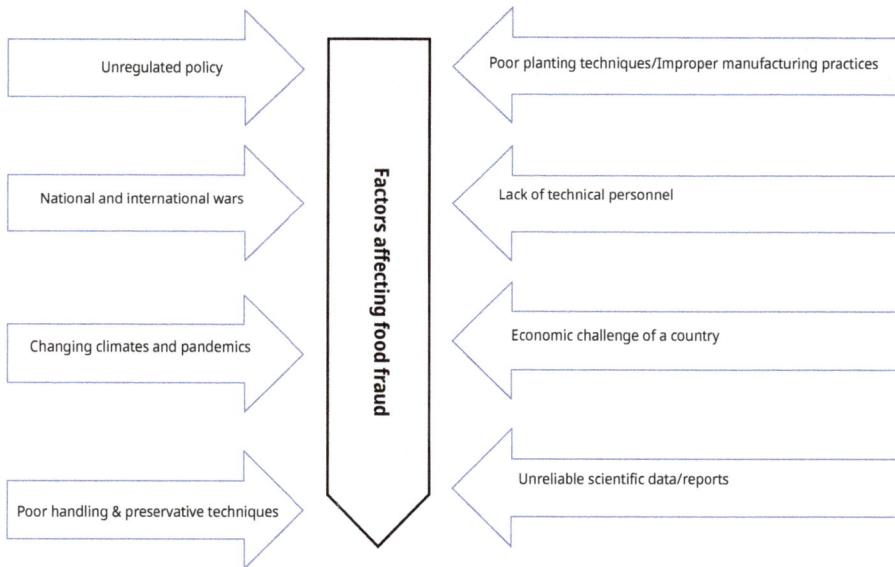

Unregulated policy

Poor planting techniques/Improper manufacturing practices

National and international wars

Lack of technical personnel

Factors affecting food fraud

Changing climates and pandemics

Economic challenge of a country

Poor handling & preservative techniques

Unreliable scientific data/reports

Figure 18.1: Factors affecting food vulnerability and fraud.

The Food and Agriculture Organization estimates that the level of undernourishment and places where moderate or severe food challenges/insecurity are rampant are gradually increasing, and recent projections from food insecurity show that hunger may persist until the year 2030 if adequate measures are not taken toward food access and inequality [39]. Although some developed nations may hold ample ability to feed some other developing or underdeveloped nations, a wide range of factors and critical circumstances such as ecological variation, supply chain disruptions, economic instability, and pandemic influence food supply [40]. For example, COVID-19 has acted as an avenue in changing the global food systems because the pandemic led to global lockdown that significantly increased violence and food fraud, food adulterations, falsification, and increased food crime [39, 41]. In addition, insurgence as well as lack of political focus for the nation may also be a contributing factor. Other key drivers challenging food insecurity in terms of availability (food supply), accessibility (food reaching the populace), consumption (nutrients intake), and stability (future needs' fulfillment) include changing weather, natural disasters, economic imbalances, lack of policy and proper check of food importers, and lack of trained personnel and public awareness of what consumers should know [39–41]. In areas where the level of awareness is high, pandemics have resulted in a large drop in people's income levels, putting food access at danger, particularly for the poor in most countries. Food security is also threatened by food safety and hygiene rules for processing, retailing, and distribution. Supply chain disruptions and the consequent policy interventions are becoming more common, increasing worries about food security [40].

Emerging risks affecting food and food produce are not limited to the supply chain but encompass other factors that do not have food origin. These challenges have greatly impacted human health, resulting in various illnesses and diseases. The emergence of these risk drivers has highlighted the need for surveillance to monitor temporal and geographical shifts in the occurrence and re-emergence patterns of these food challenges.

References

[1]   May, J. (2018). Keystones affecting sub-Saharan Africa's prospects for achieving food security through balanced diets. *Food Research International, 104*, 4–13.

[2]   Santeramo, F. G., Carlucci, D., De Devitiis, B., Seccia, A., Stasi, A., Viscecchia, R., et al. (2018). Emerging trends in European food, diets and food industry. *Food Research International, 104*, 39–47.

[3]   Hassoun, A., Måge, I., Schmidt, W. F., Temiz, H. T., Li, L., Kim, H., et al. (2020). Fraud in animal origin food products: Advances in emerging spectroscopic detection methods over the past five years. *Foods, 9*(8), 1069.

[4]   Bhat, R., & Jõudu, I. (2019). Chapter 2 – Emerging issues and challenges in agri-food supply chain. In R. Accorsi & R. Manzini (Eds.), *Sustainable food supply chains* (pp. 23–37). Academic Press, USA.

[5]   King, T., Cole, M., Farber, J., Eisenbrand, G., Zabaras, D., Fox, E., et al. (2017). Food safety for food security: Relationship between global megatrends and developments in food safety. *Trends in Food Science and Technology, 68*, 160–175.

[6]   Badimon, L., Chagas, P., & Chiva-Blanch, G. (2019). Diet and cardiovascular disease: Effects of foods and nutrients in classical and emerging cardiovascular risk factors. *Current Medicinal Chemistry, 26*(19), 3639–3651.

[7]   Yousuf, B., Deshi, V., Ozturk, B., & Siddiqui, M. W. (2020). 1 – Fresh-cut fruits and vegetables: Quality issues and safety concerns. In M. W. Siddiqui (Ed.), *Fresh-cut fruits and vegetables* (pp. 1–15). Academic Press, USA.

[8]   Melo, R., Mendonça, E., Monteiro, G., Siqueira, M., Pereira, C., Peres, P., et al. (2017). Intrinsic and extrinsic aspects on Campylobacter jejuni biofilms. *Frontiers in Microbiology, 8*, 1332.

[9]   Deng, L., Mujumdar, A. S., Pan, Z., Vidyarthi, S. K., Xu, J., Zielinska, M., et al. (2020). Emerging chemical and physical disinfection technologies of fruits and vegetables: A comprehensive review. *Critical Reviews in Food Science and Nutrition, 60*(15), 2481–2508.

[10]   Jung, S., Kim, S., Bishop, A., & Hermann, J. (2019). Poor nutritional status among low-income older adults: Examining the interconnection between self-care capacity, food insecurity, and depression. *Journal of the Academy of Nutrition and Dietetics, 119*(10), 1687–1694.

[11]   Trüeb, R. (2020). *Nutrition basics. Nutrition for healthy hair* (pp. 17–35). Springer.

[12]   Akhtar, S. (2015). Food safety challenges – A Pakistan's perspective. *Critical Reviews in Food Science and Nutrition, 55*(2), 219–226.

[13]   Kirk, M. D., Pires, S. M., Black, R. E., Caipo, M., Crump, J. A., Devleesschauwer, B., et al. (2015). World Health Organization estimates of the global and regional disease burden of 22 foodborne bacterial, protozoal, and viral diseases, 2010: A data synthesis. *PLoS Medicine, 12*(12), e1001921.

[14]   Nieuwenhuijse, D., & Koopmans, M. (2017). Metagenomic sequencing for surveillance of food- and waterborne viral diseases. *Frontiers in Microbiology, 8*, 230.

[15]   Di Marco, M., Baker, M., Daszak, P., De Barro, P., Eskew, E., Godde, C., et al. (2020). Opinion: Sustainable development must account for pandemic risk. *Proceedings of the National Academy of Sciences, 117*(8), 3888–3892.

[16]   Odeyemi, O., Sani, N., Obadina, A., Saba, C., Bamidele, F., Abughoush, M., et al. (2019). Food safety knowledge, attitudes and practices among consumers in developing countries: An international survey. *Food Research International, 116*, 1386–1390.

[17]   Kleter, G., & Marvin, H. (2009). Indicators of emerging hazards and risks to food safety. *Food and Chemical Toxicology, 47*(5), 1022–1039.

[18]   Kirk, M., Angulo, F., Havelaar, A., & Black, R. (2017). Diarrhoeal disease in children due to contaminated food. *Bulletin of the World Health Organization, 95*(3), 233–234.

[19]   Verhaelen, K., Bauer, A., Günther, F., Müller, B., Nist, M., Ülker Celik, B., et al. (2018). Anticipation of food safety and fraud issues: ISAR – A new screening tool to monitor food prices and commodity flows. *Food Control, 94*, 93–101.

[20]   Hadjigeorgiou, E., Clark, B., Simpson, E., Coles, D., Comber, R., Fischer, A. R. H., et al. (2022). A systematic review into expert knowledge elicitation methods for emerging food and feed risk identification. *Food Control, 136*, 108848.

[21]   Rubio-Armendáriz, C., Alejandro-Vega, S., Paz-Montelongo, S., Gutiérrez-Fernández, Á., Carrascosa-Iruzubieta, C., & Hardisson-de la Torre, A. (2022). Microplastics as emerging food contaminants: A challenge for food safety. *International Journal of Environmental Research and Public Health, 19*(3), 1174.

[22]   Nilsen, E., Smalling, K. L., Ahrens, L., Gros, M., Miglioranza, K., Picó, Y., et al. (2019). Critical review: Grand challenges in assessing the adverse effects of contaminants of emerging concern on aquatic food webs. *Environmental Toxicology and Chemistry, 38*(1), 46–60.

[23]   Rai, P. K., Lee, S. S., Zhang, M., Tsang, Y. F., & Kim, K. (2019). Heavy metals in food crops: Health risks, fate, mechanisms, and management. *Environment International, 125*, 365–385.

[24]   Nielsen, C., Rahman, A., Rehman, A. U., Walsh, M. K., & Miller, C. D. (2017). Food waste conversion to microbial polyhydroxyalkanoates. *Microbial Biotechnology, 10*(6), 1338–1352.

[25]   Rahman, M. T., Sobur, M. A., Islam, M. S., Ievy, S., Hossain, M. J., El Zowalaty, M. E., et al. (2020). Zoonotic diseases: Etiology, impact, and control. *Microorganisms, 8*(9), 1405.

[26]   Belluco, S., Losasso, C., Maggioletti, M., Alonzi, C., Ricci, A., & Paoletti, M. (2015). Edible insects: A food security solution or a food safety concern? *Animal Frontiers, 5*(2), 25–30.

[27]   Chhabra, M., Pathak, K., & Muraleedharan, K. (2017). Food-borne parasitic zoonoses: Status, emerging risk factors and issues: An overview. *Journal of Foodborne and Zoonotic Diseases, 5*(2), 16–31.

[28] Tabanelli, G. (2020). Biogenic amines and food quality: Emerging challenges and public health concerns.

[29] Weber, J. V., & Sharypov, V. I. (2009). Ethyl carbamate in foods and beverages: A review. *Environmental Chemistry Letters, 7*(3), 233–247.

[30] Gowd, V., Su, H., Karlovsky, P., & Chen, W. (2018). Ethyl carbamate: An emerging food and environmental toxicant. *Food Chemistry, 248*, 312–321.

[31] Gwenzi, W., Mangori, L., Danha, C., Chaukura, N., Dunjana, N., & Sanganyado, E. (2018). Sources, behaviour, and environmental and human health risks of high-technology rare earth elements as emerging contaminants. *Science of the Total Environment, 636*, 299–313.

[32] Baken, K., Sjerps, R., Schriks, M., & van Wezel, A. (2018). Toxicological risk assessment and prioritization of drinking water relevant contaminants of emerging concern. *Environment International, 118*, 293–303.

[33] Oplatowska-Stachowiak, M., & Elliott, C. T. (2017). Food colors: Existing and emerging food safety concerns. *Critical Reviews in Food Science and Nutrition, 57*(3), 524–548.

[34] McGrath, T., Haughey, S., Patterson, J., Fauhl-Hassek, C., Donarski, J., Alewijn, M., et al. (2018). What are the scientific challenges in moving from targeted to non-targeted methods for food fraud testing and how can they be addressed? – Spectroscopy case study. *Trends in Food Science and Technology, 76*, 38–55.

[35] Robson, K., Dean, M., Haughey, S., & Elliott, C. (2021). A comprehensive review of food fraud terminologies and food fraud mitigation guides. *Food Control, 120*, 107516.

[36] Spink, J., Fortin, N., Moyer, D., Miao, H., & Wu, Y. (2016). Food fraud prevention: Policy, strategy, and decision-making-implementation steps for a government agency or industry. *CHIMIA International Journal for Chemistry, 70*(5), 320–328.

[37] Millar, S. (2015). Meeting industry's science and technology needs. *Food Technology, 69*(7), 56–+.

[38] van-Ruth, S., Huisman, W., & Luning, P. (2017). Food fraud vulnerability and its key factors. *Trends in Food Science and Technology, 67*, 70–75.

[39] Ahn, S., & Norwood, F. (2021). Measuring food insecurity during the COVID-19 pandemic of spring 2020. *Applied Economic Perspectives and Policy, 43*(1), 162–168.

[40] Panghal, A., Mor, R., Kamble, S., Khan, S., Kumar, D., & Soni, G. Global food security post COVID-19: Dearth or dwell in the developing world? *Agronomy Journal*, n/a(n/a).

[41] Khan, S., Razzaq, A., Yu, Z., Shah, A., Sharif, A., & Janjua, L. (2021). Disruption in food supply chain and undernourishment challenges: An empirical study in the context of Asian countries. *Socio-Economic Planning Sciences, 82*, 101033.

Ajibola B. Oyedeji*, Ezekiel Green, Olufemi P. Sotayo,
Celestina Omohimi, Olalekan J. Odukoya and Oluwafemi A. Adebo**

Chapter 19
Predicting food safety using systems approach

Abstract: In this chapter, we revisited the different system approaches to predicting the growth, survival and enumeration of microorganisms that have food safety implications. Different established mathematical and statistical approaches, including kinetic modeling, response surface methodology as well as emerging predictive approaches such as artificial neural networks and Bayesian network are being used to understand the effect of pathogenic microorganisms on foods and predict their effects in foods. Since these applications are mathematical in nature, it is important to maintain constant environmental conditions and also validate the models to confirm their applicability.

19.1 Introduction

The different approaches employed in ascertaining food safety and overall quality could be precautionary, preventive, and corrective. Preventive measures, involving the use of predictive mechanisms, seem to be the "cheapest." Food safety stakeholders can assess risks and set food safety laws, research methods and quality assurance mechanisms based on forecasts. Predictive modeling can be used to determine how different

Acknowledgments: Special thanks to the Faculty of Science, University Research Committee (URC), the Global Excellence and Stature (GES 4.0) Catalytic Initiative Grant, both of the University of Johannesburg and the Thuthuka Grant (121826) of the National Research Foundation (NRF), South Africa, for partly sponsoring this work and authors.

**Corresponding author: Oluwafemi A. Adebo, Department of Biotechnology and Food Technology, Faculty of Science, University of Johannesburg, P.O. Box 17011, Doornfontein Campus, Gauteng, South Africa, e-mail: oadebo@uj.ac.za
*Corresponding author: Ajibola B. Oyedeji, Research and Development (Commercial Product Development), Labstat International Inc., Edmonton, Alberta, Canada, e-mail: jibanky2@gmail.com
Ezekiel Green, Department of Biotechnology and Food Technology, Faculty of Science, University of Johannesburg, P.O. Box 17011, Doornfontein Campus, Gauteng, South Africa
Olufemi P. Sotayo, Department of Microbiology, Faculty of Science, University of Lagos, Akoka-Yaba, Lagos, Nigeria
Celestina Omohimi, Olalekan J. Odukoya, Department of Food Science and Technology, College of Food Sciences and Human Ecology, Federal University of Agriculture, Abeokuta, Nigeria

https://doi.org/10.1515/9783110748345-019

processing parameters affect microorganisms and their behavior. Enterprises and institutions also employ microbiological models to refine the time-consuming and costly process of traditional microbial testing [1].

Mathematical models can be used to predict microbial behavior and how different conditions influence their growth in diverse environments. These conditions can be divided into two groups: inherent and external factors. The intrinsic (inherent) factors (such as pH, redox potential, and water activity) are the physicochemical parameters of food, while factors such as relative humidity, temperature, and atmospheric conditions are the extrinsic factors [1]. Setting maximum and minimum permissible tolerance levels is an important aspect of the development of models. However, such values must be set using the offered mathematical models' comprehensive and realistic approaches [2]. In addition, detailed account of upcoming threats, such as climate change effects and outbreak of foodborne diseases that can disrupt the present production chain, could be predicted through modeling via improved and adapted experimental techniques [3].

Such models which are adopted in predictive microbiology can be broadly classified based on probabilistic and kinetic properties [4]. With respect to modeling approach being used, they can be classified into mechanistic and empirical models and in terms of variables being considered, such models are classified into primary, secondary, and tertiary variables [5]. In this chapter, we discussed different predictive modeling approaches used in forecasting the behaviors of microorganisms in food and other biological systems and forecasting food safety.

19.2 Kinetic model approach to food safety prediction

Kinetics is defined as the examination of a reaction with the purpose of determining the rate of variation of the connected parameters like time, temperature, concentration, and pH [6–8]. A model is referred to as the description of a theory or phenomenon which accounts for its inferred features and capable of being used for further study. According to Smith and Schaner [9], a simplified representation of the connections between system responses and the elements thought to give rise to the responses observed is referred to as a model. The outcome of the microbiological interactions within a food systems can be predicted with the kinetic models. There are two major techniques generally used for kinetics modeling of microorganisms' activities. The growth is modeled, then predictions are made based on exponential increase [10]. After this, a sigmoid curve is used to fit observed growth curve, and the effect of external factors on the values of the parameters of that fitted curve is then modeled [11].

There are three types of microbial kinetic models: primary, secondary, and tertiary [12]. Primary models explain the evolution pattern of microorganisms over time [13]. The secondary models are concerned with conditions relating to the environmental conditions on germination and growth of microorganisms. The conditions are water activity, pH, temperature, and others. The combination of primary and secondary models with a computer interphase leads to creation of tertiary models, which give a detailed prediction tool [13].

In tertiary models, the growth rate is modelled in one of the two combined models (primary and secondary) and then utilized to produce forecasts based on exponential population increase in the other. Gibson et al. [11] first proposed this method. The pace of the microbial growth could be used to distinguish these two distinct techniques to kinetic modeling. Both approaches are applied in building models by taking into consideration, the increase in the biomass over population growth curve for a variety of levels and combinations of extrinsic parameters of importance. Consequently, the information about the lag phase duration, maximum population density, and growth rate recorded under the prevailing conditions is recorded. Some of the recognized models are discussed, and their specific application in previous studies, especially on mycotoxin production and degradation, is presented in Table 19.1.

19.2.1 Belehnidek/square root model types

In accordance with the findings of Ratkowsky et al. [71], a simple model was developed to describe microbial growth rates as a function of temperature. Using the square root model, the square root of nucleotide breakdown rate in carp muscle is proportional to temperature as reported by Ohta and Hirahara [24]. They also found that this connection accurately reflects the rate of growth of numerous bacteria in response to temperature. The form of the model is presented in the following equation:

$$K^{1/2} = b \ (T - T_{\min}) \tag{19.1}$$

where K is the growth rate, T is the temperature, $T_{\min}$ is the notional minimum temperature for growth, and b is the unknown coefficient.

The "square root" model was later demonstrated to be a particular example of the widely utilized Belehnidek equation in other biological disciplines [25]. However, with consideration to temperature, water activity, and pH, many adjustments have been made to the model equations. The proposed adjusted equation is

$$K^{1/2} = b(T - T_{\min}) \left(1 - e^{(c(T - T_{\max}))}\right) \tag{19.2}$$

$T_{\max}$ is the maximum temperature and c is the coefficient to be determined. k, T, $T_{\min}$, and b have been defined in eq. (19.1).

19.2.2 The modified Gompertz model

The modified Gompertz model is an empirical model that has been examined by several researchers and is used in programs such as the Pathogen Modeling Programs (https://pmp.errc.ars.usda.gov/pmponline.aspx) [26]. The growth kinetics characteristics of *Pseudomonas* were evaluated using the modified Gompertz method in a study of microorganism interactions during natural rotting of pork at 5 °C [27].

19.2.3 The Baranyi model

To anticipate microbial growth, this model was proposed by Baranyi and Roberts [28], along with an adjustment function $A(t)$ that was dependent on the physiological condition of the microbial cells. As reported by many researchers, application of Baranyi model in certain microbiological processes produced excellent results compared to both Gompertz function and other models [29–34].

19.3 The Arrhenius-type models

In predictive microbiology, the classical Arrhenius equation is the most basic "Arrhenius-type" model applied. When used to model microbial growth

$$\ln k = \ln A - \frac{E_a}{RT} \tag{19.3}$$

where K is the rate of growth, A is the fitting parameter, R is the gas constant ($8.314 \ \text{K}^{-1} \ \text{mol}^{-1}$), T is the temperature, and E_a is the activation energy.

19.4 Response surface methodology

Response surface methodology is a widely used model in the field of optimization, according to McMeekin et al. [35]. When processing conditions or other elements change in a predictable pattern, this model can be adjusted. It allows for the measurement of a system's output and the creation of experimental designs that are reliant on a variety of input variables. Each variable is dependent on its complicated equation; hence, the relationship is not linear.

Table 19.1: Some reports on the use of kinetics modeling in predicting mycotoxin detoxification.

Food samples	Mycotoxins	Input conditions/ detoxification methods	Model	Results obtained	References
Corn, rice, and peanut	Aflatoxin B1	Thermal treatments	Arrhenius equation	Aflatoxin B1 conversion was simulated	Zhang et al. [14]
Rice and wheat	Aflatoxin B1 and G1	UV-Vis	Second-order	Speed of toxin degradation was estimated	Liu et al. [15]
Water solution	Aflatoxin B1	UV irradiation	First-order reaction kinetics	The pathway for the degradation of toxin was proposed	Liu et al. [16]
Water	Aflatoxin B1, B2, G1, and G2	Ozone	First-order kinetic equation	Ozone degradation was higher for both AFB1 and AFG1 despite AFB2 and AFG2	Agriopoulou et al. [17]
Liquid culture	Aflatoxin B1	Biodegradation: *Rhodococcus erythropolis*	Second-order model	Efficiency of degradation recorded was 95.8%	Kong et al. [18]
Meal extracts peanut, maize kernels, dried grapes, and coffee beans	Production of ochratoxin A by eight strains of *Aspergillus*	(a) Varying water activity (a_w) (0.995–0.85) (b) Incubation temperature (15, 25, and 30 °C) (c) Incubation time (7, 14, and 21 days)	–	Highest OTA levels were achieved after 7 days of incubation	Astoreca et al. [19]
Wheat grain	Production of deoxynivalenol by *F. culmorum* and *F. graminearum*	Overtime (up to 40 days)	–	Increased toxin concentration (mg/g wheat) found with increasing incubation time, a_w, with higher level achieved at 25 °C than at 15 °C	Hope et al. [20] and Hope and Magan [21]

Table 19.1 (continued)

Food samples	Mycotoxins	Input conditions/ detoxification methods	Model	Results obtained	References
Grapes	Ochratoxin A production	Production profile of ochratoxin A by *Aspergillus* section Nigri between 5 and 20 days	Multiple linear regression and response surface predictive models	Max production of OTA occurred at the earlier growth stages (5 days for *A. carbonarius* and 7–13 days for *A. niger* aggregate) The amount of produced OTA detected diminished with time, minimal found after 20 days	Bellí et al. [22]
Synthetic grape medium	Kinetics of ochratoxin A production and accumulation by *A. carbonarius*	Varying temperature levels (7–42 °C)	Sigmoidal Gompertz model	At considerably higher temperature, OTA was synthesized after 2–4 days, whereas below 25 °C, maximum OTA producing capacity was detected only after 10 days	Marin et al. [70]
Agar media	Aflatoxin B1, B2, G1, and G2	Varying pH (5.5 and 5.9)	Arrhenius- like temperature function	–	Pitt [23]

19.5 Artificial neural network approach to predicting food safety

Artificial neural network (ANN) is a mathematical modeling tool capable of dealing with linear and nonlinear multivariate regression problems, without the need for a clear expression of the physical structure of the system or process under study [36, 37]. It is an information processing structure that was developed to mimic the composition and use of the biological nervous system, and is composed of the input layer, hidden layer, and output layer neurons [38].

The inspiration for the ANN development was based on the idea of neural networks simulating the structure, significance and transformation in the human nervous system [39, 40]. Similar to a biological neuron, ANN models consist of layers of simple computing nodes that are connected by links of different weights which serves as the connection strength between neurons. When data are presented to the network during a training process, the weight inputs are summed by the neuron and the results are passed to a transfer function to generate an output [39, 41, 42]. One of the efficient ways to generate an ideal output or obtain a practical accurate solution is the use of weights that correspond to the importance of each input signal, as some input signals might be more important than others [39]. Also, ANN can be programmed to function by adjusting the magnitudes of connections (weights) between the processing elements [37].

The use of ANN in various fields of emerging technologies has been reported to be on the increase due to their unique characteristics in solving problems that analytical methods find difficult to apply [43]. Some authors have highlighted some of these unique characteristics to include adaptive learning capacity; ability to establish a model without the specific knowledge of systems structure; huge fault or failure tolerance to establish a complex nonlinear relation model; resistance to noisy data; nonreliance on expert experience; operation in a real-time environment with minimal training or need to change weights connection; easy insertion of a network into existing technology through rapid training, testing, verification and transferring of a network into a hardware implementation [43–46].

As emerging computer technology, it is being used in various field applications for mapping, regression, monitoring and modeling, recognition, clustering, image processing, optimization and signal processing, and multivariate data analysis [43, 47]. Hence, due to its adaptability, ANN has been successfully applied in all aspects of food science and technology to model and predict processes in the food industry such as food process optimization, food engineering, food properties and quality, food sensory analysis, classification, and microbial prediction [39, 48], although the introduction of its application has been relatively slow [49].

ANN is one of the several models used in predictive microbiology to give the quantitative microbial ecology of foods by predicting the microbial growth inactivation and probability of survival as affected by complicated environmental factors [50, 51]. Its application in modeling microbial growth and survival as a better alternative to traditional time-consuming microbiological quantitative techniques has made it a useful tool in predicting food safety [39, 52].

Its application in food safety has been attributed to its ability to deal with a high level of variability and uncertainty that is typically associated with microbial responses; ability to offer alternative and powerful techniques for accurately modeling microbial survival and growth; presentation of high accuracy in predicting the nonlinear relationship between input (e.g., temperature, pH, and initial bacteria load) and output in food microbial system [52].

Various types of neural networks have been previously described for predictive microbiology. Some of the commonly used ANNs are multilayer perceptron neural network, backpropagation (BP) neural network, and radial basis function (RBF) neural network. The BP neural network is a commonly used feed-forward ANN. This is because it is easy to learn and, hence, useful for many food safety applications [43, 53]. However, according to Han et al. [69], it has a slow convergence speed and easy trapping of its training process in local optimum could give a poor training effect. The RBF neural network is also a feed-forward ANN with a faster training advantage and faster approximation abilities with its application as the activation function of the hidden layer being a major factor that differentiates it from other feed-forward NN [69].

Multilayer perceptron neural network is made up of orderly interconnected neurons arranged in layers. For this type of ANN, many neurons can be activated simultaneously by one signal and can all be relevant in the calculation of network output [40, 54].

Presented in Table 19.2 are some works in the literature on the application of some of the commonly used ANN-based networks to assess their predictive ability and efficacy for early warning of food safety risk. According to Gosunkonda et al. [44], it was concluded from their work that artificial neuron network-based models possess the characteristic to handle high biological differences linked with the survival of microorganisms. All the authors reported better predictive performance of the different neural networks, however, limitation on the use of ANN due to its complexity and high cost of learning was reported [56].

19.6 Bayesian network

A Bayesian network is a graphical model which represents knowledge and reasoning under various uncertainties, bringing about informed conclusions based on available information by presenting probabilistic correlations among a set of variables [57]. Simply said, Bayesian networks are a complex data analysis method. A Bayesian network transforms enormous amounts of frequently divergent, sometimes non-quantitative data into a common representation that can be interpreted consistently and expressed clearly. A Bayesian network resembles a flow chart, complete with labeled nodes and directed links, and it depicts a complicated system of interconnected variables known as a model domain [58].

The many sources of uncertainty are treated equally and consistently in Bayesian network modeling to provide an overall level of confidence linked with the values of certain variables. It was recently proved that the Bayesian network methodology can be used to reveal interactions between food fraud and drivers of change, and forecast food fraud, therefore allowing for a systems approach to achieving food safety [59, 60].

Expert knowledge, measurement data, and feedback from experience are all sources and types of data that Bayesian networks may integrate. Bayesian networks are also used in a variety of fields, including hazard and risk assessment [68], food fraud prediction [61], food safety of nanomaterials [62], food safety hazards prediction in herbs and spices [63], and sample size optimization for food allergen monitoring [64].

Table 19.2: Predictive abilities of artificial neural network-based models for food safety risk.

Objective of investigation	Experimental material	Type of model	Model input	Model output	Performance indices	Results	References
Ochratoxin A (OTA) accumulation in the grape juice-based medium by aspergillus carbonarius	Juice	MLP-ANN RBFNs	1. Temp. (20–28 °C) 2. A_w (0.94–0.98) 3. Carbonarius level (0–450 ng/mL) 4. Time (3–15 days)	Ochratoxin A level	MSE RMSE SEP	MLP-ANN, with Bayesian algorithm (BP), gave the minimum error, and highest mean square root (MSE) to estimate OTA accumulation by A. carbonarius, hence giving the best predictive ability.	Mateo et al. [42]
Prediction efficacy of sanitizers in the reduction of coliforms and Escherichia coli	Tomato and lettuce leaves	ANN	1. Hypochlorous – 25, 50, and 75 ppm 2. Peracetic acids – 30, 40, and50 ppm	Residual coliform and Escherichia coli levels after treatment	SSE R^2	Both sanitizers effectively reduced the microorganisms; hypochlorous acid treatment was shown to be more effective in reducing the viable coliforms and E. coli load while, on the contrary, peracetic acid was more effective in reducing E. coli than coliforms. However, in reducing both coliforms and E. coli on tomatoes and lettuce, peracetic acid was more effective than hypochlorous acid, at increased concentration. This was attributed to the different bactericidal mechanisms of the two sanitizers and food types. Out of the two models developed, the ANN model with three hidden networks was the best to predict E. coli reduction while the network with five neurons predicted coliform reduction efficiently.	Keeratipibul et al. [53]

(continued)

Table 19.2 (continued)

Objective of investigation	Experimental material	Type of model	Model input	Model output	Performance indices	Results	References
Quantification of *E. coli* O157:H viability responses due to electrical stresses on beef surfaces	Beef	MLP-ANN with BP learning algorithm MLP-ANN with KF learning algorithm	1. Current (300, 600, and 900 mA) 2. Frequency (1, 10, and 100 kHz) 3. Duty cycles (30%, 50%, and 70%) 4. Time (each with three levels of treatment 2, 8, and 16 min)	*% E. coli* reduction	R^2	A positive quadratic relationship between the treatments and the % E. coli was observed, i.e., as the time for the three treatments increased, % reduction also increased. For current versus time treatment, the network with the KF learning algorithm gave a better prediction than that of BP, with a higher R^2 value. However, both KF and BP gave a more significant prediction than the polynomial regressions for both frequencies versus time and duty cycle versus time treatments. Hence, both BP and KF learning algorithms presented a superior correlation over the polynomial regression models for this assessment.	Gosukonda et al. [44]
Residual *Listeria monocytogenes* ATCC 7644 growth on fresh-cut produce after treatment with a combination of bacteriophage and sucrose monolaurate	Tomato and carrot	MLP-ANN	1. Type of fresh produce 2. Temperature 3. pH 4. Initial bacterial load 5. Antimicrobial concentration 6. Storage period	Residual *Listeria monocytogenes*	MSE R^2	Actual values from the experiment showed a significant reduction in *L. monocytogenes* population on both fresh produce, however, a more reduction was observed in carrot than tomato. Predicted values from the MLP-ANN gave a positive correlation with the observed values. Comparison with linear regression model gave a lower MSE and moderate R^2 values, indicating MLP-ANN has more predictive ability than the linear regression model.	Oladunjoye et al. [54]

| ANN predictive model of S. Typhimurium in the processing of surimi with citric acid | Surimi (fish myofibrillar protein concentrated from raw minced fish) | PSO-based BP-ANN | 1. Citric acid concentration (0.5%, 1%, and 2%) 2. Temperature (4 and 25 °C) 3. Time (1–15 min) | S. Typhimurium population | MSE R^2 | Actual data from the results showed a faster reduction in the S. Typhimurium population as the concentration of citric acid increased with time. Better predictive ability of PSO-based BP-ANN than BP-ANN was established with a higher correlation coefficient and lower MSE. | Qin et al. [48] |
| Assessment of fungal population in the bulk of rapeseed stored under different conditions | Rapeseeds | MLP RBF | 1.Temperature (T = 12–30 °C) 2.Water activity (a_w = 0.75–0.90) 3. Storage time | Fungal population level | R^2 RMSE MAE | Determination of the more suitable architecture for the two neural networks tested was first established. For MLP, an increase in the neurons of the hidden layer from 2 to 8 gave decreased mean values for the training and test errors, while an increase from 2 to 20 was observed for RBF. Comparison of the predictive capability of the two neural networks revealed that MPL gave a better capability than RBF based on the higher R^2, lower RMSE and MAE values generated by MPL. | Wawrzyniak [40] |

(continued)

Table 19.2 (continued)

Objective of investigation	Experimental material	Type of model	Model input	Model output	Performance indices	Results	References
Early warning and control of food safety	Meat	BP RBF AHC-RBF	Sampling time	1. Heavy metal (cadmium, chromium, lead) 2. Microbiology (coliform and total colony count) 3. Food additive (nitrite and pentadiene carboxylic acid)	ARGE RMSE	Observed values from the result showed that the different sampling times gave different risk levels. Based on the weights of the metal, microbiological, and food additive indexes (73.11%, 15.89%, and 11% respectively), the highest value from the metal index indicates that it has the highest risk value for the food sample. Of the three neural networks tested, AHC-RBF, with the smallest ARGE and RMSE, gave the best generalization and prediction accuracy while BG gave the worst prediction, having the largest ARGE and RMSE.	Geng et al. [55]

ANN, artificial neural network; MLP, multilayer perceptron; RBFN, radial basis function network; BP, backpropagation; KF, Kalman filter; PSO, particle swarm optimization; AHC, agglomerative hierarchical clustering; MSE, mean square error; RMSE, root mean square error.

Bayesian networks are a type of probabilistic graphical model that combines graph theory, probability theory, computer science, and statistics to create cause-and-effect prediction models. It shows probable correlations between the chosen factors based on what is known. This enables bidirectional thinking in the face of ambiguity, as well as the drawing of conclusions based on given data [57].

The first technique in the Bayesian network is the identification of the primary elements that can affect the occurrence of food safety hazards in foods (directly or indirectly). There are three distinct processes in the Bayesian network approach, which are data gathering and processing, Bayesian network model creation and Bayesian network model validation [60]. Nodes, arcs, and probabilities make up a Bayesian network structure. The nodes' values might be discrete or continuous, but discrete nodes are the most common. The conditional probabilities are derived using Bayes' theorem, which states that the likelihood of event A under the condition of event B is

$$P(A|B) = \frac{P(B|A) \times P(A)}{P(B)} \tag{19.4}$$

where $P(A)$ is the prior probability of A, $P(B|A)$ is the probability of B when a known event A occurs, and $P(B)$ is the prior probability of B [65].

The product of all conditional probabilities defined in the Bayesian network yields a unique joint probability distribution of all variables $P(U) = P(A_1, \ldots, A_n)$, which is produced by the product of all conditional probability specified in the Bayesian network:

$$P(U) = \prod_{i=1}^{n} P\left(A_i | pa(A_i)\right) \tag{19.5}$$

where $pa(A_i)$ is the variable's parents. A conditional probability distribution is defined by A_i and $P(A_i | pa(A_i))$ [65].

Sensitivity analysis is the common method used to determine how much an uncertain change in input will change the outcome, or which of the inputs will have the greatest impact on the output [63]. The entropy function and parameter sensitivity analysis are the two most used sensitivity methodologies.

Entropy is a measure of how evenly a variable's probability mass is distributed throughout its states. Entropy is used to discover and rank the most significant parameters in a network since it can be used to measure the uncertainty in the distribution of a variable. It entails determining the function $H(X)$ of a node X [66]:

$$P(X) = \sum_{X} P(X) \log P(X) H(X) \tag{19.6}$$

where $P(X)$ is the probability distribution of X.

Parameter sensitivity analysis is a type of one-factor-at-a-time method, in which sensitivity measurements are generated when only one component is altered while all other variables remain constant [67]. The goal of this method is to explain how the selection of a given condition (state of a parameter) affects the direction of change in the probability of the output parameter in greater detail [66]. Food safety analyses benefit from the Bayesian network method's efficiency and consistency, and the Bayesian network makes it easier to describe many indications consistently related to food hazard domains.

19.7 Conclusion

This chapter highlights the mathematical modeling approaches to predict food safety in multiple applications. Due to the limits of the methodologies employed in mathematical models, data collected for modeling must be given from a consistent environment, to give reliable predictions. Also, other extrinsic parameters affecting the data used in predictive modeling must be considered. The model must be validated after development by comparing the values it generates with sufficient experiments, to affirm their dependability. Again, because microbial growth is often influenced by the nature or state of foods, mathematical models are useful in forecasting the behavior of the microbial population in food matrices, allowing for reasonable projections in the lifespan of foods and assurance of supply of safe foods.

References

[1]   Stavropoulou, E., & Bezirtzoglou, E. (2019). Modeling of microbial behavior in food. *Foods, 8,* 654.
[2]   Safe Foods Report. Available online: https://www.safefood.eu/Publications/Corporate-publications /Annual-Reports.aspx (accessed on 24 October 2019).
[3]   Valdramidis, V. P., Geeraerd, A. H., Gaze, J. E., Kondjoyan, A., Boyd, A. R., Shaw, H. L., & van Impe, J. F. (2006). Quantitative description of *Listeria monocytogenes* inactivation kinetics with temperature and water activity as the influencing factors: Model prediction and methodological validation on dynamic data. *Journal of Food Engineering, 76,* 79–89.
[4]   Roberts, T. A. (1989). Combinations of antimicrobials and processing methods. *Food Technology, 43,* 156–163.
[5]   Whiting, R. C., & Buchanan, R. L. (1993). A classification of models for predictive microbiology. *Food Microbiology, 10,* 175–177.
[6]   Hashemi, S. M., & Roohi, R. (2019). Ohmic heating of blended citrus juice: Numerical modeling of process and bacterial inactivation kinetics. *Innovative Food Science and Emerging Technologies, 52,* 313–324.
[7]   Hashemi, S. M. B., Mahmoudi, M. R., Roohi, R., Torres, I., & Saraiva, J. A. (2019b). Statistical modeling of the inactivation of spoilage microorganisms during ohmic heating of sour orange juice. *LWT, 111,* 821–828.

[8] Hashemi, S. M. B., Roohi, R., & Mahmoudi, M. R. (2019a). Modeling inactivation of *Listeria monocytogenes*, *Shigella sonnei*, *Byssochlamysfulva* and *Saccharomyces cerevisiae* and ascorbic acid and β-carotene degradation kinetics in tangerine juice by pulsed thermosonication. *LWT*, *111*, 612 621.

[9] Smith, S., & Schaner, D. W. (2004). Evaluation of a *Clostridium perfringens* predictive model isothermal conditions in broth, to predict growth in G cooling. *Applied Environmental Microbiology*, *70*, 2728–2733.

[10] McMeekin, T. A., Olley, J., Ross, T., & Ratkowsky, D. A. (1993). *Predictive microbiology theory and application*. Taunton, U.K: Research Studies Press.

[11] Gibson, A. M., Bratchell, N., & Roberts, T. A. (1988). Predicting microbial growth: growth responses of *Salmonellae* in a laboratory medium as affected by pH, sodium chloride and storage temperature. *International Journal of Food Microbiology*, *6*, 155–178.

[12] Whiting, R. C. (1995). Microbial modelling in foods. *Critical Reviews in Food Science and Nutrition*, *6*, 467–494.

[13] Marks, B. P. (2007). Status of microbial modelling in food process models. *Comprehensive Reviews in Food Science and Food Safety*, *7*, 137–143.

[14] Zhang, C., Ma, Y., Zhao, X., Zeng, Y., & Wang, F. (2011). Kinetic modelling of aflatoxins B1 conversion and validation in corn, rice, and peanut during thermal treatments. *Food Chemistry*, *129*, 1114–1119.

[15] Liu, Z., Zhao, H., Yang, G., He, K., Sun, X., Wang, Z., Wang, D., & Qui, J. (2020). Study of photo-degradation kinetics of aflatoxins in cereals using trilinear component modeling of excitation-emission matrix fluorescence data. *Spectrochimica Acta Part A: Molecular and Biomolecular Spectroscopy*, *235*, 118 266.

[16] Liu, R., Jin, Q., Tao, G., Shan, L., Huang, J., Liu, Y., Wang, X., Mao, W., & Wang, S. (2010). Photo-degradation kinetics and byproducts identification of the Aflatoxin B1 in aqueous medium by ultra-performance liquid chromatography quadrupole time-of-flight – mass spectrometry. *Journal of Mass Spectrometry*, *45*, 553–559.

[17] Agriopoulou, S., Koliadima, A., Karaiskakis, G., & Kapolos, J. (2016). Kinetic study of aflatoxins' degradation in the presence of ozone. *Food Control*, *61*, 221–226.

[18] Kong, Q., Zhai, C., Guan, B., Li, C., Shan, S., & Yu, J. (2012). Mathematic modeling for optimum conditions on aflatoxin B1 degradation by the aerobic bacterium *Rhodococcus erythropolis*. *Toxins*, *4*, 1181–1195.

[19] Astoreca, A., Magnoli, C., Barberis, C., Chiacchiera, S. M., Combina, M., & Dalcero, A. (2007). Ochratoxin A production in relation to ecophysiological factors by *Aspergillus* section Nigri strains isolated from different substrates in Argentina. *Science of the Total Environment*, *388*, 16–23.

[20] Hope, R., Aldred, D., & Magan, N. (2005). Comparison of environmental profiles for growth and deoxynivalenol production by *Fusarium culmorum* and *F. graminearum* on wheat grain. *Letters of Appied Microbiology*, *40*, 295–300.

[21] Hope, R., & Magan, N. (2003). Two-dimensional environmental profiles of growth, deoxynivalenol and nivalenol production by *Fusarium culmorum* on a wheatbased substrate. *Letters of Appied Microbiology*, *37*, 70–74.

[22] Bellí, N., Ramos, A. J., Sanchis, V., & Marın, S. (2004b). Incubation time and water activity effects on ochratoxin A production by *Aspergillus* section Nigri strains isolated from grapes. *Letters of Applied Microbiology*, *38*, 72–77.

[23] Pitt, R. E. (1993). A descriptive model of mold growth and aflatoxin formation as affected by environmental conditions. *Journal of Food Protection*, *2*, 139–146.

[24] Ohta, F., & Hirahara, T. (1977). Rate of degradation in cool-stored carp muscle. Memoirs of the Faculty of Fisheries, Kagoshima University, *26*, 97–102.

[25] Ross, T. (1987). Belehradek temperature functions and growth of organisms. CSIRO-DSIR Joint Workshop on Seafood Processing, Nelson, New Zealand, April 1986. CSIRO Tasmanian Regional Laboratory Occasional Paper No. 18.

[26] McClure, P. J., Blackburn, C. W., Cole, M. B., Curtis, P. S., Jones, J. E., Legan, J. D., Ogden, I. D., Peck, M. W., Roberts, T. A., Sutherland, J. P., & Walker, S. J. (1994). Modeling the growth, survival and death of microorganisms in foods: The UK Food Micro-Model approach. *International Journal of Food Microbiology, 23*, 265–275.

[27] Liu, F., Guo, Y. Z., & Li, Y. F. (2006). Interactions of microorganism during natural spoilage of pork at 5 °C. *Journal of Food Engineering, 72*, 24–29.

[28] Baranyi, J., & Roberts, T. A. (1994). A dynamic approach to predicting bacterial growth in food. *International Journal of Food Microbiology, 23*, 277–294.

[29] Dalgaard, P. (1995). Modeling of microbial activity and prediction of shelf life for packed fresh fish. *International Journal of Food Microbiology, 26*, 305–317.

[30] Fernandez, P. S., George, S. M., Sills, C. C., & Peck, M. W. (1997). Predictive model of the effects of CO_2, pH, temperature and NaCl on the growth of *Listeria monocytogenes*. *International Journal of Food Microbiology, 37*, 37–45.

[31] McClure, P. J., Baranyi, J., Boogard, E., Kelly, T. M., & Roberts, T. A. (1993). A predictive model for the combine effect of pH, sodium chloride and storage temperature on the growth of *Brochothrix thermosphacta*. *International Journal of Food Microbiology, 19*, 161–178.

[32] McClure, P. J., Beaumont, A. L., Sutherland, J. P., & Roberts, T. A. (1997). Predictive modeling of growth of *Listeria monocytogenes*: The effects on growth of NaCl, pH, storage temperature and sodium nitrate. *International Journal of Food Microbiology, 34*, 221–232.

[33] Sutherland, J. P., Bayliss, A. J., & Braxton, D. S. (1995). Predictive modeling of growth of *Escherichia coli* O157:H7 the effects of temperature, pH and sodium chloride. *International Journal of Food Microbiology, 25*, 29–49.

[34] Sutherland, J. P., Bayliss, A. J., Braxton, D. S., & Beaumont, A. L. (1997). Predictive modeling of *Escherichia coli* O157:H7: Inclusion of carbon dioxide as a fourth factor in a pre-existing model. *International Journal of Food Microbiology, 37*, 113–120.

[35] McMeekin, T. A., Brown, J., Krist, K., Miles, D., Neumeyer, K., Nichols, D. S., Olley, J., Presser, K., Ratkowsky, D. A., Ross, T., et al. (1997). Quantitative microbiology: A basis of food safety. *Emerging Infectious Diseases, 3*, 541549.

[36] Khayet, M., Cojocaru, C., & Essalhi, M. (2011). Artificial neural network modeling and response surface methodology of desalination by reverse osmosis'. *Journal of Membrane Science, 368*(1–2), 202–214.

[37] Mashhadban, H., Kutanaei, S. S., & Sayarinejad, M. A. (2016). Prediction and modeling of mechanical properties in fiber reinforced self-compacting concrete using particle swarm optimization algorithm and artificial neural network. *Construction and Building Materials, 119*, 277–287.

[38] Gevrey, M., Dimopoulos, I., & Lek, S. (2003). Review and comparison of methods to study the contribution of variables in artificial neural network models. *Ecological Modelling, 160*(3), 249–264.

[39] Huang, Y., Kangas, L. J., & Rasco, B. A. (2007). Application of artificial neural networks (ANN) in food science. *Critical Reviews in Food Science and Nutrition, 47*, 113–126.

[40] Wawrzyniak, J. (2020). Application of artificial neural networks to assess the mycological state of bulk stored rapeseeds. *Agriculture, 10*, 567.

[41] Basheer, I. A., & Hajmeer, M. (2000). Artificial neural networks: Fundamentals, computing, design, and application. *Journal of Microbiological Methods, 43*, 3–31.

[42] Mateo, F., Gadea, R., Medina, Á., Mateo, R., & Jiménez, M. (2009). Predictive assessment of ochratoxin A accumulation in grape juice based-medium by *Aspergillus carbonarius* using neural networks'. *Journal of Applied Microbiology, 107*, 915–927.

[43] Funes, E., Allouche, Y., Beltrán, G., & Jiménez, A. (2015). A review: Artificial neural networks as tool for control food industry process. *Journal of Sensor Technology, 5*, 28–43.

[44] Gosukonda, R., Mahapatra, A. K., Liu, X., & Kannan, G. (2015). Application of artificial neural network to predict Escherichia coli O157:H7 inactivation on beef surfaces. *Food Control, 47*, 606–614.

[45] Popescu, O., Popescu, D., Wilder, J., & Karwe, M. (2001). A new approach to modelling and control of a food extrusion process using artificial neural network and an expert system. *Journal of Food Process Engineering, 24*, 17–36.

[46] Song, Q., Zheng, Y. J., Xue, Y., Sheng, W. G., & Zhao, M. R. (2017). An evolutionary deep neural network for predicting morbidity of gastrointestinal infections by food contamination. *Neurocomputing, 226*, 16–22.

[47] Guiné, R. (2019). The use of artificial neural networks (ANN) in food process engineering. *International Journal of Food Engineering, 5*(1), 15–21.

[48] Qin, T., Liu, S., Mao, Y., Liu, X., Tang, X., Li, R., & Cai, W. (2017). PSO-based BP-ANN predictive model of S. Typhimurium in processing of surimi with citric acid. *Journal of Food Safety, 38*, e12420.

[49] Dębska, B., & Guzowska-Świder, B. (2011). Application of artificial neural network in food classification. *Analytica Chimica Acta, 705*(1), 283–291.

[50] Pérez-Rodríguez, F., & Valero, A. (2013). *Predictive models: Foundation, types, and development, predictive microbiology in foods* (pp. 25–55). New York, NY: Springer.

[51] Ross, T., & McMeekin, T. A. (2003). Modeling microbial growth within food safety risk assessment. *Risk Analysis, 23*, 179–197.

[52] Jeyamkondan, J., Jayas, D. S., & Holley, R. A. (2001). Microbial growth modeling with artificial neural networks. *International Journal of Food Microbiology, 64*, 343–354.

[53] Keeratipibul, S., Phewpan, A., & Lursinsap, C. (2011). Prediction of coliforms and Escherichia coli on tomato fruits and lettuce leaves after sanitizing by using Artificial neural networks'. *LWT – Food Science and Technology, 44*, 130–138.

[54] Oladunjoye, A. O., Oyewole, S. A., Singh, S., & Ijabadeniyi, O. A. (2017). Prediction of *Listeria monocytogenes* ATCC 7644 growth on fresh cut produce treated with bacteriophage and sucrose monolaurate by using artificial neural network. *LWT-Food Science and Technology, 76*, 9–17.

[55] Geng, Z., Liu, F., Shang, D., Han, Y., Shang, Y., & Chu, C. (2021). Early warning and control of food safety risk using an improved AHC-RBF neural network integrating AHP-EW. *Journal of Food Engineering, 292*, 110239.

[56] Mahdinia, E., Liu, S., Demiciri, A., & Puri, V. M. (2020). Modelling and process design. In A. Demiri, H. Feng & K. Krishnamurthy (Eds.), *Food engineering series* (pp. 357–421). Switzerland: Published by Springer.

[57] Cheng, J., Greiner, R., Kelly, J., Bell, D., & Liu, W. (2002). Learning Bayesian networks from data: An information-theory based approach. *Artificial Intelligence, 137*(1), 43–90.

[58] Barker, G. C (2004). Application of Bayesian belief network models to food safety science In MAJS Van Boekel, A Stein, AHC Van Bruggen (eds). *Bayesian Statistics and Quality Modelling in the Agro-Food Production Chain.* (Pp. 117–130). The Netherlands: Published by Springer Dordrecht.

[59] Bouzembrak, Y., & Marvin, H. J. P. (2016). Prediction of food fraud type using data from rapid alert system for food and feed (RASFF) and Bayesian network modelling. *Food Control, 61*, 180–187.

[60] Marvin, H. J. P., Bouzembrak, Y., Janssen, E. M., van der Fels-Klerx, H. J., van Asselt, E. D., & Kleter, G. A. (2016). A holistic approach to food safety risks: Food fraud as an example. *Food Research International, 89*, 463–470.

[61] Ngai, E. W. T., Hu, Y., Wong, Y. H., Chen, Y., & Sun, X. (2011). The application of data mining techniques in financial fraud detection: A classification framework and an academic review of literature. *Decision Support Systems, 50*, 559–569.

[62] Marvin, H. J. P., Bouzembrak, Y., Janssen, E. M., van der Zande, M., Murphy, F., Sheehan, B., et al. (2017). Application of Bayesian networks for hazard ranking of nanomaterials to support human health risk assessment. *Nanotoxicology, 11*, 123–133.

[63] Bouzembrak, Y., Camenzuli, L., Janssen, E., & van der Fels-Klerx, H. J. (2018). Application of Bayesian Networks in the development of herbs and spices sampling monitoring system. *Food Control, 83*, 38–44.

[64] Elegbede, C. F., Papadopoulos, A., Gauvreau, J., & Crepet, A. (2015). A Bayesian network to optimise sample size for food allergen monitoring. *Food Control*, *47*, 212–220.

[65] Bouzembrak, Y., & Marvin, H. J. P. (2019). Impact of drivers of change, including climatic factors, on the occurrence of chemical food safety hazards in fruits and vegetables: A Bayesian network approach. *Food Control*, *97*, 67–76.

[66] Bouzembrak, Y., & Marvin, H. J. P. (2020). A system approach towards prediction of food safety hazards: Impact of climate and agrichemical use on the occurrence of food safety hazards. *Agricultural Systems*, *178*, 102760.

[67] Saltelli, A. (1999). Sensitivity analysis: Could better methods be used?. *Journal of Geophysical Research*: *Atmospheres*, *104*, 3789–3793.

[68] Qazi, A., Dickson, A., Quigley, J., & Gaudenzi, B. (2018). Supply chain risk network management: A Bayesian belief network and expected utility based approach for managing supply chain risks. International Journal of Production Economics, *196*, 24–42.

[69] Han, Y., Wu, H., Jia, M., Geng, Z., & Zhong, Y. (2019). Production capacity analysis and energy optimization of complex petrochemical industries using novel extreme learning machine integrating affinity propagation. Energy Conversion and Management, *180*, 240–249.

[70] Marın, S., Bellı´, N., Lasram, S., Chebil, S., Ramos, A.J., Ghorbel, A. and Sanchis, V. (2006a). Kinetics of ochratoxin a production and accumulation by Aspergillus carbonarius on synthetic grape medium at different temperature levels. Journal Food Sci. *71*, 196–200.

[71] Ratkowsky, D. A., Olley, J., McMeekin, T. A. and Ball, A. (1982). Relationship between temperature and growth rate of bacterial cultures. Journal of Bacteriology, *149*(1), 1–5.

Yemisi Adefunke Jeff-Agboola*, Rebecca Olajumoke Oloniyo
and Tayo Funmilola Imafidon

Chapter 20
Food safety legislation and Food Safety Modernization Act

Abstract: In this chapter, the food safety laws and its Acts were outlined and discussed. Food Safety Acts are tools used for the implementation of food laws and this has been the bedrock to control and monitor safety in order to present food in a safe and hygienic state for human consumption, The Food Safety Act discussed includes Food Safety Act 1990, Food Safety Act 2006, Food Information Regulation 2014 and Natasha's Law. The role of key agencies of food laws, Food and Agriculture Organization (FAO) and Food and Drug Administration (FDA) through its Center for Food Safety and Applied Nutrition, include to control, monitor and regulate foods and food-related laws. The other topics discussed include Food laws and Regulations in Nigeria (National Agency for Food and Drug Administration and Control - NAFDAC) and Food Safety Modernization Act.

20.1 Introduction

Food is anything that is used for the maintenance of life, growth, and development of the body when consumed [1, 2]. It provides nutritional support to organisms (plants and animals). The two main food sources are plants and animals [3]. Food mainly consists of carbohydrates, proteins, lipids, and micronutrients [4, 5]; They are used in the body to support life, growth and development. The absorption and utilization of food in the body are the basic values of human nutrition facilitated by consumption. Food provides nutrients [6]. Nutrients are substances that provide energy for the body to maintain vitality, growth, and all other bodily functions such as movement, respiration, tissue repair, and immune system management. The quality of food is determined by its value [7]. Nutritional value is a measure of the quality of food for a healthy life; Food value can be good or bad, good when food is considered safe and bad when it is considered unsafe. The Food Act provides the rules and regulations that ensure that food offered for

*Corresponding author: Yemisi Adefunke Jeff-Agboola, Department of Biological Sciences, University of Medical Sciences, Ondo, Nigeria, e-mail: yjeffagboola@yahoo.com
Rebecca Olajumoke Oloniyo, Department of Food Nutrition and Home Sciences, Prince Abubakar Audu University, Anyigba, Kogi State, Nigeria
Tayo Funmilola Imafidon, Chemical Evaluation and Research, National Agency for Food and Drug Administration and Control (NAFDAC), Lagos State, Nigeria

https://doi.org/10.1515/9783110748345-020

sale is safe and suitable for human consumption [8]. The Act also applies to the Food Standards Code (the Code). The law details food violations, orders that can be made against a business, and safety programs. The Food Act is food safety standards that reflect policy decisions regarding acceptable risk and the cost of avoiding risk [9, 10]. Food standards are criteria that food must meet in order to be fit for human consumption. This standard checks for food source, ingredients (including contaminants and residues), appearance, freshness, permitted additives, and maximum bacterial content [11]. The Food Standards Code are standards for food additives, food safety, labeling and foods that require prior approval. Food standards are used to ensure that the food consumed is safe to maintain a healthy life [12].

20.2 Food safety

Food safety is the protection of food, from raw material to finished product, from foodborne illnesses and infections while ensuring that the food contains sufficient nutrients for a healthy diet. Aspects of food control that deal with the handling, processing and storage of food is known as Food Safety [13–15]. Food safety literally means food that is safe to eat [16]. Food safety is also called food hygiene. It is a scientific method (discipline) to ensure that a product is safe to consume, but if not properly controlled, consumers can be exposed to foodborne illness, foodborne infection, foodborne illness, and even the disease can be fatal [16, 17]. Food is said to be "safe" if it is properly handled, processed, stored, and distributed to the final consumer [16, 18].

Training on food safety practices should be encouraged in all food industries to provide food processors in production and decision making with sufficient knowledge to understand the importance of these (food safety) practices [15, 19]. When people are properly trained, less food is wasted, contaminated, or lost due to improper handling [20].

20.2.1 Food safety laws in Nigeria

The board of directors for Food Safety and Applied Nutrition (FSAN) make sure that all foods produced, distributed, sold, imported, advertised, and exported from Nigeria adhere to the necessary requirements for food safety.

Sections 5 and 30 of the National Agency for Food and Drug Administration and Control Act Cap NI Laws of the Federation of Nigeria (LFN) 2004 grant the National Agency for Food and Drug Administration and Control (NAFDAC) and the Honorable Minister of Health the authority to make the following regulations:

1. Cocoa and cocoa products regulations 2021
2. Food fortification regulations 2021
3. Food grade table or cooking salt regulations 2021
4. Food additives regulations 2021
5. Food irradiation regulations 2021
6. Food products advertisement regulations 2021
7. Fruit juice and nectar regulations 2021
8. Milk and dairy products regulations 2021
9. Non-nutritive sweeteners in food products regulations 2021
10. Soft drinks regulations 2021
11. Spirit drinks regulations 2021
12. Wine regulations 2021

20.2.2 Importance of food regulations

1. To safeguard consumers from eating tainted or unhealthy food
2. To promote trade and avoid rejecting agricultural goods at borders
3. Aiming to stop food fraud
4. To prevent misleading packaging or advertising

20.3 Food law

Food law is the legal framework that underlines an effective food control system [21]. Globally, there are specific rules and regulations established to ensure food safety that food chain operators and producers or manufacturers must achieve to ensure that food is safe for consumption. It outlines basic government standards that must be met [22, 23]. "Food Law" governs all aspects of food processing, manufacturing, storage and distribution, and governs food control, food safety, food quality, and all other aspects of food distribution from raw materials to end users. Rules and regulations to supervise aspects are also included [16, 24].

20.3.1 The activities of food law

The activities of food law included:
(i) how hygienically the food is handled;
(ii) safe food preparation, cooking, re-heating, cooling; and
(iii) storage.

Food laws are the rules and regulations that ensure that food can be safely consumed by all food businesses [16, 25, 26]. The purpose of the Food Act is to protect the health of individual consumers. The Food Act should first create a legal basis and an appropriate legal framework for various food control measures. The Food and Drug Administration (FDA), through the Center for Food Safety and Applied Nutrition (CFSAN), manages, monitors, and regulates food and food-related laws [27, 28].

The Food and Agriculture Organization (FAO) assists governments in drafting, amending, and updating food safety and quality legislation and implementing regulations to strengthen food control systems in accordance with international food law and practice for:

- Alignment of National legislative frame-works with Codex-Alimentarius standards, guidelines for relevant documents that establish the benchmark for maintaining food safety and quality on a global scale [29, 30].
- Establishing National policies among interested parties to facilitate international food safety regulations by establishing a direct line of communication with officials and legal counsel [31, 32].
- Assisting nations with the creation of scientifically based food control systems to safeguard consumers' health, hence enabling access to food markets and averting food safety emergencies [33].
- Creating tools and manuals for various food control administrative and technical issues.
- Encouraging the use of the risk-analysis method to enhance food control systems.

20.3.2 Brief history of food laws and regulations in Nigeria

In Nigeria, various governments have attempted to impose welfare and health requirements on the country's food supply.

- Public health laws were established in 1917 and currently known as Public Health Ordinance (PHO) cap 165 of 1958
- The Standards Organization of Nigeria (SON) decree no. 56 of 1971
- The Food and Drug (FAO) Act no. 35 of 1974 (currently known as Food and Drug (FAO) Act Cap F32 Laws of the Federal Republic of Nigeria, 2004)
- The Animal Disease Control (ADC) Decree No. 10 of 1988
- The Marketing of Breast Milk Substitute Decree No. 41 of 1990 (now Marketing (Breast Milk) Act Cap M5 LFN 2004
- The National Agency for Food and Drugs Administration and Control Decree No. 15 of 1993 (now NAFDAC Act Cap N1 Laws of the Federal Republic of Nigeria, 2004)

Others are:
- The Food, Drugs & Related Product (Registration, etc.) Act Cap F33 Laws of the Federal Republic of Nigeria (LFN), 2004 replaces the Food, Drug and Related Products (Registration etc.) Decree No. 19 of 1993
- The Counterfeit and Fake Drugs and Unwholesome Processed Foods (Miscellaneous Provisions) Act Cap C34 LFN 2004 replaced the Counterfeit & Fake Drugs and Unwholesome Processed Foods Act No. 25 of 1999
- Bylaws passed by several L.G. across the nation. Decree 66 of the Consumer Protection Council from 1992
- 1992's Inland Fisheries Decree No. 108

20.4 Food safety act

Food safety legislation is a tool for implementing food legislation, which is the basis for controlling and monitoring food safety for healthy consumption [34].

The Food Safety Act includes: labeling, storing, transporting, importing or exporting food [16, 24]. Under the Food Safety Act 1990, food consumers are guaranteed the practice of eating safe and quality foods at all times [35].

1. The Food Safety Act 2006

   The Food Safety Act 2006 is a law unifying food laws. It establishes the Food Safety and Standards Authority for Food and regulates the manufacturing process, storage, distribution and sale of food to ensure the availability of food that is safe and healthy for humans [36]. The Food Sanitation Ordinance 2006 requires food businesses to operate their food businesses safely and hygienically so that the health of consumers is not compromised. The Food Hygiene Regulations 2006 make it a crime for any food company to supply unhealthy food [11].

   The two main laws relevant to food retailers are the Food Safety Act 1990 and the Food Sanitation Regulations 2006 [37, 38].

2. Food Information Regulation 2014

   The Food Information Regulation 2014 (FIR) is the UK implementation of European Union (EU) Regulation 1169/2011 [39, 40]. The aim is to provide consumers with clearer information about the ingredients of the food they eat.

3. Natasha's law

   The Natasha's law is a significant food law introduced on October 1, 2001. The law requires all grocery stores (companies) to provide a complete ingredient list with clear allergen labeling on their direct-to-sale packaged foods (PPDS) [41]. This is peculiar to food which is prepared on the same platform, packaged and sold to consumers. The law is well known in England, Wales, Scotland and Northern Ireland. This law helps fight allergic reactions [42].

20.5 Food Safety Modernization Act

The Food Safety Modernization Act (FSMA) is the latest food safety law signed into law by former U.S. President Barack Obama on January 4, 2011. The primary purpose of this law is to ensure that the U.S. food supply is safe by shifting the focus of federal agencies from responding to instances of microbial contamination to preventing a contaminated food supply [43]. This law applies to all human food as well as animal food, including pets. FSMA protects public safety by implementing mandatory food safety training and risk prevention controls for all food processors. It is very important that all food processors fully understand their responsibilities and comply with all relevant international and local food safety regulations. The FSMA has given the FDA new authority to regulate the way foods are grown, harvested, processed, stored, and distributed to the end users [27, 44].

The general purposes of U.S. food laws are to minimize the risk of unsafe food (it is too costly to assure that food is safe) and that consumers have the information needed to make an informed decision.

FSMA is not different from HACCP: At the elementary level, HACCP focuses on preventing post-process contamination, whereas FSMA food safety plan takes a more preventive focus, identifying potential risks and implementing appropriate controls to proactively prevent contamination. Food Safety Modernization Act (FSMA) has some other units that work together as one out of which is "HARPC" which means "Hazard Analysis and Risk-Based Preventive Controls." The term comes from the FSMA; it is a sweeping food safety legislation amending the Food, Drug and Cosmetic Act that was enacted in January of 2011. HARPC is an upgrade to HACCP [27, 45].

HACCP involves a multidisciplinary team for implementation and follows prescriptive steps to ensure safety, while HARPC is an aspect of food safety beyond critical control points (CCPs) and is mandated by the FDA for most food facilities, with some exemptions [27].

20.5.1 The seven major rules for FSMA compliance

The seven major rules for FSMA compliance are:
1. Preventive Controls Rules for Human and Animal Food
2. Produce Safety Rule
3. Foreign Supplier Verification Program (FSVP) Rule
4. Accredited Third-Party Certification
5. Sanitary Transportation Rule
6. Intentional Adulteration Rule
7. Complying with FSMA

1. Preventive Controls Rules for Human and Animal Food

The Preventive Controls for Human and Animal Food Rule was proposed in January 2013 and confirmed in October 2013. They are intended for very small businesses (less than $1,000,000 in annual revenue). All food companies must have a written food safety policy that identifies known and unknown physical, biological, and chemical hazards associated with food produced by the company. For each identified hazard, the food safety plan should determine whether the hazard requires a preventive control and, if necessary, outline a preventive control to minimize or prevent the hazard. The preventive controls rule also considered approval of raw materials and other production materials from suppliers. When approving a supplier's goods, manufacturers must consider several factors, including the supplier's implementation (that is, compliance with FDA regulations, FDA warning letters, import alerts, etc.). The FDA has acceptable standards for monitoring the status of all suppliers. This rule depends on the size of your company [46–48].

2. Produce Safety Regulation

The Produce Safety Rule was proposed in January 2013 and approved in November 2015. The FDA's Produce Safety Regulations are established science-based minimum rules and regulations for the safe growing, harvesting, packing, and holding of food. This emphasizes farmers' responsibility to protect their crops from contamination by setting standards such as water testing, raw fertilizer application, pasture surveys, and employee health and hygiene training. The rule pays special attention to sprouts, as they are frequently associated with outbreaks of foodborne illness. Product safety rules vary depending on the size of the company [49, 50].

3. Foreign Supplier Verification Program (FSVP) Rule

The Foreign Supplier Verification Program (FSVP) Rule was proposed in July 2013 and approved in November 2015. The FSVP regulation ensures that all importers certify that their foreign food suppliers produce their food in accordance with German FDA regulatory requirements. If a foreign food supplier identifies and evaluates the hazards present in the particular food that such food importer intends to import, thereby determining the potential hazards associated with such food. Additionally, the importer must implement her FSVP for each approved food and its suppliers. FSVP provides importers with flexible means to use appropriate verification measures for each food and its suppliers. Possible verification activities include product sampling and testing, inspection of supplier facilities, and review of supplier records [51, 52].

4. Third-Party Certification

Third-party certification groups primarily focus on food for the purposes of safety inspection and certification of food products in foreign food facilities. Under this program, the FDA will identify accrediting organizations that can accredit third-party certification groups. The audit process, consists of both advisory and regulatory audit laws. An advisory review is the first step in the review process, followed by a regulatory review. Regulatory review is the basis for certification [53].

5. Sanitary Transport Regulations

Sanitary Transport Regulations were proposed in February 2014 and confirmed in April 2016. It contains important and useful requirements for shippers, carriers, and recipients of food by road, to ensure that food is adequately secured throughout the transportation system [54].

6. Intentional Adulteration Rule

The Intentional Adulteration Rule was proposed in December 2013 and confirmed in May 2016. The regulation includes the use of food safety plans (documents containing vulnerability analysis, mitigation strategy implementation, and food safety monitoring procedures, corrective actions and reviews) to identify mitigation strategies for various vulnerable groups and used for monitoring. The prove of the efficiency and effectiveness of applied mitigation strategies are procedures for food defense monitoring, corrective actions, and verification [55–57]. The rule "prevents acts of intentional adulteration intended to harm public health, including acts of terrorism aimed at supplying food."

7. Compliance with FSMA

To comply with FSMA rules and regulations, you must arbitrarily comply with all FSMA rules and regulations. Registrar Corp's FSMA Assistant is one of the tools used to monitor a food supplier's compliance with her FSMA requirements. The FSMA Wizard is an unlimited device that helps you control the requirements of the FSMA. It also helps develop new requirements under FSMA, food safety, and food safety plans.

20.6 Conclusions

Food hygiene and safety are two fundamental tools of global-health policy aimed at good health for all. The continuous supply of safe, nutritious, and healthy food to the population is an important component of global health and well-being maintained at the international level. Good global food hygiene and safety guidelines improve the safety of available food, produce better nutrition, improve health outcomes, improve food trade and economic development, and ensure that food is the lifeblood of life. As it is an important factor and a necessity of life, there is a need to continuously develop maintenance mechanisms to ensure a stable supply of healthy and nutritious food. National food security structures have a positive impact on consumers and high standards of living for national economies, to reduce the burden and image of the world from disease, war, famine, famine, malnutrition, and short life expectancy. This should be strengthened, rebuilt, and managed commendably. The importance of national standardization bodies as catalysts in the prevention and control of food-borne diseases should be recognized and promoted for the definition and realization of industrial and economic development.

References

[1] Di Francesco, A., Di Germanio, C., Bernier, M., & De cabo, R. (2018). A time to fast. *Science*, *362*(6416), 770–775.

[2] Pajno, G. B., Fernandez-Rivas, M., Arasi, S., Roberts, G., Akdis, C. A., & Alvaro-Lozano, M. (2018). EAACI Guidelines on allergen immunotherapy: IgE-mediated food allergy. *Allergy*, *73*(4), 799–815. EAACI Allergen Immunotherapy Guidelines Group.

[3] Upadhyay, M. K., Shukla, A., Yadav, P., & Srivastava, S. (2019). A review of arsenic in crops, vegetables, animals and food products. *Food Chemistry*, *276*, 608–618.

[4] Kusmayadi, A., Leong, Y. K., Yen, H. W., Huang, C. Y., & Chang, J. S. (2021). Microalgae as sustainable food and feed sources for animals and humans–Biotechnological and environmental aspects. *Chemosphere*, *271*, 129800.

[5] Roleda, M. Y., Lage, S., Aluwini, D. F., Rebours, C., Brurberg, M. B., Nitschke, U., & Gentili, F. G. (2021). Chemical profiling of the Arctic sea lettuce Ulva lactuca (Chlorophyta) mass-cultivated on land under controlled conditions for food applications. *Food Chemistry*, *341*, 127999.

[6] Comerford, K. B., Papanikolaou, Y., Jones, J. M., Rodriguez, J., Slavin, J., Angadi, S., & Drewnowski, A. (2021). Toward an evidence-based definition and classification of carbohydrate food quality: An expert panel report. *Nutrients*, *13*(8), 2667.

[7] De Pasquale, I., Verni, M., Verardo, V., Gómez-Caravaca, A. M., & Rizzello, C. G.(2021). Nutritional and functional advantages of the use of fermented black chickpea flour for semolina-pasta fortification. *Foods*, *10*(1), 182.

[8] Arnold, M., Wickramatilake, S., Fernando, D., Sampath, R., Karunapema, P., Mahesh, B., . . . Denawaka, C. (2020). Health and nutrition related claims of non-alcoholic beverage labels in supermarkets: their compliance with sri lanka food labelling and advertising regulations. *Journal of the College of Community Physicians of Sri Lanka*, *26*(1).

[9] Lau, H., Nakandala, D., & Shum, P. K. (2018). A business process decision model for fresh-food supplier evaluation. *Business Process Management Journal*.

[10] Gray, A., & Hinch, R. (Eds.). (2019). *A handbook of food crime: Immoral and illegal practices in the food industry and what to do about them*. Policy Press.

[11] King, T., Cole, M., Farber, J. M., Eisenbrand, G., Zabaras, D., Fox, E. M., & Hill, J. P. (2017). Food safety for food security: Relationship between global megatrends and developments in food safety. *Trends in Food Science & Technology*, *68*, 160–175.

[12] Lee, C. L., Liao, H. L., Lee, W. C., Hsu, C. K., Hsueh, F. C., Pan, J. Q., . . . Chen, M. J.(2018). Standards and labeling of milk fat and spread products in different countries. *Journal of Food and Drug Analysis*, *26*(2), 469–480.

[13] Farooq, U., Shafi, A., Shahbaz, M., Khan, M. Z., Hayat, K., Baqir, M., & Iqbal, M. (2021). Food quality and food safety: an introduction. In *Sequencing Technologies in Microbial Food Safety and Quality* (pp. 3–24). CRC Press.

[14] Feng, Y., & Archila-Godinez, J. C.(2021). Consumer knowledge and behaviors regarding food safety risks associated with wheat flour. *Journal of Food Protection*, *84*(4), 628–638.

[15]   Kwol, V. S., Eluwole, K. K., Avci, T., & Lasisi, T. T. (2020). Another look into the Knowledge Attitude Practice (KAP) model for food control: An investigation of the mediating role of food handlers' attitudes. *Food Control*, *110*, 107025.

[16]   Kamboj, S., Gupta, N., Bandral, J. D., Gandotra, G., & Anjum, N.(2020). Food safety and hygiene: A review. *International Journal of Chemical Studies*, *8*(2), 358–368.

[17]   Zhao, Y., & Talha, M. (2021). Evaluation of food safety problems based on the fuzzy comprehensive analysis method. *Food Science and Technology*.

[18]   Yousuf, B., Deshi, V., Ozturk, B., & Siddiqui, M. W. (2020). Fresh-cut fruits and vegetables: Quality issues and safety concerns. In *Fresh-cut fruits and vegetables* (pp. 1–15). Academic Press.

[19]   Lin, N., & Roberts, K. R. (2020). Using the theory of planned behavior to predict food safety behavioral intention: A systematic review and meta-analysis. *International Journal of Hospitality Management*, *90*, 102612.

[20]   Behera, B. C. (2021). Challenges in handling COVID-19 contaminated waste material and its sustainable management mechanism. *Environmental Nanotechnology, Monitoring & Management*, 100432.

[21]   Faour-Klingbeil, D., & CD Todd, E.(2020). Prevention and control of foodborne diseases in middle-east north african countries: review of national control systems. *International Journal of Environmental Research and Public Health*, *17*(1), 70.

[22]   Ehuwa, O., Jaiswal, A. K., & Jaiswal, S.(2021). Salmonella, Food Safety and Food Handling Practices. *Foods*, *10*(5), 907.

[23]   Okpala, C. O. R., & Korzeniowska, M. (2021). Understanding the relevance of quality management in agro-food product industry: From ethical considerations to assuring food hygiene quality safety standards and its associated processes. *Food Reviews International*, 1–74.

[24]   Shukla, S., Shankar, R., & Singh, S. P. (2014). Food safety regulatory model in India. *Food Control*, *37*, 401–413.

[25]   Eriksson, M., Giovannini, S., & Ghosh, R. K.(2020). Is there a need for greater integration and shift in policy to tackle food waste? Insights from a review of European Union legislations. *SN Applied Sciences*, *2*(8), 1–13.

[26]   Kapala, A. (2020). Agricultural Retail Trade Regulation as a Legal Instrument to Support Local Food Systems. *European Food and Feed Law Review*, *15*, 231.

[27]   Malik, S., Krishnaswamy, K., & Mustapha, A.(2021). Hazard analysis and risk-based preventive controls (HARPC): Current food safety and quality standards for complementary foods. *Foods*, *10*(9), 2199.

[28]   Verrill, L., Boyer, M., Williams, L., Otto, J., Lando, A., Dawood, N., & Liggans, G.(2021). Hand washing observations in fast-food and full-service restaurants: Results from the 2014 US food and drug administration retail food risk factors study. *Journal of Food Protection*, *84*(6), 1016–1022.

[29]   Bansal, R., & Dhiman, A.(2020). Nutraceuticals: a comparative analysis of regulatory framework in different countries of the World. *Endocrine, Metabolic & Immune Disorders-Drug Targets (Formerly Current Drug Targets-Immune. Endocrine & Metabolic Disorders)*, *20*(10), 1654–1663.

[30]   Molnar, G., & Godefroy, S. B. (2020). Review of mechanisms for food safety-related SPS measures within African regional Economic Communities (RECs): Paving the way for a continent-wide food safety coordination effort. *Food Control*, *115*, 107206.

[31]   Dupouy, E., & Gurinovic, M. (2020). Sustainable food systems for healthy diets in Europe and Central Asia: Introduction to the special issue. *Food Policy*, *96*, 101952.

[32]   Lu, H., Mangla, S. K., Hernandez, J. E., Elgueta, S., Zhao, G., Liu, S., & Hunter, L.(2021). Key operational and institutional factors for improving food safety: a case study from Chile. *Production Planning & Control*, *32*(14), 1248–1264.

[33]   Laar, A., Barnes, A., Aryeetey, R., Tandoh, A., Bash, K., Mensah, K., . . . Holdsworth, M. (2020). Implementation of healthy food environment policies to prevent nutrition-related non-

communicable diseases in Ghana: National experts' assessment of government action. *Food Policy*, *93*, 101907.

[34] Gunnarsdottir, M. J., Gardarsson, S. M., Schultz, A. C., Albrechtsen, H. J., Hansen, L. T., Bergkvist, K. S. G., . . . Bartram, J. (2020). Status of risk-based approach and national framework for safe drinking water in small water supplies of the Nordic water sector. *International Journal of Hygiene and Environmental Health*, *230*, 113627.

[35] Nagyová, Ľ., Andocsová, A., Géci, A., Zaják, P., Palkovič, J., Košičiarová, I., & Golian, J. (2019). Consumers'awareness of food safety. *Potravinarstvo*, *13*(1).

[36] Prashar, D., Jha, N., Jha, S., Lee, Y., & Joshi, G. P.(2020). Blockchain-based traceability and visibility for agricultural products: A decentralized way of ensuring food safety in india. *Sustainability*, *12*(8), 3497.

[37] Elizondo, C. J. F., Lord, N., & Spencer, J. (2019). Food fraud and the Fraud Act 2006: Complementarity and limitations. In *Financial crime and corporate misconduct: A critical evaluation of fraud legislation* (pp. 48–62). Routledge.

[38] Nayak, R., & Waterson, P. (2019). Global food safety as a complex adaptive system: Key concepts and future prospects. *Trends in Food Science & Technology*, *91*, 409–425.

[39] Leialohilani, A., & de Boer, A. (2020). EU food legislation impacts innovation in the area of plant-based dairy alternatives. *Trends in Food Science & Technology*, *104*, 262–267.

[40] Thrasyvoulou, A., Tananaki, C., Goras, G., Karazafiris, E., Dimou, M., Liolios, V., . . . Gounari, S.(2018). Legislation of honey criteria and standards. *Journal of Apicultural Research*, *57*(1), 88–96.

[41] Vaque, L. G. (2020). New Publications. *European Food and Feed Law Review*, *15*, 165.

[42] Soon, J. M., & Wahab, I. R. A. (2021). Global food recalls and alerts associated with labelling errors and its contributory factors. *Trends in Food Science & Technology*.

[43] Pineda, C. E. (2018). The Food Safety Modernization Act: A summary of the act, education, and implementation.

[44] LaBorde, L. F. (2020). The hazard analysis risk-based preventive controls. In *Food safety engineering* (pp. 205–226). Cham: Springer.

[45] Nicolau, A. I., & Oddgeirsson, O. (2017). Food Safety Management in Fish Processing Units. *Trends in Fish Processing Technologies*, 241–254.

[46] Davies, R. H., Lawes, J. R., & Wales, A. D.(2019). Raw diets for dogs and cats: a review, with particular reference to microbiological hazards. *Journal of Small Animal Practice*, *60*(6), 329–339.

[47] Jones, S. L., Ricke, S. C., Keith Roper, D., & Gibson, K. E.(2020). Swabbing the surface: critical factors in environmental monitoring and a path towards standardization and improvement. *Critical Reviews in Food Science and Nutrition*, *60*(2), 225–243.

[48] Vermaak, S. (2020). Worker Health, Hygiene, and Training to Satisfy the Requirements of the FSMA Produce Safety Rule.

[49] Tack, D. M., Ray, L., Griffin, P. M., Cieslak, P. R., Dunn, J., Rissman, T., . . . Payne, D. C.(2020). Preliminary incidence and trends of infections with pathogens transmitted commonly through food – Foodborne Diseases Active Surveillance Network, 10 US Sites, 2016–2019. *Morbidity and Mortality Weekly Report*, *69*(17), 509.

[50] Torres-Tiji, Y., Fields, F. J., & Mayfield, S. P. (2020). Microalgae as a future food source. *Biotechnology Advances*, *41*, 107536.

[51] Ansah, E. O., Kaplowitz, M. D., Lupi, F., & Kerr, J.(2020). Smallholder participation and procedural compliance with sustainable cocoa certification programs. *Agroecology and Sustainable Food Systems*, *44*(1), 54–87.

[52] Bailey, R. L.(2020). Current regulatory guidelines and resources to support research of dietary supplements in the United States. *Critical Reviews in Food Science and Nutrition*, *60*(2), 298–309.

[53] Lamin, A., & Livanis, G. (2020). Do third-party certifications work in a weak institutional environment? *Journal of International Management*, *26*(2), 100742.

[54] Telesetsky, A. (2020). Food safety law in the united states: risk management in the organic food supply chain. In *Regulatory issues in organic food safety in the Asia Pacific* (pp. 217–228). Singapore: Springer.

[55] Choudhary, A., Gupta, N., Hameed, F., & Choton, S.(2020). An overview of food adulteration: Concept, sources, impact, challenges and detection. *International Journal of Chemical Studies*, *8*(1), 64–2573.

[56] Sharma, S. D., Bhagat, A. R., & Parisi, S. (2019). Food adulteration episodes. The impact of frauds in the american market of dairy raw materials. In *Raw material scarcity and overproduction in the food industry* (pp. 43–54). Cham: Springer. Lamin, A., & Livanis, G. (2020). Do third-party certifications work in a weak institutional environment? *Journal of International Management*, *26*(2), 100742.

[57] Ulberth, F. (2020). Tools to combat food fraud–a gap analysis. *Food Chemistry*, *330*, 127044.

[58] Overdiep, J. L., & Shaw, A. M.(2019). Assisting Food Processors with Food Safety Modernization Act Compliance. *Journal of Extension*, *57*(3), v57–3tt4.

[59] Han, S., Roy, P. K., Hossain, I., Byun, K. H., Choi, C., & Ha, S. D. (2021). COVID-19 pandemic crisis and food safety: Implications and inactivation strategies. *Trends in Food Science & Technology*.

[60] Ruth, S. M., Huisman, W., & Luning, P. A. (2017). Food fraud vulnerability and its key factors. *Trends in Food Science & Technology*, *67*, 70–75.

Abiola Folakemi Olaniran*, Abiola Ezekiel Taiwo,
Yetunde Mary Iranloye and Clinton Emeka Okonkwo

Chapter 21
The role of good agricultural practices (GAPs) and good manufacturing practices (GMPs) in food safety

Abstract: Ensuring safe food for consumers is becoming a persistent issue in developing and developed countries. When consumers' demands and perceptions are not integrated into industrial operations and manufacturing processes, conflicts of interest arise. High society's expectations have led to intense competition among food producers and industrial investors. However, these contentions should not influence the growers' and manufacturers' decision-making in safeguarding the safety of the final food product either raw or processed; because violations of food safety is not only costly to the industry, it also plays a significant role in human lives. This is the primary motivation for implementing specified guidelines in agricultural practices, which helps in the establishment of a balance food chain spanning production to distribution. Good agricultural practices (GAPs) acquaint farmers with the prerequisites for sustainable agricultural approach without jeopardizing the safety of consumers by creating a good consumption and production pattern. Good manufacturing practices are crucial links to all the elements which ensures food's quality, safety, and efficacy in relation to employees, facilities, environment, process control, and equipment meet up its specifications and purposes. Therefore to boost consumer-producer understanding and enhance positive dynamism an adaptation and implementation of good agricultural and manufacturing practices is paramount.

*Corresponding author: Abiola Folakemi Olaniran, Department of Food Science and Nutrition, College of Pure and Applied Sciences, Landmark University, Omu-Aran, Kwara State, Nigeria, e-mail: abiolaolaniran@gmail.com
Abiola Ezekiel Taiwo, Faculty of Engineering, Mangosuthu University of Technology, Durban, South Africa
Yetunde Mary Iranloye, Department of Food Science and Nutrition, College of Pure and Applied Sciences, Landmark University, Omu-Aran, Kwara State, Nigeria
Clinton Emeka Okonkwo, Department of Food Science, College of Food and Agriculture, United Arab Emirates University, Al Ain, United Arab Emirate

https://doi.org/10.1515/9783110748345-021

21.1 Introduction

The importance of food safety has grown over the years as a result of its impact on consumer health and the expansion in trading zones for several food products both domestically and internationally. Food has a shelf life, whether nonperishable or perishable, and the quality can be influenced by a wide range of factors such as temperature, mechanical static, light, and stresses [1]. Food hazards can arise at any point in supply chain, starting from primary production (planting, harvesting, topsoil contamination), secondary (washing, trimming), tertiary processing (transformation to products), and packaging; storage and distribution stages. Thus, the need to use suitable methods during on-farm production and postproduction operations to ensure a safe food supply [2–4] and reduce consumer susceptibility to foodborne illnesses [5]. Foodborne illness has a huge social and financial impact on communities and health systems. Hence, food security must be considered across the entire agri-food production chain, from seed selection to harvesting, processing, manufacturing, distribution, and retailing to consumption [6].

Also, the introduction and adoption of good practices in agricultural and manufacturing sectors have progressively contributed to the improvement in food safety over the centuries which allowed integration of specific activities in the food industry [7]. Implementing these food safety measures at the farm level served as a decisive phase in the paradigm shift in food production [8]. According to the FAO, good agricultural practice (GAP) is gathering diverse principles to utilize on-farm production and post-production processes, resulting in nutritious food that are safe products, while considering environmental sustainability, social, and economic aspects [9]. Farmers and manufacturers need to apply quality control ethics in definite phases and activities throughout the production chain and this has resulted in several inventions to achieve good manufacturing practices (GMPs) [10]. Considering the divergence in researches, ecological affluence, culture, belief, and techniques in breeding, production, preservation and employment scheme are crucial factors in good practices. There is need for proper understanding of relevant concepts in good practices. In addition to designing distinct tasks that form the integral part of the enterprise which cut across numerous facets of operations that runs simultaneously in a well-functioning agricultural/manufacturing system to inculcate systematic hygiene, processing, engineering, information and technology update mechanisms contribute significantly to the successful attainment of food safety and overall growers and producer's integrity while saving consumers lives.

21.2 The role of implementation of good agricultural practices in attaining food safety

Agriculture plays a significant role in daily lives and is indispensable in providing food for human survival. Consumer perceptions of food and agriculture are changing as a result of cultural, regional, and religious influences [11]. In this era of globalization, ensuring safe food for consumers is a persistent issue for both developed and developing countries [12]. GAPs are a set of rules to follow before, during, and after the production of agricultural commodities, resulting in safe and nutritious food while considering economic, social, and environmental sustainability of the process [13]. Consumer health is protected by GAP rules, which are based on three key principles: risk prevention, risk analysis, and sustainable agricultural practices [2] followed during farm cultivation, harvesting, and production processes [11].

GAP is well-defined on international level by the *Codex Alimentarius* Commission's code of practice for different product like fresh fruits and vegetables. The code of practice applies to all operations in and near fields before, during, and after production and harvest (water quality, personal hygiene of the workers, and manure composting) [14]. GAPs are effective in minimizing the risk of food contamination in agricultural production when they are well managed. GAP documentation describes the program and farm components targeted in helping farmers to conduct self-audits of their plant and animal-based product production, processing, and delivery processes to protect food, environment, and the consumer [15]. Soil, water, lands, and surfaces are the four fundamental components of production and processing that GAPs drives. There are different types of GAP programs, each with its unique set of requirements. GAPs guidelines are designed for enhancing agricultural product quality and safety and any production system can benefit from these general recommendations [16]. Providing healthy food and supporting environmental safety involves all steps in the manufacturing chain, from farmer to consumer [4]. Application of the standardized tools and guiding principle for the farmers, stakeholders, and people involved in commerce and trade are germane in realization of non-hazardous food supplies to consumers throughout the world [8]. GAPs were also created to address potential causes of pre-harvest microbial contamination, although certification is still in short supply [17]. Processes, ingredients, and their sources must have traceability, clarity, and control, and directives must be written in clear terms [18]. GAP guidelines ensure that agricultural goods meet regulatory safety and quality requirements. All farmers are encouraged to make implementation a priority because it benefits both consumers and growers. GAPs can help trace infected foods back to the "handlers and growers" responsible for contaminated agricultural products. Failure to follow GAP guidelines in a farm's operation increases the risk of contamination [19]. Also, consumers' trust in food systems are jeopardized sequel to reports of food poisoning incidences and mortalities triggered by contamination of fresh, minimally processed, and processed food product by different pathogens [8]. This has led to sporadic surge in healthcare overheads in testing and controlling outbreaks, loss of proceeds in businesses, and legal expenses connected litigations/lawsuits [20]. Therefore, self-audits have to be conducted consistently to ascertain if the farm meets up with the set individually outline and government GAP standard rules. Iowa State University Extension recommended five steps for a successful GAP adoption as shown in Figure 21.1 [16]. These recommended steps include the development of strategic plans that allow farmers and growers to specify all of the actions that will be carried out across the production chain, identification of potential critical spots with a high risk of contamination during processing, followed by careful monitoring of the indicated points/areas and, if necessary, modification of operations. This is followed by prompt and proper documentation of every change made to the typical routine to help in future decisions making with assurance.

Planning

Identification of
potential critical spots

Monitoring of the indicated
points/areas

Modification of previous operations

Document all of the changes

Figure 21.1: Recommended steps for successful GAP adoption [16].

Agricultural production is a series of events that starts with sowing (or perhaps before that) and ends with the consumer. In terms of quality, safety, and value for money, it must fulfill consumer expectations [21]. Handling, storage, and transportation across national boundaries or continents are part of numerous intermediate operations. Automation will play a large role in agricultural and biological production systems. Growing concern about preserving biodiversity and genetic resources compels food producers and consumers to be aware of relevant elements of goods, procedures, and process environments before conclusion [22]. Everyone in the food chain believes that if manufacturing is done following appropriate agricultural uses, their expectations may be met. The GAP guidelines are the cornerstones of food safety. Consumers expect food that is safe and provides accurate information; suitable for the task. For a specific reason, information must be accurate, relevant, exact, timely, and comprehensive [23]. Reconsideration ease in availability and applicability of familiar methods in quality assurance systems along a value chain in developing countries can positively change the business model connecting the stakeholders resulting in economic boom. GAP criteria are divided into four sections: (i) food safety; (ii) environmental management; (iii) workers' health, safety, and welfare; and (iv) product quality on the farm and in the packing house, this includes production, harvesting, and postharvest handling [9]. Individual GAP standards differ in details of practices, levels of implementation, and enforcement, although the underlying concept is identical. GAPs cover a wide range of procedures and activities aimed at reducing microbial contamination in farm-fresh produce [24]. Providing bathroom and hand-washing facilities for pickers in the field are examples of simple GAPs practices, cleaning of the packing shed or cooling pad, and, where applicable, the use of single-use trays, monitoring irrigation water, and developing a crisis management strategy for the firm in the case of a food safety epidemic are all examples of hygiene training [25]. Application of diverse sustainable agricultural techniques such as integrated pest management, integrated nutrient management, and conservation agriculture was developed to result in economic and efficient production of adequate (food security), harmless (food safety), and nourishing foods (food quality); sustains and enhances natural resources; finally preserves the viable farming enterprises by contributing to sustainable livelihoods in terms of occupational health, safety, and growers welfare [26]. In the bid to increase the sustainability of food production, enhancement of food security and safety in different countries [27], the Malaysian Ministry of Agriculture in the year 1999 formulated NAP3 (Third National Agricultural Policy) covering pre- and post-agricultural production [28]. Likewise, the Australian Center of International Agricultural Research codified standards for GAP to be observed by many South Asian countries who wish to trade their agricultural commodities globally during Asian Initiatives in Vietnam Conference in the year 2000 [29]. Additionally, the US Food and Drug Administration (FDA) worked with the packaging industry to identify GAP violations and develop food safety regulations unique to sprouts, tomatoes, lettuce, melons, and leafy greens. FDA findings were divided into eight categories: manure, and municipal bio-solids, packing facility sanitation, water, sanitary facilities, worker health and hygiene, field sanitation, transportation, and trace-back, all of which have improved food safety and shelf life of packaged produce significantly [4].

Food safety is becoming a primary priority for all parties engaged in the food supply chain, and food manufacturers' production processes are being scrutinized more than ever. At this time, in the food supply chain, there is also a need for enhanced quality assurance, transparency, and traceability [30]. Human health and the environment, food safety, and access to sustainable agricultural development (achieve environmental, economic, and social sustainability) are all covered by the Good Agriculture Practices standard, which focuses on the long-term viability of on-farm activities while also ensuring the safety and quality of food and non-food crops [31]. Due to recurrent food contamination, food safety has become a serious issue in the society. The risk of contaminated food products during the manufacturing process increases if a GAP program is not implemented [32]. Olutegbe et al. [33] approached sustainable agriculture, food safety, and high-quality food that enable farmers in developing countries to take advantage of new market opportunities by improving supply chain control, natural resource utilization, worker health and working conditions, consumer and farmer family health, and opening up new markets. Microorganisms are natural habitat in growing media of plants such as soil, solid substrates, and nutrient solutions, especially when biological wastes are used as a nutrient source. Pathogens may come in contact with edible parts of crops and even be internalized in vegetable organisms [34]. Common associated pathogens are Shiga toxin-producing *Escherichia coli* (STEC), *Salmonella enterica*, and *Listeria monocytogenes* [35, 36]. Growers have also progressively taken measures to prevent contamination of produce from sources of animal contamination like those identified by the Centers for Disease Control and Prevention (CDC). Twenty percent of food safety expenses were attributed to un-harvested produce near animal tracks or feces and to the time spent by harvest foremen inspecting for animal tracks or feces prior to harvest. Hence, the usage of fencing around the production areas, as well as surface water sources for irrigation in order to keep out domestic and wild animals as preventive measures [37]. Contamination of farm items has an impact on both consumers and farmers. Microbial contamination of produce can occur from fields or surfaces of tools and equipment when it touches contaminated surfaces [38]. Therefore, growers' implementation of food safety practices such as regular cleaning and sanitizing surfaces of harvesting tools and equipment that come in contact with large quantities of produce has become a key element to reduce the risk of microbial contamination and increase their sales [22]. The risk of microbial contamination had been reported to be higher for surface water sources than ground (well) water sources, and municipal water sources carry the least risk. It has been reported that growers have made clear shifts toward adoption of less risky types of irrigation systems and sources of irrigation water. Changing the type or source of irrigation such as installing drip irrigation or drilling a groundwater well though are more costly; installation of new irrigation infrastructure has resulted in reduction in risk of microbial contamination [37]. Implementation of agricultural practice (GAPs) guidelines established by the FDA reduced microbial food safety hazards in fresh fruits and vegetables. The guidelines were used by shippers, producers, and packers to address frequent microbiological dangers in their operations [39]. At the production sites, the workers must be trained to follow prescribed procedures. Storage areas must be kept clean and prepared to reduce the possibility of cross-contamination [11]. Government and nongovernmental organizations together with farmers' groups, producers' companies, and private companies have developed GAP applications to meet the different cultural and social demands of society [40]. GAP standard developed for beef cow farming are used to help farmers increase their livestock output while maintaining high quality, safety, effective resource usage, long-term beef cattle production, and a clean environment [41].

21.3 Collaboration of manufacturers in implementation of good manufacturing practices to achieve production of safe foods

Food production, breeding, and preservation technologies have evolved in response to variances in environmental wealth, cultural differences, and disciplines, providing people with a generally safe way to subsist [42]. Manufacturing takes place in the food supply chain and is usually done in a factory [43]. Industry has made significant progress in the adoption of best practices regarding food safety. GMPs are regulations that are put in place to ensure that there are no hazards in the overall practices that are used to ensure product quality, safety, and standards. GMPs must be followed for various product testing, manufacturing, storage, handling, and distribution practices [44]. GMPs must meet the following criteria: safety, integrity, purity, quality, and composition [45]. The FDA developed voluntary produce food safety guidelines, known as GAPs and GMPs [37]. GMPs are intrinsically tied to food technology. GMPs connect everything together all of the elements that ensure food's quality, safety, and efficacy under its specifications and purposes [11]. Developing a food safety plan is critical for reducing the risks that could affect the safety of food products [14]. While some food sector investors are afraid that GMP could jeopardize the effectiveness of an established hazard analysis and critical control point (HACCP) system, others are completely unaware of what defines good practice [46]. Hence to achieve food system quality procedures, food items and related processes must be consistent and managed. The World Health Organization (WHO), defines standard operating procedures as pertaining to dealing with people, quality control, documentation, equipment, manufacturing, and building from its fundamental requirements and principles [44].

GMP implementation is a continuous process based on the PDCA (plan, do, check, act approach) cycle management concepts. The PDCA cycle may be used to divide the implementation of GMPs into four phases: initial diagnosis, road map building, nonconformity resolution, and re-evaluation of corrective measures taken. Initial diagnosis and re-evaluation of corrective measures are frequently accomplished by inspecting processing facilities using a checklist based on the country's GMP regulations [47]. Following the audit, road maps will be created, while the execution of corrective actions will necessitate the determination of resource priorities and efforts. Indicators are used to measure the true advantages and efficacy of GMP implementation in terms of food safety and quality. Several indicators, such as microbiological indicators, pre- and post-implementation expenses, among other things, can be used to assess how well GMPs have been implemented [48]. According to studies, the biggest obstacles to implementing GMP in the cheese processing factory were the difficulty in modifying food handlers' hygienic habits and some of their unwillingness to compromise on filling out forms and spreadsheets for verification and record management [49]. Continuous training and multiple seminars were arranged to address these difficulties, demonstrating to food handlers the necessity of their support in the success of GMP execution. In addition, food handlers were reminded in the briefings that failing to follow hygienic measures could result in advertisements resulting in tiny salary decreases [50]. All of the structural adjustments made in the cheese processing plant were only conceivable after the food handlers had been completely conceded by GMP enactment, according to the authors' findings [51].

Food must be extremely safe and hygienic because it is directly consumed by living beings. Hence, in designing a food production process or pattern there will be need to consider series of decision-making on both physical facilities and the manufacturing system which comprises the building, unit operations, equipment, factory logistic and material handling [52]. Researchers have shown that several factors such as adequate facilities and infrastructure, strategic location, potential market opportunity changes, consumers' consumption patterns due to healthy life awareness, and advances in technology with good information management influenced the rate of GMP implementation in small and medium enterprises [53, 54]. The internal control system is linked to entities adhering to specific procedures, such as HACCP, GMP, and Good Hygienic Practice, among others, ISO 22000 (Food Safety Management System), and others [55]. Hygiene in primary production, hygienic design of equipment and facilities, control of operations, maintenance and sanitation practices, personal hygiene, transportation, product information, and consumer awareness and training are some of the general measures to be implemented by food industries to achieve GMP as described by *Codex Alimentarius* [56].

GMPs facilitate better learning, core technical skills, and increased executive performance; and promote organizational learning, the development of internal competencies, and increased organizational performance [57]. As reported by Abbas [58], waste and resource expenditures, which are primarily achieved by GMPs, enable higher production speed and shorter time to market while simultaneously enhancing issues of social obligation and basic rights. This good manufacturing perspective draws more consumers, allowing manufacturers to enhance their financial and in-use performance [59]. Good hygienic measures and following the guidelines set forth by the HACCP food safety management system are essential for food safety [55]. Prerequisite programs such as GMPs and standard operating procedures, programs aimed at improving employee hygiene practices, cleaning and sanitation programs, proper facility-design practices, equipment maintenance, and supplier selection and specification programs (cross-contamination co-ordination) all should be implemented before the HACCP system can be implemented [60]. Food safety has received a lot of attention in recent years as a result of a series of food scandals, including several outbreaks of *E. coli* contamination in fresh produce, such as fresh spinach in the United States in 2006 [2] and bean sprouts in Germany in 2011 [61]. Food safety violations not only cost the industry a lot of money, but they also have a significant impact on people's lives. Several efforts have been made to achieve the policy strategic goals, including the adoption of required standards such as minimum residue levels based on the *Codex Alimentarius* and safety requirements for agricultural commodities and food items, advocating for a national "good agricultural practice" (GAP) program to reduce pesticide consumption [32].

In addition, various initiatives are underway to build food safety assurance systems, which are seen as a useful tool for managing, monitoring, and controlling the quality and safety of food production [53]. Many governmental and commercial organizations have launched voluntary certification programs as a method of establishing effective food quality and safety management systems in the food production, processing, preservation, and distribution processes, intending to assist adopters in more than merely meeting requirements [62]. GAP practices address environmental, economic, and social sustainability for on-farm operations, resulting in safe and high-quality food and non-food agricultural products, according to the FAO definition. GAP is currently often utilized in worldwide regulatory frameworks as commercial or public voluntary standards such as GLOBALGAP, FAO GAP, and USDA GAP [63].

GLOBALGAP is a nonprofit multinational organization that establishes voluntary certification standards and procedures for GAPs. It was founded in 1997 as EurepGAP by a group of European supermarket chains who were members of the Euro-Retailer Produce Working Group (EUREP) and changed its name to GLOBALGAP in 2007. GLOBAL GAP aspires to create a single GAP standard for a variety of product uses, allowing for worldwide coverage of agricultural output. It largely focuses on food safety and traceability, with some provisions for worker safety, health, and welfare, as well as environmental considerations tossed in for good measure [9]. GLOBALGAP is a pre-farm-gate or on-farm standard that certifies a product's complete agricultural production process from the moment the seed is sown until it leaves the farm. The standard also incorporates traceability and segregation control points, allowing certified products to be distinguished from others [62]. However, because it was created primarily as a business-to-business label, this norm is not explicitly communicated to consumers. GLOBALGAP is made up of many control points and compliance criteria, which are separated into three categories: (1) the entire farm support; (2) crop support; and (3) fruit and vegetable cultivation. This covers both on-farm and post-harvest and handling operations, as well as record-keeping and tracking. Its objective is to put up a comprehensive measurement and control system that will allow for efficient item back-tracing [64]. HACCP and GMPs programs give consumers trust and faith that adequate testing consistency, safety, and quality checks have been maintained throughout product manufacturing, packaging, and distribution [49]. GMPs are used by businesses to ensure that the science- and technology-based norms, regulations, and standards are met. An integrated systems approach includes quality, facilities, and equipment, as well as materials, production, packaging, labeling, and laboratory control. It preserves accurate records for planned validation and cross-contamination changes [65].

GMP specifies the procedures to be utilized, as well as the equipment or controls that will be utilized, in the creation, processing, packaging, or holding of a food or drug to ensure that it is safe, has the correct distinctiveness, limpidness, and power, and fulfills excellence and limpidness requirements [66]. Food safety and quality control are essential for ensuring that food assistance supplies are wholesome, of excellent value, and easy to get sufficient volumes, on time, and at reasonably priced rates to guarantee that all population groups have a nutritional and acceptable health status. HACCP, GMP, ISO, WHO, and *Codex Alimentarius* are major food safety and quality systems [49]. HACCP is a system for ensuring food safety that can be used in conjunction with GMP. GMP ensures that food, pharmaceuticals, and medicines are made in a way that ensures their safety, efficacy, and quality, as well as providing marketing authority with consistency [66].

All food processors, medicinal drug manufacturers, food product manufacturers, packagers, labelers, and distributors, warehouse/storage facilities keepers should strictly follow GMP and be given different regulations of food safety and standards [1]. Figure 21.2 presents the five significant (P) factors that affect the quality and safety of food products throughout production and processing while adhering to GMPs [65].

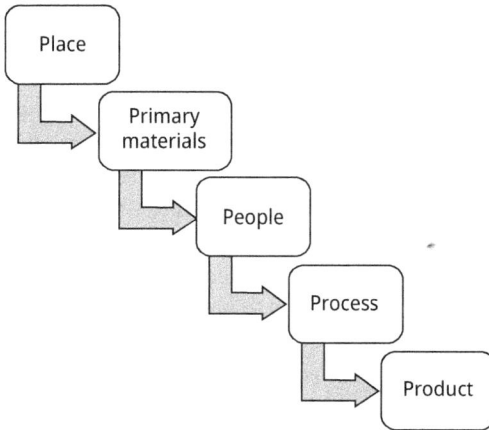

Figure 21.2: GMP control factors for determining food quality and safety throughout production and processing stages [65].

Places and premises must be clean and well organized. All surfaces and roofs, sidewalls should be well cleaned and designed to prevent any kind of contamination. People who are working in the industry must be well qualified with the help of education, training, workshops, experience, and seminars. To avoid food contamination, cleaning, disinfection, and maintenance equipment and containers should be easy to transport and handle [67]. Wearing of light-colored outerwear is appropriate for the job and protects us from food contamination, food-contact exteriors, and food containers or food packaging materials. Documented methods for proper cleaning of the premises, equipment, material handling, and employee health and are all part of Process Sanitation programs. Every product has specifications, its effectiveness and test techniques inspection and issuing Certificate for Analysis and actual test findings [68].

Equipment product design and construction should be well planned, developed, and positioned to accommodate the product's manufacturing process. Equipment surfaces that come into contact with any food in-process goods or ingredients should not react or absorb the processed materials [69]. Leaking valves, lubrication leaks, and inappropriate alterations or adaptations should not negatively impact the product. Equipment should be built in such a way that it is simple to clean. Explosion-proof equipment should be utilized with combustible substances. To reduce congestion, equipment should be strategically placed and easily identifiable. Water, steam, and pressure or vacuum pipes must be installed wherever they are required throughout all phases of operation. They must be simple to clean, handle, and identify. Space heating, aeration, air cooling, water (fit to drink, filtered, distilled), steam, air pressure, and gases (such as nitrogen) should all work as intended and be easily recognized [70].

Food industry employees were asked whether the food would also be served to their households. Is it safe to utilize on one's own family if so? Participants in GMP training learn about the processes that are unique to their facility, as well as the reasons behind them [71]. This helps to ensure the safety of the product. In terms of resources such as time away from production or other responsibilities, creation of training materials, and documentation of staff training, GMP training is a required burden for the industrial sector. GMP training should be part of every supervisor's and manager's job description because managers wear multiple hats [72].

Food producers and industrial investors who are now aware of the issue of food safety will undoubtedly play an essential role in making operational decisions [73]. Purchases of new equipment or supplies will be made with safety in mind, and hence materials that are easy to clean and disinfect will be chosen. When planning to extend an operation or packaging line, the same priorities should be applied. Choosing sterilized building materials and constructing buildings with sufficient ventilation, floor drains, walls, and ceilings that promote simple cleaning and sanitation are all part of developing new facilities with sanitation in mind [74]. To meet commodity needs, several farmers have created innovative modifications to existing equipment or constructed their own equipment. In other circumstances, such as the invention of coring machines and stainless-steel harvesting equipment that can be cleaned and disinfected in the field at the end of each harvest day, this inventiveness is employed in the field. Although existing technologies cannot entirely eliminate all potential food safety issues linked with raw food products, the focus should be on prevention [19].

Purchases of new equipment or supplies should be made with safety in mind, and hence materials that are easy to clean and disinfect must be chosen. When planning to extend an operation or packaging line, the same priorities should be applied. Choosing sterilized building materials and constructing buildings with sufficient ventilation, floor drains, walls, and ceilings that promote simple cleaning and sanitation are all part of developing new facilities with sanitation in mind [74]. To meet commodity needs, several farmers have created innovative modifications to existing equipment or constructed their own equipment. The invention of coring machines and stainless-steel harvesting equipment that can be cleaned and disinfected in the field after every harvest is recommended. Annually millions of people become sick, hospitalized, and some die from foodborne illnesses [75]; hence, improving agricultural and manufacturing practices produce safe is a key step and neglecting it is not an option to protecting public health. There are formal and informal channels of food flow which impose a check in quality and safety of foods. Proactive measures using random sampling, testing, and quarantine of products at ports entry are good practices. While capacity building initiatives can increase food safety along the food production chain without disrupting the import food supply, they cannot completely eliminate reactive actions like product recalls in response to an outbreak of a foodborne illness. In developing countries, different international and national aids which come as provision of training material, instructors, cost-free food safety training, cost-sharing agreements, loans, and grants have helped in dissemination of new food safety techniques among grassroots who are the major growers in these countries. Non-implementation of the duo (GAP and GMP) to achieve food safety will affect all the stakeholders, hence the need for collaboration to save lives.

Therefore, the role of implementation of good agricultural and manufacturing practices cannot be overemphasized, but it is a continuous clarion call with rooms for improvements and inventions of several practice without generating conflict among the growers, manufacturers, and the consumer.

References

[1] Olaniran, A. F., Abiose, S. H., & Adeniran, A. H. (2015 Nov). Biopreservative effect of ginger (Zingiber officinale) and garlic powder (Allium sativum) on tomato paste. *Journal of Food Safety*, *35*(4), 440–452.

[2] Elik, A., Yanik, D. K., Istanbullu, Y., Guzelsoy, N. A., Yavuz, A., & Gogus, F. (2019). Strategies to reduce post-harvest losses for fruits and vegetables. *Strategies*, *5*(3), 29–39.

[3] Olaniran, A. F., Afolabi, R. O., Abu, H. E., Owolabi, A., Iranloye, Y. M., & Okolie, C. E. (2020 Dec). Lime potentials as biopreservative as alternative to chemical preservatives in pineapple, orange and watermelon juice blend. *Food Research*, *4*(6), 1878–1884.

[4] Cole, M. B., Augustin, M. A., Robertson, M. J., & Manners, J. M. (2018 Aug 6). The science of food security. npj. *Science of Food*, *2*(1), 1–8.

[5] Jackson, P., & Meah, A. (2018 Jan 1). Re-assessing vulnerability to foodborne illness: Pathways and practices. *Critical Public Health*, *28*(1), 81–93.

[6] Horton, P., Banwart, S. A., Brockington, D., Brown, G. W., Bruce, R., Cameron, D., Holdsworth, M., Koh, S. L., Ton, J., & Jackson, P. (2017 Apr 1). An agenda for integrated system-wide interdisciplinary agri-food research. *Food Security*, *9*(2), 195–210.

[7] Weaver, P., Jansen, L., Van Grootveld, G., Van Spiegel, E., & Vergragt, P. (2017 Sep 8). *Sustainable technology development*. London, Routledge.

[8] FAO. (2016a). Influencing food environments for healthy diets. Food and Agriculture Organization of the United Nations Rome, 2016.

[9]   FAO. (2016b). A scheme and training manual of good agricultural practices for fruits and vegetables. Volume 2 training manual. Food and agriculture organization of the United Nations. Regional Office for Asia and the Pacific Bangkok, 2016.

[10]   Zeghibe, A. C. (2021). Growing into their Own? Plant Molecular Farming and the Pursuit of Biotechnological Sovereignty for Lower and Middle Income Countries.

[11]   Alzeer, J., Rieder, U., & Abou Hadeed, K. (2020 Aug 1). Good agricultural practices and its compatibility with Halal standards. *Trends in Food Science and Technology, 102*, 237–241.

[12]   Raspor, P. (2008 Aug 1). Total food chain safety: How good practices can contribute? *Trends in Food Science and Technology, 19*(8), 405–412.

[13]   Leong, W. H., Teh, S. Y., Hossain, M. M., Nadarajaw, T., Zabidi-Hussin, Z., Chin, S. Y., Lai, K. S., & Lim, S. H. (2020 Apr 15). Application, monitoring and adverse effects in pesticide use: The importance of reinforcement of Good Agricultural Practices (GAPs). *Journal of Environmental Management, 260*, 109987.

[14]   Rodrigues, R., Loiko, M. R., de Paula, C. M., Hessel, C. T., Jacxsens, L., Uyttendaele, M., Bender, R. J., & Tondo, E. C. (2014 Aug 1). Microbiological contamination linked to implementation of good agricultural practices in the production of organic lettuce in Southern Brazil. *Food Control, 42*, 152–164.

[15]   Johnson, L. K., Dunning, R. D., Bloom, J. D., Gunter, C. C., Boyette, M. D., & Creamer, N. G. (2018 Oct 1). Estimating on-farm food loss at the field level: A methodology and applied case study on a North Carolina farm. *Resources, Conservation and Recycling, 137*, 243–250.

[16]   Reyes, R., & Evans, E. (2008 Oct 23). Good agricultural practices (GAPs). *EDIS, 2008*(7) 1–4.

[17]   Ramos, T. M., Jay-Russell, M. T., Millner, P. D., Shade, J., Misiewicz, T., Sorge, U. S., Hutchinson, M., Lilley, J., & Pires, A. F. (2019 Sep 6). Assessment of biological soil amendments of animal origin use, research needs, and extension opportunities in organic production. *Frontiers in Sustainable Food Systems, 3*, 73.

[18]   Malo, M. (2020 Oct 29). Good agricultural practices: A working concept. *Biotica Research Today, 2*(10), 1075–1077.

[19]   Schmit, T. M., Wall, G. L., Newbold, E. J., & Bihn, E. A. (2020 Jul 2). Assessing the costs and returns of on-farm food safety improvements: A survey of good agricultural practices (GAPs) training participants. *PloS One, 15*(7), e0235507.

[20]   Yu, H., Neal, J. A., & Sirsat, S. A. (2018 Apr 1). Consumers' food safety risk perceptions and willingness to pay for fresh-cut produce with lower risk of foodborne illness. *Food Control, 86*, 83–89.

[21]   Handayati, Y., Simatupang, T. M., & Perdana, T. (2015 Dec). Agri-food supply chain coordination: The state-of-the-art and recent developments. *Logistics Research, 8*(1), 1–5.

[22]   Astill, J., Dara, R. A., Campbell, M., Farber, J. M., Fraser, E. D., Sharif, S., & Yada, R. Y. (2019 Sep 1). Transparency in food supply chains: A review of enabling technology solutions. *Trends in Food Science and Technology, 91*, 240–247.

[23]   De Baerdemaeker, J., & Saeys, W. (2016). Good agricultural practices, quality, traceability, and precision agriculture. *Precision Agriculture Technology for Crop Farming*, 1st edition, CRC Press Boca Raton, Florida, USA, 279–298.

[24]   Kouhizadeh, M., Zhu, Q., & Sarkis, J. (2020 Sep 9). Blockchain and the circular economy: Potential tensions and critical reflections from practice. *Production Planning and Control, 31*(11–12), 950–966.

[25]   Gole, S., & Dave, S. (2018). *Block-4 other food safety practices* (pp. 6–30). Indira Gandhi National Open University School of Agriculture.

[26]   Paudel, S., Sah, L. P., Devkota, M., Poudyal, V., Prasad, P. V., & Reyes, M. R. (2020 Jan). Conservation agriculture and integrated pest management practices improve yield and income while reducing labor, pests, diseases and chemical pesticide use in smallholder vegetable farms in Nepal. *Sustainability, 12*(16), 6418.

[27] Pawlak, K., & Kołodziejczak, M. (2020 Jan). The role of agriculture in ensuring food security in developing countries: Considerations in the context of the problem of sustainable food production. *Sustainability*, *12*(13), 5488.

[28] Da Cruz, A. G., Cenci, S. A., & Maia, M. C. (2006 Aug 1). Quality assurance requirements in produce processing. *Trends in Food Science and Technology*, *17*(8), 406–411.

[29] Burange, P. S., Wargantiwar, R. K., Gonde, A. D., & Rajendran, T. P. (2015). Good agricultural practices-relevance for crop health management. *Ecology, Environment and Conservation*, *21*(1),155–164.

[30] Adam-Bradford, A., El-Kahlout, G., Byrne, R., Wright, J., & Rahman, M. (2020 Jun 26). Stabilisation agriculture: Reviewing an emerging concept with case studies from Afghanistan and Iraq. *CAB Reviews: Perspectives in Agriculture, Veterinary Science, Nutrition and Natural Resources*, *15*(42), 1–19.

[31] Panahzadeh, P. M., Razzaghi, B. F., Shabanali, F. H., Motiee, N., & Hosseinpoor, A. (2015 Sep 1). Major barriers to application of good agricultural practices (GAPs) technologies in sustainability of livestock units. *International Journal of Agricultural Management and Development*, *5*(3), 169–178.

[32] de Quadros Rodrigues, R., Loiko, M. R., de Paula, C. M., Hessel, C. T., Jacxsens, L., Uyttendaele, M., Bender, R. J., & Tondo, E. C. (2014 Aug 1). Microbiological contamination linked to implementation of good agricultural practices in the production of organic lettuce in Southern Brazil. *Food Control*, *42*, 152–164.

[33] Olutegbe, N. S., & Sanni, A. O. (2021). Determinants of compliance to good agricultural practices among cocoa farmers in Ondo State, Nigeria. *Caraka Tani: Journal of Sustainable Agriculture*, *36*(1), 123–134.

[34] Riggio, G. M., Wang, Q., Kniel, K. E., & Gibson, K. E. (2019 Feb 2). Microgreens – A review of food safety considerations along the farm to fork continuum. *International Journal of Food Microbiology*, *290*, 76–85.

[35] Buscaroli, E., Braschi, I., Cirillo, C., Fargue-Lelièvre, A., Modarelli, G. C., Pennisi, G., Righini, I., Specht, K., & Orsini, F. (2021 Aug 1). Reviewing chemical and biological risks in urban agriculture: A comprehensive framework for a food safety assessment of city region food systems. *Food Control*, *126*, 108085.

[36] Marshall, K. E., Nguyen, T. A., Ablan, M., Nichols, M. C., Robyn, M. P., Sundararaman, P., Whitlock, L., Wise, M. E., & Jhung, M. A. (2020 Nov 13). Investigations of possible multistate outbreaks of Salmonella, Shiga Toxin-producing Escherichia coli, and Listeria monocytogenes infections – United States, 2016. *MMWR Surveillance Summaries*, *69*(6), 1.

[37] Astill, G., Minor, T., & Thornsbury, S. (2019 Oct 1). Changes in US produce grower food safety practices from 1999 to 2016. *Food Control*, *104*, 326–332.

[38] Julien-Javaux, F., Gérard, C., Campagnoli, M., & Zuber, S. (2019 Jun 1). Strategies for the safety management of fresh produce from farm to fork. *Current Opinion in Food Science*, *27*, 145–152.

[39] FDA, C. (2007). *Guidance for industry: Use of mechanical calibration of dissolution apparatus 1 and 2 – Current good manufacturing practice*. Rockville, MD: Office of Regulatory Affairs.

[40] Zhu, X., Huang, I. Y., & Manning, L. (2019 Feb 1). The role of media reporting in food safety governance in China: A dairy case study. *Food Control*, *96*, 165–179.

[41] Tricarico, J. M., Kebreab, E., & Wattiaux, M. A. (2020 Nov 1). Milk Symposium review: Sustainability of dairy production and consumption in low-income countries with emphasis on productivity and environmental impact. *Journal of Dairy Science*, *103*(11), 9791–9802.

[42] Garcia, S. N., Osburn, B. I., & Jay-Russell, M. T. (2020 Jan 28). One health for food safety, food security, and sustainable food production. *Frontiers in Sustainable Food Systems*, *4*, 1.

[43] Hasnan, N. N., Aziz, N. A., Zulkifli, N., & Taip, F. S. (2014 Jan 1). Food factory design: Reality and challenges faced by Malaysian SMEs. *Agriculture and Agricultural Science Procedia*, *2*, 328–336.

[44] World Health Organization. (2007). *Quality assurance of pharmaceuticals: A compendium of guidelines and related materials. Good manufacturing practices and inspection*. World Health Organization.

[45] Manning, L. (2018 Oct 22). *Food and drink-good manufacturing practice: A Guide to its responsible management (GMP7)*. Hoboken, NJ, United States: John Wiley & Sons, 1–6.

[46] Al-Busaidi, M. A., Jukes, D. J., & Bose, S. (2017 Mar 1). Hazard analysis and critical control point (HACCP) in seafood processing: An analysis of its application and use in regulation in the Sultanate of Oman. *Food Control, 73*, 900–915.

[47] Jani, U. K. (2016). Good manufacturing practices (GMP):"Planning for quality and control in microbiology". In *Frontier discoveries and innovations in interdisciplinary microbiology* (pp. 71–77). New Delhi: Springer.

[48] Godoy, R., & Bessas, C. (2020 Jul 27). *Plan, do, check, act-a practical guide for managers: Understanding results-focused management*. Escola de Gestão Aquila.

[49] De Oliveira, C. A., Da Cruz, A. G., Tavolaro, P., & Corassin, C. H. (2016 Jan 1). Food safety: Good manufacturing practices (GMP), sanitation standard operating procedures (SSOP), hazard analysis and critical control point (HACCP). In *Antimicrobial food packaging* (pp. 129–139). Academic Press.

[50] Hasnan, N. N., Aziz, N. A., Zulkifli, N., & Taip, F. S. (2014 Jan 1). Food factory design: Reality and challenges faced by Malaysian SMEs. *Agriculture and Agricultural Science Procedia, 2*, 328–336.

[51] Ahmad, G. A. (2019). Evaluation of good manufacturing practices in small dairy plants in Jordan. *Evaluation, 84*, 37–43.

[52] Abd Rahman, A., Singhry, H. B., Hanafiah, M. H., & Abdul, M. (2017 Mar 1). Influence of perceived benefits and traceability system on the readiness for Halal Assurance System implementation among food manufacturers. *Food Control, 73*, 1318–1326.

[53] Umesha, S., Manukumar, H. M., & Chandrasekhar, B. (2018 Jan 1). Sustainable agriculture and food security. In *Biotechnology for sustainable agriculture* (pp. 67–92). Woodhead Publishing.

[54] Chávez-Dulanto, P. N., Thiry, A. A., Glorio-Paulet, P., Vögler, O., & Carvalho, F. P. (2021 Feb). Increasing the impact of science and technology to provide more people with healthier and safer food. *Food and Energy Security, 10*(1), e259.

[55] Panghal, A., Chhikara, N., Sindhu, N., & Jaglan, S. (2018 Aug). Role of Food Safety Management Systems in safe food production: A review. *Journal of Food Safety, 38*(4), e12464.

[56] Chen, H., Liu, S., Chen, Y., Chen, C., Yang, H., & Chen, Y. (2020 Feb). Food safety management systems based on ISO 22000: 2018 methodology of hazard analysis compared to ISO 22000: 2005. *Accreditation and Quality Assurance, 25*(1), 23–37.

[57] Benton, H., Binder, M., & Egel-Hess, W. (2007 Jan 1). Benchmarking operations to promote learning: An internal supply chain perspective. *International Journal of Learning and Change, 2*(2), 145–169.

[58] Abbas, J. (2020 Jan 1). Impact of total quality management on corporate green performance through the mediating role of corporate social responsibility. *Journal of Cleaner Production, 242*, 118458.

[59] Chaudhari, V., Yadav, V., Verma, P., & Singh, A. (2014 Sep 1). A review on good manufacturing practice (GMP) for medicinal products. *PharmaTutor, 2*(9), 8–19.

[60] Jali, M. B., Ghani, M. A., & Nor, N. M. (2016 Aug 3). The confusion of design and facilities in good manufacturing practice requirements among industries in Malaysia. *Environment-Behaviour Proceedings Journal, 1*(3), 156–167.

[61] Goetz, G. (2011 Jun). Germany's E. coli outbreak: A global lesson. *Food Safety News, 18*, 1–2.

[62] Wongprawmas, R., Canavarpi, M., & Waisarayutt, C. (2015 Sep 7). A multi-stakeholder perspective on the adoption of good agricultural practices in the Thai fresh produce industry. *British Food Journal. 117*(9), 2234–2249.

[63] Asogwa, E. U., Dongo, L. N., Ndubuaka, T. C., & Hammed, L. A. (2021 Jun 22). The sustainability of Nigerian cocoa sector through good agricultural practices. *African Journal of General Agriculture, 5*(2) 59–69.

[64] Halabi, S. F., & Lin, C. F. (2017). Assessing the relative influence and efficacy of public and private food safety regulation regimes: Comparing codex and GlobalG. AP standards. *Food and Drug Law Journal, 72*, 262.

[65] Meghwal, M., Heddurshetti, U., & Biradar, R. (2017 Aug 22). Good manufacturing practices for food processing industries: Principles and practical applications. In *Food technology* (pp. 3–28). Apple Academic Press.

[66] Ricke, S. C. (2016 Jun 15). Enhancing food safety, product quality, and value-added in food supply chains using whole-chain traceability. *International Food and Agribusiness Management Review, 19* (1030–2016–83149), 191–213.

[67] Mohammad, Z. H., Yu, H., Neal, J. A., Gibson, K. E., & Sirsat, S. A. (2020 Jan). Food safety challenges and barriers in southern United States farmers markets. *Foods, 9*(1), 12.

[68] Marriott, N. G., Schilling, M. W., & Gravani, R. B. (2018 Mar 30). *Principles of food sanitation*. Berlin/ Heidelberg, Germany: Springer.

[69] Levy, S. M. (2018). *Project management in construction*. New York City: McGraw-Hill Education.

[70] Singapurwa, N. M., Semariyani, A. M., & Candra, I. P. (2017). Identification of implementation of GMP and SSOP on processing of Balinese traditional food sardine pedetan. *International Research Journal of Engineering, IT and Scientific Research, 3*(3), 20–30.

[71] Konecka-Matyjek, E., Turlejska, H., Pelzner, U., & Szponar, L. (2005 Jan 1). Actual situation in the area of implementing quality assurance systems GMP, GHP and HACCP in Polish food production and processing plants. *Food Control, 16*(1), 1–9.

[72] Yaskun, M., Pambudy, A. P., & Hidayat, K. (2020 Nov 1). Development strategy of WingkoBabat Industry with good manufacturing practices (GMP) Method. *Jurnal Mantik, 4*(3), 1929–1934.

[73] Jaffee, S., Henson, S., Unnevehr, L., Grace, D., & Cassou, E. (2018 Dec 11). *The safe food imperative: Accelerating progress in low-and middle-income countries*. NW, Washington, D.C.: World Bank Publications.

[74] Clark, J. P. (2008 Sep 12). *Practical design, construction and operation of food facilities*. Amsterdam, Netherlands: Elsevier.

[75] Narrod, C., Dou, X., Chfadi, T., & Miller, M. (2021 Jul 1). Participant characteristics and learning outcomes: Lessons from international food safety capacity building. *Food Policy, 102*, 102105.

Christiana Eleojo Aruwa*, Frank Abimbola Ogundolie
and Olabisi Omowumi Adebisi

Chapter 22
Food fraud: causes and control

Abstract: Globally and annually food fraud (FF) causes major losses in the food indus-try. Food fraud can be perpetrated at any stage along the food chain including pro-duction, processing, transportation, storage, packaging, product advertisement and delivery, and/or distribution points. Food fraud is one of the four elements of food protection, with others being food crime, food safety and food defence, all of which may or may not be interconnected. The differentiation between these terms has been proposed to be dependent on the motivation, intent and aim, and have been discussed in this chapter using varying models and theories. Food fraud types vary and occur at varied points with certain foods being more susceptible to fraud or adulteration with adverse consequences for public health, the consumer, food business owner, other in-dustry stakeholder and governments. Asides the points highlighted above, this chap-ter further covers an in-depth overview of factors encouraging FF, FF control and preventive strategies, FF study and monitoring gaps, as well as means to stem the FF tide. Overall, mitigating FF would require the evolution of detection methods, as well as the vigilance and cooperation of all food industry actors.

22.1 Background

The annual global costs of food fraud to the food industry has been estimated to be-tween $30 and $40 billion [1]. Food fraud (FF) is said to have occurred when the inten-tion, presence of a law infringement, economic advantage, and consumer deception have been identified. According to PAS 96 [2], a threat arises from the malicious intent to cause harm or loss, while a hazard is that which arises from ignorance, negligence

Acknowledgment: Authors acknowledge the support of the South African National Research Founda-tion's (NRF) Innovation Postdoctoral Grant awarded to Dr. C. E. Aruwa, and tenable at the Department of Biotechnology and Food Science, Durban University of Technology (DUT), South Africa.

*Corresponding author: Christiana Eleojo Aruwa,** Faculty of Applied Sciences, Department of Bio-technology and Food Science, Durban University of Technology (DUT), Durban 4000, South Africa, e-mail: ChristianaA@dut.ac.za
Frank Abimbola Ogundolie, Faculty of Computing and Applied Sciences, Department of Biotechnology, Baze University, Abuja, Nigeria
Olabisi Omowumi Adebisi, Molecular Biology & Interdisciplinary Life Sciences (MBIL), New Mexico University, Las Cruces, PO Box 30001, New Mexico, United States.

https://doi.org/10.1515/9783110748345-022

or incompetence, and can be accidental or natural. The four elements of food protection – FF, food crime, food safety, and food defense – could be intertwined based on the determination of unintentionality/intentionality of incidents or the ability to cause harm [3]. Nonetheless, the uniqueness of each element may be maintained while accepting their interconnectivity depending on the assessment of specific FF scenarios [4]. Opportunities to commit FF is enhanced by flux in food and food products availability, supply, and demand. These changes result in monetary gains to fraudsters, opportunists, and perpetrators [5]. A crime implies illegality or committing an offence subject to existing laws [4, 6]. Food crime aims at knowingly deceiving or causing injury to the target consumer [7]. Hence, food crime and FF for economic gains could lead to both food quality and food safety issues [5]. FF can arise from misrepresentation, positional abuse or prejudice of consumer rights for illicit and unlawful gains. FF becomes food crime when acts are no longer random and become more intentional and organized to injure or deceive consumers [7]. Thus, mitigating FF impact calls for measures that involve disciplines like food science and technology, criminology, behavioral and social sciences [8], which are useful in propounding theories and approaches surrounding motivations for FF [5, 9–10].

FF aims at illegally and purposefully tampering with foods for selfish gains [11]. It involves the purposeful substitution, misrepresentation, alteration or mislabeling of a food product at any point within a food supply chain [12]. In addition, the three elements of FF, that is, pressure or motivation (e.g., presence of debt, unrealistic production targets), opportunity (e.g., loopholes in quality control checks) and rationalization (gains outweigh the probability of getting caught in the fraudulent act) of the individual or group engaging in fraud must be identified for FF to take place. Nevertheless, even when some food products like meat, milk and seafoods are more prone to FF, about 10% of commercially available foods are still affected by FF. FF may also involve contamination with foreign substances or inclusion of unlisted/blacklisted additives in food products [11]. The inclusion of hazardous food fillers to increase product weight also constitutes a major public health concern. When sizeable amounts of food are produced, imported or exported there is an increased probability of FF incidence. Production of high value products also induces FF and adulterations [11]. Moyer et al. [13] reported that FF in whatever form constitute profitable crimes. The United States Pharmacopeia [14, 15] further stated that the most prone geographic regions to FF are usually developing regions characterized by a growing population, and social and political instability. In Brazil, counterfeiting, artificial enhancement, substitution and dilution events constituted 7%, 14%, 22%, and 50%, respectively, of FF types recorded [16]. In Africa, in 2018 adulterated/tainted biscuits were reported in Abuja, Nigeria, while cooking oil repackaged from recycled oil were found in Kenya [17]. In order to stem the rising FF tide, scientists and food industry stakeholders continue to collaborate and evolve new measures based on varying indices. This chapter therefore discussed the causes/motivators, prevention, and control methods of FF. In addition, it expands on types of FF, foods most prone to FF, FF scientific theories, detection methods, and impacts of FF.

22.2 Food fraud types and points of occurrence

The various types of FF and the points at which they happen are itemized below:

1. **Tampering fraud**: This is when a food product or packaging is purposely modified. Such modifications cause adverse effects to consumers. For example, in 2018, strawberries in Australia were sabotaged through the insertion of sewing needles in the fruits. This spurred similar occurrence within the fruit industry with deleterious effects to strawberry farmers and producers in Australia. The intentional practice of extending products best before/expiry dates beyond acceptable industrial standards for meat and dairy products also constitute food tampering [11].

2. **The unauthorized production of food (overrun)**: In this scenario, excessive number of legitimate products are produced beyond legal limits stated in a production agreement. Excess produce then flood the supply chain but are marketed with alternative labels that deceive consumers. The products are sold outside the regulated supply chain and therefore constitute a threat to public health [11].

3. **Food/product adulteration**: This is the addition of foreign components like sugars, melamine, formalin, among others, into food products to increase quantity and profit. This often leads to use of false labels. Oregano and spices are mostly prone to adulteration with olive and sumac leaves which were reported in Australia in 2016. These adulterants constituted 50–90% of adulterated samples [11].

4. **Simulation**: This involves making a fake product look like the legitimate product. It is similar to making "knock offs," imitations, or copycats. Honey product copycats to which sugar syrup had been added have been reported in Australia in 2018. The fraudulent manufactures came under serious scrutiny by regulatory agencies following detection of the fake products which were sold at exorbitant prices [11].

5. **Food theft**: Here, a known, legal and original product is stolen and criminally sold at unregulated market sections. This can be observed when products are smuggled illegally and redistributed to the target market. Since 2008, the International Union of Food Science and Technology (IUFoST) reported that food and drink are the most prone to theft [11].

6. **Diversion fraud**: This is when legal and original products fail to reach the end user or target market. Diversion occurs often with relief foods which are re-distributed to areas that do not really require them. The manipulation of food aid lists and illicit removal of food items in 2019 by local officials in charge of distributing United Nations (UN) relief materials to needy populations in Yemen and other countries constitute an example under food diversion fraud [11, 18].

7. **Counterfeiting**: This is when substandard goods are produced outside regulatory boundaries and passed off as established and trusted brands. The most common and

prone counterfeited products include spices, honey and olive oil. Counterfeiting occurs in both developed and developing regions of the world [11].

In addition to the above, any of the FF types may happen at the production, processing, transportation, storage, packaging, product advertisement and delivery, and/or distribution points, as depicted in Figure 22.1. The potential FF incidence points are not exhaustive and may grow as the global FF trend develops. This could also remain the case given the complex nature of the food industry and its array of stakeholders/ actors. Likewise, the food production, supply and distribution chains are complex. FF vulnerability occurs when presence of an opportunity meets motivation, and the likelihood of an incident increases when control measures are non-existent. FF may therefore most likely occur at the beginning of a food supply chain since more opportunities exist to manipulate a product [18].

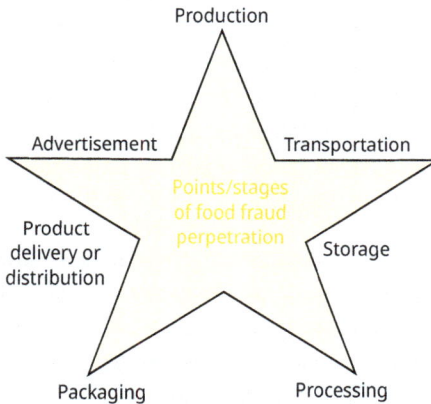

Figure 22.1: A simplistic view of points or stages of food fraud occurrence.

22.3 Foods prone to fraud

1. **Seafood**: The concept of "mock meat" or "imitation seafoods" is a fast-growing trend in seafood restaurants. Adulterated seafoods like shrimps and crabs are unknowingly supplied to restaurant kitchens, and in certain cases the kitchens are aware that seafoods supplied have been tampered with. In America, of hundreds of shrimp products tested, about 30% were misrepresented, for example, farmed shrimps sold as wild-caught, or many shrimp varieties used in one product. Restaurants pur-

chase these foods because they are cheaper. Consumers are also often unaware of their purchases or provided little information on where and how the foods were farmed or caught, thus making it difficult to make informed choices [12].

2. **Milk and milk products**: Milk and milk products FF increased due to increased demand for milk products. Increased demands exposes a product to adulteration for illicit financial profit. The most common fraud in milk products is the addition of water to increase volume. Also, starch, salts and flour are added to milk, leading to loss of nutritional value. Adulterated milk added to infant formula foods constitute a public health concern and cause malnutrition, weight loss and stunted growth in infants [12]. Milk powder is also filled with salt, detergent, sugar, melamine, urea and caustic soda. In 2009, about 60 Chinese infants were killed after consumption of adulterated milk powder [19]. Milk powder may also be diluted with non-milk fats like palm and soybean oil, and animal fat such as pork lard and cow tallow [20]. Some ultra-high temperature (UHT) milk products have been artificial enhanced with hydrogen peroxide, urine, chlorine, and formaldehyde addition [21], as well as the inclusion of sodium hydroxide and starch [22, 23]. The dilution of bulk goat milk with cow milk has also been reported [24].

3. **Fat and oils**: Truffle oils and many olive oil product varieties have been subject to fraud. The popular extra-virgin olive oil is the concentrated derivative of a first-pressed olive harvest with a luminescent green color and no additives. Unfortunately, the versions sold are highly substandard. Some adulterated versions are a mixture of seed oil and additives like chlorophyll and beta-carotene to mimic the smell and look of pure extra virgin olive oil. Again, within the gourmet industry, truffle oil is highly priced, but low-quality and scam truffle oils now fill most marketplaces. Additions to fraudulently packaged truffle oils include synthetic ingredients, artificial flavorings and chemical additives. Sunflower seed oil is also used to dilute olive oil. The "extra virgin" label is also too often misused [19]. The addition of soybean oil to coconut, linseed, safflower and evening primrose oils has also been practiced in countries like Brazil [25]. Intellectual property right (IPR) infringements, verification documents forgery, manipulation or absence of data, invoices, and other documents are also practiced by FF actors [26].

4. **Meat and fish**: Fish and meats are prone to mislabeling, substitution or counterfeit frauds. For example, not all "tuna" is actual tuna fish. Cow beef is also often misrepresented by other meats such as horse meat, especially in the burger industry [19]. Pricy fishes like the cod, pink cusk-eel and flounder have been replaced with inexpensive basa and Alaska pollock fishes [27]. Croaker (Sciaenidae) fillets brands have been substituted with a variety of fish species in the families Serranidae, Lutjanidae, and Sciaenidae, and other taxa which were not identified [28]. Meat products diluted with chicken DNA have been reported in pork sausages sold in Brazil [29]. Beef hamburgers are also adulterated with soy protein [30]. Other meat product frauds include

manipulation, falsification, or absence of invoices and other verification documents and IPR breaches [26, 31].

5. **Syrups and honey**: Fraudulent maple syrups have been found in Quebec and the United States. In Quebec, an $18 million loss was recorded following theft of over 3,000 maple syrup barrels. These were restocked and repackaged with water to yield outrageous profit margins. Within the United States, maple syrup may have a close semblance with adulterated honey such that the most reliable means of purchase would be directly from trusted producers or subsistent farms to ensure quality of the syrup is maintained. Surprisingly, not all honey is bee derived. Food coloring and honeysuckle fragrance added to sugar are mislabeled and advertised as honey by dubious producers. The probability of purchasing impure honey can be reduced if sourced from local trusted beekeepers. Many honey producers with honey products in the global marketplace that are exported from China usually dose producing bees with antibiotics [19]. Corn syrup may also be added to honey to sweeten and thicken it.

6. **Coffee and tea**: Many persons have fallen victim to diluted or mislabeled coffee and tea products. Coffee labels mislead buyers and are often not truthful about the source of the coffee. Coffee and coffee products may likewise be filled with non-coffee ingredients (fillers) like grains, starch, figs, among others. Instant coffee is also often adulterated and has resulted in coffee lovers purchasing whole coffee beans for subsequent processing for their home use. This helps with reducing the risk of consuming unwanted coffee fillers in marketed "impure" coffee products [12]. The presence of sucrose in coffee and tea products has been recorded in Saint Catarina state of Brazil [32]. Soil and sand have also been detected in tea [33]. The presence of coffee husk, barley, corn, and rice was detected in coffee using hydrogen nuclear magnetic resonance (H-NMR) [34].

7. **Juices and extracts**: Ascorbic acid, corn sugar and potassium sulfate are often added to orange and other juices to improve and protect their color and flavor. In other instances, lemon juice and high fructose corn syrup are added to orange juice. Some juices are actually stale juices kept in vats for months, which then have the taste improved prior to sale with flavorings and fragrances. Fruit extracts and chemical-based concentrates have also been used in juices which are mislabeled as "fresh" or "whole" squeezed to entice customers to purchase them. Given these practices, food experts have advised the consumption of only fresh, whole fruits. Vanilla extracts are commonly used in the pastry industry to subtly enhance smell, flavor and sweetness in pastry products. Unfortunately, vanilla extracts in the United States, Mexican restaurants and malls may contain coumarin from other sources besides vanilla beans and pose a health risk since it is a known blood thinning chemical. Coumarin is also toxic to the liver. Sugar alcohols and synthetic vanillin could also be added to vanilla extracts [35].

8. **Saffron and other spices**: Saffron is well known for its earthy flavor and ability to add rich color to dishes. Although considered an exquisite spice in some regions, chemical analysis and identification of some saffron products revealed mislabeling fraud. The actual contents were mismatched with label and origin claims [12]. Cheaper saffron products have been offered to consumers, but are made from sandalwood, poppy petals, turmeric, and gypsum. Powdered saffron is also usually contaminated with paprika. In order to increase weight of saffron fibers, vegetable oil may be added. However, as a test, fake saffron has been known to turn red when dipped in water [35]. Other spices have been found to be contaminated following their dilution with lead [19]. The presence of unsavory levels of toxic lead chromate was reported in cumin, curry powder, and turmeric and cinnamon spices [36].

9. **Wine, beer, and other alcoholic beverages**: Fraud in the form of mislabeling, adulteration, substitution and misrepresentation thrives in the wine and beverage industry. Common diluents and additives used in wines include flavoring, colorants and elderberry juice. Fake labels which mimic expensive wine brands are common in cheaper adulterated versions. Counterfeit beer products with fake labels and caps from cheaper brands switched with caps and labels of highly priced leading brands are also common [37]. IPR violations, documents falsification, subjection of products to unapproved processing, content replacement, dilution, removal, or addition are also practiced in the beverage producing sector [26].

10. **Other foods**: Grains, cereals, probiotic products and formulations, and organic foods are some food products that have also been exposed to FF [38]. Dietetics, fortified foods and food supplements are also subject to unapproved processes, mislabeling, documents falsification, IPR theft and breach, content dilution, removal, addition, and replacement [26].

22.4 Food fraud as an overlap of food protection elements

There are four food protection elements which comprise food quality/safety, FF, food crime, and food defense. Food safety points to the unintended introduction of contaminants that can make a food or food product hazardous to health. On the other hand, food quality targets the delivery of food product characteristics which inform the value placed by consumers on products [39]. Food defense is intentional food adulteration that is ideologically motivated. The motivator for FF is economic gains (Table 22.1) which may or may not lead to public health concerns arising from consumption of injurious foods. Hence, there is an overlap of FF with some food defense issues, as well as food quality and food safety. Spink and Moyer [39] identified these elements as distinct with no interlink with FF, but the Global Food Safety Initiative (GFSI) [4] identified all elements as overlapping. Nevertheless, given the motivators, intent, and aim linked with each element as depicted in Table 22.1, one may submit that they are both distinct and overlapping. Hence, it is important to note that in considering the creation of organizational or regional food control and protection frameworks, all elements need to be adequately covered for wholesome coverage.

Table 22.1: Food protection elements, their motivators, intent, and aim.

Food protection element	Motivators	Intent	Aim
Food fraud	Economic gain-driven	Intentional	Profit, consumer deception
Food defense	Ideologically driven		
Food crime	Economic gain	Intentional	Consumer deception and harm
Food safety and food quality	Accidental	Unintentional	–

22.5 Food fraud causes and motivation theories

22.5.1 Food fraud causes/motivators

The central characteristic of FF is the intention behind it, the economic gain and consumer or target market deception (Table 22.1) [40]. The perpetrators of FF have been classified under headings which include disgruntled/irrational individuals, opportunists, extortionists, and extremists. Business operators can also perpetrate fraud to gain the upper hand against competitors and/or consumers. At the center of FF is the derivation of profit in form of monetary benefits which is considered the sole motivator. To proffer solutions, there needs to be a full grasp of this motivator as a cause and driver for FF. Depending on the adulterant used and how the fraudster operates, either unnoticed as an individual within or outside an organization or in collaboration with an organization, the severity and likelihood (Figure 22.2) to cause harm which their deliberate adulterations cause would increase.

Two models are known which explain and improve comprehension on FF motivator and impact [40]. In both high- and low-profit models, there is a low- and high-likelihood scenario with the detection of FF. However, the risk to consumers range from medium to high in the high profit model compared to the low profit model where risk remains low (Figure 22.2). The high-profit model can be majorly observed in highly organized FF incidents involving larger group of persons and corporations within the food industry. The low-profit model is common among individuals or smaller group of fraudsters. On another hand, the digital sale of foods which constitute e-commerce has also not helped to reduce fraud incidents but created more avenues for fraudulent practices while having a wider reach. Since there is a complexity of the global food supply chain and economic need to produce inexpensive foods for

FOOD FRAUD

Motivator

ECONOMIC GAINS/PROFIT
Impact

Profit impact on food
fraud detection and risk

| low likelihood of detection with high risk | **HIGH PROFIT MODEL** | high likelihood of detection with medium risk |
| high likelihood of detection with low risk | **LOW PROFIT MODEL** | low likelihood of detection with low risk |

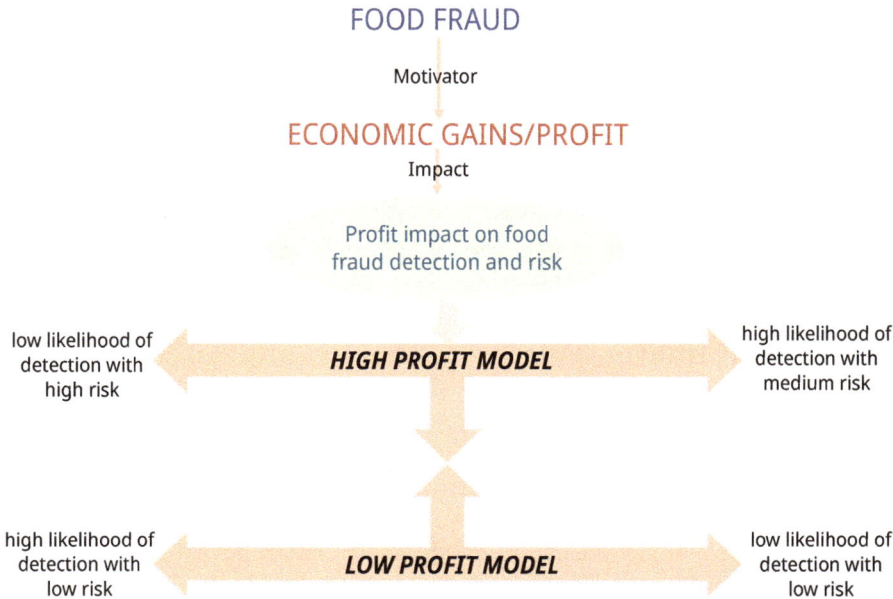

Figure 22.2: An overview of food fraud main motivator and profit models.

teeming populations, FF possibilities have increased [40]. Other FF motivators and causes are the increasing demand for cheap goods, insatiable market demands, and challenges with food proofing and detection of fraud [41]. These causes have been further expanded to include lockdowns in the event of a disease outbreak, raw materials shortage, purchasing power loss and globalization [42].

22.5.2 Food fraud motivation theories

FF motivation theories aid in the assessment of FF cases and in determining the probability of an FF occurring (FF vulnerability). Three major elements within FF vulnerability are motivations, opportunities, and control measures. Hence, FF theories also aid in the comprehension of these underlying elements [18]. Some of these theories are herein discussed.

1. In the *rational choice theory*, a choice is only made after the fraudster matches benefits to the cost of perpetrating and committing the act. Economic gains are considered, and a decision is then made based on access to important information, ability and time [43].

2. In the *routine activity theory*, a decision to commit an unlawful act is made based on location, time and targeted victims. This approach is three-pronged such that there is a motivated perpetrator, a potential victim, and the presence/absence of a guardian. The offender is also aware of the target's behavior or routine [44].

3. The *social control theory* takes advantage of four social control and fear levels. First, the attachment level where strong social expectations equate with stronger attachment and increased likelihood of consumer or victim conformity. Second is the commitment level, where increased commitment equates with reduced likelihood of deviation from the product one is committed to. The third stage is involvement; that is, the more time an individual spends being engaged in an activity, the less time they devote to other things. The fourth bond is the belief level which states that if an individual is trained or groomed in a certain right and lawful way, they are less likely to deviate into unlawful acts [45].
4. The *common-sense motivation approach* simply puts forward FF as being driven by greed or financial needs [46].
5. The *relative deprivation approach* takes advantage of persons who feel they have been deprived. It is especially linked to material scenarios that are economically driven [46].
6. The *game-theoretic approach* reconstructs the monetary incentives of profit-oriented actors. The ability of perpetrators to commit unlawful acts increase with increased probability of making profit and reduces with the anticipation of loss from detection of their fraudulent activities [47].

22.6 Food fraud control and prevention

Some measures which could help in the control and prevention of FF are discussed in this section.

22.6.1 Development of strong risk management and control strategies

This includes food safety programs like the Hazard Analysis and Critical Control Point (HACCP) techniques and the Six Sigma (standard deviation) method which mathematically means a "six-standard-deviation event from the mean for an error to happen," and is dependent on consistent practice and enforcement of Sanitation, Standard Operating Procedures (SOP), Good Manufacturing Practices (GMP) and Good Agricultural Practices (GAP) throughout the supply and production chains [48].

22.6.2 Setting certification standards for industries

Initiatives like the Global Food Safety Initiative (GFSI) set certification standards for stakeholders within the food safety industry [48]. Data from incident outcomes could also inform benchmark setting when inputted into databases such as the USP [49]. Local, regional, geographical, national, and international and/or border certifications following food and food products batch sampling and testing should be the norm for passage of foods within and across regions and areas. The Association of Official Agricultural Chemists International (AOAC) have also floated the Food Authenticity and Fraud (FAF) program for developing and identifying standard reference methods for targeted and non-targeted approaches to food authenticity testing. This ensures food materials compliance and quality control (QC). Where the identified contaminant is known the targeted approach is applicable, but when the adulterant is novel and not previously identified the non-targeted approach applies. Some effective food compliance requirements and certifications are GFSI certification, Food Safety Modernization Act (FSMA), Food Drug and Cosmetic Act (FDCA), and Codex Alimentarius (CODEX), among others [8, 48, 50].

22.6.3 Review of vulnerability, occurrence predictive factors and monitoring systems

Within the European Media Monitor (EMM) infrastructure system is the MedISys database which birthed MedISys-FF tool developed by Bouzembrak et al. [51]. MedISys-FF is capable of text mining published media reports on FF, globally, and is updated around the clock. Compared to known FF reporting platforms such as Economically Motivated Adulteration (EMA), Rapid Alert for Food and Feed (RASFF) and Horizon-Scan databases, and after text mining for a period of sixteen months, MedISys-FF showed improved capacity to collect and report FF incidences of greater relevance. The platform also stood out in terms of report frequency. Such systems enhance traceability of FF along the food supply chain, globally.

Traceability systems achieve target market and end user protection, removal of non-consumable foods, accuracy in recalls targeting. It also enhances investigations into FF incidents [52]. The utilization of Internet of Things (IoT) and blockchains to implement traceability programs has offered an information database for all participants within a supply chain. A blockchain is a chain of traceable transactions that serves as a digital register of records of transactions between parties and is advantageous when protected and is not easily manipulated. It ensures neutrality, transparency, security and reliability [53]. Traceability systems can also be enhanced with the incorporation of Threat Assessment Critical Control Point (TACCP) which serves to prevent malicious threats to food, and Vulnerability Assessment Critical Control Point (VACCP) for preventing FF [53].

The continued assessment of factors which make a food supply system vulnerable to fraud could also be targeted to enhance monitoring and reduce FF incidents. Some of these food system vulnerability assessors include possible threat size to public health, tourism, animal welfare, food trade and export, food supply chain, animal health, the effectiveness of the region's food safety surveillance mechanisms and infrastructure, and effective emergency response capabilities. Other assessors are the motivation for perpetrators of food terrorism, availability of potential food contaminants, potential for transmission of contaminants between humans, and an agent's potential to contaminate foods produced in large quantities and gain wide coverage and distribution [5]. Certain databases like Eurofins and Decernis have combined resource and expertise to assist with risk-based decisions on FF control and vulnerability. The collaborative databases also give annual access to thousands of updated potential adulterants and the related report linked to FF. The platform is user-friendly, customizable, and automated for incidence analytics and alerts. Records are kept and displayed as surveillance records, analytical methods, and incident and inference reports [54]. Some FF predictive factors include raw material or food origin assessment [55, 56], food product category, and food price fluctuations [57].

22.6.4 Creation, review, and update of FF classification databases

Several databases exist that classify FF and adulteration types, including The European Union Rapid Alert System for Food and Feed (RASFF), the USP Food Fraud Database, and the Economically Motivated Adulteration (EMA) Incidents databases also exist for FF classification [13, 58]. Common FF incidents grouping such as dilution, theft and resale, mislabeling, artificial enhancement, substitution, origin masking or transhipment, counterfeiting and intentional distribution of contaminated products are done in the EMA incidents database [16]. The USP collates incidents data for open-source platforms and gives an appropriate technique for the authentication of food ingredients and products. It therefore covers a range of adulterants identified over time using different detection methods [14]. These databases contribute to the enhancement of FF monitoring systems, as well as the spatial, time-dependent evaluation of food adulteration and fraud.

22.6.5 Partnerships

Public–private and industries-driven partnerships could contribute immensely to stem the tide of FF globally. Given the increased sensitization on the economic impact of FF on producers and consumers, many industrial stakeholders have become more protective of their brands. They form associations and groups which prioritize on-the-ground monitoring and safe delivery of their original brands to the target market. These groups and partnerships further improve communication. Association with legal and jurisdictional support can also recommend standard assay techniques and carry out periodic analysis of food products distributed within the marketplace. Such industry-led partnerships help to regain consumer trust and confidence [59]. Nestle's

FF prevention document described measures and SOP for preventing FF for the timely detection of potential emerging threats/adulterants [60]. The Brazilian Coffee Industry Association (ABIC) creates and implements programs which target coffee sustainability, quality and purity maintenance, and the integration of industry, retail, distribution and purchase points. The ABIC also has a certified purity seal for ground roasted coffee to curb products adulteration [61]. Food production companies could also improve relations with other companies and product distributors, carry out periodic auditing measures that cover fraudulent activities, espionage and is anti-fraud, and encourage whistle blowing among customers. Companies could also enforce the need for supplier auditing and certifications for packages and raw materials received. The sharing of information between industry stakeholders is also encouraged [41].

22.6.6 Increase consumer awareness and public perception

There still exist consumers who are unaware of FF. A widespread dearth of concern with the purchase of beverages and foods has also been reported. Many individuals do not know what to do when FF is noticed and instead believe that regulatory bodies within the food industry are responsible for their protection [62]. Some consumers accede to being aware of FF and what it may entail [63]. How the public perceives potential risks of FF varies. In order to improve consumer awareness and perceptions of FF and its impact, the utilization of trusted information outlets cannot be overstated [64]. Alongside consumer and stakeholder education, the regular dissemination and publication of FF contact lists, as practiced in some European Union (EU) and non-EU countries, would be beneficial in reducing FF.

22.6.7 Protection of the food supply chain integrity

This approach involves the core elements of a food chain integrity which are the product, process, persons and data integrity. Data integrity protection (DIP) involves all forms of data entered into documents that could be utilized and required at all stages in a food supply system. Examples covered under DIP include forged, absent, or missing certifications, unlawful imports, missing, fraudulent, expired or improper import declarations and entry documents, mislabeling, among others. Checking the activities of FF perpetrators falls under persons/people integrity protection. Process integrity covers products' diversions, theft, overrun, illegal imports, etc. Product integrity is covered under examples such as tampering, simulation, counterfeiting, and expiration date. If food chain integrity can be assured at national and international levels, it would aid in the management of FF risks [65]. Studies which update, educate, and inform on food safety and potential public health risks also add to knowledge on all four food chain integrity strategies [54, 66].

22.6.8 Other control and prevention strategies

The re-evaluation of testing procedures and possible replacement with new and more sensitive technologies can aid the early detection and reduction of FF cases. Examples of innovative assay techniques can be observed in the European Commission's Food Integrity Project and USP. The provision and introduction of new laws and extending enforcement activities within and across borders is another strategy worth sustaining in the fight against FF. The Chinese Food Safety Laws and US Food Safety Modernization Act (FSMA) are among the new regulations targeted toward checking FF and food safety [14, 48]. The extent to which FF is controlled or prevented depends on the proper implementation of adequate techniques and measures. The measures employed would vary with the type of fraudulent act engaged in and relies on organizational commitment. Preventive approaches must implement risk evaluation and management programs to aid the classification of FF vulnerabilities and threats. The utilization of probability and impact tables help to better classify risks and establish appropriate countermeasures [67].

22.7 Food fraud detection methods

The prevention of FF requires the enforcement of regulatory systems. This must also be supported by food handlers and manufacturers training, increase in monitoring and sampling frequency. Detection methods that are inexpensive and rapid, but do not undermine precision and accuracy, should be developed. Efficient techniques could achieve early detection and prevention of FF especially in producing regions [16]. A robust global system for authenticating food products cannot be overemphasized in protecting consumers and the food industry at all stages of production and supply [68]. The efficacy of these systems can be enhanced by the regular enforcement of authentication standards throughout the supply and production chain [69].

Adulterants and levels of purity of foods have been evaluated using identity, authenticity and ingredient purity testing. Specific methods like the isotopic ratio mass spectrometry have been used to test meat products for soy protein inclusions [30], as well as the presence of extraneous sucrose in juices [70]. In the detection of nitrogen-rich contaminants like cyanuric acid and melamine in food product matrices, liquid chromatography tandem mass spectrometry has been used [71]. Chemo-metrics in combination with near-infrared spectroscopy have also proved to be powerful techniques for detecting fraud in milk [72, 73]. NMR [74], molecular identification (polymerase chain reaction – PCR), and chromatographic (gas and liquid) methods are also commonly used for food adulteration detection. However, a combination of techniques may be necessary to assay for food contaminants and adulterants [75, 76]. These methods also aid with the validation and verification of preventive measures introduced in food manufacturing process. The coupling of chromatographic methods with MS continues to be developed as a reproducible and accurate technique for food adulterants detection. Furthermore, methods like DNA barcoding [77, 78], SDS-PAGE [79], two-dimensional electrophoresis (2-DE) [80], isoelectric focusing (IEF) [81], and enzyme-linked immunosorbent assay (ELISA) [82, 83] have been useful for species identification (for example in fish and seafood product) and mislabeling detection [84]. Point of purchase detection using advanced technological tools like food scanners could also be exploited to detect foreign nutrients/components, microorganisms, mislabeling, and other FF types [85]. Methods must also be regularly evaluated and evolved for detection of specific contaminants [76].

22.8 Impacts of food fraud on public health and economy

The direct and global impact of FF on the economy is loss of confidence by customers, authorities, and investors (Figure 22.3). FF directly threatens economic sustainability

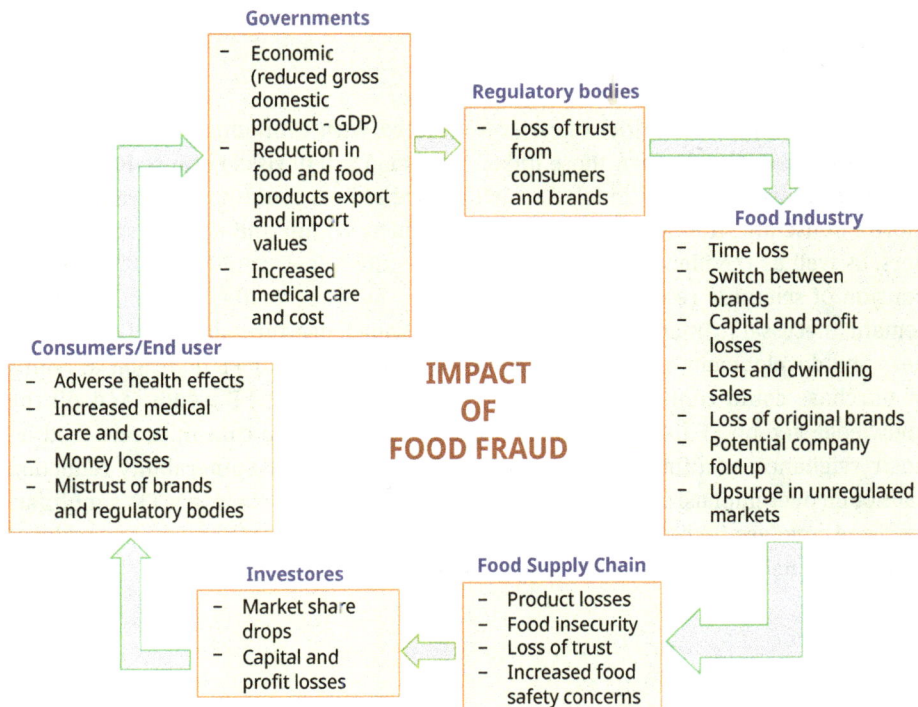

Figure 22.3: Food fraud impact on various stakeholders.

and health [16]. Monetary and nutritional losses also accrue to consumers of foods subjected to fraud. More important are the health risks posed from the consumption of additives which make food products impure. Consumer and investor trust in markets, food regulators, and food protection agencies are significantly undermined by FF. Financial losses to consumers lead to an instant switch to other products and brands, and the consumer may no longer return to purchase the original food(s) [69].

Potential health risks of FF to consumers have been reported. The unauthorized use of di(2-ethylhexyl) phthalate (DEHP) plasticizer in food clouding agents was linked to reproductive health disorders and cancer [86]. Natural sweeteners and honey products containing heavy metal contaminants and pesticides also induce cancerous growths [87]. Another health threat is the presence of melamine in dairy products [88]. The use of nut and legume oils in the adulteration of olive oil has also been shown to cause allergies [89]. Some seafood substitutes are also allergenic and lead to fish poisoning [90]. The inability to guarantee that assurance specifications for food safety are kept also remains a present potential threat [91, 92]. Health impacts spiral into huge sums of money to cover medical care and costs (Figure 22.3). Other impacts include drops in market share, loss of time in building brands [59], loss in industry profit, loss of sales to companies, and lost wages [39, 93].

22.9 Concluding remarks

In conclusion, public health concerns associated with economically motivated FF come with greater risks than those linked with food safety risks because FF contaminants are more unconventional. As such, future efforts could be tailored toward a more intense and frequent testing of food products within and across regional borders, as well as the identification of other unidentified food adulterants [39]. The expansion of scientific research into methods used by fraud mafias to avoid detection remains necessary. In addition, increasing our knowledge through use of food scanners and development of new DNA sequencing techniques for FF detection at points of purchase could contribute immensely to reduce current FF trends, and fill the knowledge gap on FF [94]. Yet, while encouraging increase in consumer and food industry vigilance, as a first step, continuous efforts toward FF vulnerability reduction cannot be overemphasized in the food supply chain [95]. It is clear that FF is multifaceted and involves both applied and social science disciplines. Thus, a cross-sectional approach is necessary to stem the global tide of FF.

References

[1] Arnold, E. (2020). Preventing food fraud and adulteration. Food Safety Magazine. BNP Media. United States.

[2] PAS 96. (2014). *Guide to protecting and defending food and drink from deliberate attack*. London: BSI.

[3] FSIS Food Safety and Food Defense – Information for In-Commerce Firms, USDA, 2014. (Accessed October 15, 2021, at http://www.fsis.usda.gov/wps/wcm/connect/86cc4d8e-c963-4be2-936a-745e2da9063a/food-defense-incommerce-firms.pdf?MOD=AJPERES.)

[4] GFSI. (October 2013). The Global Food Safety Initiative GFSI Guidance Document Version 6.3.

[5] Manning, L., & Soon, J. M. (2016). Food safety, food fraud, and food defense: A fast-evolving literature. *Journal of Food Science, 81*, R823–34.

[6] Rapalje, S., & Lawrence, R. L. (1997). *A dictionary of American and English law*. The Lawbook Exchange Ltd. vol. 2, New Jersey, USA.

[7] Elliott Review. (July 2014). *Elliott review into the integrity and assurance of food supply networks – Final report a national food crime prevention framework*. London: HM Government.

[8] Spink, J., Bedard, B., Keogh, J., Moyer, D. C., Scimeca, J., & Vasan, A. (2019). International survey of food fraud and related terminology: Preliminary results and discussion. *Journal of Food Science, 84*, 2705–2718.

[9] Manning, L. (2016). Food fraud: Policy and food chain. *Current Opinion in Food Science, 10*, 16–21.

[10] Manning, L., Baines, R. N., & Chadd, S. A. (2005). Deliberate contamination of the food supply chain. *British Food Journal, 107*, 225–245.

[11] Anthill Magazine. (2020). Expert identifies 7 acts of food fraud that every Australian food brand should watch out for. (Accessed May 15, 2021, at https://www.anthillonline.com/expert-identifies-7-acts-of-food-fraud-that-every-australian-food-brand-should-watch-out-for/.)

[12] Balance team. (2021). 10 common food frauds to avoid. (Accessed May 11, 2021, at https://blog.mybalancemeals.com/health/10-common-food-frauds-to-avoid/.)

[13] Moyer, D. C., DeVries, J. W., & Spink, J. (2017). The economics of a food fraud incident – Case studies and examples including melamine in wheat gluten. *Food Control, 71*, 358–364.

[14] United States Pharmacopeia. (2016a). *Food fraud mitigation guidance. Food chemicals codex* (10th Edition). Appendix XVII. (Accessed June 12, 2021, at https://www.usp.org/foods/food-fraudmitigation-services.)

[15] United States Pharmacopeia. (2016b). Food fraud database. (Accessed June 12, 2021, at https://www.foodfraud.org/.)

[16] Tibola, C. S., da Silva, S. A., Dossa, A. A., & Patrício, D. I. (2018). Economically motivated food fraud and adulteration in Brazil: Incidents and alternatives to minimize occurrence. *Journal of Food Science, 83*, 2028–2038.

[17] Nwuneli, N. (2018). Fake processed food is becoming an epidemic in African urban life. (Accessed October 2, 2021, at https://qz.com/africa/1226112/fake-food-or-fraud-food-in-nigeria-kenya-and-other-african-countries/.)

[18] Guzder, K. (2019). What is Food Fraud? – Types of EMA (Economically Motivated Adulteration). (Accessed May 20, 2021, at https://www.highspeedtraining.co.uk/hub/what-is-food-fraud/.)

[19] Organic authority. (2018). 8 food frauds: From horse meat to olive oil. (Accessed May 05, 2021, at https://www.organicauthority.com/buzz-news/8-food-frauds-horse-meat-olive-oil.)

[20] Garcia, J. S., Sanvido, G. B., Saraiva, S. A., Zacca, J. J., Cosso, R. G., & Eberlin, M. N. (2012). Bovine milk powder adulteration with vegetable oils or fats revealed by MALDI-QTOF MS. *Food Chemistry, 131*, 722–726.

[21] Souza, S. S., Cruz, A. G., Walter, E. H. M., et al. (2011). Monitoring the authenticity of Brazilian UHT milk: A chemometric approach. *Food Chemistry, 124*, 692–695.

[22] Rosa, L. S., Garbin, C. M., Zamboni, L., & Bonacina, M. S. (2015). Avaliac͂ao da qualidade f´ısicoqu ´ımica do leite ultra pasteurizado comercializado no munic´ıpio de Erechim–RS. *Vigilˆancia Sanit´aria Em Debate: Sociedade, Ciˆencia & Tecnologia, 3*, 99–107.

[23] Buzzo, M. L., Carvalho, M. D. F. H., Arakaki, E. E. K., Matsuzaki, R., Oliveira, C. C. de, & Kira, C. S. (2015). Teores de s´odio em leites industrializados consumidos no Brasil. *Revista Do Instituto Adolfo Lutz, 74*, 12–20.

[24] Rodrigues, N. P. A., Givisiez, P. E. N., Queiroga, R. C. R. E., Azevedo, P. S., Gebreyes, W. A., & Oliveira, C. J. B. (2012). Milk adulteration: Detection of bovine milk in bulk goat milk produced by smallholders in northeastern Brazil by a duplex PCR assay. *Journal Dairy Science, 95*, 2749–2752.

[25] Silva, S. A., Torres, E. A. S., Almeida, A. P. de, & Sampaio, G. R. (2018). Polycyclic aromatic hydrocarbons content and fatty acids profile in coconut, safflower, evening primrose, and linseed oils. *Food Chemistry, 245*, 798–805.

[26] European Union Annual Report. (2018). The EU food fraud network and the system for administrative assistance – Food fraud. European Commission, 1–13.

[27] Carvalho, B. M. A. de, Carvalho, L. M. de, Coimbra, R. J. dos, Minim, L. A., Barcellos, S. E., de Silva W. F., et al. (2015). Rapid detection of whey in milk powder samples by spectrophotometric and multivariate calibration. *Food Chemistry, 174*, 1–7.

[28] Brito, M. A., Schneider, H., Sampaio, I., & Santos, S. (2015). DNA barcoding reveals high substitution rate and mislabelling in croaker fillets (Sciaenidae) marketed in Brazil: The case of "pescada branca" (*Cynoscion leiarchus* and *Plagioscion squamosissimus*). *Food Research International, 70*, 40–46.

[29] Felkl, G. S. (2014). Autenticidade molecular de produtos c´arneos. Master's dissertation, Universidade Tecnol´ogica Federal do Paran´a, Ponta Grossa. (Accesed Sepember 17, 2021, at https://repositorio.utfpr.edu.br:8080/jspui/bitstream/1/1550/2/PG_PPGEP_M_Felkl%2c%20Gabriela %20Sartori_2014.pdf.)

[30] Ducatti, R., Pinto, J. P. D. A. N., Sartori, M. M. P., & Ducatti, C. (2016). Quantification of soy protein using the isotope method (δ 13 C and δ 15 N) for commercial brands of beef hamburger. *Meat Science, 122*, 97–100.

[31] Jack, L., & Krzyzaniak, S. A. (2020). Food fraud – tackling the food industry's adulterous underbelly. (Accessed May 29, 2021, at https://www.aweimagazine.com/article/food-fraud/.)

[32] Mendes, R. M. D. O., Quadri, M. B., & Quadry, M. G. N. (2007). Mat´erias estranhas emer va-mate (Ilex paraguariensis St Hil) beneficiada no estado de Santa Catarina, Brasil. *Revista Do Instituto Adolfo Lutz, 66*, 103–107.

[33] Santos, J. M. D., & Abrantes, S. D. M. P. (2015). Presenc,a de mat´erias estranhas em ervadoce, *Pimpinella anisum* L. Vigilˆancia Sanit´aria em Debate: Sociedade. *Ciˆencia and Tecnologia, 3*, 120–127.

[34] Moura Ribeiro, M. V. de, Boralle, N., Pezza, H. R., Pezza, L., & Toci, A. T. (2017). Authenticity of roasted coffee using 1 H NMR spectroscopy. *Journal of Food Composition and Analysis, 57*, 24–30.

[35] Cathe Friedrich Newsletter. (2021). 10 most common food frauds that can affect your health. (Accessed May 17, 2021, at https://cathe.com/10-most-common-food-frauds-that-can-affect-your-health/.)

[36] Food Fraud Advisors. (2019). Latest news in food fraud. (Accessed May 17, 2021, at https://www.food fraudadvisors.com/food-fraud-news-oct-19/.)

[37] Pereira, H. V., Amador, V. S., Sena, M. M., Augusti, R., & Piccin, E. (2016). Paper spray mass spectrometry and PLS-DA improved by variable selection for the forensic discrimination of beers. *Analytica Chimica Acta, 940*, 104–112.

[38] Hellberg, R. S., Sklare, S. A., & Everstine, K. (2021). *Food fraud – A global threat with public health and economic consequences* (pp. 401). United States: Elsevier Academic Press.

[39] Spink, J., & Moyer, D. C. (2011). Defining the public health threat of food fraud. *Journal of Food Science, 769*, 157–163.

[40] European Commission. (2021). Agri-food fraud/Food Safety. (Accessed June 3, 2021, at https://ec.eu ropa.eu/food/safety/food-fraud_en.)

[41] Lau, J.-E. (2021). Food fraud – Part 1 – What is food fraud? (Accessed June 7, 2021, at https://sciencemeetsfood.org/food-fraud-what-is-food-fraud/.)

[42] Spielman, S. (2020). Food fraud is not only an economic drain but also a supply chain safety concern. BNP Media. Accessed May 14, 2021.

[43] Pease, K. (2006). Rational choice theory. In E. McLaughlin & J. Muncie (Eds.), *The sage dictionary of criminology*. London: Sage.

[44] Cohen, L., & Felson, M. (1979). Social change and crime rate trends: A routine activity approach. *American Sociological Review, 444*, 588–608.

[45] Hirschi, T. (1969). *Causes of delinquency*. Berkeley: University of California Press.

[46] Walklate, S. (2007). *Understanding criminology*. Buckingham: Open University Press.

[47] Hirschauer, N., & Zwoll, S. (2008). Understanding and managing behavioral risks: the case of malpractice in poultry production. *European Journal of Law and Economics, 26*, 27–60.

[48] Aruwa, C. E. (2020). Food regulations and governance. In *Food Science and Technology: Trends and Future Prospects*, Part VI, Food Business: entrepreneurship and regulation (pp. 485–510). Germany: DeGruyter.

[49] Spink, J., Moyer, D. C., & Speier-Pero, C. (2016). Introducing the food fraud initial screening model (FFIS). *Food Control, 69*, 306–314.

[50] Danezis, G. P., Tsagkaris, A. S., Camin, F., Brusic, V., & Georgiou, C. A. (2016). Food authentication: Techniques, trends & emerging approaches. *Trends in Analytical Chemistry, 85*, 123–132.

[51] Bouzembrak, Y., Steen, B., Neslo, R., Linge, J., Mojtahed, V., & Marvin, H. J. (2018). Development of food fraud media monitoring system based on text mining. *Food Control, 93*, 283–296.

[52] Badia-Melis, R., Mishra, P., & Ruiz-García, L. (2015). Food traceability: New trends and recent advances. A review. *Food Control, 7*, 393–401.

[53] Tian, F. (2017). A supply chain traceability system for food safety based on HACCP, blockchain, & internet of things. Proceedings of the 2017 International Conference on Service Systems and Service Management, Dalian, China.

[54] Eurofins. (2021). Food assurance – The Food Fraud Database. (Accessed March 25, 2021, at https://reg.abcsignup.com/reg/event_page.aspx?ek=0055-0021.)

[55] Aruwa, C. E., Akindusoye, A. J., & Awala, S. I. (2017). Socio-demographic characteristics and food hygiene level assessment of food handlers in cafeterias around a Federal University in Nigeria. *Journal of Scientific Research and Reports, 14*, 1–9.

[56] Aruwa, C. E., & Ogundare, O. (2017). Microbiological quality assessment of pupuru and plantain flours in an urban market in Akure, Ondo State, Southwestern Nigeria. *Open Access Library Journal, 4*, e3783.

[57] Wisniewski, A., & Buschulte, A. (2019). How to tackle food fraud in official food control authorities in Germany. *Journal of Consumer Protection and Food Safety, 14*, 319–328.

[58] Zhang, W., & Xue, J. (2016). Economically motivated food fraud and adulteration in China: An analysis based on 1553 media reports. *Food Control, 67*, 192–198.

[59] Everstine, K., Spink, J., & Kennedy, S. (2013). Economically motivated adulteration (EMA) of food: Common characteristics of EMA incidents. *Journal of Food Protection, 76*, 723–735.

[60] Nestle. (2016). Food Fraud Prevention – Economically-motivated adulteration. (Accessed June 22, 2021, at https://www.nestle.com/asset-library/documents/library/documents/suppliers/food-fraudprevention.)

[61] Associacao Brasileira da Indústria de Café-ABIC. Selo de pureza (2017). (Accessed June 20, 2021, at https://abic.com.br/certificacao/pureza.)

[62] Giovannucci, D., Scherr, S., Nierenberg, D., et al. (2012). *Food and agriculture: The future of sustainability. A strategic input to the sustainable development in the 21st century (SD21) project.*

New York (USA): United Nations Department of Economic and Social Affairs, Division for Sustainable Development.

[63] Fikselová, M., Benešová, L., Zajác, P., Golian, J., & Čapla, J. (2020). Food adulteration and safety regarding detected market cases and consumer opinions. *Potravinarstvo Slovak Journal of Food Sciences, 14*, 417–428.

[64] Food Insight Newsletter. Consumer perceptions about food fraud – food insight (2021). (Accessed March 21, 2021, at https://foodinsight.org/consumer-perceptions-about-food-fraud/.)

[65] Bouzembrak, Y., & Marvin, H. J. P. (2016). Prediction of food fraud type using data from Rapid Alert System for Food and Feed (RASFF) and Bayesian network modelling. *Food Control, 61*, 180–187.

[66] Aruwa, C. E., & Farotimi, A. R. (2019). Microbiological and proximate analyses of lebanese bread (Pita) from Akure metropolis. *South Asian Journal of Research in Microbiology, 3*, 1–11.

[67] Christino, C. (2020). Do you know the difference between food fraud and food defense?. (Accessed February 11, 2021, at https://blog.softexpert.com/en/difference-between-food-fraud-food-defense/.)

[68] Dong, H., Luo, D., & Luo, H. (2016). Effective use of food traceability in cereal grain food supply chains. In M. Espi~neira & F. J. Santaclara (Eds.), *Advances in food traceability techniques and technologies: Improving quality throughout the food chain* (pp. 303–319). Cambridge: Woodhead Publishing.

[69] Food and Agriculture Organization of the United Nations – FAO, World Trade Organization-WHO. (2017). Trade and food standards. (Accessed May 11, 2021, at https://www.wto.org/english/res_e/pub lications_e/tradefoodfao17_e.htm.)

[70] Diniz, A. P. C. (2010). Aplicac,~ao da raz~ao isot´opica do carbono (13C/12C) para detecc,~ao de adulterac,~ao em bebidas a base de maracuj´a–Passiflora edulis Sims. f. flavicarpa Deg. Master's dissertation, Faculdade de Ci^encias Agron^omicas, UNESP, Botucatu. (Accessed May 17, 2021, at https://200.145.6.238/handle/11449/90497.)

[71] Frank, N., Bessaire, T., Tarres, A., Goyon, A., & Delatour, T. (2017). Development of a quantitative multi-compound method for the detection of 14 nitrogen-rich adulterants by LC-MS/MS in food materials. *Food Addit Contam: Part A, 34*, 1842–1852.

[72] Nascimento, C. F., Santos, P. M., Pereira, E. R., & Rocha, F. R. P. (2017). Recent advances on determination of milk adulterants. *Food Chemistry, 221*, 1232–1244.

[73] Indelicato, S., Bongiorno, D., Pitonzo, R., et al. (2017). Triacylglycerols in edible oils: Determination, characterization, quantitation, chemometric approach, and evaluation of adulterations. *Journal of Chromatography A, 1515*, 1–16.

[74] Chauhan, A. (2020). Food fraud – an evolving crime with profit at its heart. (Accessed June 1, 2021, at https://www.newfoodmagazine.com/article/109059/food-fraud-an-evolving-crime-with-profit-at-its-heart/.)

[75] Moore, J. C., Spink, J., & Lipp, M. (2012). Development and application of a database of food ingredient fraud and economically motivated adulteration from 1980 to 2010. *Journal of Food Science, 77*, 118–126.

[76] Aruwa, C. E., & Ogunlade, S. T. (2016). Classical identification, 16S rDNA sequencing, and molecular characterization of *Bacillus* species from convenience foods. *Current Journal of Applied Science and Technology, 15*, 1–11.

[77] Yan, S., Lai, G., Li, L., Xiao, H., Zhao, M., & Wang, M. (2016). DNA barcoding reveals mislabeling of imported fish products in Nansha new port of Guangzhou, Guangdong province, China. *Food Chemistry, 202*, 116–119.

[78] Ha, T. T. T., Huong, N. T., Hung, N. P., & Guiguen, Y. (2018). Species identification using DNA barcoding on processed panga catfish products in Viet Nam revealed important mislabeling. *Turkish Journal of Fisheries and Aquatic Sciences, 18*, 457–462.

[79] Chen, T.-Y., & Hwang, D.-F. (2002). Electrophoretic identification of muscle proteins in 7 puffer species. *Journal of Food Science, 67*, 936–942.

[80]   Martinez, I., ˇSliˇzyte, R., & Daukˇsas, E. (2007). High resolution two-dimensional electrophoresis as a tool to differentiate wild from farmed cod (*Gadus morhua*) and to assess the protein composition of klipfish. *Food Chemistry, 102*, 504–510.

[81]   Berrini, A., Tepedino, V., Borromeo, V., & Secchi, C. (2006). Identification of freshwater fish commercially labelled "perch" by isoelectric focusing and two-dimensional electrophoresis. *Food Chemistry, 96*, 163–168.

[82]   Taboada, L., S´anchez, A., Velasco, A., Santaclara, F. J., P´erez-Mart´ın, R. I., & Sotelo, C. G. (2014). Identification of Atlantic cod (*Gadus morhua*), ling (*Molva molva*), and Alaska pollock (*Gadus chalcogrammus*) by PCR-ELISA using duplex PCR. *Journal of Agricultural and Food Chemistry, 62*, 5699–5706.

[83]   Carrera, E., Terni, M., Montero, A., Garc´ıa, T., Gonz´alez, I., & Mart´ın, R. (2014). ELISA-based detection of mislabeled albacore (*Thunnus alalunga*) fresh and frozen fish fillets. *Food and Agricultural Immunology, 25*, 569–577.

[84]   Fiorino, G. M., Garino, C., Arlorio, M., Logrieco, A. F., Losito, I., & Monaci, L. (2018). Overview on untargeted methods to combat food frauds: a focus on fishery products. *Journal of Food Quality, 2018*, 1–13.

[85]   Saheed, S., Charlene, P., Taofeeq, G., & Mamosa, N. (2020). *18 Food scanners: applications in the food industry* (pp. 447–466). Food Science and Technology, De Gruyter.

[86]   Sreenath, H. K., Crandall, P. G., & Baker, R. A. (1995). Utilization of citrus by-products and wastes as beverage clouding agents. *Journal of Fermentation and Bioengineering, 80*, 190–194.

[87]   Arvanitoyannis, I. S. (2016). *Authenticity of foods of animal origin*. Abingdon: CRC Press.

[88]   Handford, C. E., Campbell, K., & Elliott, C. T. (2016). Impacts of milk fraud on food safety and nutrition with special emphasis on developing countries. *Comprehensive Review in Food Science Food Safety, 15*, 130–142.

[89]   P´erez-Jim´enez, M., Besnard, G., Dorado, G., Hernandez, P., & Bakkali, A. (2013). Varietal tracing of virgin olive oils based on plastid DNA variation profiling. *PLoS ONE, 8*, e70507.

[90]   Leschin-Hoar, C. Specious species: Fight against seafood fraud enlists DNA testing. Scientific American, 2011. (Accessed May 22, 2021, at https://www.scientificamerican.com/article/dna-testing-for-seafood-fraud/.)

[91]   Hohmann, M., Christoph, N., Wachter, H., & Holzgrabe, U. (2014). 1H NMR profiling as an approach to differentiate conventionally and organically grown tomatoes. *Journal of Agricultural and Food Chemistry, 62*, 8530–8540.

[92]   Esteki, M., Regueiro, J., & Simal-G´andara, J. (2019). Tackling fraudsters with global strategies to expose fraud in the food chain. *Comprehensive Review in Food Science Food Safety, 18*, 425–440.

[93]   Tola, A. (2018). Global food fraud trends and their mitigation strategies: The case of some dairy products: A review. *Food Science and Quality Management, 77*, 30–42.

[94]   GFSI. (2014). GFSI Position on mitigating the public health risk of food fraud. (Accessed October 15, 2021, at http://www.mygfsi.com/news-resources.html.)

[95]   van Ruth, S. M., Huisman, W., & Luning, P. A. (2017). Food fraud vulnerability and its key factors. *Food Science and Technology, 67*, 70–75.

James A. Elegbeleye, Victor Ntuli, Wisdom Selorm Kofi Agbemavor
and Adeyemi Ayotunde Adeyanju

Chapter 23
Enhancement of food safety through technological innovations

Abstract: Myriads are the number of people at risk of consuming unsafe food. Many likewise have become sick because of unsafe foods, whereas thousands die annually for the same reason. Therefore, food safety is vital for several reasons, such as enhancing population health and the economic growth of a nation. Consumers' increasing desire for minimally processed foods has created an equal exposure to potential food hazards such as pathogenic microbes and pesticides. These varied foodborne have fueled the search for innovative solutions to guarantee safe, nutritious, and healthy food products. These technological advances are new opportunities to ensure a continuing supply of safe, nutritious food products with extended shelf-life to the population and expansion of product distribution beyond a specific region, thus promoting global trade. This chapter explores these state-of-the-art technologies and their superior advantages over the conventional methods used to inactivate potential hazards in food.

23.1 The general perspective on food safety

Food safety is essential for several reasons, such as enhancing a nation's population health and economic growth [1]. The WHO estimated that about 600 million globally became sick after consuming contaminated food, with an annual death of 420,000. These altogether have occasioned approximately 33 million disability-adjusted life years (DALYs), causing extended hospitalization of the affected people [2]. Children below the age of 5 bear 40% of the foodborne disease burden, with an estimated death of 125,000 annually. The estimated cost of foodborne illnesses is approximately $90.2 billion and $60.9 billion globally and in the USA [3, 4]. Hence, the issues around food safety are

James A. Elegbeleye, Department of Microbiology, Faculty of Pure and Applied Sciences, Southwestern University Nigeria, Kilometre 20 Sagamu-Benin Expressway, Okun-Owa, Ogun State, Nigeria
Victor Ntuli, Department of Food Science and Technology, University of Venda, Private Bag X5050 Thohoyandou, Limpopo, South Africa
Wisdom Selorm Kofi Agbemavor, Radiation Technology Centre, Biotechnology and Nuclear Agriculture Research Institute, Ghana Atomic Energy Commission, P. O. Box LG 80, Legon Accra, Ghana
Adeyemi Ayotunde Adeyanju, Department of Food Science and Microbiology, Landmark University PMB 1001, Omu-Aran, Kwara State, Nigeria

https://doi.org/10.1515/9783110748345-023

crucial to public health because of the devastating impacts on the economy within a geographical location and even beyond the boundaries where such food is produced.

Food is considered safe when it is produced, processed, transported, stored, and prepared in a manner that the consumers are not harmed upon consumption. In contrast, a food product is considered unsafe when contaminated with any physical, biological or chemical hazard, which may cause debilitating effects in humans when consumed [5]. These hazards or contaminants in food can also lead to food recalls and untold economic losses to the food industry and, in turn, affect a country's economy, leading to job losses and even bankruptcy [6, 7]. Hence, all stakeholders in the food industry, government, and (inter)national agencies saddled with the responsibility of food safety must ensure only safe food is displayed in the market and consumed [8].

Food safety involves protecting the food chain from potential hazards from the farm to the fork and ensuring optimum food quality. The potential hazards of food are primarily chemical and biological agents, which may harm the consumers. Among the biological hazards, *Salmonella, Clostridium perfringens, Campylobacter,* and *Staphylococcus aureus* are the leading causes of outbreaks in the USA [9]. The other biological hazards are yeasts, molds, and viruses exhibiting different susceptibility to the various operations during food processing [10]. Besides the unit operations, the persistence of these pathogens in the food chain is determined by other factors such as improper handling/storage of raw material, unhygienic food processing facilities, poor application of good manufacturing practices and food safety standards [11]. More often than not, the failure to reduce the metabolic activity of the contaminating microbes, especially after harvesting, slaughtering, or milking, compromises the food product's safety and quality. This failure or nonadherence provides a favorable environment for contaminating spoilage and pathogenic bacteria, yeasts, and molds to thrive. Some of these process microbes may establish their dominance in the food industry by evolving adaptive re-

Table 23.1: Common foodborne and their medical and economic impacts.

Foodborne hazards	Infectious/toxic agents	Incidence of foodborne illness	Death	Total DALYs
Bacteria	*Salmonella, Vibrio, E. coli, Shigella, Listeria, Brucella, Listeria, Campylobacter*	359,747,420	272,554	20,188,792
Virus	Noro virus, hepatitis A	138,513,782	120,814	3,849,845
Protozoa	*Entamoeba, Giardia, Cryptococcus, Toxoplasma*	77,462,734	6,242	1,311,435
Worms	Cestodes (tapeworms), nematodes (roundworms), trematodes (flatworms); helminths (parasites)	26,063,664	90,261	11,599,735
Chemicals	Aflatoxins, cyanogenics, dioxins, heavy metals	217,632	19,712	908,356

Source: [1].

sponses to the stresses by altering their genes and physiology, making them more resistant to processing conditions [12, 13].

Besides the biological agents compromising food safety, chemical toxins, environmental pollutants, and residues from pesticides have been implicated in many outbreaks. Some chemical toxins in food include naturally occurring mycotoxins, marine biotoxins, cyanogenic glycosides, and poisonous mushrooms, as shown in Table 23.1. Among the chemical contaminants of food are mycotoxins such as aflatoxin and ochratoxin. Mycotoxins are often found in different concentrations in staples such as corn meals and breakfast cereals, which may cause debilitating conditions such as cancer immune disorder after prolonged and repeated exposures (14). Besides the mycotoxins, persistent organic pollutants (POPs) such as dioxins and polychlorinated biphenyls (PCBs) are also gaining attention in public health. These environmental pollutants, alongside heavy metals, accumulate in the water, animals, and soil, where they gain entry into the food chain.

Another aspect of chemical contamination may arise from the chemical substances used to produce the packaging material. Food may also interact with the packaging materials, which may cause the migration of the chemical substances from the packaging film or material into the food upon contact. Several factors regulate the extent of this migration, such as the type of food, polymer, physicochemical properties of the migrant, period and temperature of storage, and proportion of food in the pack [15].

Unlike the biological and physical contaminants of food products which can either be inactivated or removed during unit operations, chemical contaminants are often tricky to manage. These chemical contaminants may persist to the food at the point of consumption, causing immediate or remote health challenges such as bioaccumulation in vital tissues and causing immune and reproductive problems [1].

23.2 Overview of some of the early food preservation methods

The traditional concepts of food safety involved the application of salting, blanching, fermentation, drying, pasteurization, canning, chemical additives, ultrafiltration, and low-temperature storage. These operations are either used as a stand-alone preservation process or, in combination, commonly referred to as hurdle technology [16]. Figure 23.1 shows the different conventional food preservation methods adopted by the food industry.

Almost all the food products are treated using a thermal application, thus making the food susceptible to the Maillard reaction, which occurs in the presence of amino acids and reducing sugars. Maillard reaction causes an undesirable sensory change in the food and loss of some essential nutrients due to the extreme amount of energy

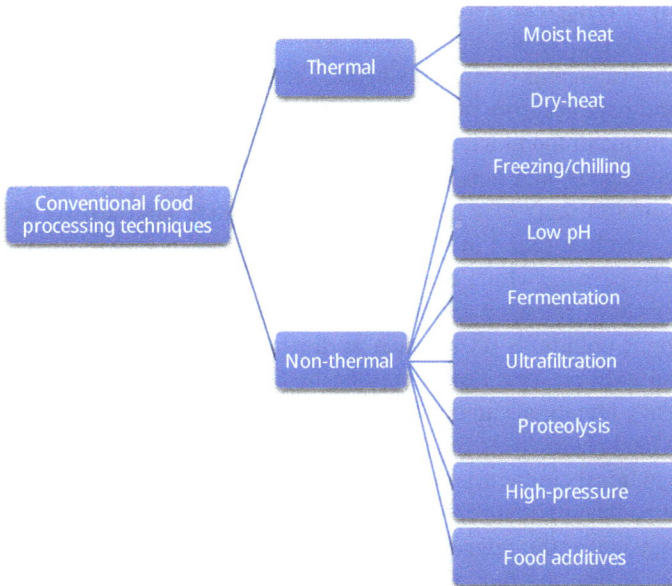

Figure 23.1: Conventional food preservation methods adopted by the food industry [17].

applied, which induces a chemical change in food products. Nonthermal processing such as high hydrostatic pressure (HPP] has been used since the 1980s. It employs high pressure above 350 MPa to inactivate spoilage, pathogenic microorganisms, and their enzymes at low temperatures. As in the case of thermal processing, there are some noticeable differences in the resistance of microorganisms to HHP, as observed in some studies. For example, yeasts and molds are more susceptible to HHP than vegetative bacteria cells; Gram-positive bacteria are generally more resistant than Gram-negative bacteria, while bacterial spores are highly resistant to HHP treatment [18, 19]. These conventional methods of processing foods are not often enough to decontaminate certain food products and ensure complete inactivation of the microbial contaminants. More so, applying them at high intensity or dosage may affect the organoleptic properties of the food product. Hence, they are often combined as an intervention strategy or hurdle targeting specific microbes while minimizing the impacts on the food. Hence, there is a need for innovative technologies that can effectively inactivate the contaminants without negatively impacting the food. These nonthermal innovative technologies are recognized as sustainable alternatives to conventional food processing because of the low energy and water consumption, solid waste reduction, and valorization potential [20]. These technologies may be applied in food ingredient modification, processing, or enhancing traceability throughout the food chain. These technologies are discussed below.

23.3 Selected innovative technologies for food ingredient modifications

23.3.1 Decaffeination

Caffeine, also known as 1, 3, 7-trimethylxanthine (Figure 23.2), is a naturally occurring chemical (an alkaloid) found in coffee, kola nuts, tea leaves, and cocoa beans. The overall conclusions about its health implications are still up for debate. Low to moderate caffeine consumption has been linked to improved alertness, learning capacity, and athletic performance [22], among other benefits. It is an essential antioxidant source because of the presence of polyphenols, which can protect against various ailments [21, 23, 24]. However, excessive use of caffeine-containing beverages has a connection with various health issues, including worsened heartburn and acid indigestion [25–27]. Caffeine consumption in excess has been linked to mutations, DNA repair inhibition, fetus deformity during pregnancy, lower fertility rates, and hastened onset of osteoporosis [26]. Given these negative consequences, much work has gone into developing beverages with no caffeine and other goods derived from caffeine-containing essential ingredients. Removing caffeine from these beverages will eliminate the health risks associated with caffeine and provide certain health benefits to the general public.

Various methods can be used to remove caffeine, including water, organic solvent extraction, and supercritical CO_2 extraction. Decaffeination techniques based on the solubility of caffeine in water, which varies with temperature, are typically used to remove caffeine from coffee without employing organic reagents. Coffee beans are steeped in hot water to dissolve both caffeine and flavor components in this approach. The water, which includes most of the flavor components, is then sprayed on the extracted beans, allowing them to absorb the flavor while eliminating the caffeine [28]. Solvent extraction, which involves either direct solvent extraction of the beans or indirect solvent extraction of the caffeine from a water extract, is purely dependent on the ability of caffeine to solubilize in various organic solvents.

The recent introduction of supercritical CO_2 extraction allows for removing only the caffeine while leaving the other flavor components intact. Compressing CO_2 under extremely high pressure is first changed from a gaseous to a liquid state. To extract the caffeine, pre-moistened beans are treated with liquid CO_2. The benefits of this technique include the absence of dangerous chemicals, higher product quality, and little extraction of other coffee soluble components in addition to caffeine [29].

Other decaffeination methods have also been created because traditional procedures have some drawbacks. Water decaffeination, for example, is ineffective due to caffeine's limited solubility in water, but solvent extraction processes suffer from the

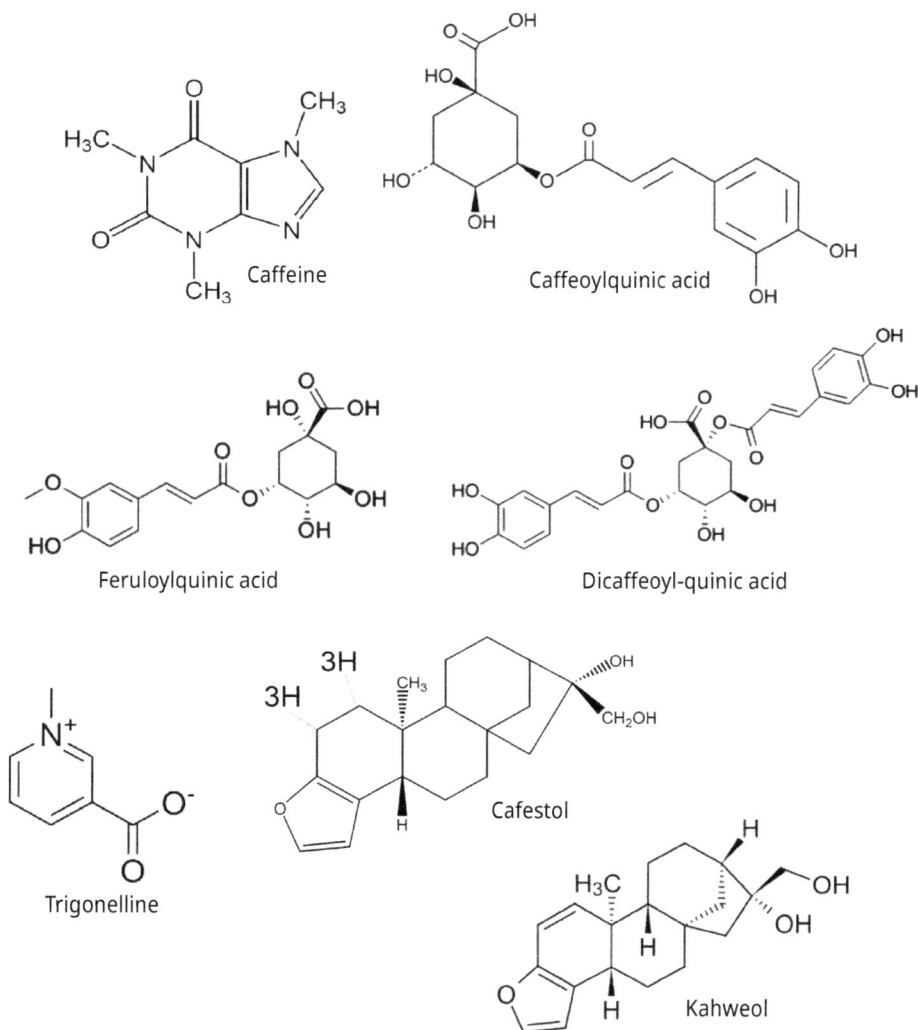

Figure 23.2: Main bioactive compounds of green coffee [21].

decaffeinating agent's side effects and high cost. Despite its various benefits, the supercritical carbon dioxide technique is capital-intensive [28].

As a result, microbial decaffeination has been developed as a viable alternative. Caffeine decomposition by bacteria is faster than fungus, and unlike fungal degradation, it is not hampered by an external nitrogen supply. Caffeine is efficiently degraded by the bacteria *Pseudomonas* and the fungus *Aspergillus* and *Penicillium* [30]. Enzymes, specifically demethylases and oxidases, are responsible for the caffeine degrading ability of these microorganisms. Because of the health hazards associated with excessive caffeine levels and its addictive character, the beverage industry is interested in discovering better procedures for removing caffeine from beverages at a low cost.

23.3.2 Fat replacement

Today, overweight and obesity have become a significant health issue in most countries. Obesity and a high fat intake have been linked to chronic diseases such as diabetes, hypertension, and cardiovascular disease [31, 32]. Therefore, lowering overall dietary fat intake and progressively substituting it with other ingredients appears to be a viable strategy for reducing heart disease, obesity, and other metabolic disorders, all of which are on the rise [31]. However, removing lipids from food may negatively impact product quality because fat contributes to the physiochemical qualities of final food items, such as viscosity, texture, appearance, and flavor [33]. As a result, fat substitutes in foods have emerged as a promising option for the food sector to lower fat content. These substances should be able to supply fat's flavor, mouthfeel, and other desirable properties of fat without excessive calories and hazards known with traditional fat. Fat replacers are classified as fat substitutes or fat mimetics depending on their chemical structure and activity [33]. Fat substitutes are designed to replace fat "one-to-one" and have structures similar to fats, but fat mimetics do not. Fat replacements are lipid-based fat substitutes, while fat mimetics are carbohydrate- and protein-based fat substitutes [33]. Carbohydrate-based replacers are derived from cereal grains and are modified to give food products a fat-like texture while offering a reduced-calorie product due to their lower energy density than fats. Protein-based replacers, like carbohydrate-based replacers, come from protein sources like egg and milk and provide lower-energy foods. Fat-based replacers offer similar characteristics to the natural fats they replace, such as texture and flavor.

In various foodstuffs, starch as a fat substitute has proven successful in lowering overall fat and calorie content. For example, 20% of the fat in mayonnaise has been successfully replaced with native or cross-linked rice starch gels, with the results indicating that mayonnaises with both types of fat replacers had superior sensory quality than those with both types of fat replacers the full-fat control. Using native and cross-linked rice starch also improved the freezing stability of mayonnaise [31]. The effects of using sweet potato starch treated with citric acid as a fat substitute on the overall quality of low-fat ice cream were also investigated. Although modified starch fat replacers affected the freezing ability of ice cream, low-fat ice cream with 1% fat replacer had the same sensory acceptability as full-fat control ice cream and did not influence microbiological stability [34].

Due to their unique functional features, such as nongelling behavior, high viscosity, and good plasticity, starch–lipid complexes can also be employed as an ingredient in industrial applications. For example, starch–lipid and starch–lipid–protein complexes could be used as fat substitutes in developing lower-calorie foods [35, 36]. 36 made a low-calorie mayonnaise using stearic acid complexes with teff or maize starch as fat substitutes. The results showed that replacing 80% of the oil with a starch–lipid complex produced a product with physical attributes similar to full-fat mayonnaise. Furthermore, as types of resistant starch, starch–lipid or starch–lipid–protein complexes can be used as functional components in food formulations, which is of nutritional relevance in light of the rising incidence of obesity and diet-related disorders [37, 38]. The addition of amylose–lipid complexes to starch gels, for example, raised the resistant starch content while lowering the hydrolysis index value, with type II complexes having a more considerable effect than type I complexes [39]. In another study, compared with control white bread, bread prepared from 60% (dry basis) starch–lipid complexes significantly reduced postprandial glucose levels and insulin response in people [40].

To reduce total fat and energy intake, consuming low-fat, low-calorie foodstuffs has emerged as treatment for preventing cardiovascular disease and obesity. Despite the benefits of synthetic fat substitutes, questions have been raised concerning its potential influence as a dietary change method. Concerns include the potential for affecting fat-soluble micronutrient bioavailability, gastrointestinal (GI) or other particular adverse effects, and growing food prices. More research is needed to solve these challenges to make fat replacement technology more relevant and acceptable.

23.3.3 Enzyme technology

Enzymatic biotechnology has made significant progress in developing novel industrial processes in recent years, particularly in health-related pharmaceuticals and food items. Enzymes offer several benefits over nonbiological catalysts, including operating with food materials, efficient catalytic activity under mild reaction conditions (ambient temperature and air pressure), and high substrate specificity. Enzymes are employed as direct food additives in the food processing industry to adjust food's physicochemical and rheological qualities and manage the quality of food product [41]. Furthermore, enzyme catalysis may result in "greener processes" because it uses less energy and produces less waste and enzymes can be reused when immobilized on supports [42, 43]. As a result, enzymes' catalytic activity has been used in large-scale food processing alternatives to chemical processing. Enzymes used in food can range from highly refined commercial enzymes to primitive preparations like plant exudates or chopped fruits. They can be directly injected into food systems or immobilized on inert supports, allowing the enzyme to interact with food systems during processing [42]. Most industrial enzymes are found in foods and beverages, with over 500 goods manufactured in various industries using enzyme-based procedures [41, 44].

Recently, enzymes in food processing have shifted away from their conventional purposes and toward offering health and nutritional benefits to food items. Oxidoreductases are used in food systems to transform cholesterol into coprostanol, a nontoxic chemical [26]. This technology reduces cholesterol's adverse effects while avoiding steam distillation and supercritical fluid extraction procedures, which are both costly and time-consuming [26]. Food is also supplemented with enzymes to improve its nutritional content. Phytic acid, for example, is an antinutritional component found in many plant-based meals and binds minerals like iron and zinc, lowering their bioaccessibility. Exogenous phytase addition enhances the micronutrient bioaccessibility of these foods. Other enzymes with nutritional and physiological benefits have also been suggested for food preparation. Antinutritional compounds in legumes, including raffinose, stachyose, and verbascose, are not digested by humans and cause gas, diarrhea, and indigestion. These oligosaccharides are connected by α-D-galactosidic linkages, which are resistant to heating and other processing procedures but can be hydrolyzed by α-D-galactosidases. As a result, α-D-galactosidases have been used in food preparations to hydrolyze heat-resistant oligosaccharides in processed legume-based meals.

Enzymes have increased process simplification, overall competitiveness, and product innovation in the food industry by serving as an environmentally friendly processing aid or additive. It is a tested and accurate approach to using a biocatalyst. However, there is still much potential for development regarding overall enzyme qualities with multiprocessing capabilities. The scientific and industrial research groups are working hard to develop more robust, effective, and diverse enzymes, as well as to ensure that the structure-function activities of enzymes can endure the bioreactor's fluctuating and severe environmental circumstances (such as pH, temperature, organic solvents, and inhibitory compounds).

23.3.4 Fermentation

Fermentation converts macromolecules such as polysaccharides, proteins, and lipids in food into products with improved flavor, fragrance, and texture using microbial enzymes [45, 46]. Fermentation is a low-cost technique of making and preserving food that boosts its nutritional and health benefits. In Africa, the method is widely used in industrial and domestic sectors. Traditional fermentation methods include yeast fermentation, *Acetobacter* fermentation, and *Lactobacillus* fermentation for alcoholic beverages, vinegar, yoghurt, and pickles, respectively [47]. Lactic acid bacteria are widely used as a natural preservative in traditionally fermented meals because the lactic acid produced by the microbes improves the food's digestibility while inhibiting the growth of harmful organisms. Studies show that fermented milk provides health benefits, including decreasing blood cholesterol and anticancer activity [47]. Fermentation achieves the desired result by producing significant levels of certain helpful bioactive chemicals in situ, removing undesirable compounds, or transforming undesirable compounds into desirable ones.

The modern application of fermentation in food processing has highlighted the production of foods with improved nutritional quality and health benefits, enhancing the commercial worth of food products. Fermentation is currently utilized to minimize anti-nutritive chemicals like tannin and phytate and improve the bioavailability of crucial elements like iron and zinc [48]. Fermentation also aids food safety by preventing pathogenic germs from growing [49]. Fermented foods are thus intended to be significant elements in producing novel meals with improved qualities in various food industries [49, 50]. Fermentation also significantly reduces the levels of substances like cyanide in cassava and nondigestible carbohydrates in legumes and thus reduces adverse side effects like abdominal distention and flatulence associated with nondigestible carbohydrates like raffinose and stachyose [46]. Despite the apparent benefits of fermentation to human and animal health, fermented foods have been connected to several health hazards. Fermentation has been connected to pathogenic and dangerous compounds such as mycotoxins and biogenic amines [26]. More research into creating nontoxic starters is needed to overcome some of these safety issues to get the most out of fermented foods.

23.4 Modern food biotechnology

The fundamental requirement for genetically modified foods is to feed the world's growing population and meet all of humanity's food-related needs. The loss of fertile land due to anthropological activities such as urbanization and industrialization is the primary cause of decreased food production, which can be offset by adopting high-yielding genetically engineered foods. Biotechnology involves using biological systems, living creatures, or organism components to create or change goods or processes for specialized applications. Farmers have used traditional breeding techniques to improve crop plants for centuries. However, because several genes are jumbled every time two plants are crossed, cross-breeding is unpredictable and challenging to manage, as undesired qualities are passed on to the next crop alongside favorable traits.

Furthermore, traditional cross-breeding can only occur between closely related species [51]. On the other hand, modern crop breeders can use modern biotechnology to choose a specific genetic feature from any plant or even an animal source and transfer it into the genetic code of another plant. By overcoming species' biological boundaries, we can employ the gene-editing technique to modify the plant's genetic features precisely, rapidly, and regulated.

The introduction of genetically modified organisms (GMOs) into food has the potential to boost crop yield and nutritional content, both of which are beneficial to human health and well-being. Indirectly, genetically modified (GM) crops promote human health by reducing the environmental effect of food production. Insect-resistant GM crops require fewer pesticides, lowering the risk of chemical residues for farmers and the environment [52]. Crops more tolerant to environmental stress factors such as drought, viruses, and herbicides are also developed using biotechnology. The cost savings from fewer chemicals and improved agricultural sustainability strengthen food security and, as a result, have a global influence on human health and development, particularly in developing nations. According to estimates, GM crops already span around 4% of the world's arable land. Between 2006 and 2012, worldwide agriculture income climbed by $116 billion [53]. The enhanced genetics and resilience to pests and weeds are responsible for the growth in productivity and profitability. GM foods have improved the quality of the food and quantity. Compared to natural foods, GM foods have a higher chemical composition of vitamins, fatty acids, and cellulose. GM foods have the potential to revolutionize food processing. Scientists were able to create insect-resistant crops expressing the Bt gene from *Bacillus thuringiensis* through genetic alterations in tomato, potato, and other food crops, better shelf life, and appeal in food products. It is also accomplished in animals, mainly fish, by enhancing development and body bulk. For example, genetically modified salmon's natural full-size growth period is three years, whereas GM fish reach full size in only 18 months. GM foods may potentially be supplemented with therapeutic bacterial and viral antigens, making them suitable for use as oral vaccinations [53].

While increasing the production and nutritional potential of plants, GM foods have specific adverse side effects, such as allergies in some people [51]. Concerns have been raised about GMOs' toxicity, susceptibility to cause allergic reactions, stability of the inserted gene, nutritional effects related to the specific genetic modification, and any unanticipated effects associated with the gene insertion. However, the magnitude of the benefits outweighs the magnitude of the drawbacks. Some negative impacts can be easily avoided with enough research investigation.

23.5 Selected innovative technologies for protecting ingredients and controlling delivery

23.5.1 Encapsulation

In recent years, the nutraceutical and functional food industries have focused on creating food matrices containing bioactive elements such as lipids, antioxidants, proteins, vitamins, minerals, phytochemicals, and probiotics. On the other hand, bioactive substances are susceptible to process and environmental factors, have poor stability during food preparation and storage and have reduced bioavailability when exposed to conditions in the upper GI tract. As a result, incorporating bioactive chemicals while maintaining their capabilities has proven difficult for the business. Therefore, they require new and innovative approaches to introduce them to various food products. Encapsulation is a popular method for encasing and protecting unstable active substances (core material), delivering them to a specific location, and extending the shelf life of a product. Bioactive substances will be more efficient if they are delivered in a highly controlled manner, allowing for determining their optimal dose [54]. This approach is feasible to protect chemicals against degradation during food preparation and digestion, thereby improving their bioavailability, preservation, and stability. This technique shields active components from adverse conditions that affect their stability, as well as light, heat, water, pH, enzymes, and oxygen [54, 55]. It can also improve the product's organoleptic properties by concealing flavor without modifying taste, fragrance, color, or texture and increasing the release of constituent characteristics [56].

For many years, encapsulation has been widely employed in the pharmaceutical and chemical sectors, and it is now finding new uses in the food industry. Food components, enzymes, cells, nutrients, and/or other bio-ingredients are encapsulated in miniature capsules (microcapsules) in the food industry, allowing the active ingredients to be inserted into food items and released at a controlled time and rate. The enclosed core material is insulated from moisture, heat, and other harsh circumstances to improve its stability and practicability. Several strategies for encapsulating bioactive substances for food and nutraceutical purposes have been reported. Chemical, physical, and physicochemical procedures are the three types. In situ polymerization and liposome, entrapment is chemical procedures; physical techniques include spray-drying, spray chilling, extrusion, and fluidized bed coating; physicochemical approaches include coacervation and sol-gel encapsulation). Spray-drying is the most used method for encapsulating active ingredients in the food business [57]. Low cost, high loading encapsulation, and ease of industrial scale-up are just a few of the benefits of this approach. As a downside, the high temperature during the process may cause the bioactive components to degrade [56]. Microcapsules can be as small as a few millimeters or as large as several millimeters, but most are between 5 and 300 microns in diameter and exist in various shapes and sizes. Encapsulating agents include emulsifiers, cellulose, carbohydrates, gums, lipids, and fibers. Typically, a combination of these encapsulating agents is utilized. Controlling the active substance's release can be done in various ways.

Active substances are encapsulated and added to food in a liquid or solid form to improve sensory qualities (for example, boost flavor or hide odor and taste), nutritional quality, or both. Epidemiological studies show that n-3 polyunsaturated fatty acids (PUFA) protect against coronary heart disease (CHD). Fish oil, a significant source of PUFA, is used in dietary supplements to protect against CHD. PUFA, on the other hand, is prone to oxidation due to its high degree of unsaturation, resulting in an off-flavor. Fish oils that have been encapsulated are protected from oxidation and can be used on various dishes. Probiotics are live bacteria that benefit the gut by lowering the number of undesirable germs in the intestine and urogenital tract [26]. However, the stress experienced during the manufacturing, storage, and consumption of probiotic-containing foods makes adding probiotics to foods problematic. It has been suggested that they can be replaced with probiotic isolates and extracts to avoid this problem. However, while these nonviable probiotics have some therapeutic benefits, they are not as potent as the living microbes Efforts to encapsulate considerable demand preparation. The encapsulating material's composition and structure, the production conditions (temperature, pH, pressure, and humidity), and the encapsulated particles' effectiveness are all critical factors in protecting the ingredients during processing and then delivering and releasing them in a highly complex food matrix. Further research is needed to improve and expand the technology's use to encapsulate a broad spectrum of beneficial substances.

23.5.2 Nanotechnology

Nanotechnology is a new revolution with immense potential in various sectors, including mechanics, medicine, and food production. It is the study of altering and regulating matter on an atomic and molecular scale having at least one characteristic dimension in nanometers, often between 1 and 100 nm [58]. Due to their higher surface area and mass transfer rates, nanoparticles appear to have more significant chemical and biological activity, enzymatic reactivity, penetrability, catalytic behavior, and quantum features than big particles with a similar composition [59]. Nanomaterials are categorized according to their size, structure, and properties. The food industry now demands technology to produce convenient, authentic, and delicious food products to continue its leadership in food marketing and processing [60]. Nanotechnology is an example of a technology with a wide range of applications in the food industry. Food additives containing nanoparticles are commonly used to prevent food from infection and extend its shelf life. Preservatives, antimicrobial sensors, flavoring agents, packaging materials, encapsulated food components, nanomaterials, and nanoscale food additives affect the nutrient composition and improve product shelf life, texture, flavor, and other characteristics [61].

Nanotechnology brings up new avenues for developing unique products and expanding food-related applications such as bioactives, nutraceuticals, functional foods, medicinal foods, etc. It can even detect food pathogens that act as quality and safety indicators [62]. In food processing, food (nano-sized) ingredients, nutritional supplements (e.g., proteins and antioxidants), and additives (e.g., flavor and color) are nano-encapsulated, resulting in nano-capsules that can be used in functional foods. This approach has various advantages, including enhanced transport and dispensability for water-insoluble chemicals, protective barriers, controlled release, increased bioavailability of several vitamins and their precursors, and improved taste and masking of the off-flavor. Nanotechnology intends to solve food-related ailments (such as diabetes and obesity), create individualized nutritious meals for a wide range of target groups, ageing populations, and lifestyles, and ensure the sustainability of food production. This technology enables the creation of nutrition nano-therapy devices that deliver nutrients to strategic parts of the body. It has the potential to develop smart/intelligent systems that allow for the controlled release of nutrients through nanoencapsulation. Food fortification can be used to construct nanoscale enzymatic reactors to develop novel products [62].

23.6 Artificial intelligence and predictive analytics

Even though AI and PA demonstrate numerous potentials in the food industry, their actual contributions and competencies are yet to be revealed. They depend on massive data collections about food production, manufacturing, imports, exports, etc. [63, 64]. However, the process of the data collection and its subsequent application in productive ways is a massive effort through which AI can aid the food industry. Accordingly, AI and PA enable the food industry to take proactive actions against potential food safety threats [65, 66]. Even though preventive activities are undertaken within the food industry, the food supply chain mostly gets affected before the threat becomes known for remedial action. However, with PA and AI analytics, there is a great potential for such food safety (supply) threats indicative factors to be determined and even prevented before any incident occurs [66, 67]. Therefore, AI and PA allow the food industry to act against potential threats proactively. Besides protection, PA and AI improve the food supply chain [66, 68]. AI and PA can also determine factors that boost crop production and animal health, improving the food supply chain to become more productive and sustainable.

23.6.1 Blockchain system

Blockchain technology is an indisputable shared (distributed) ledger database that transaction partners can use to transmit information securely, the details (events and transactions) of which cannot be deleted or modified [69, 70]. The blockchain enhances transparency by ensuring that captured data is accessible to all business partners and the distribution network [71, 72]. Blockchain, as a decentralized data exchange, guarantees that information is shared and synchronized across several locations enabling the data to be more trustworthy, resilient, and secure [71, 73]. This system allows for product information, including lot and batch numbers. Expiry dates and supply chain events (shipping and delivery) to be correctly taken at every point of the product journey and shared with stakeholders in real time and allowing backward tracing to be more accessible [74–76]. It also enables defective, expired, or contaminated products to be quickly traced within the supply chain and allows for recalls much more efficiently to prevent a great deal of waste. This procedure is conducted by laser-focused identification of target products for withdrawal and saving unnecessary disposal of others [77–79].

23.6.2 Automated labeling

Labeling is an indispensable part of any food packaging, ensuring the food complies with food safety laws. It is a critical access point for consumers watching out to comprehend its nutritional information [80–82]. For the past few years, labeling has seen increased regulation, as well as helping consumers confirm whether a food product meets their nutritional requirements [83, 84]. However, quality assurance personnel still struggle to inspect data-rich labels with the needed accuracy [85–87]. This section of the article discusses the advantages of visual examination systems and the need for their introduction and implementation before they become a regulatory obligation. To meet regulatory requirements for customer traceability and information, it is vital for the information printed on films and labels to be accurate and legible [88, 89]. The continuously increasing demand in food labeling contributed to the inevitable more significant complexities within the (design of) labels.

All required product information must be printed in line with the EU Food Information Regulation, with at least a specified minimum font size [90, 91]. Besides the list of constituents, the compulsory details must entail expiry (best before) date, possible allergy information, and a consumer-friendly nutritional table in a specified structure [92, 93]. The label may consist of a broad range of other vital information, including bar codes for scanners, lot and batch numbers for traceability, several seals, and logos of approval, as well as QR codes (that enable consumers to contact the manufacturer for additional information using smartphones) [94–96]. The complexity of food labels is further complicated by diversity in product types, different flavors, and regulations for displaying product information in various languages, especially for products intended for multiple export markets. Labeling errors, including incorrect best before dates, illegible barcodes, or entirely wrong product labels, may return entire deliveries [97, 98]. This subjects product manufacturers to significant direct and indirect charges spanning from considerable costs for logistics and disposal to the applicable charges on product replacement or even the cost of damage to the company's reputation.

Severe product label violations mostly lead to contractual penalties or a complete product delisting. Graphical errors, including misaligned labels, can result in adverse effects due to damage to brand image, suggesting to the consumer that such a product is of poor quality due to carelessness in processing and packaging. These problems can completely be circumvented by adopting a well-planned automated labeling system.

23.7 Wearable technologies (smart sensors) with industrial IoT

Appliances including smart helmets, goggles and wrist tags for machinists are progressively gaining acceptance due to the numerous advantages and higher returns regarding efficiency, hygienic practices, and time management [99, 100]. Most food production units are equipped with wash-down environments, implying that extreme precaution must be exercised to circumvent minor food contaminants. Accordingly, food handlers should be prohibited from wearing watches and jewelry [101]. However, some of these devices should not be compromised for technological wearables, which can be used to impact significantly on specific areas (including monitoring of food hygiene and improving safety standards) of the supply chain and food production. For example, machinists use wearables to detect machinery malfunction, spilt fluids, and scorching surfaces, which helps avoid accidents on the factory floor. Additionally, smart glasses enable in-built cameras to allow quality control personnel to scrutinize the hygiene standards along the production process [99, 102–104].

23.7.1 Mobile solutions

Food safety, handling, preparation, and storage of food to prevent foodborne illnesses is of utmost importance to any food processing industry. Traditionally, food safety assessments are done using clipboards and paper forms coupled with data writing with the hand to compile various reports [105]. Accordingly, this could lead to overextension of this method due to the numerous activities that usually take place in factories, production kitchens, warehouses, and shops. More so, the manual completion of such documents is mainly saddled with inaccuracies and errors, contributing to work complexities and its concomitant health hazards for people [106]. However, with the advent of appropriate modern technologies, these associated problems are circumventable whilst conducting such important food safety tasks. Specialized smartphone-based apps can be developed to undertake such operations, including faster, more detailed, and adequate safety inspections than the traditional pen and paper system [107, 108].

Snappii is an institution for developing health assessment, food safety and other fast mobile business applications for tracking documentation involving food handling, processing, storage, and presentation, cleaning, and maintenance equipment, kitchen practices, and hygienic status of food handlers. Some apps allow pictorial attachments, PDF creation, and instant generating checklists and completed forms in excel rapidly, which are sharable via email and other mobile application platforms. Simply put, such mobile apps allow compliance with food safety regulations conveniently and easily [109, 110].

23.7.2 Boosting operational systems

The food and beverage industry is developing at a progressively fast pace. New government regulations and consumer and stakeholder anticipations compel food industries, eateries, and grocery shops to focus more on food safety [111]. These veracities enhance consumer awareness and sensitivity to foodborne diseases. Meanwhile, the food safety risks associated with new and innovative operative, yet occasionally troublesome, models and preventive techniques are everyday struggles for several companies. The structure and management of food safety programs should reflect today's food safety landscape applicable to food and beverage industries, restaurants, and grocery shops.

Food safety is an organizational-level risk, so companies can no longer depend on a sole task to conquer food safety and quality [112]. Food industries must embrace the scientific, operational, and consumer perceptions of food safety to identify and moderate their organization's numerous challenges. In the face of increasing government regulations, the continual redefinition of food safety by consumers, coupled with evolving operational schemes to aid companies in maintaining significance in today's competitive world, it is undeniable that the food industry is undergoing continuous evolution [113, 114]. Consumers factor possible long-term effects of foods and ingredients into their purchasing decisions. They are also concerned about the immediate impacts of physical contaminations or foodborne illnesses.

Additionally, consumer trends bring manufacturers' focus to health and wellness, fresh and minimally processed foods, social impact, local sourcing, supply chain transparency, health, and fitness [1, 115]. Partakers of the farm-to-fork food supply chain are presented with a massive burden to cut costs while addressing consumer requests for more variety and customization. This demand compels food manufacturers to outsource critical operations to reduce costs whilst maximizing efficiencies. This situation calls for restaurants and food retailers to adapt smartly to the rapidly changing consumer demands. Thus, efficiency improvement and cost reduction have become necessary and standard operational requirements [5, 116].

23.7.3 Improving the bottom line

An organizational ethos consists of written and unwritten rules and the values that influence the behavior of everyone in such an institution. One of the essential things a food business can do is to foster a positive food safety culture [117]. Besides determining how effective the food safety program will be, it also affects the business's bottom line. Food safety errors are costly and damage the company's reputation [106]. A company's business may collapse if it is infested with pests, dirty, or lackadaisical about food safety [118–120]. In this modern age of information sharing via social media, one unfortunate incident can quickly destroy one's years of toiling to earn customer trust. This reason underscores that food manufacturers must deem food safety and hygiene to be the company's priority. The returns of investing time and resources into one's food safety training and promoting positive food safety culture far outweigh ways of neglecting such.

It enables the effective management of risks associated with food allergies. It protects consumers from health risks, including food poisoning [121]. It creates the confidence that the business is always inspection ready. It protects the company's brand and reputation from media scandals, negative online reviews, and complaints. It protects the company against costly fines and eventual closures from regulators [122]. It enables the company to save costs on food wastage via pest control and eradication services, improper stock control, and other unnecessary operational costs. Recruiting qualified staff to manage and take care of the daily operations of a food (retail) service (restaurant) enables top and middle management to focus on long-term strategies to develop the business [123]. It should always be acknowledged that employees are the heart and soul of every business. Investing them with requisite skills through training and short courses will enable them to diligently carry on with their job. They will contribute to the success story of the business and all other stakeholders [124].

23.7.4 Fraud reduction

The definition of food fraud varies depending on the opinion of an individual or an institution. The Food and Drug Administration (FDA) defined the term as economically driven food tempering, leading to food safety–related issues [125]. It is a deceitful exclusion act involving food production and or supply for personal gains at the expense of the consumer [126]. It is also defined as a thoughtful and intentional addition, substitution, misrepresentation, or tempering of food, food constituents, food packing, and misleading (dishonest) declarations involving food products for financial gains [127].

Food fraud involves any alleged deliberate action by any individual or business. Thus, food fraud is a deliberate act to acquire economic advantage whilst adulterating the food to mislead the consumer [128]. Food fraud is a potential hazard that negatively affects food safety and can have dire consequences on plants, animals, and even human health (life) [129]. In Spain, fraud involving rapeseed oil meant for industrial use (1981) affected about 20,000 whilst causing from 370 to 835 fatalities in Spain [130]. In Belgium (1999), dioxin resulted in substantial economic losses [131]. Milk adulteration with melamine in China (2008) caused 50,000 sicknesses among babies, with six fatalities and fifty-nine (59) casualties recorded in Poland and the Czech Republic (2012–2014) [132, 133]. Horse meat was detected among beef products (2013) [134], fipronil was detected in eggs (2017), and slaughtering of ailing cows (2019) was also reported [135–137].

Alcoholic beverages, cereals, coffee, dairy products, fish, fruit juices, grain-based foods, honey, meats and meat products, milk and dairy products, oils and fats, organic foods, pulses, seafood, spices, sweeteners, tea, wine, and alcoholic beverages together with some highly processed foods are often associated with food fraud. Unfortunately, the extent of food fraud is difficult to be estimated both at national and global levels, mainly due to the complexity of its nature.

Many analytical techniques used in the detection of food adulteration and fraud include chromatographic analysis [gas chromatography (GC), elemental/isotopic analysis, isotope ratio mass spectrometry (IRMS), high-performance liquid chromatography (HPLC), thin-layer chromatography], electrophoretic methods (amplified fragment length polymorphism, polymerase chain reaction (PCR), PCR-restriction fragment length polymorphism, random amplified polymorphic DNA, simple sequence repeats, sodium dodecyl sulfate-polyacrylamide gel electrophoresis). Others are immunoassays (enzyme-linked immunosorbent assay), mass spectrometry (MS)-based techniques [GC-mass selective detector, molecular composition (spectroscopic techniques), liquid chromatography with tandem MS (LC-MS-MS)], Raman, nuclear magnetic resonance (NMR), site-specific natural isotope fractionation NMR, inductively coupled plasma (ICP)-atomic emission spectrometer, ICP-MS], and spectroscopic methods Fourier transform infrared (FTIR), infrared (IR), near-IR (NIR), mid-IR (MIR)] as well as human sensory panels. The following are a few cited cases involving different analytical techniques to detect adulteration in other foods. The real-time PCR technique has been used to detect adulteration involving different milk species in cheese (Goncalves, Pereira, Amorim, & van Asch, 2012) and horse and donkey meats as adulterants in raw minced meat [138]. IRMS was used to detect milk differences in cheese from different geographical locations [139]. The visible NIR technique was also used to detect thawed chicken as an adulterant in pork [140]. Castro et al. [141] also used reverse-phase HPLC with ultraviolet radiation (RP-HPLC-UV) technique to detect soybean proteins as adulterants from beef, pork, chicken, or complex mixture meat products.

The application of analytical techniques to detect contaminants in food and products cannot be overemphasized. McGrath et al. [142] employed a two-tiered system of techniques to test rice fraud in selected countries, including China, India, Vietnam, and Ghana. The first tier involved a screening procedure based on a small hand-held (and field-portable] molecular spectroscopic technique that enables end-users with affordable, easy-to-use, and rapid "food-fingerprinting." Due to its simplistic and convenient nature, this technology can be used by stakeholders along the food supply (value) chain to quickly screen food samples to detect anomalies based on their established fingerprints. The second tier involves a particular laboratory-based analytical procedure like GC-MS, ICP-MS, and liquid chromatography quadrupole time-of-flight

MS (LC-QTOF-MS) to identify, quantify and confirm the abnormal screened results. The question at hand informs the choice of analytical technique(s) to be followed. Although this approach was followed to detect anomalies involving adulteration(s) in rice, this procedure can conveniently be adopted for other foods and products.

The consequences of falling short with regulators involving food fraud may not only affect one's reputation but can lead to the complete closure of the company. The losses incurred in food fraud far outweigh the gains, suggesting that various stakeholders along the food value chain should discourage and avoid the practice.

23.8 Infrared thermal processing technologies

Before using electronics and manufacturing industries, automotive curing applications initially utilized IR heating industrially during the 1930s [143, 144]. Following substantial progress in comprehending its operational mechanism over the past two decades, its application was extended to the food industry [145, 146]. Currently, the food industry applies the IR heating technique for roasting nuts, baking bread and other foods, and browning meats and other foods [147, 148]. The following are selected latest IR applications currently undergoing further improvements.

23.8.1 Infrared blanching

IR processing provides a better substitute for traditional blanching, saving water and energy [149, 150]. Blanching helps fruits and vegetables to attain suitable attributes whilst acquiring superior qualities [151, 152]. The produce is treated with steam or hot water (90–100 °C) and held for 1–10 min. The blanching is required to disable enzymes, like polyphenol oxidase (PPO), responsible for the senescence and subsequent spoilage of fruits and vegetables through food preparations, treatments and stowage [153]. Conventional food blanching involves substantial volumes of water, leading to large amounts of wastewater [154]. The process leaches out significant water-soluble nutrients, including minerals, ascorbic acids, and other vitamins, along with the wastewater. This setback is overcome by developing and using novel IR dry-blanching technology, which simultaneously offers the opportunity to achieve drying [149, 150].

23.8.2 Infrared drying

IR heating technology is noted for effective drying, especially high-moisture foods, which can be distributed in shallow layers using hot air-supported IR technology. This technology dries food products much faster than conventional hot air drying. Additionally, there is much improvement in the final product quality because of the shorter total processing time involved. Combining the IR with hot air drying facilitates initial moisture removal, followed by traditional drying to slowly remove the remaining moisture from the food. The IR drying technology has been demonstrated to be effective for bananas, onions, pineapples, apples, herbs, potatoes, blueberries, walnuts, rice, cashews, barley, and shrimp.

23.8.3 Infrared peeling

The food processing industry uses two techniques: hot lye and steam to peel fruits and vegetables [155, 156]. The main concerns confronting the food industry are the low quality of steam-peeled products and a large amount of wastewater. The ability to transfer intensive heat with its characteristic low penetration powers renders IR an appropriate technique to loosen the skin and promote peeling for fruits and vegetables [156, 157]. This technique requires no heating medium like water to deliver heat to the produce. This technique was patented and termed dry peeling in the United States.

IR peeling technology demonstrated significantly higher peelability and substantially reduced peeling loss compared to lye and steam peeling techniques [158, 159]. IR peeling technology was demonstrated to increase the peelability significantly whilst substantially reducing peeling losses, required amount of caustic lye and generated wastewater when peaches and white potatoes were subjected to IR dry-caustic peeling [160]. Two tested tomato processing factories successfully demonstrated superior quality performance using a designed and constructed pilot dry-peeling system equipped with IR emitters. The equipment with the technology was said to be undergoing further optimization and consideration to cater for different fruits and vegetables.

23.8.4 Infrared for microbial inactivation

IR is a physical (nonchemical) technique that can effectively decontaminate foodborne microbial pathogens and food spoilage microorganisms whilst maintaining good product quality [161]. The potency of this technique for microbial deactivation is dependent on IR heating bandwidth, intensity power, peak wavelength, food temperature, microbial growth phase, microbial types, size, and type (matrix) of food (product) [162]. The competence of using IR heating has been studied for various objectives involving fruit surface decontamination, milk sterilization, rice disinfestation, and almond pasteurization.

The applications of IR (heating) technology for treating foods and food products have been widely studied to achieve different objectives. A study to investigate microbial decontamination of black pepper and onion flakes using a novel method of combining infra-red (IR), ultraviolet (UV) and ozone demonstrated different efficacies against the two spices [163]. The treatment was generally more effective for the black pepper than the onion flakes. Treatment with ozone and UV ultimately reduced the artificial bio-burden of the two spices in less than 20 min compared to IR treatment in 30 min. Sequential ozone and combined IR and UV yielded improved results by inactivating 99.99% of *E. coli* over shorter periods. In another study, IR proved very effective in decontaminating eggs to prevent Salmonellosis [164]. In another study involving the detoxification of hazelnuts from aflatoxins, a traditional static hot air roasting was compared to an IR roasting method in terms of efficiency and their effects on the product quality [165].

The results indicated that both treatments were influential in detoxifying the aflatoxins from the hazelnuts without compromising nutritional quality. However, testing the two methods under similar conditions revealed that roasting with the IR method induced higher decontamination of the different aflatoxins (aflatoxins B1, G1, B2, and G2) than the traditional static hot air roasting method. These results render the IR roasting method a preferred choice over the other. However, in another study to evaluate the resistance of parental and an antibiotic (nalidixic acid) adapted bacteria like Shiga toxin-producing *E. coli* (STEC) strains and serotypes to IR and pulsed ultraviolet treatments (PUVs) [166], showed that while some *E. coli* O157:H7 parental strain 4 was the most resistant to IR radiation, *E. coli* O157:H7 parental strain 5 was most resistant against PUV out of the several serogroups tested. Some of the nalidixic acid-adapted strains demonstrated less resistance to PUV and IR treatments than their parental strains.

A study to evaluate a novel nonthermal combined technology using modulated IR light, ozone, and UV-C radiation to decontaminate *E. coli*-challenged chili flakes demonstrated that IR and ozone treatment ultimately reduced the artificial bioburden in less than 20 min, unlike UV-C treatment that took about 40 min [167]. The combination treatment, however, yielded a better treatment order of UV-C and IR and then ozone than ozone, UV followed by IR. Thus, the combined therapy of chilis with IR, ozone and UV-C showed a better performance than their treatments alone. *Bacillus cereus* in cardamom was decontaminated to acceptable levels in another study [168] using IR treatment modeled by a genetic algorithm-artificial neural network (GA-ANN). Thus, the application of IR, ozone, and UV has been exploited for the treatment of different foods, including onion and onion products [169], food grains [170], water [171] peanut [165], corn [172], etc. has been extensively discussed [169]. The purpose is to decontaminate (disinfect) microbial bioburden [171], detect microbial (fungal), toxin (mycotoxin) contamination [173], and reduce (inactivate) toxins (mycotoxin) in foods with the aim of safety assurance and shelf-life extension [172, 174].

23.8.5 Prospects of IR thermal processing technologies in food safety

Undeniably, IR heating might not be appropriate for the thermal processing of all foods due to its low penetration power. However, it offers considerable gains in specific applications [175, 176]. The technology is expected to grow with more comprehensive applications within the food industry, with future processing techniques requiring surface heating. Additionally, the technology is expected to be controlled mainly in terms of heating to meet specific food processing requirements in the years ahead. This technique should enable the provision of selective heating with the best possible outcome.

23.9 Next-generation sequencing

Next-generation sequencing (NGS) involves various methods for sequencing novel DNA, which profoundly impacts multiple areas of applications in the agricultural, food, and health sectors. There is a continuous increase in applying NGS-based techniques for food safety applications to detect food pathogens, trace the origin of disease outbreaks, and test food authenticity (fraud) [177, 178]. Considering the significance of NGS to food safety, numerous emerging and established players are venturing into this landscape. Hardware inventions in NGS are mostly restricted to incremental advancements in scale and speed but are dominated mainly by a few big players [179, 180]. Among the NGS technologies, whole-genome sequencing (WGS), which detects the presence of a single cultured and sequenced isolate, is remarkable within the food safety framework. As of 2020, the USA and a handful of other European countries demonstrated higher adoption rates of the WGS for food safety [181, 182]. The technique will likely expand further in the coming years because of strict food safety regulations.

The food sector importantly requires vigorous testing techniques for food pathogen detection considering recent pathogen outbreaks, for example, the *Escherichia coli* outbreak in romaine lettuce in the Canada and USA (2018), during which the WGS was predominantly used for the outbreak source detection [183, 184]. Also, enacting the Food Safety Modernization Act (FSMA) in 2011 catalyzed the growth of WGS technology in the USA [185, 186]. The FDA's Genome Trakr is a prominent example of a distributed network of laboratories utilizing WGS for pathogen detection and monitoring for the attainment of food safety. Several (~ 250,000) uploads of raw sequences of *E. coli*, *Salmonella*, *Listeria*, and *Campylobacter* have been made available for sharing [187, 188].

The essence of streamlining the collection and interpretation of WGS data is under development in the EU. A scientific opinion was published (December 2019) by the European Food Safety Authority (EFSA) on the applicability of NGS in food-borne investigations [189]. The report demands capacity building regionally and globally grounded on coordinated, quality-controlled operational systems among European research labs as a critical requirement for the investigations of cross-border epidemics. It was anticipated that the EU would drive the broader adoption of NGS, including WGS. However, the adoption of the WGS in developing countries is impeded by a gross lack of supporting infrastructure and investment in these technologies [190]. This situation is worsened by inadequate (lack) of the requisite expertise for WGS data storage, handling, and interpretation [191].

23.9.1 Prospects of NGS/WGS in food safety

There will be increased interest in food safety applications, especially in the post-CoViD era considering the necessity to mitigate the hazards related to foodborne illnesses [183]. This interest will offer substantial opportunities to improve food pathogen detection and traceability technologies, further boosting the interest and broader adoption of NGS/WGS in food safety management schemes [192]. Bioinformatics will be pivotal to interpreting and applying biological data from NGS techniques, providing a robust and combined approach. Interested individuals and institutions should be capitalized on the steady increase of NGS techniques in food safety. When exploring these techniques, it is imperative to consider factors including ease of use, speed, the cost of food pathogen, and accuracy.

23.10 The impact of some technological innovations on food safety

Recent foodborne outbreaks, the need for contemporary food preservation approaches and consumer demand for fresh, healthy, convenient, diverse, and safe food have prompted some collaborative efforts by the food industry, regulators, and researchers. These efforts are geared toward proactively improving safety strategies and preservation techniques that can be applied successfully throughout the food supply chain [193, 194]. A range of factors along the farm-to-fork chain, including quality of raw material, processing techniques, packaging, transportation, storage, and cooking, contribute immensely to a food product's nutritional and safety quality. Raw materials are transformed into value-added food products through unit operations. The final quality of food, such as appearance, texture, taste, and nutritional content, is strongly impacted by how it is processed. Innovative food processing technologies have recently been widely investigated in food processing research. Within the arsenal of preservation techniques, innovative thermal and nonthermal technologies for food processing, depending on their primary principle of action, have been investigated and developed to modify the prevailing food processing techniques or replace the traditional techniques. This section discusses some prominent technologies that have either found commercial application in the food industry or are still under investigation.

Indisputably, research has shown that some novel food processing techniques have increased production and process efficiency with minimal or no changes in the nutritional properties of foods. They have further reduced energy consumption, improved product safety, and reduced food wastage by improving shelf life and meeting other essential consumer drivers such as convenience. Although efforts have been made to improve food processing techniques, there is no ultimate technology that eliminates pathogens from the food supply. Depending on the food product, a combination of multiple intervention technologies (thermal and nonthermal) to inactivate food pathogens, also known as the hurdle approach, has shown some potential in improving food safety. The food processing business is still hampered by persister organisms (such as *Listeria monocytogenes*) in the food processing environments, inactivating bacterial spores in minimally processed or ready-to-eat foods and mycotoxin contamination in a specific type of food. Mycotoxins such as aflatoxin, ochratoxin A, fumonisins deoxynivalenol, and zearalenone are not destroyed during food processing operations. Consequently, they have the potential to cross-contaminate finished processed foods [195]. On the other hand, removing naturally occurring chemicals toxins in food and chemical contaminants from food processing is also a daunting task for the food industry [196].

23.11 Novel thermal technologies and food safety

Thermal processing relies on heat to reduce or destroy microbial and enzymatic activities and which result in physical or chemical changes to make the food meet a certain quality standard. Food heating, mainly through cooking, is arguably the oldest means of processing foods and has been the principal method of eliminating pathogens in food [197]. Heat safely preserves food by pasteurization, sterilization, hot-air drying, and ultrahigh temperature. In essence, blanching, cooking, baking, roasting, and frying processes have improved food quality and safety. The challenge of thermal processing is that the extreme temperatures that destroy microorganisms also cause the loss of nutrients, bioactive compounds, and sensory food properties, resulting in low-quality food. However, a range of novel thermal processing techniques have been developed to improve the quality of foods while minimizing the processing impacts such as thermal degradation [194].

Numerous investigations in dielectric heating (including microwave and radio-frequency heating), ohmic heating, inductive heating and IR heating have demonstrated their effectiveness in ensuring product safety, quality and acceptability compared to conventional thermal processing techniques. These techniques have replaced, at least partially, the heating technologies that relied predominantly on conductive, convective,

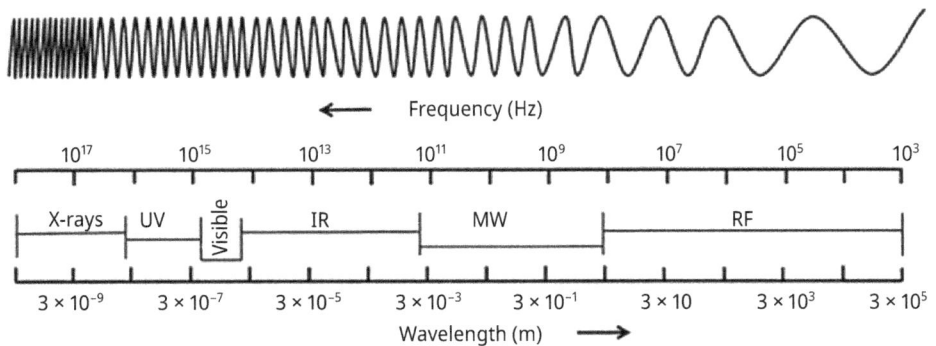

Figure 23.3: Representative scheme of the entire distribution of electromagnetic radiation according to frequency or wavelength [199].

and radiative heat transfer [198, 199]. Microwave, radio-frequency and IR heating are based on the electromagnetic spectrum (Figure 23.3). Each technology covers a different frequency and wavelength range to deliver heat to foods selectively and efficiently.

Novel thermal techniques have a standard feature, which involves heat being generated directly inside the food by molecular dipole rotation and ionic migration by alternating the electrical field or electric conductivity. This directly affects energy and heating efficiency [199]. The advantages of thermal technologies that employ an electromagnetic field or electric conductivity are the partial retention of the food product's original nutritional and sensory attributes and the inactivation of unwanted microbes in the food product [200].

23.11.1 Microwave heating (MH)

Microwave heating (MH) is a process involving electromagnetic radiation or a high-frequency alternating electric field that is transformed into thermal energy by affecting the polar molecules [201]. The rapidly varying electric and magnetic fields generate heat (volumetric heating) inside a food product, and any food material exposed to the radiations is heated up. This process generates heat throughout the material, leading to faster heating rates over a short contact time. MH is dependent on the dielectric (electric properties) and magnetic properties of the heated material. The dielectric properties of the food material influence how the food interacts with electromagnetic energy [194]. When a food material (dielectric material with poor electrical conductors) is brought into a rapidly alternating electric field, the material absorbs electromagnetic energy very rapidly, generating heat inside the material, a process known as dielectric heating. Microwave frequencies range between 300 megahertz (MHz) and 300 gigahertz (GHz) or 300×10^6 to 300×10^9 Hz (cycles per second) (Figure 23.3), but the FDA has approved two microwave frequencies (915 and 2,450 MHz) for application to foods; with most domestic microwave ovens using 2,450 MHz [202].

MH is a popular and well-studied novel food processing technology applied domestically and industrially. MH has long been applied to food processing activities such as drying, pasteurization, sterilization, thawing, tempering, precooking, and baking food products. Its main advantage is high heating rates, which reduces processing time and saves energy (Table 23.1). Because of reduced processing time (high-temperature short-time), microwave processed foods have better sensory and nutritional quality compared to products processed by conventional heat treatment. However, the technology has not entirely replaced the traditional thermal processes because of the inhomogeneity of the electromagnetic field and low food penetration depth if 2,450 MHz microwave systems are used, resulting in nonuniform heating of food [202].

The nonuniform heating of food has implications on product safety (the incomplete kill of microbes due to uneven temperature distribution) and quality. Apart from nonuniform heating, studies have revealed that MH is a more negligible effect on bacterial spores due to shorter times [201, 202]. Time is a critical factor affecting the effectiveness of MH when considering the safety of processed food products. However, continuous microwave systems have been designed to offer a uniform solution to heat food products potentially. Research has also revealed that MH has been successfully combined in batch and continuous forms with RF heating to obtain the benefits of both dielectric and conduction forms of heating [194]. Several commercial microwave instruments are currently used in Belgium, Holland, Italy, Japan, and the USA for multiple food sterilization and application.

23.11.2 Radiofrequency heating (RFH)

Radiofrequency electric fields as a pasteurization method have been studied for more than six decades [203]. Radiofrequency heating (RFH) applies the same dielectric heating principles as MH. However, its application in food processing is relatively less common. Although radio frequencies range between 3 kiloHertz (kHz) and 300 MHz (Figure 23.3), only the frequencies of 13.56, 27.12, and 40.68 MHz are permitted for food processing. RFH relates to the fact that the molecules within a food product placed in a radio frequency environment re-orient themselves (27 million times/s at 27 MHz) continuously in response to the applied field (i.e., dipole heating) [203]. RFH has a deeper penetration due to its lower frequencies, longer wavelength, and better uniform electric field distribution than MH. The main applications of RFH are blanching, thawing, drying and processing foods as supplementary heat in drying pasta, crackers and snacks [204]. Radiofrequency heating can improve the food quality and reduce the surface overheating experienced with MH [199].

The effect of radiofrequency energy on the inactivation of microorganisms is due to heat. However, research has discovered that time–temperature regimes are much milder than those required with conventional heating techniques [204]. During radio-frequency processing, dielectric materials are placed in an alternating electric field. The benefit of using this technology is the potential selective destruction effect on microorganisms other than that attributable to a heating effect or targeted, localized heating of the microorganisms [203]. Because of this, a cell might be destroyed thermally at a comparatively low heating rate if heat can be generated faster in the microbial cell than in the suspension medium. RFH has been reported to control pathogens in foods successfully (*Bacillus cereus* and *Clostridium perfringens* in pork luncheon meat, *E. coli* and *Listeria innocua* in milk and *Clostridium sporogenes* in scrambled eggs) because of fast and volumetric heating, as well as reducing loss in food quality [204]. Radiofrequency use is reported to give better results than MH due to the better homogeneity of heating, greater penetration depth, and more stable control of the product temperature. However, due to technical problems such as dielectric breakdown and thermal runaway, which can damage the product and the packaging, the technology has not been fully adopted by the food industry. Moreover, full commercial adoption is hampered by the high operational costs of using radiofrequency systems [199].

23.11.3 Ohmic heating (OH)

Ohmic heating, also known as Joule heating, electro heating or electro-conductive heating, is an advanced thermal processing method where an electric current is passed through a food material with the primary purpose of rapidly increasing the internal temperature of the food for either cooking or sterilization purposes [205]. OH applies the principle of dissipation of electrical energy in the form of heat using an electrical conductor. Therefore, the electric conductivity of foods determines the heat generated in the food material due to resistance [199]. Usually, the frequency employed in OH is 50 or 60 Hz, although it can vary depending on the food material's electric conductivity. The advantage of the ohmic thermal process over MH and RFH is its ability to rapidly and uniformly heat liquid food products, reduce fouling, and low cost and energy efficiency.

Moreover, OH treatment has no penetration depth limitation, can avoid hot surfaces, and reduce the temperature gradients, resulting in improved product quality and safety [194]. Nonetheless, OH technology provides post-processing product quality like MH and RFH compared to conventional heat treatment methods. OH is mainly applied in blanching, evaporation, thawing, sterilization, pasteurization, and pre-treatment on mass-transfer operations such as extraction and dehydration of foods [194, 199]. In addition to advantages of OH such as high-temperature short time and ultra-high temperature technique on solids or suspended materials, ohmically processed foods have extended shelf life comparable to that of canned and sterile processed products. Thermal and nonthermal (electroporation) cell damage are the two primary microbial inactivation mechanisms. OH technology has been applied in food processing on a commercial scale across the globe in fruits and vegetable processing, particularly for liquid and viscous products and pumpable multiphase foods [194, 199].

23.11.4 Infrared heating

Infrared heating (IRH) is a short-time high-temperature process that uses electromagnetic waves like MH and RFH. IR radiation lies between ultraviolet and microwave energy regions, and the frequencies are above 300 GHz or in the wavelength range from 0.7 to 1,000 mm (Figure 23.3). IRH spectral ranges are generally categorized into NIR (700–1,400 nm), MIR (1400–3,000 nm), and far-IR (3,000–10,000 nm) regions. However, far-IR is suitable mainly for food processing because most food constituents absorb radiation in that region [206]. According to Priyadarshini et al. [194], IRH (radiant heating) is an indirect heating mode wherein electromagnetic energy penetrates the food, gets adsorbed on the surface and then converts to heat. The heat adsorbed on the food surface is mostly by radiation but by convection and conduction mechanism to a lesser extent. Because of the food surface characteristics and color, IRH is also applied to alter food quality. Among the advantages of IRH is rapid and uniform heating, reduced processing time and energy costs, and improved product quality.

IRH has been commercialized for cooking, frying, drying, dehydration, roasting, baking, peeling, blanching, and surface pasteurizing ham and other meat products. IRH has shown great potential in the decontamination of pathogenic microorganisms in food. The technology is used to reduce bacteria and its spores and yeast on the surface of RTE products while retaining product quality. Microorganism inactivation depends on thermal effects and damage to the DNA, RNA, and proteins of microbial cells due to the radiation [199, 207]. Despite the promises, the use of the IRH process has not yet been successfully marketed as a single technology. IRH, combined with other intervention technologies, provide a promising tool for the inactivation of microorganisms on the post-packaged product. Studies have suggested that IR surface pasteurization is an effective technology to decontaminate *L. monocytogenes* on cooked meat products' surfaces before final packaging with minimal quality deterioration [207]. Furthermore, IRH technology has been successfully employed to inactivate lipooxygenase, lipases, α amylases and other enzymes responsible for developing off-flavor and deterioration of fruits and vegetables. The major setback of IRH technology is the poor penetration capacity, which can affect product safety, thus the need to use the system with other novel technologies [208].

23.12 Novel nonthermal technologies and food safety

Due to the recurrent consumer-driven demands for foods that are minimally processed fresh like food products, there is an increased interest in developing alternative nonthermal or cold pasteurization techniques in food processing. Applying thermal technologies in food processing undoubtedly reduces or eliminates microorganisms contaminating food but often adversely affects many of its organoleptic characteristics and nutritional properties. Nutrients, including vitamins, minerals, and nutrients with functional properties such as pigments, antioxidants, and bioactive compounds, are sensitive to heat [198, 209]. Although cold pasteurization techniques can preserve the product's quality attributes while lowering foodborne levels, they cannot presently be trusted to inactivate bacterial spores [209]. To achieve the food safety goals, nonthermal techniques must be combined with an additional preservation hurdle such as refrigeration or acid formulation to prevent spore outgrowth. Nonthermal processing technologies such as high hydrostatic pressure (HHP) and ultrasound (US), pulsed electric fields (PEF), ultraviolet light (UV), pulsed light (PL), and cold plasma (CP) allow the processing of foods below temperatures used during thermal pasteurization, so flavors, essential nutrients, and vitamins undergo minimal or no changes.

23.12.1 High hydrostatic pressure (HHP)

The most well-developed and studied nonthermal technology is HHP. The technology has probably advanced further than any alternative physical food processing technology. HHP applies a mechanical ultra-high hydrostatic pressure of around 100–800 MPa to a packaged product into a handling basket that is placed into a vessel, and the pressure is transmitted uniformly and instantaneously throughout the pre-and post-packaged food product [193, 210]. The process of HHP pressurization takes between 2 and 5 min, and during the process, there is adiabatic heating, causing the water temperature to increase by 3 °C for every 100 MPa [211]. HHP is applied chiefly to moisture foods and acts instantaneously and uniformly throughout a mass of food independent of size, shape and composition, which means that the food retains its shape and texture despite undergoing extreme pressure. Therefore, there is no gradient of effectiveness from outside to inside as there is with thermal processing [199]. The critical parameters that need process optimization are pressure, temperature, and time.

Microorganism inactivation by high pressure has been reported, and the mechanism of destruction is caused by structural changes in the cell membrane without heat [212]. Again, some chemical reactions induced by HHP can cause microbial damage. The compression of the food product may alter the pH of the food as a function of imposed pressure and this inactivation of microbes. Successful inactivation of vegetative microorganisms by HHP in plant-based products, egg products, dairy, seafood, meat and beverages have been reported [212]. Microorganisms are pretty variable concerning their sensitivity to HPP. However, research suggests that the order of sensitivity is Gram-negative bacteria > yeasts > Gram-positive bacteria > bacterial spores [193]. One challenge of HHP is its inefficiency in inactivating bacterial spores at high pressures of up to 1,000 MPa. Therefore, combined treatment of parameters such as pressure, mild heat and low pH is typically required for inactivation and control of outgrowth. Notably, HHP has the advantage of pasteurizing food products without modifying the nutritional and sensory properties of the product in contrast to conventional processing methods [213, 214].

Since the commercialization of HHP in the mid-1990s, the process has been successfully applied to heat-sensitive liquid and solid foods currently on the market, such as guacamole, jams and jellies, fruit juices, and tomato salsas, and applesauce. Of importance to note is the application of HPP to inactivate microbes in ham, other cooked RTE meat products and seafood products. HPP is effective when applied to high-acid foods to extend shelf life or improve food safety. Today, high-pressure pasteurization has become a commercial reality in many commercial operations worldwide [17, 194].

23.12.2 Pulsed electric fields (PEF)

Due to the growing demand for safe food with nutritional qualities, PEF has been a commercialized nonthermal technology since 2005. The process involves the application of high voltage pulses (15–80 kV/cm) for a short duration (μs to ms) to liquid or semi-solid foods that can flow easily and pass between two electrodes in a PEF chamber [214]. Critical parameters which need process optimization during PEF treatment are electric field strength, treatment time, pulse shape, pulse width, pulse frequency, pulse polarity, and temperature. Among these parameters, the electric field strength is most important when considering cell damage and microbial inactivation – increasing the strength of the electric field and treatment time results in higher microbial inactivation [193]. PEF processing is unsuitable for solid food products with no air bubbles and low electric conductivity. However, PEF effectiveness is also influenced by the microorganism and medium characteristics [194, 215].

The mechanism of microbial deactivation in food treated with PEF is electropermeabilization (electroporation- which is the formation of pores on the cellular membrane) of the cells without thermal effects [193, 194, 209]. Other studies have also reported the structural arrangement of microbial enzymes, electromechanical compression, and osmotic imbalance as an explanation for microbial death [216]. Food is exposed to PEF for a concise duration with no undesirable changes in food due to high temperature. High safety levels have been proven in PEF-treated foods such as fruit juices, milk, yoghurt, soups, cooked meats, liquid eggs and other pumpable food products. Successful elimination of *E. coli*, *L. innocua*, *S. aureus*, Enterobacteriaceae, and *P. fluorescens* has been reported [193]. Just like HHP, sterility levels could not be achieved with PEF since the treatment cannot destroy spores on its own [217]. However, PEF, combined with other technologies, can produce shelf stabile foods even at room temperature.

23.13 Ultraviolet (UV) light and pulsed light (PL)

Light treatments are innovative minimal food processing technologies that improve the safety of food products and maintain their appearance and nutrient content while extending their shelf life [18]. UV (the most effective light treatment in killing microbes in food) processing uses radiation (at 254 nm) from the ultraviolet region of the electromagnetic spectrum (Figure 23.3). PL involves the application of intense and short pulses (100–400 μs) of light (200–1,110 nm) [218]. Parameters to consider in the process optimization of light treatments are the transparency of the medium, energy dose, pulses number, the depth of the samples for PL and the power, wavelength and treatment time for UV [219].

UV's antimicrobial properties are believed to be due to nucleic acid damage, cytoplasmic membrane integrity damage, and cellular enzyme activity without thermal effects. PL uses the exact mechanism of microbial deactivation, although it also causes protein denaturation and other photothermal and photophysical effects. UV and PL have been used for years to decontaminate food factories' air, water and surfaces. Recently, light treatments have been applied to decontaminate surfaces of fruits, vegetables, liquid food, chicken meat, baked products, water for food and beverage formulation, and wash water [193, 220].

UV and PL significantly eliminated pathogens and spoilage microflora in transparent and translucent food products, including fruit and vegetable juices, particularly as some change in flavor and color when thermally treated [193]. Although UV and PL have been approved for the treatment of fruit juices because of their low capital cost and energy efficiency, the technology is less effective to dense and opaque due to poor penetration power. Regarding spore inactivation, UV has little effect. Hence, the need to combine the technology with other intervention methods reap higher benefits, particularly for products that cannot be treated thermally. On the contrary, studies have reported that PL can inactivate bacterial spores [221].

23.13.1 Ionizing radiation (IR)

Radiation is a nonthermal food preservation technology that applies a low IR level (gamma rays, x-rays, or electron beams) to inactivate microorganisms, spores, naturally occurring chemical toxins, yeasts, molds and parasites in foods [193]. Radiation is carried out using gamma rays emitted from radioactive forms of the element cobalt

60 or cesium 137, X-rays produced by reflecting a high-energy stream of electrons (high energy of up to 5 MeV) off a heavy metallic substance or electron beam wherein the high-energy electrons (high energy of up to 10 MeV) are propelled from an electron accelerator into food [222]. Relative to microwaves, gamma and X-rays are high-frequency irradiation with more penetration power (Figure 23.3). They penetrate food and inactivate microorganisms rapidly without thermal emission; therefore, the food remains intact. The process uses very little energy to inactivate microorganisms without increasing temperature, and because of this, there is no chance of damage to components in food that are heat sensitive. Some studies have shown that up to one kiloGray (kGy) of ionizing radiation inactivates *E. coli* inside or on food sample surfaces to about a 5-log reduction without meaningfully changing the quality, thus making the technology a promising solution for the produce industry [223, 224]. The mechanism of microbial inactivation is damage to the nucleic acid and the ionization of water molecules, resulting in oxidative damage to microbial cells. No radioactive waste is produced at food processing facilities. However, the process optimization parameters to consider when using IR are irradiation dose and the time duration the food is in contact with the radiation source [193, 194].

Although the use of IR on food to preserve food has since been approved, the regulations for how food is irradiated and the foods allowed to be irradiated vary significantly from country to country [209]. Recent developments include regulatory approval of irradiation for RTE meat products and juices. Irradiation is mainly employed in the food processing sector for the preservation of food products where the technology is used to minimize the post-harvest loss, retain the color of fresh meat, inhibit sprout formation in products such as potatoes and control post-packaging contamination in a range of food products including cereals, legumes, spices, poultry, fish, seafood, meat, fruits vegetables, tubers and dried vegetable seasonings [214]. Although the technology is effective against foodborne pathogens, its use is limited to food types such as milk and high lipid and vitamin content foods. This is because radiation causes peroxidation of unsaturated bonds present in the polyunsaturated fatty acids (especially omega 3, C22.5, and C22.6 fatty acids), increases the onset of oxidative rancidity in milk and high lipid foods, and also destroys vitamins, in particular vitamin B1 [194].

23.13.2 Ultrasonication processing (USP)

Ultrasound processing has since been used in other processing sectors but is an emerging nonthermal technology in the food sector [209]. Ultrasound technology is based on a series of compression and rarefaction cycles induced by sound waves on the molecules of the medium they pass through at a frequency above the threshold of human hearing, i.e., sound wave bearing a specific frequency that is more than the regular human hearing frequency (>20 kHz) [209, 214]. Mechanical waves travel through the material or on its surface, leading to cavitation bubbles forming. These bubbles distribute throughout the liquid at a high ultrasound power, and at high acoustic pressure, they grow to a critical size throughout a few cycles and violently collapse. This leads to energy accumulations in hot spots and generates extremely high pressure (up to 100 MPa) and temperature (up to 5,000 K), producing shear energy shock waves and turbulence in the cavitation zone. Combining these microevents can induce various physical and chemical properties (such as breakdown of the water molecules, disruption of the cell wall of biological tissue or polymeric chain of biomolecules), which can be harnessed in food processing [194]. During food processing, frequency in the range of 20–100 kHz is used to extract bioactives, emulsification, cooking, debittering and intensified synthesis [209].

USP has been used to ensure food safety by inactivating pathogens/spoilage microorganisms in foods such as milk, dairy products and fruit juices: blueberry, orange and strawberry. The primary mechanism of microbial destruction is cavitation. Though USP-assisted processing has advantages, studies have shown that it does not affect reducing microorganisms, including spores in food, when applied alone. Therefore, combining USP with other novel food preservation technologies can enhance its effectiveness [193].

23.13.3 Cold plasma (CP)

CP is the most recent nonthermal intervention method that uses high voltage electricity to ionize a gas (such as air) to generate a plasma field. Plasma is the fourth state of matter after solid, liquid and gas. It is an ionized gas that consists of a large number of differently charged species (such as the electron, ions, photons and free radicals, as well as gas atoms and molecules in their actual or excited states) which are produced by providing energy to a neutral gas causing the production of these charged carriers [209]. Microorganisms, including spores exposed to this nonthermal gaseous energy in the form of reactive chemical species, are eliminated on contact, and the plasma dissipates immediately when the electrode is turned off [225].

CP was initially developed to enhance polymer surface energy and sterilize medical equipment in hospitals. However, the technology is now employed in the food industry for in-package and post-packaging decontamination of food products, including the dry disinfection of solid and liquid food surfaces like dried milk, meat, poultry, fish, herbs, sprouted seeds, grains, spices, and fresh produce [193, 194].

23.14 Conclusion

The supply of food products that are wholesome, nutritious, and safe is a crucial process from farm to fork. Each processing technology utilizes different means of bacterial decontamination, which may be combined or applied as a stand-alone to effect injury or stresses on contaminating bacterial cells, eventually leading to metabolic exhaustion and death. Nevertheless, some stress-resistant bacteria and bacterial spores such as *Bacillus* and *Clostridium* are still of concern. Regardless of the effectiveness of these technologies, it is pertinent to monitor food ingredients for chemical contaminants and water used in food production that may put the consumers in harm's way. Everyone, including the consumers, is responsible for ensuring food safety, with the greater responsibility on the food industry, regulators and government. Various food technologies must interact to produce safe and nutritional foods that retain their sensory characteristics. Because of continual advances in food technology and the seemingly unlimited supply of newly discovered functional ingredients, the sky is the limit for producing novel food items with health advantages. The innovation process might take two directions: from research to practice or from practice to research. Nutritional data is being translated into consumer items faster because of advancements in food technology. As a result, more research will be required in several key areas, including ingredient synergy, biological efficacy, and the safety aspects of long-term usage of value-added food made using new food technology. Developing new healthy foods promises to be an exciting endeavor that will almost surely surpass our wildest hopes.

References

[1]   Fung, F., Wang, H. S., & Menon, S. (2018a). Food safety in the 21st century. *Biomedical Journal, 41*, 88–95. https://doi.org/10.1016/j.bj.2018.03.003

[2]   WHO estimates of the global burden of foodborne diseases. (2015). https://apps.who.int/iris/handle/10665/199350

[3]   Devleesschauwer, B., Haagsma, J. A., Mangen, M.-J.-J., Lake, R. J., & Havelaar, A. H. (2018). In The global burden of foodborne disease. *Food safety economics* (pp. 107–122). Cham: Springer International Publishing. https://doi.org/10.1007/978-3-319-92138-9_7

[4]   Scharff, R. L. (2018). *The economic burden of foodborne illness in the United States* (pp. 123–142). In Food Safety Economics. Springer International Publishing. https://doi.org/10.1007/978-3-319-92138-9_8

[5]   Focker, M., & Van der fels-klerx, H. J. (2020). Economics applied to food safety. *Current Opinion in Food Science*. https://doi.org/10.1016/j.cofs.2020.10.018

[6]   Gorton, A., & Stasiewicz, M. J. (2017). Twenty-two years of u.s. meat and poultry product recalls: Implications for food safety and food waste. *Journal of Food Protection, 80*, 674–684. https://doi.org/10.4315/0362-028X.JFP-16-388

[7]   Taylor, M., Klaiber, H. A., & Kuchler, F. (2016). Changes in U.S. consumer response to food safety recalls in the shadow of a BSE scare. *Food Policy, 62*, 56–64. https://doi.org/10.1016/j.foodpol.2016.04.005

[8]   Hussain, M. A., & Dawson, C. O. (2013). Economic impact of food safety outbreaks on food businesses. *Foods, 2*, 585–589. https://doi.org/10.3390/foods2040585

[9]   CDC. (2016). Foodborne Germs and Illnesses. www.cdc.gov/foodsafety. https://www.cdc.gov/foodsafety/foodborne-germs.html (accessed 10.5.20).

[10] Alvarenga, V. O., Campagnollo, F. B., Do prado-silva, L., Horita, C. N., Caturla, M. Y. R., Pereira, E. P. R., Crucello, A., & Sant'Ana, A. S. (2018). Impact of unit operations from farm to fork on microbial safety and quality of foods. *Advances in Food and Nutrition Research*, 131–175. https://doi.org/10.1016/bs.afnr.2018.02.004

[11] Stein, R. A., & Chirilă, M. (2017). Routes of transmission in the food chain. In *Foodborne diseases* (pp. 65–103). Elsevier. https://doi.org/10.1016/B978-0-12-385007-2.00003-6

[12] Begley, M., & Hill, C. (2015). Stress adaptation in foodborne pathogens. *Annual Review of Food Science and Technology*. https://doi.org/10.1146/annurev-food-030713-092350

[13] Liao, X., Ma, Y., Daliri, E. B.-M., Koseki, S., Wei, S., Liu, D., Ye, X., Chen, S., & Ding, T. (2020). Interplay of antibiotic resistance and food-associated stress tolerance in foodborne pathogens. *Trends in Food Science and Technology, 95*, 97–106. https://doi.org/10.1016/j.tifs.2019.11.006

[14] De Ruyck, K., Huybrechts, I., Yang, S., Arcella, D., Claeys, L., Abbeddou, S., De Keyzer, W., De Vries, J., Ocke, M., Ruprich, J., De Boevre, M., & De Saeger, S. (2020). Mycotoxin exposure assessments in a multi-center European validation study by 24-hour dietary recall and biological fluid sampling. *Environment International, 137*. https://doi.org/10.1016/j.envint.2020.105539

[15] Alamri, M. S., Qasem, A. A. A., Mohamed, A. A., Hussain, S., Ibraheem, M. A., Shamlan, G., Alqah, H. A., & Qasha, A. S. (2021). Food packaging's materials: A food safety perspective. *Saudi Journal of Biological Sciences, 28*, 4490–4499. https://doi.org/10.1016/j.sjbs.2021.04.047

[16] Leistner, L., & Gorris, L. G. M. (1995). Food preservation by hurdle technology. Trends in Food Science and Technology. https://doi.org/10.1016/S0924-2244(00)88941-4

[17] Pou, K. R. J., & Raghavan, V. (2020). Recent advances in the application of high pressure processing-based hurdle approach for enhancement of food safety and quality. *Journal of Biosystems Engineering, 45*, 175–187. https://doi.org/10.1007/s42853-020-00059-6

[18] Elegbeleye, J. A., Gervilla, R., Roig-Sagues, A. X., & Buys, E. M. (2021). Ultraviolet-C inactivation and hydrophobicity of Bacillus subtilis and Bacillus velezensis spores isolated from extended shelf-life milk. *International Journal of Food Microbiology, 349*, 109231. https://doi.org/10.1016/j.ijfoodmicro.2021.109231

[19] Pinela, J., Ferreira, I. C. F. R., & Pinela, J. E. (2017). Critical Reviews in Food Science and Nutrition Nonthermal physical technologies to decontaminate and extend the shelf-life of fruits and vegetables: Trends aiming at quality and safety Nonthermal physical technologies to decontaminate and extend the shelf-life of fruits and vegetables: Trends aiming at quality and safety. https://doi.org/10.1080/10408398.2015.1046547

[20] Hefferon, K. L., Feei Ma, Z., Picart-Palmade, L., Cunault, C., Chevalier-Lucia, D., Belleville, M.-P., & Marchesseau, S. (2019). Potentialities and limits of some non-thermal technologies to improve sustainability of food processing. *Frontiers in Nutrition* www.frontiersin.org, *1*, 130. https://doi.org/10.3389/fnut.2018.00130

[21] Bosso, H., Barbalho, S. M., De alvares goulart, R., & Otoboni, A. M. M. B. (2021). Green coffee: economic relevance and a systematic review of the effects on human health. *Critical Reviews in Food Science and Nutrition, 0*, 1–17. https://doi.org/10.1080/10408398.2021.1948817

[22] Klingel, T., Kremer, J. I., Gottstein, V., De Rezende, T. R., Schwarz, S., & Lachenmeier, D. W. (2020). A review of coffee by-products including leaf, flower, cherry, husk, silver skin, and spent grounds as novel foods within the European Union. *Foods 9*. https://doi.org/10.3390/foods9050665

[23] Asbaghi, O., Kashkooli, S., Mardani, M., Rezaei, K., Fry, M., Kazemi, H., & Kaviani, M, M. (2021). Effect of green coffee bean extract supplementation on liver function and inflammatory biomarkers: A meta-analysis of randomized clinical trials. *Complementary Therapies in Clinical Practice, 43*, 101349. https://doi.org/10.1016/j.ctcp.2021.101349

[24] Nogaim, Q. A., Bugata, L., Prabhakar, P., Reddy, U. A., Mangala, G. P., Indu, K. S., & Mahboob, M. (2020). Protective effect of Yemeni green coffee powder against the oxidative stress induced by Ochratoxin A. *Toxicology Reports, 7*, 142–148. https://doi.org/10.1016/j.toxrep.2019.11.015

[25] Butt, M. S., Ahmed, A., Imran, S. A., Yasin, M., & Imran, M. (2011). Evaluating the effect of decaffeination on nutritional and antioxidant status of different coffee brands. *Internet Journal of Food Safety, 13*, 150–156.

[26] Hsieh, Y. H. P., & Ofori, J. A. (2007). Innovations in food technology for health. *Asia Pacific Journal of Clinical Nutrition, 16*, 65–73.

[27] Temple, J. L., Bernard, C., Lipshultz, S. E., Czachor, J. D., Westphal, J. A., & Mestre, M. A. (2017). The safety of ingested caffeine: A comprehensive review. *Frontiers in Psychiatry, 8*, 1–19. https://doi.org/10.3389/fpsyt.2017.00080

[28] Franca, A. S. (2016). Coffee: Decaffeination. *Encyclopedia of food and health* (1st ed). Elsevier Ltd. https://doi.org/10.1016/B978-0-12-384947-2.00183-5

[29] Gokulakrishnan, S., Chandraraj, K., & Gummadi, S. N. (2005). Microbial and enzymatic methods for the removal of caffeine. *Enzyme and Microbial Technology, 37*, 225–232. https://doi.org/10.1016/j.enzmictec.2005.03.004

[30] Ibrahim, S., Shukor, M. Y., Syed, M. A., Arina, N., Rahman, A., Khalil, K. A., Khalid, A., & Ahmad, S. A. (2014). Bacterial degradation of caffeine: A review. *Asian Journal of Plant Biology, 2*, 18–27. https://doi.org/10.54987/ajpb.v2i1.84

[31] Chen, Y., She, Y., Zhang, R., Wang, J., Zhang, X., & Gou, X. (2020). Use of starch-based fat replacers in foods as a strategy to reduce dietary intake of fat and risk of metabolic diseases. *Food Science and Nutrition, 8*, 16–22. https://doi.org/10.1002/fsn3.1303

[32] Paglarini, C. D. S., Vidal, V. A. S., Martini, S., Cunha, R. L., & Pollonio, M. A. R. (2020). Protein-based hydrogelled emulsions and their application as fat replacers in meat products: A review. *Critical Reviews in Food Science and Nutrition, 0*, 1–16. https://doi.org/10.1080/10408398.2020.1825322

[33] Peng, X., & Yao, Y. (2017). Carbohydrates as fat replacers. *Annual Review of Food Science and Technology, 8*, 331–351. https://doi.org/10.1146/annurev-food-030216-030034

[34] Babu, A. S., Parimalavalli, R., & Mohan, R. J. (2018). Effect of modified starch from sweet potato as a fat replacer on the quality of reduced fat ice creams. *12*, 2426–2434. https://doi.org/10.1007/s11694-018-9859-4

[35] Agyei-Amponsah, J., Macakova, L., DeKock, H. L., & Emmambux, M. N. (2019). Sensory tribological, and rheological profiling of"clean label" starch-lipid complexes as fat Replacers. *Starch, 71*(9–10), 1180340.

[36] Teklehaimanot, W. H., Duodu, K. G., & Emmambux, M. N. (2013). Maize and teff starches modified with stearic acid as potential fat replacer in low calorie mayonnaise-type emulsions. *Starch, 65* (9–10), 773–781.

[37] Zhao, Y., Ai, Y., Li, L., Jane, J.-L., Hendrich, S., & Birt, D. F. (2014). Inhibition of azoxymethane-induced preneoplastic lesions in the rat colon by a stearic acid complexed high-amylose cornstarch using differentcooking methods and assessing potential gene targets. *Journal of Functional Foods, 6*, 499–512.

[38] Zhao, Y., Hasjim, J., Li, L., Jane, J.-L., Hendrich, S., & Birt, D. F. (2011). Inhibition of azoxymethane-induced preneoplastic lesions in the rat colon by a cooked stearic acid complexed high-amylose cornstarch. *Journal of Agricultural and Food Chemistry, 59*(17), 9700–9708.

[39] Putseys, J. A., Derde, L. J., Lamberts, L., Goesaert, H., & Delcour, J. A. (2009). Production of tailor made short chain amylose-lipid complexes using varying reaction conditions. *Carbohydrate Polymers, 78*(4), 854–861.

[40] Hasjim, J., Lee, S. O., Hendrich, S., Setiawan, S., Ai, Y., & Jane, J. L. (2010). Characterization of a novel resistant-starch and its effects on postprandial plasma-glucose and insulin responses. *Cereal Chemistry, 87*(4), 257–262.

[41] Bilal, M., & Iqbal, H. M. N. (2020) State-of-the-art strategies and applied perspectives of enzyme biocatalysis in food sector – current status and future trends. *Critical Reviews in Food Science and Nutrition, 60*, 2052–2066. https://doi.org/10.1080/10408398.2019.1627284

[42] Chandra, P., Enespa, S., & Arora, P.K, R. (2020). Microbial lipases and their industrial applications: A comprehensive review. *Microbial Cell Factories. BioMed Central.* https://doi.org/10.1186/s12934-020-01428-8

[43] Hasan, F., Shah, A., & Hameed, A. (2006). Industrial applications of microbial lipases. *Enzyme and Microbial Technology, 39,* 235–251. https://doi.org/10.1016/j.enzmictec.2005.10.016

[44] Patel, A. K., Singhania, R. R., & Pandey, A., (2016). Novel enzymatic processes applied to the food industry. *Current Opinion in Food Science, 7,* pp. 64–72. https://doi.org/10.1016/j.cofs.2015.12.002

[45] Garrido-Galand, S., Asensio-Grau, A., Calvo-Lerma, J., Heredia, A., & Andrés, A. (2021). The potential of fermentation on nutritional and technological improvement of cereal and legume flours: A review. *Food Research International, 145.* https://doi.org/10.1016/j.foodres.2021.110398

[46] Sharma, R., Garg, P., Kumar, P., Bhatia, S. K., & Kulshrestha, S. (2020). Microbial fermentation and its role in quality improvement of fermented foods. *Fermentation, 6,* 1–20. https://doi.org/10.3390/fermentation6040106

[47] Xu, L. N., Guo, S., & Zhang, S. W. (2019). Effects of solid-state fermentation on the nutritional components and antioxidant properties from quinoa. *Emirates Journal of Food and Agriculture, 31,* 39–45. https://doi.org/10.9755/ejfa.2019.v31.i1.1898

[48] Adeyanju, A. A., Kruger, J., Taylor, J. R. N., & Duodu, K. G. (2019). Effects of different souring methods on the protein quality and iron and zinc bioaccessibilities of non-alcoholic beverages from sorghum and amaranth. *International Journal of Food Science and Technology, 54,* 798–809. https://doi.org/10.1111/ijfs.13998

[49] Xiang, H., Sun-Waterhouse, D., Waterhouse, G. I. N., Cui, C., & Ruan, Z. (2019). Fermentation-enabled wellness foods: A fresh perspective. *Food Science and Human Wellness, 8,* 203–243. https://doi.org/10.1016/j.fshw.2019.08.003

[50] Xing, Q., Dekker, S., Kyriakopoulou, K., Boom, R. M., Smid, E. J., & Schutyser, M. A. I. (2020). Enhanced nutritional value of chickpea protein concentrate by dry separation and solid state fermentation. *Innovative Food Science and Emerging Technologies, 59,* 102269. https://doi.org/10.1016/j.ifset.2019.102269

[51] Chavre, B. W. (2020). Positive and negetive role of biotechnology in human life : A Review. *International Journal of Researches in Biosciences, Agriculture and Technology,* II, 107–113.

[52] Mathur, R. (2018). Genetic Engineering and Biosafety in the use of genetically modified foods, *International Journal of Advanced Scientific Research and Management, 1,* 76–82.

[53] Zhang, C., Wohlhueter, R., & Zhang, H. (2016). Genetically modified foods : A critical review of their promise and problems. *Food Science and Human Wellness, 5,* 116–123. https://doi.org/10.1016/j.fshw.2016.04.002

[54] Ghorani, B., & Tucker, N. (2015). AC SC. *Food Hydrocoll.* https://doi.org/10.1016/j.foodhyd.2015.05.024

[55] Bamidele, O. P., & Emmambux, M. N. (2020). Encapsulation of bioactive compounds by "extrusion" technologies : A review. *Critical Reviews in Food Science and Nutrition, 0,* 1–19. https://doi.org/10.1080/10408398.2020.1793724

[56] Drosou, C. G., Krokida, M. K., & Biliaderis, C. G. (2016). Encapsulation of bioactive compounds through electrospinning / electrospraying and spray drying : A comparative assessment of food related applications 3937. https://doi.org/10.1080/07373937.2016.1162797

[57] Estevinho, B. N., & Rocha, F. (2017). Kineics models applied to soluble vitamins delivery systems prepared by spray drying. *Drying Technology, 35,* 1249–1257.

[58] Chellaram, C., Murugaboopathi, G., John, A. A., Sivakumar, R., & Ganesan, S. (2014). Significance of nanotechnology in food industry. *Procedia – Social and Behavioral Sciences, 8,* 109–113. https://doi.org/10.1016/j.apcbee.2014.03.010

[59] Avella, M., Bruno, G., Errico, M. E., Gentile, G., Piciocchi, N., Sorrentino, A., & Volpe, M. (2007). Innovative packaging for minimally processed fruits. https://doi.org/10.1002/pts.761

[60] Bajpai, S. K., Chand, N., & Chaurasia, V. (2012). Nano zinc oxide-loaded calcium alginate films with potential antibacterial properties 1871–1881. https://doi.org/10.1007/s11947-011-0587-6

[61] Samal, D. (2017). Use of nanotechnology in food industry : A review. https://doi.org/10.22161/ijeab/2.4.90

[62] Sahoo, M., Vishwakarma, S., Panigrahi, C., & Kumar, J. (2021). Nanotechnology : Current applications and future scope in food 3–22. https://doi.org/10.1002/fft2.58

[63] Kumar, I., Rawat, J., Mohd, N., & Husain, S. (2021). Opportunities of artificial intelligence and machine learning in the food industry. *Journal of Food Quality, 2021.*

[64] Misra, N. N., Dixit, Y., Al-Mallahi, A., Bhullar, M. S., Upadhyay, R., & Martynenko, A., 2020. IoT, big data and artificial intelligence in agriculture and food industry. *IEEE Internet of Things Journal, 9*, 6305–6324.

[65] Friedlander, A., & Zoellner, C. (2020). Artificial intelligence opportunities to improve food safety at retail. *Food Protection Trends, 40*, 272–278.

[66] Nychas, G.-J. E., Panagou, E. Z., & Mohareb, F. (2016). Novel approaches for food safety management and communication. *Current Opinion in Food Science, 12*, 13–20.

[67] Marvin, H. J. P., Janssen, E. M., Bouzembrak, Y., Hendriksen, P. J. M., & Staats, M. (2017). Big data in food safety: An overview. *Critical Reviews in Food Science and Nutrition, 57*, 2286–2295. https://doi.org/10.1080/10408398.2016.1257481

[68] Deng, X., Cao, S., & Horn, A. L. (2021). Emerging applications of machine learning in food safety. *Annual Review of Food Science and Technology, 12*, 513–538. https://doi.org/10.1146/annurev-food-071720-024112

[69] Campanile, L., Iacono, M., Marulli, F., & Mastroianni, M. (2021). Designing a GDPR compliant blockchain-based IoV distributed information tracking system. *Information Processing and Management, 58*, 102511.

[70] Perwej, Y. (2018). A pervasive review of blockchain technology and its potential applications. **Open Science Journal** of **Electrical** and **Electronic** Engineering (OSJEEE), New York, USA, *5*, 30–43.

[71] Özyilmaz, K. R., Doğan, M., & Yurdakul, A., 2018. IDMoB: IoT data marketplace on blockchain. In *2018 Crypto Valley Conference on Blockchain Technology (CVCBT)*. IEEE, pp. 11–19.

[72] Wan, P. K., Huang, L., & Holtskog, H. (2020). Blockchain-enabled information sharing within a supply chain: A systematic literature review. *IEEE Access [Internet], 8*, 49645–49656.

[73] Lu, Q., & Xu, X. (2019). Adaptable blockchain-based systems. *Ieee Software, 6*, 21–27. https://doi.org/10.1109/MS.2017.4121227

[74] Galvez, J. F., Mejuto, J. C., & Simal-Gandara, J. (2018). Future challenges on the use of blockchain for food traceability analysis. *TrAC Trends in Analytical Chemistry, 107*, 222–232.

[75] Liao, D.-Y., & Wang, X. (2018). Applications of blockchain technology to logistics management in integrated casinos and entertainment. *Informatics, 4*, 44.

[76] Reda, M., Kanga, D. B., Fatima, T., & Azouazi, M. (2020). Blockchain in health supply chain management: State of art challenges and opportunities. *Procedia Computer Science, 175*, 706–709.

[77] Kshetri, N. (2018). 1Blockchain's roles in meeting key supply chain management objectives. *International Journal of Information Management, 39*, 80–89.

[78] Kshetri, N., & Loukoianova, E. (2019). Blockchain adoption in supply chain networks in Asia. *IT Professional, 21*, 11–15.

[79] Rajput, R. K., Pandey, A., Kumar, D., & Wairiya, M. (2021). Permissioned Blockchain Model to Strengthen Food Supply Chain during pandemic (Covid-19). Available SSRN 3882509.

[80] Dainelli, D., Gontard, N., Spyropoulos, D., Zondervan-van den beuken, E., & Tobback, P. (2008). Active and intelligent food packaging: Legal aspects and safety concerns. *Trends in Food Science and Technology, 19*, S103–S112.

[81] Tonkin, E., Meyer, S. B., Coveney, J., Webb, T., & Wilson, A. M. (2016). The process of making trust related judgements through interaction with food labelling. *Food Policy, 63*, 1–11.

[82]   Yam, K. L. (2012). Intelligent packaging to enhance food safety and quality. In *Emerging food packaging technologies* Yam K.L and Lee D.S (eds) (pp. 137–152). Elsevier.

[83]   Campos, S., Doxey, J., & Hammond, D. (2011). Nutrition labels on pre-packaged foods: a systematic review. *Public Health Nutrition, 14*, 1496–1506. https://doi.org/10.1017/S1368980010003290

[84]   Huang, L., Li, N., Barzi, F., Ma, G., Trevena, H., Dunford, E., Land, M.-A., & Neal, B. (2014). A systematic review of the prevalence of nutrition labels and completeness of nutrient declarations on pre-packaged foods in China. *Journal of Public Health (Bangkok), 37*, fdu091. https://doi.org/10.1093/pubmed/fdu091

[85]   Ahmed, I., Lin, H., Zou, L., Li, Z., Brody, A. L., Qazi, I. M., Lv, L., Pavase, T. R., Khan, M. U., & Khan, S. (2018). An overview of smart packaging technologies for monitoring safety and quality of meat and meat products. *Packaging Technology and Science, 31*, 449–471.

[86]   Kneafsey, M., Venn, L., & Bos, E. (2017). Consuming rural connections: Tracing leeks back to their roots. In *Transforming the rural*. Emerald Publishing Limited.

[87]   Soon, J. M., & Saguy, I. S. (2017). Crowdsourcing: A new conceptual view for food safety and quality. *Trends in Food Science and Technology, 66*, 63–72.

[88]   Ribeiro, F. D. S., Caliva, F., Swainson, M., Gudmundsson, K., Leontidis, G., & Kollias, S., 2018. An adaptable deep learning system for optical character verification in retail food packaging. In *2018 IEEE Conference on Evolving and Adaptive Intelligent Systems (EAIS)*. IEEE, pp. 1–8.

[89]   Thota, M., Kollias, S., Swainson, M., & Leontidis, G. (2020). Multi-source domain adaptation for quality control in retail food packaging. *Computers in Industry, 123*, 103293.

[90]   Canali, M., Amani, P., Aramyan, L., Gheoldus, M., Moates, G., Östergren, K., Silvennoinen, K., Waldron, K., & Vittuari, M. (2017). Food waste drivers in Europe, from identification to possible interventions. *Sustainability, 9*, 37.

[91]   Miernicki, M., Hofmann, T., Eisenberger, I., Von der kammer, F., & Praetorius, A. (2019). Legal and practical challenges in classifying nanomaterials according to regulatory definitions. *Nature Nanotechnology, 14*, 208–216. https://doi.org/10.1038/s41565-019-0396-z

[92]   Ballco, P., De-Magistris, T., & Caputo, V. (2019). Consumer preferences for nutritional claims: An exploration of attention and choice based on an eye-tracking choice experiment. *Food Research International, 116*, 37–48. https://doi.org/10.1016/j.foodres.2018.12.031

[93]   Yue, D. A. I., Qiang, L. I., Wen, L. I. U., Hongrui, Z., Jia, H. A. O., Ye, D., Peng, L. I. U., Min, D., & Xiaoran, F. A. N., 2020. Research on consumers' cognition and demand for food label information. In *IOP conference series: Earth and environmental science* (p. 12071). IOP Publishing.

[94]   Chowdhury, E. U., & Morey, A. (2019). Intelligent packaging for poultry industry. *Journal of Applied Poultry Research, 28*, 791–800.

[95]   Deng, M., & Feng, P. (2020). A food traceability system based on blockchain and radio frequency identification technologies. *Journal of Computer and Communications, 8*, 17–27.

[96]   Tan, A., & Ngan, P. T. (2020). A proposed framework model for dairy supply chain traceability. *Sustainable Futures, 2*, 100034.

[97]   Kendall, H., Clark, B., Rhymer, C., Kuznesof, S., Hajslova, J., Tomaniova, M., Brereton, P., & Frewer, L. (2019). A systematic review of consumer perceptions of food fraud and authenticity: A European perspective. *Trends in Food Science and Technology, 94*, 79–90.

[98]   Wilson, N. L. W., Rickard, B. J., Saputo, R., & Ho, S.-T. (2017). Food waste: The role of date labels, package size, and product category. *Food Quality and Preference, 55*, 35–44.

[99]   Bader, F., & Jagtap, S. (2020). Internet of things-linked wearable devices for managing food safety in the healthcare sector. In *Wearable and implantable medical devices* (p. 229–253). Elsevier.

[100] Griesche, C., & Baeumner, A. J. (2020). Biosensors to support sustainable agriculture and food safety. *TrAC Trends in Analytical Chemistry, 128*, 115906.

[101] Nyawo, T., Kesa, H., & Onyenweaku, E. (2020). Food safety and hygiene: Knowledge, attitude and practices among food handlers. *African Journal of Hospitality, Tourism and Leisure, 10*, 547–558. https://doi.org/10.46222/ajhtl.19770720-117

[102] Clark, J., Crandall, P. G., Pellegrino, R., & Shabatura, J. (2019). Assessing smart glasses-based foodservice training: An embodied learning theory approach. *Canadian Journal of Learning and Technology Review/Can. L'apprentissage la Technol, 45*, 2.

[103] Kang, L., Wu, C., & Wang, B. (2019). Principles, approaches and challenges of applying big data in safety psychology research. *Frontiers in Psychology, 10*, 1596.

[104] Pierdicca, R., Prist, M., Monteriù, A., Frontoni, E., Ciarapica, F., Bevilacqua, M., & Mazzuto, G., 2020. Augmented reality smart glasses in the workplace: Safety and security in the fourth industrial revolution era. In *Augmented Reality, Virtual Reality, and Computer Graphics: 7th International Conference, AVR 2020, Lecce, Italy, September 7–10, 2020, Proceedings, Part II 7*, pp. 231–247. Springer.

[105] Lavau, S., & Bingham, N. (2017). Practices of attention, possibilities for care: Making situations matter in food safety inspection. *Sociological Review, 65*, 20–35.

[106] Walsh, C., & Leva, M. C. (2019). A review of human factors and food safety in Ireland. *Safety Science, 119*, 399–411.

[107] Nelis, J. L. D., Tsagkaris, A. S., Dillon, M. J., Hajslova, J., & Elliott, C. T. (2020). Smartphone-based optical assays in the food safety field. *TrAC Trends in Analytical Chemistry, 129*, 115934. https://doi.org/10.1016/j.trac.2020.115934

[108] Sivakumar, R., & Lee, N. Y. (2021). Recent progress in smartphone-based techniques for food safety and the detection of heavy metal ions in environmental water. *Chemosphere, 275*, 130096. https://doi.org/10.1016/j.chemosphere.2021.130096

[109] Bland, C., Dalrymple, K. V., White, S. L., Moore, A., Poston, L., & Flynn, A. C. (2020). Smartphone applications available to pregnant women in the United Kingdom: An assessment of nutritional information. *Maternal and Child Nutrition, 16*, e12918. https://doi.org/10.1111/mcn.12918

[110] Choi, J. R. (2017). Smartphone-based sensing in food safety and quality analysis. *Sensing Techniques for Food Safety and Quality Control, 2*, 332–358.

[111] Boyd, M. C. (2017). Cricket soup: a critical examination of the regulation of insects as food. *Yale Law and Policy Review, 36*, 17.

[112] Nyarugwe, S. P., Linnemann, A. R., Ren, Y., Bakker, E. J., Kussaga, J. B., Watson, D., Fogliano, V., & Luning, P. A. (2020). An intercontinental analysis of food safety culture in view of food safety governance and national values. *Food Control, 111*, 107075. https://doi.org/10.1016/j.foodcont.2019.107075

[113] Jaffee, S., Henson, S., Unnevehr, L., Grace, D., & Cassou, E. (2018). *The safe food imperative: Accelerating progress in low-and middle-income countries*. Washington, D.C., World Bank Publications.

[114] Tse, D., Zhang, B., Yang, Y., Cheng, C., & Mu, H., 2017. Blockchain application in food supply information security. In *2017 IEEE International Conference on Industrial Engineering and Engineering Management (IEEM)*. IEEE, pp. 1357–1361.

[115] Käferstein, F., & Abdussalam, M. (1999). Food safety in the 21st century. *Bulletin of the World Health Organization, 77*, 347–351.

[116] Siu, J., Jackson, L. J., Bensassi, S., Manjang, B., & Manaseki-Holland, S. (2021). Cost-effectiveness of a weaning food safety and hygiene programme in rural Gambia. *Tropical Medicine and International Health, 26*, 1624–1633. https://doi.org/10.1111/tmi.13691

[117] Sharman, N., Wallace, C. A., & Jespersen, L. (2020). Terminology and the understanding of culture, climate, and behavioural change–Impact of organisational and human factors on food safety management. *Trends in Food Science and Technology, 96*, 13–20.

[118] Adewunmi, A. A., & Fapohunda, S. O. (2018). Pesticides and food safety in Africa. *European Journal of Biological Research, 8*, 70–83.

[119] Mézes, M. (2018). Food safety aspect of insects: A review. *Acta Alimentaria, 47*, 513–522.

[120] Rortais, A., Arnold, G., Dorne, J.-L., More, S. J., Sperandio, G., Streissl, F., Szentes, C., & Verdonck, F. (2017). Risk assessment of pesticides and other stressors in bees: principles, data gaps and perspectives from the European Food Safety Authority. *Science of the Total Environment, 587*, 524–537.

[121] Lehotay, S. J. (2018). Food safety analysis. *Analytical and bioanalytical chemistry, 410*, 5329–5330.

[122] Meltzer, R., Rothbart, M. W., Schwartz, A. E., Calabrese, T., Silver, D., Mijanovich, T., & Weinstein, M. (2019). What are the financial implications of public quality disclosure? Evidence from New York City's restaurant food safety grading policy. *Public Finance Review, 47*, 170–201.

[123] De andrade, M. L., Stedefeldt, E., Zanin, L. M., & da Cunha, D. T. (2020). Food safety culture in food services with different degrees of risk for foodborne diseases in Brazil. *Food Control, 112*, 107152.

[124] Zanin, L. M., da Cunha, D. T., De rosso, V. V., Capriles, V. D., & Stedefeldt, E. (2017). Knowledge, attitudes and practices of food handlers in food safety: An integrative review. *Food Research International, 100*, 53–62.

[125] Moyer, D. C., DeVries, J. W., & Spink, J. (2017). The economics of a food fraud incident–Case studies and examples including Melamine in Wheat Gluten. *Food Control, 71*, 358–364.

[126] Pendrous, R. (2012). Hunting out food fraudsters: Food safety. *South African Food Review, 39*, 30–31.

[127] Spink, J. W. (2019). The current state of food fraud prevention: Overview and requirements to address 'How to Start?'and 'How Much is Enough? *Current Opinion in Food Science, 27*, 130–138.

[128] Şerbancea, F., Belc, N., & Stănescu, A. (2018). Risk factors in the assessment of the conformity of falsified dairy products. *Quality-access to Success, 19*, 133–139.

[129] Robson, K., Dean, M., Brooks, S., Haughey, S., & Elliott, C. (2020). A 20-year analysis of reported food fraud in the global beef supply chain. *Food Control, 116*, 107310. https://doi.org/10.1016/J.FOOD CONT.2020.107310

[130] Casadei, E., Valli, E., Panni, F., Donarski, J., Gubern, J. F., Lucci, P., Conte, L., Lacoste, F., Maquet, A., & Brereton, P. (2021). Emerging trends in olive oil fraud and possible countermeasures. *Food Control, 124*, 107902.

[131] Heres, L., Hoogenboom, R., Herbes, R., Traag, W., & Urlings, B. (2010). Tracing and analytical results of the dioxin contamination incident in 2008 originating from the Republic of Ireland. *Food Additives and Contaminants: Part A, 27*, 1733–1744.

[132] Pei, X., Tandon, A., Alldrick, A., Giorgi, L., Huang, W., & Yang, R. (2011). The China melamine milk scandal and its implications for food safety regulation. *Food Policy, 36*, 412–420.

[133] Shi, X., Dong, R., Chen, J., Yuan, Y., Long, Q., Guo, J., Li, S., & Chen, B. (2020). An assessment of melamine exposure in Shanghai adults and its association with food consumption. *Environment International, 135*, 105363.

[134] Meira, L., Costa, J., Villa, C., Ramos, F., Oliveira, M. B. P. P., & Mafra, I. (2017). EvaGreen real-time PCR to determine horse meat adulteration in processed foods. *LWT, 75*, 408–416.

[135] Roche, S. M., Renaud, D. L., Genore, R., Shock, D. A., Bauman, C., Croyle, S., Barkema, H. W., Dubuc, J., Keefe, G. P., & Kelton, D. F. (2020). Canadian National Dairy Study: Describing Canadian dairy producer practices and perceptions surrounding cull cow management. *Journal of Dairy Science, 103*, 3414–3421.

[136] Stafford, E. G., Tell, L. A., Lin, Z., Davis, J. L., Vickroy, T. W., Riviere, J. E., & Baynes, R. E. (2018). Consequences of fipronil exposure in egg-laying hens. *Journal of the American Veterinary Medical Association, 253*, 57–60.

[137] Van der merwe, D., Jordaan, A., & van den Berg, M. (2019). Case report: Fipronil contamination of chickens in the Netherlands and surrounding countries. In *Chemical hazards in foods of animal origin* (pp. 363–373). Wageningen, Wageningen Academic Publishers.

[138] Walker, M. J., Burns, M., & Burns, D. T. (2013). Horse meat in beef products – species substitution 2013. *Journal of the Association of Public Analysts, 41*, 67–106.

[139] Drivelos, S. A., & Georgiou, C. A. (2012). Multi-element and multi-isotope-ratio analysis to determine the geographical origin of foods in the European Union. *TRAC: Trends in Analytical Chemistry, 40*, 38–51. https://doi.org/10.1016/j.trac.2012.08.003

[140] Liu, Y., Barton, F. E., Lyon, B. G., Windham, W. R., & Lyon, C. E. (2004). Two-dimensional correlation analysis of visible/near-infrared spectral intensity variations of chicken breasts with various chilled and frozen storages. *Journal of Agricultural and Food Chemistry, 52*(3), 505–510. https://doi.org/10.1021/jf0303464

[141] Castro, F., García, M. C., Rodriguez, R., Rodríguez, J., & Marina, M. L. (2007). Determination of soybean proteins in commercial heat-processed meat products prepared with chicken, beef or complex mixtures of meats from different species. *Food Chemistry, 100*(2), 468–476. https://doi.org/https://doi.org/10.1016/j.foodchem.2005.09.067

[142] McGrath, T. F., Shannon, M., Chevallier, O. P., Ch, R., Xu, F., Kong, F., . . . Elliott, C. T. (2021). Food fingerprinting: Using a two-tiered approach to monitor and mitigate food fraud in rice. *Journal of AOAC International, 104*(1), 16–28. https://doi.org/10.1093/jaoacint/qsaa109

[143] Jia, H., & Gu, S.-Y. (2020). A near infrared induced self-healable composite based on disulfide bonds for flexible electronics. *Journal of Polymer Research, 27*, 1–13.

[144] Wu, S., Li, J., Zhang, G., Yao, Y., Li, G., Sun, R., & Wong, C. (2017). Ultrafast self-healing nanocomposites via infrared laser and their application in flexible electronics. *ACS Applied Material Interfaces, 9*, 3040–3049.

[145] Ratti, C., & Mujumdar, A. S. (2020). Infrared drying. In *Handbook of industrial drying* (pp. 567–588). CRC Press.

[146] Wang, X. (2019). Near-infrared spectroscopy for food quality evaluation. In *Evaluation technologies for food quality* (pp. 105–118). Elsevier.

[147] Aboud, S. A., Altemimi, A. B., A -hiiphy, R. S., Yi-Chen, A., & Cacciola, F, L. (2019). A comprehensive review on infrared heating applications in food processing. *Molecules, 24*, 4125.

[148] Catelani, T. A., Santos, J. R., Páscoa, R. N. M. J., Pezza, L., Pezza, H. R., & Lopes, J. A. (2018). Real-time monitoring of a coffee roasting process with near infrared spectroscopy using multivariate statistical analysis: A feasibility study. *Talanta, 179*, 292–299.

[149] Feng, Y., Wu, B., Yu, X., Yagoub, A. E. A., Sarpong, F., & Zhou, C. (2018). Effect of catalytic infrared dry-blanching on the processing and quality characteristics of garlic slices. *Food Chemistry, 266*, 309–316.

[150] Wu, B., Guo, Y., Wang, J., Pan, Z., & Ma, H. (2018). Effect of thickness on non-fried potato chips subjected to infrared radiation blanching and drying. *Journal of Food Engineering, 237*, 249–255.

[151] Akomea-Frempong, S., Perry, J. J., & Skonberg, D. I. (2021). Effects of pre-freezing blanching procedures on the physicochemical properties and microbial quality of frozen sugar kelp. *Journal of Applied Phycology, 39*, 609–624

[152] Nguyen, T. V. L., Tran, T. Y. N., Lam, D. T., Bach, L. G., & Nguyen, D. C. (2019). Effects of microwave blanching conditions on the quality of green asparagus (Asparagus officinalis L.) butt segment. *Food Science and Nutrition, 7*, 3513–3519.

[153] Noreña, C. Z., & Rigon, R. T. (2018). Effect of blanching on enzyme activity and bioactive compounds of blackberry. *Brazilian Archives of Biology and Technology, 61*, 1–13.

[154] Deng, L.-Z., Mujumdar, A. S., Zhang, Q., Yang, X.-H., Wang, J., Zheng, Z.-A., Gao, Z.-J., & Xiao, H.-W. (2019). Chemical and physical pretreatments of fruits and vegetables: Effects on drying characteristics and quality attributes–a comprehensive review. *Critical Reviews in Food Science and Nutrition, 59*, 1408–1432.

[155] Pan, Z., & El-Mashad, H. (2018). Infrared peeling technology for fruits and vegetablesThe need for IR. *Resources Magazine, 25*, 4–6.

[156] Vidyarthi, S. K., El-Mashad, H. M., Khir, R., Zhang, R., McHugh, T. H., & Pan, Z. (2019a). Tomato peeling performance under pilot scale catalytic infrared heating. *Journal of Food Engineering, 246*, 224–231.

[157] Shen, Y., Khir, R., Wood, D., McHugh, T. H., & Pan, Z. (2020). Pear peeling using infrared radiation heating technology. *Innovative Food Science and Emerging Technologies, 65*, 102474.

[158] Kate, A. E., & Sutar, P. P. (2018). Development and optimization of novel infrared dry peeling method for ginger (Zingiber officinale Roscoe) rhizome. *Innovative Food Science and Emerging Technologies, 48*, 111–121.

[159] Wang, W., Wang, L., Feng, Y., Pu, Y., Ding, T., Ye, X., & Liu, D. (2018). Ultrasound-assisted lye peeling of peach and comparison with conventional methods. *Innovative Food Science and Emerging Technologies, 47*, 204–213.

[160] Farhadi, R., Afkari-Sayyah, A. H., Jamshidi, B., & Gorji, A. M. (2020). Prediction of internal compositions change in potato during storage using visible/near-infrared (Vis/NIR) spectroscopy. *International Journal of Food Engineering 16*, p.20190110.

[161] Horton, L., Torres, A. E., Narla, S., Lyons, A. B., Kohli, I., Gelfand, J. M., Ozog, D. M., Hamzavi, I. H., & Lim, H. W. (2020). Spectrum of virucidal activity from ultraviolet to infrared radiation. *Photochemical and Photobiological Sciences, 19*, 1262–1270.

[162] Yang, -Y.-Y., Feng, H.-P., Niu, C.-G., Huang, D.-W., Guo, H., Liang, C., Liu, H.-Y., Chen, S., Tang, N., & Li, L. (2021). Constructing a plasma-based Schottky heterojunction for near-infrared-driven photothermal synergistic water disinfection: Synergetic effects and antibacterial mechanisms. *Chemical Engineering Journal, 426*, 131902.

[163] El Darra, N., Xie, F., Kamble, P., Khan, Z., & Watson, I. (2021). Decontamination of Escherichia coli on dried onion flakes and black pepper using Infra-red, ultraviolet and ozone hurdle technologies. *Heliyon, 7*(6), e07259. https://doi.org/10.1016/j.heliyon.2021.e07259

[164] Keerthirathne, T., Ross, K., Fallowfield, H., & Whiley, H. (2017). Reducing risk of salmonellosis through egg decontamination processes. *International Journal of Environmental Research and Public Health, 14*(3), 335. https://doi.org/10.3390/ijerph14030335

[165] Siciliano, I., Dal Bello, B., Zeppa, G., Spadaro, D., & Gullino, M. L. (2017). Static hot air and infrared rays roasting are efficient methods for aflatoxin decontamination on hazelnuts. *Toxins (Basel), 9*(2). https://doi.org/https://doi.org/10.3390/toxins9020072

[166] Chintagari, S., Jadeja, R., & Hung, Y. C. (2019). Resistance of various shiga-toxin producing Escherichia coli (STEC) strains and serogroups to infra-red and pulsed UV radiation and effect of nalidixic acid adaptation. *LWT, 102*, 356–363. https://doi.org/10.1016/j.lwt.2018.11.094

[167] Watson, I., Kamble, P., Shanks, C., Khan, Z., & El Darra, N. (2020). Decontamination of chilli flakes in a fluidized bed using combined technologies: Infrared, UV and ozone. *Innovative Food Science and Emerging Technologies, 59*, 102248. https://doi.org/10.1016/j.ifset.2019.102248

[168] Shavandi, M., Sadeghi, A., & Sarani, A. (2020). Modeling the effect of different infrared treatment on *B. cereus* in cardamom seeds and using genetic algorithm-artificial neural network. *Journal of Food and Bioprocess Engineering, 3*(1), 29–34. https://doi.org/https://doi.org/10.22059/jfabe.2020.75644

[169] Savitha, S., Chakraborty, S., & Thorat, B. N. (2021). Microbial contamination and decontamination of onion and its products. *Applied Food Research*, 100032. https://doi.org/https://doi.org/10.1016/j.afres.2021.100032

[170] Sirohi, R., Tarafdar, A., Kumar Gaur, V., Singh, S., Sindhu, R., Rajasekharan, R., . . . Pandey, A. (2021). Technologies for disinfection of food grains: Advances and way forward. *Food Research International, 145*, 110396. https://doi.org/10.1016/j.foodres.2021.110396

[171] Dodoo-Arhin, D., Bowen-Dodoo, E., Agyei-Tuffour, B., Nyankson, E., Obayemi, J. D., Salifu, A. A., . . . Soboyejo, W. O., 2021. Modified nanostructured titania photocatalysts for aquatic disinfection applications. *Materials Today: Proceedings, 38*, 1183–1190. https://doi.org/https://doi.org/10.1016/j.matpr.2020.07.710

[172] Drishya, C., Yoha, K. S., Perumal, A. B., Moses, J. A., Anandharamakrishnan, C., & Balasubramaniam, V. M. (2022). Impact of nonthermal food processing techniques on mycotoxins and their producing fungi. *International Journal of Food Science and Technology, 57*(4), 2140–2148. https://doi.org/10.1111/ijfs.15444

[173] Rispail, N., & Rubiales, D. (2015). Rapid and efficient estimation of pea resistance to the soil-borne pathogen fusarium oxysporum by infrared imaging. *Sensors, 15*(2), 3988–4000. https://doi.org/10.3390/s150203988

[174] Adebo, O. A., Njobeh, P. B., Gbashi, S., Nwinyi, O. C., & Mavumengwana, V. (2017). Review on microbial degradation of aflatoxins. *Critical Reviews in Food Science and Nutrition, 57*(15), 3208–3217. https://doi.org/10.1080/10408398.2015.1106440

[175] Kirimtat, A., & Krejcar, O. (2018). A review of infrared thermography for the investigation of building envelopes: Advances and prospects. *Energy and Buildings, 176*, 390–406.

[176] Lu, H., Carroll, G. M., Neale, N. R., & Beard, M. C. (2019). Infrared quantum dots: Progress, challenges, and opportunities. *ACS Nano, 13*, 939–953.

[177] Cao, Y., Fanning, S., Proos, S., Jordan, K., & Srikumar, S. (2017). A review on the applications of next generation sequencing technologies as applied to food-related microbiome studies. *Frontiers in Microbiology, 8*, 1829.

[178] Jagadeesan, B., Gerner-Smidt, P., Allard, M. W., Leuillet, S., Winkler, A., Xiao, Y., Chaffron, S., Van Der Vossen, J., Tang, S., & Katase, M. (2019). The use of next generation sequencing for improving food safety: translation into practice. *Food Microbiology, 79*, 96–115.

[179] Gupta, A. K., & Gupta, U. D. (2020). Next generation sequencing and its applications. In *Animal biotechnology* (pp. 395–421). Elsevier.

[180] Mondal, S., & Khatua, S. (2019). Accelerating pairwise sequence alignment algorithm by mapreduce technique for next-generation sequencing (ngs) data analysis. In *Emerging technologies in data mining and information security* (pp. 213–220). Springer.

[181] Mutiga, S. K., Mushongi, A. A., & Kangéthe, E. K. (2019). Enhancing food safety through adoption of long-term technical advisory, financial, and storage support services in maize growing areas of East Africa. *Sustainability, 11*, 2827.

[182] Shimazawa, R., & Ikeda, M. (2020). Regulatory perspectives on next-generation sequencing and complementary diagnostics in Japan. *Expert Review of Molecular Diagnostics, 20*, 601–610.

[183] Deurenberg, R. H., Bathoorn, E., Chlebowicz, M. A., Couto, N., Ferdous, M., García-Cobos, S., Kooistra-Smid, A. M. D., Raangs, E. C., Rosema, S., & Veloo, A. C. M. (2017). Application of next generation sequencing in clinical microbiology and infection prevention. *Journal of Biotechnology, 243*, 16–24.

[184] Rantsiou, K., Kathariou, S., Winkler, A., Skandamis, P., Saint-Cyr, M. J., Rouzeau-Szynalski, K., & Amézquita, A. (2018a). Next generation microbiological risk assessment: opportunities of whole genome sequencing (WGS) for foodborne pathogen surveillance, source tracking and risk assessment. *International Journal of Food Microbiology, 287*, 3–9.

[185] Garcia, S. N., Osburn, B. I., & Jay-Russell, M. T. (2020). One health for food safety, food security, and sustainable food production. *Frontiers in Sustainable Food Systems, 4*, 1.

[186] Weinroth, M. D., Belk, A. D., & Belk, K. E. (2018). History, development, and current status of food safety systems worldwide. *Animal Frontiers, 8*, 9–15.

[187] Allard, M. W., Strain, E., Melka, D., Bunning, K., Musser, S. M., Brown, E. W., & Timme, R. (2016). The practical value of food pathogen traceability through building a whole-genome sequencing network and database. *Journal of Clinical Microbiology, 54*, 1975–1983. https://doi.org/10.1128/jcm.00081-16

[188] Brown, E., Dessai, U., McGarry, S., & Gerner-Smidt, P. (2019). Use of whole-genome sequencing for food safety and public health in the United States. *Foodborne Pathogens and Disease, 16*, 441–450.

[189] Motro, Y., & Moran-Gilad, J. (2017). Next-generation sequencing applications in clinical bacteriology. *Biomolecular Detection and Quantification, 14*, 1–6.

[190] Liu, Z., Zhu, L., Roberts, R., & Tong, W. (2019). Toward clinical implementation of next-generation sequencing-based genetic testing in rare diseases: where are we? *Trends in Genetics Tig, 35*, 852–867.

[191] Akkari, Y., Smith, T., Westfall, J., & Lupo, S. (2019). Implementation of cancer next-generation sequencing testing in a community hospital. *Molecular Case Studies, 5*, a003707.

[192] Gu, W., Miller, S., & Chiu, C. Y. (2019). Clinical metagenomic next-generation sequencing for pathogen detection. *Annual Review of Pathology: Mechanisms of Disease, 14*, 319–338.

[193] Morales-de la peña, M., Welti-Chanes, J., & Martín-Belloso, O. (2019). Novel technologies to improve food safety and quality. *Current Opinion in Food Science, 30*, 1–7. https://doi.org/10.1016/j.cofs.2018.10.009

[194] Priyadarshini, A., Rajauria, G., O'Donnell, C. P., & Tiwari, B. K. (2019a). Emerging food processing technologies and factors impacting their industrial adoption. *Critical Reviews in Food Science and Nutrition, 59*, 3082–3101. https://doi.org/10.1080/10408398.2018.1483890

[195] Bullerman, L. B., & Bianchini, A. (2007). Stability of mycotoxins during food processing. *International Journal of Food Microbiology, 119*, 140–146. https://doi.org/10.1016/j.ijfoodmicro.2007.07.035

[196] Rather, I. A., Koh, W. Y., Paek, W. K., & Lim, J. (2017). The sources of chemical contaminants in food and their health implications. *Frontiers in Pharmacology, 8*. https://doi.org/10.3389/fphar.2017.00830

[197] Robinson, R. (2000). Encyclopedia of food microbiology. *Choice Reviews Online, 38*, 38–0033-38–0033. https://doi.org/10.5860/CHOICE.38-0033

[198] Cullen, P. J., Tiwari, B. K., & Valdramidis, V. P. (2012). Status and trends of novel thermal and non-thermal technologies for fluid foods. In: *Novel thermal and non-thermal technologies for fluid foods* (pp. 1–6). Elsevier. https://doi.org/10.1016/B978-0-12-381470-8.00001-3

[199] Hernández-Hernández, H. M., Moreno-Vilet, L., & Villanueva-Rodríguez, S. J. (2019a). Current status of emerging food processing technologies in Latin America: Novel non-thermal processing. *Innovative Food Science and Emerging Technologies, 58*, 102233. https://doi.org/10.1016/j.ifset.2019.102233

[200] Van Impe, J., Smet, C., Tiwari, B., Greiner, R., Ojha, S., Stulić, V., Vukušić, T., & Režek Jambrak, A. (2018). State of the art of nonthermal and thermal processing for inactivation of micro-organisms. *Journal of Applied Microbiology, 125*, 16–35. https://doi.org/10.1111/jam.13751

[201] Vadivambal, R., & Jayas, D. S. (2010). Non-uniform temperature distribution during microwave heating of food materials – A review. *Food & Bioprocess Technology*. https://doi.org/10.1007/s11947-008-0136-0

[202] Tang, J. (2015). Unlocking potentials of microwaves for food safety and quality. *Journal of Food Science, 80*, E1776–E1793. https://doi.org/10.1111/1750-3841.12959

[203] Tănase, E. E., Miteluţ, A. C., Popa, M. E., Ştefănoiu, G. A., & Drăghici, M. (2015). Radio frequency heating for food safety and preservation – State of the art. International Nonthermal Processing Work shop, 12–13 November 2015, Athens, Greece.

[204] Altemimi, A., Aziz, S. N., Al-HiIphy, A. R. S., Lakhssassi, N., Watson, D. G., & Ibrahim, S. A. (2019). Critical review of radio-frequency (RF) heating applications in food processing. *Food Quality and Safe, 3*, 81–91. https://doi.org/10.1093/fqsafe/fyz002

[205] Schottroff, F., Pyatkovskyy, T., Reineke, K., Setlow, P., Sastry, S. K., & Jaeger, H. (2019). Mechanisms of enhanced bacterial endospore inactivation during sterilization by ohmic heating. *Bioelectrochemistry, 130*, 107338. https://doi.org/10.1016/j.bioelechem.2019.107338

[206] Wang, Y., Li, X., Sun, G., Li, D., & Pan, Z. (2014). A comparison of dynamic mechanical properties of processing-tomato peel as affected by hot lye and infrared radiation heating for peeling. *Journal of Food Engineering, 126*, 27–34. https://doi.org/10.1016/j.jfoodeng.2013.10.032

[207] Nazari, S. H. (2014). Impact of using infrared irradiation energy in food processing. *Energy Education Science and Technology Part A – Energy Science and Research, 32*, 1387–1396.

[208] Rastogi, N. K. (2012). Recent trends and developments in infrared heating in food processing. *Critical Reviews in Food Science and Nutrition, 52*, 737–760. https://doi.org/10.1080/10408398.2010.508138

[209] Jadhav, H. B., Annapure, U. S., & Deshmukh, R. R. (2021). Non-thermal technologies for food processing. *Frontiers in Nutrition, 8*. https://doi.org/10.3389/fnut.2021.657090

[210] Aganovic, K., Hertel, C., Vogel, R. F., Johne, R., Schlüter, O., Schwarzenbolz, U., Jäger, H., Holzhauser, T., Bergmair, J., Roth, A., Sevenich, R., Bandick, N., Kulling, S. E., Knorr, D., Engel, K., & Heinz, V. (2021). Aspects of high hydrostatic pressure food processing: Perspectives on technology and food safety. *Comprehensive Reviews in Food Science and Food Safety, 20*, 3225–3266. https://doi.org/10.1111/1541-4337.12763

[211] Zhang, H. Q., Barbosa-Cánovas, G. V., Balasubramaniam, V. M., Dunne, C. P., Farkas, D. F., & Yuan, J. T. C. (2010). *Nonthermal processing technologies for food, nonthermal processing technologies for food*. Oxford, UK: Wiley-Blackwell. https://doi.org/10.1002/9780470958360

[212] Buzrul, S. (2014). Multi-pulsed high hydrostatic pressure inactivation of microorganisms: A review. *Innovative Food Science and Emerging Technologies, 26*, 1–11. https://doi.org/10.1016/j.ifset.2014.07.004

[213] Hu, G., Zheng, Y., Liu, Z., Xiao, Y., Deng, Y., & Zhao, Y. (2017). Effects of high hydrostatic pressure, ultraviolet light-C, and far-infrared treatments on the digestibility, antioxidant and antihypertensive activity of α-casein. *Food Chemistry, 221*, 1860–1866. https://doi.org/10.1016/j.foodchem.2016.10.088

[214] Priyadarshini, A., Rajauria, G., O'Donnell, C. P., & Tiwari, B. K. (2019b). Emerging food processing technologies and factors impacting their industrial adoption. *Critical Reviews in Food Science and Nutrition*. https://doi.org/10.1080/10408398.2018.1483890

[215] Li, X., & Farid, M. (2016). A review on recent development in non-conventional food sterilization technologies. *Journal of Food Engineering, 182*, 33–45. https://doi.org/10.1016/j.jfoodeng.2016.02.026

[216] Fernández, A., Cebrián, G., Álvarez-Ordóñez, A., Prieto, M., Bernardo, A., & López, M. (2018). Influence of acid and low-temperature adaptation on pulsed electric fields resistance of Enterococcus faecium in media of different pH. *Innovative Food Science and Emerging Technologies, 45*, 382–389. https://doi.org/10.1016/j.ifset.2017.12.001

[217] Pillet, F., Formosa-Dague, C., Baaziz, H., Dague, E., & Rols, M.-P. (2016). Cell wall as a target for bacteria inactivation by pulsed electric fields. *Scientific Reports, 6*, 19778. https://doi.org/10.1038/srep19778

[218] Mahendran, R., Ramanan, K. R., Barba, F. J., Lorenzo, J. M., López-Fernández, O., Munekata, P. E. S., Roohinejad, S., Sant'Ana, A. S., & Tiwari, B. K. (2019). Recent advances in the application of pulsed light processing for improving food safety and increasing shelf life. *Trends in Food Science and Technology, 88*, 67–79. https://doi.org/10.1016/j.tifs.2019.03.010

[219] Martinez-Garcia, M., Sauceda-Gálvez, J. N., Codina-Torrella, I., Hernández-Herrero, M. M., Gervilla, R., & Roig-Sagués, A. X. (2019). Evaluation of continuous UVC treatments and its combination with UHPH on spores of Bacillus subtilis in whole and skim milk. *Foods, 8*, 539. https://doi.org/10.3390/foods8110539

[220] Fan, X., Huang, R., & Chen, H. (2017). Application of ultraviolet C technology for surface decontamination of fresh produce. *Trends in Food Science and Technology, 70*, 9–19. https://doi.org/10.1016/j.tifs.2017.10.004

[221] Esbelin, J., Malléa, S., Clair, G., & Carlin, F. (2016). Inactivation by pulsed light of Bacillus subtilis Spores with impaired protection factors. *Photochemistry and Photobiology, 92*, 301–307. https://doi.org/10.1111/php.12568

[222] Cleland, M. R. (2007). Advances in gamma ray, electron beam, and X-ray technologies for food irradiation. In *Food irradiation research and technology* (pp. 11–35). Ames, Iowa, USA: Blackwell Publishing. https://doi.org/10.1002/9780470277638.ch2

[223] Pinela, J., & Ferreira, I. C. F. R. (2017). Nonthermal physical technologies to decontaminate and extend the shelf-life of fruits and vegetables: Trends aiming at quality and safety. *Critical Reviews in Food Science and Nutrition, 57*, 2095–2111. https://doi.org/10.1080/10408398.2015.1046547

[224] Tejedor-Calvo, E., Morales, D., García-Barreda, S., Sánchez, S., Venturini, M. E., Blanco, D., Soler-Rivas, C., & Marco, P. (2020a). Effects of gamma irradiation on the shelf-life and bioactive compounds of Tuber aestivum truffles packaged in passive modified atmosphere. *International Journal of Food Microbiology, 332*. https://doi.org/10.1016/j.ijfoodmicro.2020.108774

[225] Olatunde, O. O., Benjakul, S., & Vongkamjan, K. (2020). Cold plasma combined with liposomal ethanolic coconut husk extract: A potential hurdle technology for shelf-life extension of Asian sea bass slices packaged under modified atmosphere. *Innovative Food Science and Emerging Technologies, 65*, 102448. https://doi.org/10.1016/j.ifset.2020.102448

Conclusion

This book is a comprehensive and thorough study of the field of food safety. Each chapter has contributed valuable insights into different aspects of hazards in food and the global state of food safety. The first few chapters focused on the identification and understanding of bacterial pathogens, viruses, and zoonoses that pose significant risks to food safety and public health. These chapters emphasized the need for rigorous control measures, surveillance, and monitoring to prevent the transmission of these pathogens from farm to fork. Additionally, the chapters on biofilm production and bacterial stress adaptation shed light on the survival mechanisms adopted by pathogens and the control strategies that can be implemented to mitigate their risks.

Prevention of foodborne illnesses was a key theme throughout several chapters. These chapters highlighted the importance of implementing preventive measures such as proper sanitation as well as hygiene practices. The application of DNA-based techniques to food safety provided a valuable tool for the rapid detection and identification of pathogens, thereby improving the efficiency and accuracy of food safety testing. The role of consumers in ensuring food safety was also highlighted in the chapter on consumer handling of food and food safety knowledge. It emphasized the need for consumer education regarding proper handling, storage, and preparation of food to minimize the risk of contamination and foodborne illnesses.

The economic cost of food safety and food recall was a significant area of discussion in this textbook. The chapters explored the financial impact of food safety breaches and emphasized the importance of effective measures and adherence to food safety protocols to avoid costly recalls and associated liabilities. The chapters on HACCP and food safety management systems, as well as food safety culture and enhancing food employee compliance, underscored the importance of establishing robust systems and a culture of food safety within organizations. These chapters emphasized the integration of food safety principles into every aspect of the food industry, so as to achieve a comprehensive and systematic approach to food safety.

This book also covered topics related to natural and synthetic toxicants, food allergies, risk assessment and analysis, emerging food safety risks, legislations, good agricultural and manufacturing practices, food fraud, and technological innovations. Each chapter provided valuable insights into these specific areas.

In conclusion, this textbook serves a valuable resource for students, researchers, and food industry professionals. The comprehensive coverage of various topics related to hazards in food and global food safety enables readers to gain a holistic understanding of the challenges and advancements in this important field. By applying the knowledge gained from this book, including implementing adequate food measures and adopting new technologies, we can all work toward ensuring a safer and more secure global food supply chain.

https://doi.org/10.1515/9783110748345-024

Index

https://doi.org/10.1515/9783110748345-025